Ordinary Differential Equations

1971 NRL–MRC Conference

Ordinary Differential Equations

1971 NRL–MRC Conference

Edited by

Leonard Weiss

University of Maryland
College Park, Maryland
and
Naval Research Laboratory
Washington, D. C.

Proceedings of a Conference on
Ordinary Differential Equations
Sponsored by the Mathematics Research
Center of the Naval Research Laboratory
Held in Washington, D. C.
June 14-23, 1971

Academic Press
New York and London 1972

ACADEMIC PRESS, INC.
111 Fifth Avenue, New York, New York 10003

United Kingdom Edition published by
ACADEMIC PRESS, INC. (LONDON) LTD.
24/28 Oval Road, London NW1 7DD

LIBRARY OF CONGRESS CATALOG CARD NUMBER: 77-187234

PRINTED IN THE UNITED STATES OF AMERICA

CONTENTS

II. SEMINAR PAPERS

PARTICIPANTS

Asterisk denotes contributor to this volume.

Agins, B., National Science Foundation, Washington, D. C.

**Antosiewicz, H. A.*, University of Southern California, Los Angeles, California

Arenstorf, R., Vanderbilt University, Nashville, Tennessee, and Naval Research Laboratory, Washington, D. C.

Bernfeld, S., University of Missouri, Columbia, Missouri

**Berger, Melvyn*, Yeshiva University, New York, New York

**Bhatia, Nam P.*, University of Maryland, Catonsville, Maryland

Bram, L., Office of Naval Research, Washington, D. C.

**Brauer, Fred*, University of Wisconsin, Madison, Wisconsin

**Brockett, Roger W.*, Harvard University, Cambridge, Massachusetts

Brodsky, S., Office of Naval Research, Washington, D. C.

**Bushaw, D.*, Washington State University, Pullman, Washington

Callas, N., Air Force Office of Scientific Research, Washington, D. C.

**Cartwright, M. L.*, Girton College, Cambridge, England

Chandra, J., U. S. Army Research-Durham, Durham, North Carolina

Chow, S., Michigan State University, East Lansing, Michigan

Coleman, C., Harvey Mudd College, Claremont, California

Comstock, C., U. S. Naval Postgraduate School, Monterey, California

**Conley, C.*, University of Wisconsin, Madison, Wisconsin

**Cooke, Kenneth L.*, Pomona College, Claremont, California

**Corduneanu, C.*, Seminarul Matematic Universitate, Iasi, Romania

**Datko, Richard*, Georgetown University, Washington, D. C.

Diliberto, S. P., University of California, Berkeley, California

Fattorini, H., University of California, Los Angeles, California

**Feldstein, Alan,* Arizona State University, Tempe, Arizona, and Naval Research Laboratory, Washington, D. C.

**Gordon, William B.,* Naval Research Laboratory, Washington, D. C.

**Gross, F.,* University of Maryland, Catonsville, Maryland, and Naval Research Laboratory, Washington, D. C.

Hahn, W., Technische Hochschule, Graz, Austria

**Halanay, A.,* Academie, Republique Socialiste de Roumanie, Bucharest, Romania

**Hale, Jack K.,* Brown University, Providence, Rhode Island

**Halkin, Hubert,* University of California, San Diego – La Jolla, California

**Harris, W. A., Jr.,* University of Southern California, Los Angeles, California

**Hartman, Philip,* Johns Hopkins University, Baltimore, Maryland

**Hautus, M. L. J.,* Stanford University, Stanford, California

Heimes, K., Iowa State University, Ames, Iowa

Henry, D., University of Kentucky, Lexington, Kentucky

Hermes, H., University of Colorado, Boulder, Colorado

**Hsieh, P. F.,* Western Michigan University, Kalamazoo, Michigan, and Naval Research Laboratory, Washington, D. C.

**Jackson, Lloyd K.,* University of Nebraska, Lincoln, Nebraska

**Jones, G. Stephen,* University of Maryland, College Park, Maryland

Junghenn, H., George Washington University, Washington, D. C.

**Kalman, R. E.,* University of Florida, Gainesville, Florida, and Stanford University, Stanford, California

Kaplan, J., Northwestern University, Evanston, Illinois

Lagnese, J., Georgetown University, Washington, D. C.

**Lakshmikantham, V.,* University of Rhode Island, Kingston, Rhode Island

**LaSalle, J. P.,* Brown University, Providence, Rhode Island

**Lasota, A.,* Jagellonian University, Krakow, Poland

**Lee, J. -S.,* Naval Research Laboratory, Washington, D. C.

Leela, S., SUNY at Geneseo, Geneseo, New York

**Leighton, Walter,* University of Missouri, Columbia, Missouri

Lepson, B., Naval Research Laboratory, Washington, D. C.

**Levin, J. J.*, University of Wisconsin, Madison, Wisconsin

Loud, W., University of Minnesota, Minneapolis, Minnesota

**Lukes, D. L.*, University of Virginia, Charlottesville, Virginia

Markus, L., University of Minnesota, Minneapolis, Minnesota

Melvin, W., College of William and Mary, Williamsburg, Virginia

Meyer, K., University of Minnesota, Minneapolis, Minnesota

Mitter, S., Massachusetts Institute of Technology, Cambridge, Massachusetts

**Morris, Grainger R.*, University of New England, Australia, and Brown University, Providence, Rhode Island

**Nohel, John A.*, University of Wisconsin, Madison, Wisconsin

Olver, F., University of Maryland, College Park, Maryland

**O'Malley, R. E., Jr.* New York University, New York, New York

Osgood, C., Naval Research Laboratory, Washington, D. C.

Pell, W., National Science Foundation, Washington, D. C.

Peterson, A., University of Nebraska, Lincoln, Nebraska

**Popov, V. M.*, University of Maryland, College Park, Maryland

**Reid, William T.*, University of Oklahoma, Norman, Oklahoma

Richards, P., Naval Research Laboratory, Washington, D. C.

Roxin, E., University of Rhode Island, Kingston, Rhode Island

**Russell, David L.*, University of Wisconsin, Madison, Wisconsin

**Sacker, Robert J.*, University of Southern California, Los Angeles, California

**Schäffer, J. J.*, Carnegie-Mellon Institute, Pittsburgh, Pennsylvania

**Schmidt, Dieter S.*, University of Maryland, College Park, Maryland

**Seibert, P.*, Universidad Catolica de Chile, Santiago, Chile

Seifert, G., Iowa State University, Ames, Iowa

**Sell, George R.*, University of Minnesota, Minneapolis, Minnesota

Shere, K., Naval Ordnance Laboratory, Silver Spring, Maryland

**Stokes, A.*, Georgetown University, Washington, D. C.

Strauss, A., University of Maryland, College Park, Maryland

Sweet, D., University of Maryland, College Park, Maryland

Swick, K., Queens College – CUNY, Queens, New York

Taam, C-T., George Washington University, Washington, D. C.

**Wasow, Wolfgang*, University of Wisconsin, Madison, Wisconsin

**Weiss, Leonard*, University of Maryland, College Park, Maryland, and Naval Research Laboratory, Washington, D. C.

Willman, W., Naval Research Laboratory, Washington, D. C.

Wilson, W., University of Colorado, Boulder, Colorado

**Wong, James S. W.*, University of Iowa, Iowa City, Iowa

**Yang, Chung-Chun*, Naval Research Laboratory, Washington, D. C.

**Yorke, J. A.*, University of Maryland, College Park, Maryland

**Yoshizawa, Taro*, Tohôku University, Sendai, Japan

PREFACE

This volume represents the Proceedings of a Conference on Ordinary Differential Equations held in Washington, D. C., June 14-23, 1971, and sponsored by the Mathematics Research Center of the Naval Research Laboratory.

The aim of this meeting was to stimulate research in ordinary differential equations by bringing together persons who were actively pursuing research in this field so they could exchange information and ideas.

Approximately 90 mathematicians representing 8 nations attended the conference, whose program consisted of 30 formal lectures and 27 seminar presentations. In addition to the regular program, a number of informal talks were given. The invited formal lectures covered geometric and qualitative theory, analytic theory, functional differential equations, dynamical systems, and algebraic theory, with applications to control theory, celestial mechanics, and biomedicine. The seminar presentations were scheduled under 6 headings: functional differential equations, oscillations and dynamical systems, analytic theory, boundary-value problems, stability and control, and differential equations on Banach spaces.

It is always difficult to capture the spirit of a meeting merely by offering a collection of technical papers. In the present case, it is impossible. The combination of concentrated mathematical talent and more time than usual to think about and discuss mathematics at a meeting with many different colleagues proved to be a potent formula for success.

My thanks go to all the authors whose papers appear here for the needed cooperation to produce this volume so relatively soon after the conference. A special word of gratitude goes to Dr. Paul B. Richards, Superintendent of the Mathematics and Information Sciences Division of NRL, who conceived of this conference and who gave unwavering support at every stage of the planning. Finally, no list of credits would be complete without noting, with appreciation, the contributions of Dr. William Gordon and Professor Philip Hsieh, who served with me on the Organizing Committee, the Office of Naval Research, for providing some financial assistance for the conference, and the staff at the Mathematics Research Center of NRL, who exhibited unfailing competence in a variety of tasks.

Leonard Weiss

I

INVITED PAPERS

A GENERAL APPROACH TO LINEAR PROBLEMS FOR NONLINEAR ORDINARY DIFFERENTIAL EQUATIONS*

H. A. Antosiewicz

This is a brief summary of a few recent results on the existence of periodic solutions of a differential equation of the form

$$\dot{x} = f(t,x) .$$

My aim is to present these results as illustrations of a broad approach to various linear problems in which the desired solutions are required to satisfy a given set of much more general linear constraints.

Throughout, I will stress basic ideas rather than utmost generality and omit all details, which may be found in the references listed at the end.

1. Let me begin with the following simple result for a differential equation

$$\dot{x} = Ax + g(t,x) , \tag{1.1}$$

where A is a linear mapping in $\mathbb{R}^n$ and g is continuous and sufficiently smooth in $\mathbb{R} \times \mathbb{R}^n$ so that the solutions of (1.1) are uniquely determined by (and hence depend continuously upon) the initial conditions.

---

*This work was done with partial support from the U.S. Army Research Office (Durham).

(1.2) Suppose A is stable. There exist positive constants μ, ρ such that, if $t \mapsto g(t,x)$ has period 1 for each $x \in \mathbb{R}^n$ and $\|g(t,x)\| < \mu$ for each $t \in [0,1]$ and $\|x\| \leq \rho$, then (1.1) has at least one solution in $\mathbb{R}$ with period 1.

Actually, (1.2) is a special case of much more general results which, in their original form, are due to Poincaré (cf. e.g., [1, 8, 10]).

(1.2) has been proved in many different ways. One method of proof applies Brouwer's fixed point theorem to the classical Poincaré mapping associated with (1.1) [1]. It hinges on the fact that there exists a real valued function V of class C^1 in $\mathbb{R}^n$ such that

$$(1.3) \qquad \lim_{\|x\| \to +\infty} V(x) = +\infty$$

and, for every $t \in [0,1]$ and every $x \in \mathbb{R}^n$ with $\|x\| \geq \rho$

$$(1.4) \qquad (DV(x), Ax + g(t,x)) < 0 .$$

Indeed, since A is stable by assumption, there is a positive definite quadratic form $V(x) = (x, Px)$ in $\mathbb{R}^n$ for which $(DV(x), Ax) = -(x,x)$ for every $x \in \mathbb{R}^n$.

Another method of proof depends upon Schauder's fixed point theorem and the admissibility, with respect to A, of the function space pair (P,P), where P is the Banach space of continuous mappings of $\mathbb{R}$ into $\mathbb{R}^n$ which are periodic with period 1 [10]. For A being stable implies that the linear differential equation $\dot{x} = Ax + b(t)$, for each $b \in P$, has precisely one solution $v(b)$ that belongs to P; in fact, v is a continuous linear mapping of P into itself. Thus, (1.1) has a solution $\phi \in P$ if and only if

$$(1.5) \qquad \phi = v \circ \omega(\phi)$$

where $\omega: P \to P$ is the substitution mapping induced by g.

Both of these methods admit various extensions to far more general settings.

2. One such extension is the basis of Krasnoselskii's method of guiding functions [12, 13] for a differential equation

(2.1) $$\dot{x} = f(t,x) ,$$

where f is continuous and sufficiently smooth in $\mathbb{R} \times \mathbb{R}^n$ so that the solutions are uniquely determined by the initial conditions, and $t \mapsto f(t,x)$ has period 1 for each $x \in \mathbb{R}^n$.

(2.2) If there exist a real valued function V of class C^1 in $\mathbb{R}^n$ and a constant $\rho > 0$ such that (1.3) holds and

(2.3) $$(DV(x),\ f(t,x)) < 0$$

for every $t \in [0,1]$ and every $x \in \mathbb{R}^n$ with $\|x\| \geq \rho$, then (2.1) has at least one solution in $\mathbb{R}$ with period 1.

Observe that the set $\{x \in \mathbb{R}^n\colon V(x) \leq \text{const.}\}$ need not be convex so that Brouwer's fixed point theorem cannot be applied directly, as in the case (1.2). Instead, the proof of (2.2) depends upon the notion of the degree of a mapping and the equivalent assertion to Brouwer's theorem that the identity mapping of the sphere in $\mathbb{R}^n$ is not null-homotopic.

A similar argument has been used by Hartman [11] to prove the existence of a solution of a general functional equation

(2.4) $$F(x) = 0$$

where F is a continuous mapping, into $\mathbb{R}^n$, of a compact convex subset K of $\mathbb{R}^n$ which contains the origin in its interior.

(2.5) Let V be a real valued positive definite function of class C^1 in K such that $DV(x) = 0$ if and only if $x = 0$. If, for every $x \in \operatorname{bd} K$,

(2.6) $$(DV(x),\ F(x)) \leq 0 ,$$

then there exists at least one point $\hat{x} \in K$ for which $F(\hat{x}) = 0$.

Hartman himself extended this result to equations in locally convex Hausdorff topological linear spaces and, in turn, deduced from this extension a general existence theorem on the solution of initial value problems for nonlinear ordinary differential equations in Hilbert space [11].

3. The solution of an equation, such as (2.4), in Hilbert space can often be accomplished by the use of projection methods which yield solutions to corresponding finite-dimensional equations (cf. [15]). Analogous techniques can be employed for the construction of fixed points.

Let H be a (real) Hilbert space with orthonormal basis (e_K), $K \geq 1$, let H_n be the (closed) linear subspace of H spanned by $e_1, e_2, \ldots, e_n$, and denote by P_n the usual projection of H onto H_n.

(3.1) Suppose f is a mapping of a closed bounded convex set $K \subset H$ into H with these properties:

(i) if (x_n) is a sequence of points of K converging weakly to a point $x_0 \in K$, then $(f(x_n))$ converges weakly to $f(x_0)$;

(ii) for each integer $n \geq 1$

$$P_n \circ f \circ P_n(K) \subset P_n(K) .$$

Then there exists at least one point $\hat{x} \in K$ such that $\hat{x} = f(\hat{x})$.

Indeed, the fixed point $\hat{x} \in K$ is the (strong) limit of a sequence (x_n) of projectional fixed points for which $(x_n, e_K) = (f(x_n), e_K)$ for $K = 1, 2, \ldots, n$.

If $K \subset H$ is simply a closed ball centered at the origin and f is defined everywhere in H, (3.2) may be satisfied by requiring that, for each $x \in \cup H_n$,

$$|(f(x), e_K)| \leq \alpha_K \|x\| + \beta_K \qquad K = 1, 2, \ldots, n ,$$

where (α_K), (β_K) are sequences with

$$\Sigma \alpha_K^2 < 1 , \qquad \Sigma \beta_K^2 < \infty .$$

This last remark yields the following generalization of a classical result of Hammerstein [9] for the scalar differential equation

$$(3.5) \qquad x'' = f(t, x, x')$$

where f is defined and continuous in $[0,\pi] \times \mathbb{R} \times \mathbb{R}$ [6].

(3.6) Suppose there are positive constants $a < 1$, b such that

$$|f(t,x,y)| \leqslant a\,|x| + b \tag{3.7}$$

holds for every $(t,x,y) \in [0,\pi] \times \mathbb{R} \times \mathbb{R}$. Then (3.5) has at least one solution ϕ in $[0,\pi]$ for which $\phi(0) = \phi(\pi) = 0$.

Originally, the condition (3.7) was required to hold with $a < \sqrt{3/\pi}$.

4. A general framework for the concept of admissibility is simple to formulate [4].

Let E, G be Banach spaces and let F be a Fréchet space which contains G algebraically and topologically (in the sense that the topology of G is stronger than the topology induced by F on G). Suppose $u\colon \mathbb{R}^n \to F$ is an injective homomorphism, $v\colon E \to F$ a continuous linear mapping, and $\omega\colon G \to E$ an arbitrary continuous mapping.

The problem of determining points $x \in \mathbb{R}^n$ and $z \in G$ such that

$$z = u(x) + v \circ \omega(z)\ , \tag{4.1}$$

is central to nearly all questions in the qualitative thoory of differential equations. Evidently, it has a solution only if there is a point $x \in \mathbb{R}^n$ and a point $y \in E$ such that $u(x) + v(y) \in G$. Thus, it is natural to include among sufficient conditions for the solution of (4.1) the requirement that, for each $y \in E$, there exist at least one point $x \in \mathbb{R}^n$ for which $u(x) + v(y) \in G$. This is the general notion of admissibility, of the pair of Banach spaces (E,G), relative to the pair of mappings (u,v) [4].

If (E,G) is admissible, there exists a constant $\mu > 0$ such that for each $y \in E$ there is a point $x \in \mathbb{R}^n$ for which $u(x) + v(y) \in G$ and

$$\mu\,\|u(x) + v(y)\| \leqslant \|y\|\ . \tag{4.2}$$

Moreover, if $X_0 = \{x \in \mathbb{R}^n\colon u(x) \in G\}$ and $\mathbb{R}^n = X_0 \oplus X_1$, there is a linear mapping $s\colon E \to X_1$ such that $v_0 = u \circ s + v$ is a linear mapping of E into G which is continuous.

Thus, if (E,G) is admissible, it is sufficient to find, for a given point $x_0 \in X_0$, a point $z \in G$ such that

$$z = u_0(x_0) + v_0 \circ \omega(z) \tag{4.3}$$

where u_0 is the restriction of u to X_0. This can be done, by use of Banach's fixed point theorem when ω is lipschitzian, and by use of Schauder's principle when v has additional properties (cf. e.g., [4, 5, 10, 14]).

In the latter case, v_0 can often be represented as the product of two suitable (continuous linear) mappings $v_0 = v_1 \circ v_2$, where v_1 maps an auxiliary space H into G and v_2 maps E into H. This device is particularly effective when x_0 is taken to be $0 \in X_0$ in (4.3) or when v_0, in fact, is identical with v (and hence (4.1) reduces to (1.5)). For, in that case, instead of determining a solution ϕ belonging to G for which

$$\phi = v_1 \circ v_2 \circ \omega(\phi) \tag{4.4}$$

one first finds a point Ψ belonging to H such that

$$\Psi = v_2 \circ \omega \circ v_1(\Psi) \tag{4.5}$$

and then obtains the solution of (4.4) as $\phi = v_1(\Psi) \in G$.

This method may be used to obtain the following improvement on (3.6) (cf. e.g., [12]).

(4.6) Suppose there are positive constants $a < 1$, b such that f in (3.5) satisfies

$$x \cdot f(t,x,y) < ax^2 + b \tag{4.7}$$

at every $(t,x,y) \in [0,\pi] \times \mathbb{R} \times \mathbb{R}$. Then (3.5) has at least one solution ϕ in $[0,\pi]$ for which $\phi(0) = \phi(\pi) = 0$.

Other illustrations of the use of (4.3) are given in [5] (cf. e.g., [2, 3, 4, 7]).

REFERENCES

[1] H.A. ANTOSIEWICZ, Forced periodic solutions of systems of differential equations, Annals Math. 57 (1953), 314-317; 58 (1953), 592.
[2] H.A. ANTOSIEWICZ, On the existence of periodic solutions of nonlinear differential equations, Proc. Colloques Internat. C.N.R.S. 148 (1965), 213-216.
[3] H.A. ANTOSIEWICZ, Boundary value problems for nonlinear ordinary differential equations, Pacific J. Math. 17 (1966), 191-197.
[4] H.A. ANTOSIEWICZ, Un analogue du principe du point fixe de Banach, Ann. Mat. Pura Appl. 74 (1966), 61-64.
[5] H.A. ANTOSIEWICZ, Linear problems for nonlinear ordinary differential equations, Proc. U.S.-Japan Sem. Diff. and Funct. Eqns. W.A. Benjamin, New York, 1967, pp. 1-11.
[6] H.A. ANTOSIEWICZ, A fixed point theorem and the existence of periodic solutions, Proc. 5th Internat. Conf. Nonlin. Oscill., Kiev, 1969, 40-44.
[7] R. CONTI, Recent trends in the theory of boundary value problems for ordinary differential equations, Bull. Un. Mat. Ital. 22 (1967), 135-178.
[8] J.K. HALE, Oscillations in Nonlinear Systems. McGraw-Hill, New York, 1963.
[9] A. HAMMERSTEIN, Die erste Randwertaufgabe für nichtlineare Differentialgleichungen 2 ter Ordnung, S.-B. Berlin Math. Ges. 30 (1932), 3-10.
[10] P. HARTMAN, Ordinary Differential Equations. John Wiley & Sons, New York, 1964.
[11] P. HARTMAN, Generalized Lyapunov functions and functional equations, Annali Mat. Pura Appl. 69 (1968), 305-320.
[12] M.A. KRASNOSELSKII, The theory of periodic solutions of nonautonomous differential equations, Russian Math. Survey 21 (1966), 53-74.
[13] M.A. KRASNOSELSKII, The operator of translation along the trajectories of differential equations, Amer. Math. Soc. Translations Math. Monographs, Vol. 19, Providence, 1968.
[14] J.L. MASSERA and J.J. SCHÄFFER, Linear Differential Equations and Function Spaces, Academic Press, New York, 1966.
[15] W.V. PETRYSHYN, Projection methods in nonlinear numerical functional analysis, J. Math. Mech. 17 (1967), 353-372.

University of Southern California, Los Angeles, California

DIFFERENTIAL RELATIONS

D. Bushaw

1. Introduction

In the early 1930's geometry, and more recently control theory, have drawn attention to conditions of the form

$$y' \in f(x,y) \,, \tag{1}$$

where f is defined on $\mathbb{R} \times \mathbb{R}^n$ and has as values subsets of $\mathbb{R}^n$. A solution of (1) is usually defined as an absolutely continuous function $\phi\colon I \to \mathbb{R}^n$ (where I is some real interval) such that

$$\phi'(x) \in f(x,\phi(x)) \tag{2}$$

almost everywhere on I. A classical solution of (1) is required to be continuously differentiable (the derivative being interpreted as the appropriate one-sided derivative at endpoints, if any, of I) and to satisfy (2) for all $x \in I$. Much is known about the existence and other qualitative properties of solutions and classical solutions of (1) under certain assumptions - typically involving continuity, compactness, and convexity - on the function f (see references).

The point of departure for this paper is a simple reinterpretation of (1)-(2). The condition (2) is equivalent to

$$(x,\phi(x),\phi'(x)) \in F \,, \tag{3}$$

where

$$F = \{(x,y,z) \in \mathbb{R} \times \mathbb{R}^n \times \mathbb{R}^n : z \in f(x,y)\} ,$$

and we may thus define solutions, or classical solutions, of the "differential relation"

$$(4) \qquad (x,y,y') \in F$$

just as before, with (3) in place of (2). In the absence of any special assumptions on f, the set F in (3) or (4) may be a perfectly arbitrary subset of $\mathbb{R} \times \mathbb{R}^n \times \mathbb{R}^n$ (or, as we shall say henceforth, of $\mathbb{R}^{2n+1}$).

For simplicity, we shall consider only classical solutions of (4).

Let $\mathcal{D}$ denote the set of all continuously differentiable functions $\phi\colon I \to \mathbb{R}^n$, where I is some interval (connected subset of $\mathbb{R}$ having at least two points) which may vary with ϕ and which, accordingly, will usually be denoted by I_ϕ. We define a map W from $\mathcal{D}$ into the collection of all subsets of $\mathbb{R}^{2n+1}$ by

$$(5) \qquad W(\phi) = \{(x,\phi(x),\phi'(x)) : x \in I_\phi\} .$$

The range of this map, $W(\mathcal{D})$, will be denoted by $\mathcal{W}$.

With these notations, a classical solution of (4) may be defined simply as a $\phi \in \mathcal{D}$ such that $W(\phi) \subset F$.

In fact, the map from $\mathcal{W}$ into the collection of subsets of $\mathbb{R} \times \mathbb{R}^n$ induced by the projection $(x,y,z) \to (x,y)$ is an inverse for W, which accordingly is one-one. Thus without real loss we may think wholly in terms of subsets of $\mathbb{R}^{2n+1}$ and regard as solutions of (4) - the word "classical" will be omitted henceforth - those $w \in \mathcal{W}$ which are contained in F.

The situation may now be described as follows. We have a collection $\mathcal{W}$ of subsets of $\mathbb{R}^{2n+1}$ which is universal in the sense that it depends on no differential equation or differential relation, but in fact only on n; we have a subset F of $\mathbb{R}^{2n+1}$ whose definition involves no differentiation; and the problem is to learn something about those members of $\mathcal{W}$ that lie in F. This problem subsumes the study of all "generalized differential equations" (1), and in particular of all real ordinary differential equations. Moreover, the pattern could be extended to problems where $\mathbb{R}^n$ is replaced by some more general space.

The rest of this paper will consist of some rudimentary and fragmentary observations based on this viewpoint.

2. The Collection $\mathcal{W}$

It is useful to have a geometrical-relational characterization of the collection $\mathcal{W}$. Such a characterization may be based on the concept of a wedge. Let $P_0 = (x_0,y_0,z_0) \in \mathbb{R}^{2n+1}$ (specifically, $x_0 \in \mathbb{R}$ and $y_0,z_0 \in \mathbb{R}^n$), and let $\varepsilon > 0$. The corresponding forward and backward wedges are:

$$V_\varepsilon^+(P_0) = \{(x,y,z)\colon x_0 < x < x_0+\varepsilon,\ |y-y_0-(x-x_0)z_0| < \varepsilon|x-x_0|,\ \text{and}\ |z-z_0| < \varepsilon\} \cup \{P_0\},$$

$$V_\varepsilon^-(P_0) = \{(x,y,z)\colon x_0-\varepsilon < x < x_0,\ |y-y_0-(x-x_0)z_0| < \varepsilon|x-x_0|\ \text{and}\ |z-z_0| < \varepsilon\} \cup \{P_0\}.$$

PROPOSITION 1. A subset w of $\mathbb{R}^{2n+1}$ belongs to $\mathcal{W}$ if and only if it is the graph of a function $\Phi\colon I_w \to \mathbb{R}^n \times \mathbb{R}^n$, where I_w is a real interval, which satisfies: for every $x_0 \in I_w$ and $\varepsilon > 0$, there exists a $\delta > 0$ such that

$$(x,\Phi(x)) \in V_\varepsilon^-(x_0,\Phi(x_0)) \cup V_\varepsilon^+(x_0,\Phi(x_0))$$

for all $x \in I_w \cap [x_0-\delta,\ x_0+\delta]$.

The proof is elementary, and is based directly on the definition of W. The concluding condition in Proposition 1 is a continuity condition; the sets $V_\varepsilon(P_0) = V_\varepsilon^-(P_0) \cup V_\varepsilon^+(P_0)$ form a basic system of neighborhoods at each $P_0 \in \mathbb{R}^{2n+1}$, and thus define a topology $\mathcal{T}_V$ on $\mathbb{R}^{2n+1}$. The above condition is that the map $x \to (x,\Phi(x))$ be continuous relative to the usual topology on $\mathbb{R}$ and the topology $\mathcal{T}_V$ on $\mathbb{R}^{2n+1}$. Note that this is not equivalent to the continuity, in some sense, of Φ; the topology $\mathcal{T}_V$ is not a product topology. In fact, this topology is rather unpleasant: it is not locally compact, for example, and a translation of $\mathbb{R}^{2n+1}$ is continuous relative to this topology if and only if its z-component vanishes.

Topologies $\mathcal{T}_{V^+}$ and $\mathcal{T}_{V^-}$ may be defined analogously in terms of the sets $V_\varepsilon^+(P_0)$ and $V_\varepsilon^-(P_0)$.

For a given n, $\mathcal{W}$ is an extremely rich cover of $\mathbb{R}^{2n+1}$, as one would expect. For example, if $P_1 = (x_1,y_1,z_1)$ and $P_2 = (x_2,y_2,z_2)$ are any two points of $\mathbb{R}^{2n+1}$ with $x_1 \neq x_2$, there exists a $w \in \mathcal{W}$ such that $\{P_1,P_2\} \subset w$; it may be found by choosing carefully the coefficients (in $\mathbb{R}^n$) in $\phi(x) = x^3a + x^2b + xc + d$ and taking $w = W(\phi)$.

Furthermore, it follows directly from the definitions that the restriction of every $w \in W$ to any subinterval of I_w again belongs to $\mathcal{W}$; and that if $w_1,w_2 \in \mathcal{W}$ and agree on $I_{w_1} \cap I_{w_2} \neq \emptyset$, then $w_1 \cup w_2 \in \mathcal{W}$.

3. Initial Value Problems

For given $F \subset \mathbf{R}^{2n+1}$ and $P_0 = (x_0,y_0,z_0) \in F$, two questions naturally arise:

The strong initial value problem. Does there exist a $w \in \mathcal{W}$ such that $P_0 \in w \subset F$?

The weak initial value problem. Do there exist $z \in \mathbf{R}^n$ and $w \in \mathcal{W}$ such that $(x_0,y_0,z_0) \in w \subset F$?

Plainly, when F is functional (i.e., the sets f(x,y) in (1) are singletons) there is no difference between the problems; so the distinction is rarely made. The following discussion will be limited mainly to the strong problem.

From Proposition 1, it is clear that the strong problem has a solution at P_0 only if P_0 is a cluster point of F relative to the topology $\mathcal{T}_V$. (Thus an F at every point of which the strong problem has a solution is perfect relative to this topology.) Similarly, if a $w \in \mathcal{W}$ starts at $P_0 \in F$, then P_0 is a cluster point relative to $\mathcal{T}_{V^+}$. The converse, as crude examples show, is false.

Clearly, the strong problem will have a solution at a point $P_0 = (x_0,y_0,z_0)$ if there exists a continuous function ψ from some neighborhood of (x_0,y_0) into $\mathbb{R}^n$ whose graph lies in F and which satisfies $z_0 = \psi(x_0,y_0)$: for such a ψ, Peano's classical existence theorem establishes the existence of a $\phi \in \mathcal{D}$ such that $\phi'(x) = \psi(x,\phi(x))$ and $\phi(x_0) = y_0$; and then $W(\phi)$ is a solution of the problem. One of Filippov's existence theorems for classical solutions works by giving sufficient conditions for the existence of such a ψ, and another adapts Peano's proof to a different set of conditions. Both are very general, but somewhat unsatisfactory because they derive conclusions that are essentially local in character (in $\mathbb{R}^{2n+1}$) from assumptions that are not.

A sufficient condition for existence which is vastly more primitive, but more natural as far as it goes, is the following.

PROPOSITION 2. If $P_0 \in F$, while $r \in \mathbb{R}^n$ and $\varepsilon > 0$ are such that

(6) $$\{(x,y,z): 0 < x-x_0 < \varepsilon, |y-y_0-(x-x_0)z_0| < \varepsilon|x-x_0|, \\ |z-z_0-(x-x_0)r| < \varepsilon|x-x_0|\} \subset F ,$$

then there exists a solution of the strong problem that starts at P_0.

Such a solution is $W(\phi)$, where ϕ is defined on a suitable interval by

$$\phi(x) = y_0 + (x-x_0)z_0 + \frac{1}{2}(x-x_0)^2 r .$$

Not every member of $\mathcal{W}$ starting at a point P_0 begins in such a pyramid (6), however. This is shown by the example defined by $n = 1$, $\phi(0) = 0$, $\phi(x) = x^{5/2}\sin(1/x)$ for $x > 0$.

It does follow from Proposition 2 (and also from several preceding remarks) that at any interior point P_0 of F, there are a great many solutions of the strong initial value problem. Thus the problem can be interesting only on the border $F \cap \partial F$ of F - interesting in the sense that there is any danger of nonexistence of solutions, or any possibility of their uniqueness. Both Proposition 2 and the Peano approach may be applicable at such points.

4. An Example

Let us choose $n = 1$ and take F to be the set of all (x,y,z) satisfying the conditions

$$y \sin x - z \cos x = 0 ,$$
$$y^2 + z^2 \leqslant 1 .$$

Geometrically, F may be described as a twisted ribbon along the x-axis. This set has no interior, but at points where $|y| < |\cos x|$ the Peano theorem may be applied and the strong problem has a (unique) solution both ways. Points where $|y| = |\cos x|$ and $0 < x < \pi/2$ (mod π) are isolated points of F relative to the topology $\mathcal{T}_{V^+}$, so no solutions of the strong problem start at such points. (It is easy to show, although it does not follow immediately from anything in this paper, that certain solutions end at such points.) Similarly, solutions start but do not end at points where $|y| = |\cos x|$ and $-\pi/2 < x < 0$ (mod π). It may be seen in various ways that there are no solutions at points $(k\pi, \pm 1, 0)$. This leaves the points where $x = \pi/2$ (mod π). The section of F perpendicular to the x-axis for such x are segments where $y = 0$, $|z| \leqslant 1$. Simple geometrical arguments show that all such points are isolated points of F relative to $\mathcal{T}_V$ except those where $z = 0$. Thus any solution $W(\phi)$ such that (say) $\pi/2 \in I_\phi$ must satisfy not only $\phi(\pi/2) = 0$ but $\phi'(\pi/2) = 0$; and this is a conclusion that may not be obvious from the statement of the problem. (Of course there exists such a solution, namely the x-axis.)

5. Conclusion

The brief and elementary discussion offered here is intended merely to suggest a way of looking at ordinary differential equations and generalized differential equations that seems to have been used little, if at all, although in some ways it is natural enough. In at least one way it is decidedly unnatural: for problems usually represented geometrically in n or $n+1$ dimensions, it uses a representation in $2n+1$ dimensions. This alone would be ample to account for its being a viewpoint toward which workers in the field would not have gravitated.

Nevertheless, it has some promise as an angle of attack in dealing with the fundamental theory of ordinary differential equations and their generalizations, such problems as approximation, stability and other kinds of limiting behavior, control problems, and perhaps more subtle matters like the existence and structure of periodic solutions. If nothing else, it provides another way of feeding and guiding the intuition about these matters.

REFERENCES

[1] T.F. BRIDGLAND, Contributions to the theory of generalized differential equations I, II, Math. Systems Theory, 3 (1969), 17-50, 156-165.

[2] A.F. FILIPPOV, Classical solutions of differential equations with multi-valued right-hand side (translation), SIAM J. Control, 5 (1967), 609-621.

[3] H. HERMES, The generalized differential equation $\mathring{x} \in R(t,x)$, Advances in Math., 4 (1970), 149-169.

Washington State University, Pullman, Washington

CONDITIONS FOR BOUNDEDNESS OF SYSTEMS OF ORDINARY DIFFERENTIAL EQUATIONS

M. L. Cartwright

1. I take as my starting point the second order equation

(1) $\ddot{x} + f(x)\dot{x} + g(x) = 0, \quad xg(x) > 0, \quad x \neq 0, \quad f(x) \geq 0, \quad |x| \geq 1$.

This equation represents most known forms of oscillation in second order equations, and it has been studied by many authors with a view to establishing that solutions are bounded as $t \to \infty$ by constants depending only on f and g and not on the initial values $x(0) = x_0$, $\dot{x}(0) = y_0$. Most authors have tried to formulate the most general conditions which will cover well known equations representing physical phenomena such as

$$\ddot{x} + k(x^2 - 1)\dot{x} + x = 0 , \qquad \text{van der Pol,}$$

$$\ddot{x} + k\dot{x} + ax + bx^3 = 0 , \qquad \text{Duffing,}$$

$$\ddot{x} + k\left(\frac{\dot{x}^3}{3} - \dot{x}\right) + x = 0 , \qquad \text{Rayleigh.}$$

It should be observed that Rayleigh's equation, if differentiated, reduces to (1) with $x = \int \xi\, dt$, and that although Duffing's equation is not of the form (1) for large x if $b < 0$, the polynomial $ax + bx^3$ is an approximation of a function $g(x)$ such that $xg(x) > 0$ valid for comparatively small x.

The most general results are obtained by expressing (1) as a system

(2) $$\dot{x} = y - F(x) , \qquad F(x) = \int_0^x f(\xi)d\xi$$

(3) $$\dot{y} = -g(x) , \qquad G(x) = \int_0^x g(\xi)d\xi .$$

One advantage of this form is that it is immediately obvious that x increases or decreases according as $y \gtrless F(x)$ while y increases or decreases according as $g(x) \lessgtr 0$. There are two main tools; first the Lyapunov function $V(x,y)$, obtained by multiplying (2) by $g(x)$ and (3) by y, adding and integrating, so that

(4) $$V(x,y) \equiv G(x) + \frac{1}{2} y^2 = V(x_0,y_0) - \int_{t_0}^{t} F(x)g(x)dt$$
$$= V(x_0,y_0) + \int_{y_0}^{y} F(x)dy$$

by (3). The second tool is the equation derived from (2) and (3)

(5) $$\frac{dy}{dx} = -\frac{g(x)}{y - F(x)} .$$

With these it can be shown that if <u>either</u> $G(x) \to \infty$ or $|F(x)| \to \infty$ as $|x| \to \infty$ and $F(x) \geq B_1 > 0$ for $x \geq B_2 > 1$, where B_1 and B_2 are constants then solutions $x(t)$ and $y(t)$ are bounded as $t \to \infty$ by constants depending only on f and g. These conditions can be relaxed by allowing $xg(x)$ to change sign for $|x| < 1$, but the complications involved are irrelevant to what I have to say here. Reuter, Bushaw, Burton and Townsend have done important work in reducing these and allied results for nonautonomous second order equations to their most general form.

2. Levinson and Smith considered the equation

(6) $$\ddot{x} + f(x,\dot{x})\dot{x} + g(x) = 0, \qquad xg(x) > 0, \quad x \neq 0 ,$$

by considering the system

(7) $\dot{x} = y, \qquad f(x,y) \geq 0, \quad |x| \geq 1, \quad |y| \geq 1,$

(8) $\dot{y} = -g(x) - f(x,y)y .$

Their additional conditions include what one might call subsidiary conditions needed to cover the strips $|x| \leq 1$ and $|y| \leq 1$ in which $f(x,y)$ may be negative. Their objective was to show that there is a periodic solution, but their proof shows that solutions are bounded by a constant independent of their initial values if $G(x) \to \infty$ as $x \to \infty$ and there exists $x_1 > x_0 > 0$ such that

(9) $$\int_{x_0}^{x_1} f(x,y(x))dx \geq Kx_0 ,$$

where $y(x)$ is an arbitrary decreasing function of x, the x_0 and K being dependent on the subsidiary conditions.

It should be observed that by putting $x = -y'$, $y = x'$ we obtain from the system (7), (8) the system

$$\dot{x}' = g'(y') - f(-y',x')x'$$

$$\dot{y}' = -x' ,$$

where $y'g(y') = -xg(-x) > 0$ if $y' \neq 0$, so that the question arises whether we can obtain a result covering both systems. This seems difficult to do partly because if $g(y')$ is not large with y' it is difficult to use the equation corresponding to (5), as in the proof for (1), to show that y' varies slowly for y' large and $|x'| < B$, but even more because the proof of Levinson and Smith seems to depend, not only on the fact that $\dot{x}$ is large when y is large, but also on the smooth and rapid rate of increase of y in $f(x,y)y$ which is involved in the use of (9). I see no obvious way of making a generalization which will cover both systems.

3. There is no doubt that the proof for the system (7) and (8) is more difficult than that for (2) and (3), and I asked Levinson whether it related to any physical system not covered by (2) and (3). He wrote that he had consulted a colleague and, so far as they knew, it

did not. Littlewood and I in proving results with $k\,f(x,\dot{x})$ in place of $f(x,\dot{x})$ were particularly interested in obtaining constants independent of k for k small or large if it was possible to do so. By imposing somewhat stronger conditions on f and g we obtained

$$|x| < B, \qquad |\dot{x}| < B(k+1), \qquad t > t_0 ,$$

and I do not think that under our stronger conditions there was much difference between the two cases. For the system (2) (3), it becomes clear that $G(x) \to \infty$ gives a better result for k small and $|F(x)| \to \infty$ is better for k large.

4. In preparing a lecture on the early history of nonlinear oscillations I looked back at a paper by Appleton and Greaves, and, as I thought, van der Pol's equation was derived from a system somewhat similar to these, but not included in either. Changing their notation for convenience, I write their system in the form

(10) $$L\dot{y} = Rz + \frac{1}{C}\int z = -x ,$$

(11) $$y + z = \psi(x) ,$$

where $\psi(x)$, or rather $\psi(x) + \alpha x$, is a function somewhat similar to $F(x)$ in (2). Let $D = d/dt$ and write (10) in the form

(12) $$LDy = -x ,$$

(13) $$(RD + \frac{1}{C})z = -Dx .$$

Their method was to eliminate both y and z by operating on (12) with $(RD + 1/C)$ and on (13) with LD and adding and using (11). This gives

$$D^2(x + R\psi(x)) + \left(\frac{\psi'}{LC} + R\right)Dx + \frac{x}{LC} = 0 .$$

They remarked that $R\psi$ is small and omitted the term $D^2(R\psi(x))$ from further consideration with apparently perfectly satisfactory results. If it is retained and $|R\psi''| < 1$, then we can divide through by $1 + R\psi''$ and we obtain an equation of the form (6). For there is a term $R\psi'\dot{x}^2$.

However if we eliminate z by operating on (11) with $RD + 1/C$, we have

$$-\frac{Rx}{L} + \frac{1}{LC}y - Dx = \left(RD + \frac{1}{C}\right)\psi(x) .$$

We now have a system

$$(1 + R\psi'(x))\dot{x} = \frac{y}{LC} - \left(\frac{\psi(x)}{C} + \frac{Rx}{L}\right) ,$$

$$\dot{y} = -\frac{x}{L} .$$

If $1 + R\psi'(x) > 0$, $w = x + R\psi(x)$ is a strictly increasing function of x, and the new system is of the form (2), (3) if $\psi(x) - Rx/C$ as a function of w, satisfies the same conditions that $F(x)$ did, which was the case in the equation considered. However it seems easier to continue working with x and not w. The Lyapunov function will be

$$V(x,y) \equiv G(x) + \frac{y^2}{2C} , \qquad G(x) = \frac{x^2}{2} + R\int_0^x \xi\psi'(\xi)d\xi \to \infty$$

as $x \to \infty$.

5. For a linear system with a, b, c, d constants,

$$(14) \qquad \begin{cases} \dot{x} = ax + by , \\ \dot{y} = cs + dy , \end{cases}$$

the necessary conditions for $x \to 0$, $y \to 0$ as $t \to 0$ are that $a + d < 0$ and $ad - bc > 0$, but it must be remembered that these imply that x and y tend to 0 exponentially.

R. Datko at a conference in Los Angeles in 1967 discussed the case in which a, b, c, d are bounded functions of x and y and for some positive constants S_1, S_2, V_1, V_2

$$(15) \qquad -S_2 \leqslant a + d \leqslant -S_1 < 0, \qquad 0 < V_2 < ad - bc < V_1 ,$$

and a further condition was imposed about a transversal. He said that the system

$$\dot{x} = -x + 2x^2y ,$$

$$\dot{y} = -y ,$$

showed that boundedness is a necessary condition. For $a = -1$, $b = 2x^2$, $c = 0$, $d = -1$ so that (15) holds.

However it seems reasonable to suppose that $\dot{x}$ is more sensitive to x than to y, and if <u>both</u> a and b are allowed to be functions of x and y the expression of the system in the form (14) is ambiguous. If we write

$$\dot{x} = (-1 + 2xy)x$$

$$\dot{y} = -y$$

$a = -1 + 2xy$, $b = d = 0$, and the conditions are only satisfied for $2xy < 1$.

If, like Datko, we abandon the attempt to cover the case of van der Pol's equation, we can build up systems with bounded solutions. Consider first the system

$$\begin{aligned} &h_1(x)\dot{x} = g_1(y), \quad h_1 > 1, \quad yg_1(y) > 0, \quad y \neq 0, \\ (16) \qquad &h_2(y)\dot{y} = -g_2(x), \quad h_2 > 1, \quad xg_2(x) > 0, \quad x \neq 0 . \end{aligned}$$

This system is conservative. For let

$$\int_0^x h_1(x)g_2(x)dx = H_1(x), \qquad \int_0^y h_2(y)g_1(y)dy = H_2(y),$$

then $H_1 > 0$, $H_2 > 0$, and

$$V \equiv H_1(x) + H_2(y) = V(x_0,y_0) = V_0 .$$

If $H_1 \to \infty$, $H_2 \to \infty$ as $x \to \infty$ and $y \to \infty$ respectively, the solutions are closed curves $V(x,y) = V_0$. Now if

$$h_1(x)\dot{x} = g_1(y) - F_1(x,y),$$

$$h_2(y)\dot{y} = -g_2(x) - F_2(x,y),$$

with the same V we have

$$V(x,y) = V_0 - \int_0^t F_1(x,y)g_2(x)dt - \int_0^t F_2(x,y)g_1(y)dt .$$

Since $xg_2(x) > 0$ for $x \neq 0$ and $yg_1(y) > 0$ for $y \neq 0$, if

$$xF_1(x,y) \geqslant 0, \quad x \neq 0, \qquad yF_2(x,y) \geqslant 0, \quad y \neq 0 \tag{17}$$

we have $V(x,y) \leqslant V_0$. Observe that these conditions eliminate the possibility of any critical point other than $x = y = 0$. For $g_1(y) \neq F_1(x,y)$ unless $x \geqslant 0$, $y \geqslant 0$ or $x \leqslant 0$, $y \leqslant 0$, and $g_2(x) + F_2(x,y) \neq 0$ unless $x \geqslant 0$, $y \geqslant 0$ or $x \leqslant 0$, $y \leqslant 0$. Some positive damping is needed in order to bring the solution to the origin, and here the formulation is necessarily different. We have no region of negative damping the effect of which has to be compensated in the region of positive damping, so that a very general condition such as either

$$F_1(x,y)g_2(x) \geqslant B_1 > 0 \quad \text{for} \quad x \geqslant B_2 > 0 ,$$

or

$$F_2(x,y)g_1(y) \geqslant B_3 > 0 \quad \text{for} \quad y \geqslant B_4 > 0 ,$$

might be sufficient to ensure boundedness with respect to initial conditions, although I have not carried through the investigation, but near the origin something like Datko's conditions might be expected to be necessary. It should be observed that in (16) the cases $g_1(y) \equiv 0$ and $g_2(x) \equiv 0$ have been excluded. If, for instance $h_1 = 1$, $g_1(y) \equiv 0$ and $F_1(x,y)/x \geqslant S_1 > 0$, and (17) holds, then

$$\dot{x} \leqslant -S_1 x , \qquad x > 0 ,$$

and it follows that $x \to 0$ as $t \to \infty$, so long as $x > 0$. If similar conditions hold for $x < 0$, we have $h_2(y)\dot{y} \to 0$, and, since the only critical point is $x = y = 0$, we have $y \to 0$. It should be observed that for the simple equation $\dot{x} = -F^*(x)$ the solution tends to 0 as $t \to \infty$ if

$$\int_x^{x_0} \frac{d\xi}{F^*(\xi)} \to \infty ,$$

so that a condition of the form $0 < F_1(x,y) \leqslant F^*(x)$, $x > 0$ with similar conditions for $x < 0$, $F_2(x,y)$ might be sufficient. The methods which I have just suggested keep the parts of the Lyapunov function depending on $\dot{x}$ and $\dot{y}$ separate and treat the damping of

$\dot{x}$ and $\dot{y}$ separately, whereas results such as Datko's and Krasovskii make provision for changing axes so as to get a better result. Krasovskii considered

$$\dot{x} = X(x,y) \ , \qquad \dot{y} = Y(x,y) \ ,$$

and assumed that

$$\frac{\partial X}{\partial x} + \frac{\partial Y}{\partial y} < 0 \ ,$$

and also that $\ell(x^2+y^2) < X^2 + Y^2$ for all small x and y. The latter condition takes the place of $ad - bc$ for the linear case, and the first takes the place of $a + d < 0$, but it requires the existence of partial derivatives. In the general case there is the possibility of finding a one to one continuous mapping which will reduce a general system to one in which x and y separate themselves conveniently into one of the forms which I have discussed, but so far I have no suggestion of how to construct such a mapping in the general case. In a particular case the nature of the system or physical applications might provide a clue, but as in the case of (1) and (6) I have little hope that any general result can be formulated to include all cases.

REFERENCES

[1] T.A. BURTON and C.G. TOWNSEND, On the generalized Liénard equation with forcing term, Journal of Differential Equations, 4 (1968), 620-633.

[2] R. DATKO, Stability properties of a second order system satisfying the Hurwitz condition, Mathematical Theory of Control (Proc. Conference, Los Angeles, 1967), ed. A.V. Balakrishan and L.W. Neustadt.

[3] N.N. KRASOVSKII, trans. T.L. Brenner, Stability of Motion. Stanford, 1963.

[4] N. LEVINSON and O.K. SMITH, A general equation for relaxation oscillations, Duke Math. Journal, 9 (1942), 382-403.

[5] G.E.H. REUTER, A boundedness theorem for nonlinear differential equations of the second order, Proc. Cambridge Phil. Soc., 47 (1951), 49-54.

Girton College, Cambridge, England

ON A GENERALIZATION OF THE MORSE INDEX

C. Conley*

A singular point of an ordinary differential equation is called elementary if the eigenvalues of the linearized equations all have non-zero real parts. In this case the set of orbits which tend to the singular point (i.e. the stable manifold) has dimension equal to the number of eigenvalues with negative real part; the unstable manifold has the complementary dimension. Either of these dimensions determines the flow near the singular point up to conjugacy; thus their importance is not overemphasized by giving them a special name. It is customary to call the dimension of the unstable manifold the Morse index (distinguishing from the Euler-Poincaré index) of the singular point.

Elementary singular points are examples of invariant sets of the flow (defined by the equations) which are "isolated" in the sense that they admit a closed neighborhood in which they are maximal. Any such neighborhood will be called an isolating neighborhood for the invariant set. The definition implies that any larger invariant set is quite a bit larger.

The purpose of this report is to describe an "index" for the isolated invariant sets of a flow on a compact metric space which, in the case of elementary singular points above contains (exactly) the information of the Morse index. This index is defined in terms of special isolating neighborhoods called isolating blocks. Roughly, B is a block

---

*The research reported on here was supported by N.S.F. GP 20858.

if there is an $\varepsilon > 0$ such that, given any $x \in \partial B$, one or the other open half (from x) of the orbit segment of length ε centered on x lies outside B. Thus, entrance and exit points are "strict" in the sense of T. Wazewski. A theorem, proved in [1] for C^1-flows and in [2] for flows on compact metric spaces, states that any isolated invariant set is realized as the maximal invariant set in some isolating block; thus it has an index as described below.

Given B, let b^+ and b^- denote respectively the entrance points (i.e. those leaving B in the negative direction) and exit points in ∂B ($\partial B = b^+ \cup b^-$). As shown in [3], the homotopy type, $[B/b^+]$, of the pointed space obtained from B by collapsing b^+ to a point depends only on the isolated invariant set and not on the block chosen. We thus define for each isolated invariant set S the "homotopy indices" $i^{\pm}(S) \equiv [B/b^{\pm}]$ where B is any block for S.

For example if S is an elementary critical point we can choose coordinates centered at S so that the equations can be written

$$\dot{u} = Au + O_2(u,v)$$

$$\dot{v} = Bu + O_2(u,v)$$

where A is negative definite and B is positive definite. A block is defined as the set $B = \{|u|, |v| \leqslant \varepsilon\}$ where ε is a small positive number. A simple computation shows, that on B, $\frac{d}{dt}(|u|^2) \simeq 2(u,Au) < 0$ and similarly $\frac{d}{dt}(|v|^2) \simeq (v,Bv) > 0$. It follows that B is a block with $b^+ = \{|u| = \varepsilon\} \cap B$ and $b^- = \{|v| = \varepsilon\} \cap B$. To calculate $i^+(S) = [B/b^+]$ one can just ignore the v coordinate and so find that $i^+(S)$ is a pointed sphere with dimension equal to that of the stable manifold. Similarly $i^-(S)$ is a pointed sphere with dimension equal to the Morse index.

There is a sum and a product formula for the homotopy index corresponding to the sum, $A \vee B$ and product $A \wedge B$ of pointed spaces. (The sum is obtained by gluing the distinguished points a_0 and b_0 together; the product is obtained on collapsing the subset $A \times b_0$ $a_0 \times B$ of $A \times B$ to a point.) Thus _if S_1 and S_2 are disjoint_ then $S = S_1 \amalg S_2$ is isolated and $i^{\pm}(S) = i^{\pm}(S_1) \vee i^{\pm}(S_2)$. Also if

S_1 is isolated for a flow on X and S_2 is isolated for a flow on Y, then $S_1 \times S_2$ is isolated for the product flow on $X \times Y$ and $i^{\pm}(S) = i^{\pm}(S_1) \wedge i^{\pm}(S_2)$.

The empty set is always an isolated invariant set; a block being provided by the empty set. Noting that the process of collapsing the empty set to a point is achieved by taking the disjoint union of the space with a point (which becomes the distinguished point) we find the index of the empty set is the pointed point which we call $\overline{0}$. Of course $\overline{0} \vee A = A$, and $\overline{0} \wedge A = \overline{0}$ for any pointed space A. The whole space X is also isolated, X now serving as a block. The index is the disjoint union of X with a (distinguished) point. The summation formula is no longer informative since S is disjoint from no other invariant set except the empty set.

To illustrate how the summation formula provides some insights; consider a Hamiltonian system with two degrees of freedom:

$$\begin{aligned} \dot{x}_i &= H_{y_i} = y_i \\ \dot{y}_i &= -H_{x_i} = -V_{x_i}(x) \end{aligned} \qquad i = 1,2$$

where $H(x_1, x_2, y_1, y_2) = \frac{1}{2}(y_1^2 + y_2^2) + V(x_1, x_2)$. Assume that the level lines of V appear as depicted in Figure 1 (the surface $z = V(x)$ is a bowl with two pouring spouts). Now consider orbits with energy $H = h$; in the figure these orbits are confined to move in the (open ended) region between the top and bottom curves. On consideration of the "acceleration" vector $-V_x$ one sees that the orbits tangent to the line segments ℓ_1 and ℓ_3 "bounce" off to the left while those tangent to ℓ_2 and ℓ_4 bounce to the right. This implies the existence of three blocks in the energy surface $H = h$; namely

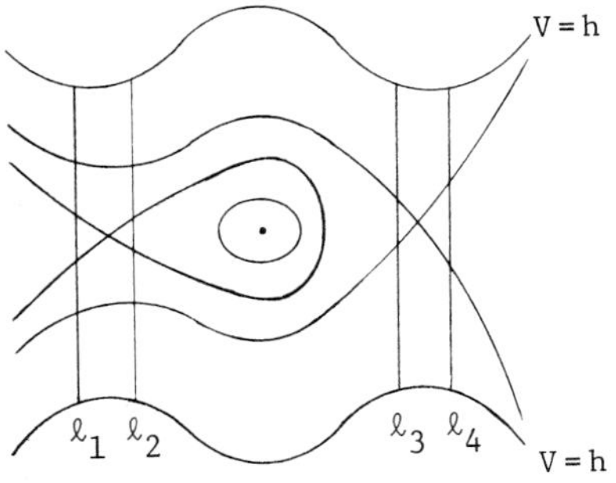

Fig. 1

$$B_1 \equiv \{(x,y) \mid x \text{ is between or on } \ell_1 \text{ and/or } \ell_2\}$$
$$B_2 \equiv \{(x,y) \mid x \text{ is between or on } \ell_3 \text{ and/or } \ell_4\}$$
$$B \equiv \{(x,y) \mid x \text{ is between or on } \ell_1 \text{ and/or } \ell_4\}$$

All of these blocks are homeomorphic to the product of a 2-sphere and an interval and the b^+ and b^- consist of two hemispheres, one on each bounding 2-sphere. Thus the indices of the isolated invariant sets S_1, S_2 and S are all $\Sigma^1 \vee \Sigma^2$ where Σ^p is the pointed p-sphere.

In particular, none of the invariant sets is empty (the index of the empty set is $\bar{0}$). Also, one concludes from the summation formula that any open set containing S_1 and S_2 must also meet orbits of S not contained in $S_1 \cup S_2$; if this were not the case, S would be the disjoint union of three isolated invariant sets $S_1 \amalg S_2 \amalg S_3$ and we would have $\Sigma^1 \vee \Sigma^2 = \Sigma^1 \vee \Sigma^2 \vee \Sigma^1 \vee \Sigma^2 \vee i^+(S_3)$ which is obviously not possible. Thus, given any $\varepsilon > 0$ there is an orbit not in $S_1 \cup S_2$ which stays in the bowl and which passes within ε of either S_1 or S_2.

If N is a closed subset of X such that the orbit through each boundary point of N meets the complement of N, then the maximal invariant set, S, in N is isolated; in fact N is an isolating neighborhood. The condition on N is preserved for all flows near the given one in the compact open topology; the corresponding isolated invariant sets are called "continuations" of S. A theorem proved in [1] for smooth flows and in [3] for flows on compact metric spaces (where it is considerably harder) states that the index is invariant under continuation. The technique of weak coupling makes use of the fact that the equations define a flow close to a product flow. Thus if one knows the indices of isolated invariant sets for each factor, one sees that for the weakly coupled systems there is an isolated invariant set with index equal to the product of the indices.

An algebraic index for isolated invariant sets is obtained on taking the cohomology of the homotopy index. Since for blocks, b^+ is always a strong deformation retraction of a neighborhood of itself in B, this algebraic index is just $H^*(B, b^+)$ and so sits in the exact sequence for the pair. Given B, one defines $a^+ \subset b^+$ to be the set of boundary points whose positive half orbit stays in B (hence these points tend to S as time increases). Similarly $a^- \subset b^-$ is defined as the set of points in B whose negative half orbit is contained in B. As shown in [2] one then has (using Čech groups) the commutative diagram

$$\begin{array}{ccccccc}
\xrightarrow{i} & H^*(b^+) & \xrightarrow{\delta} & H^*(B,\, b^+) & \xrightarrow{j} & H^*(B) & \xrightarrow{i} \\
 & \downarrow & & \downarrow & & \downarrow & \\
\xrightarrow{i'} & H^*(a^+) & \xrightarrow{\delta'} & H^*(i^+(S)) & \xrightarrow{j'} & H^*(S) & \xrightarrow{i'}
\end{array}$$

where the rows are exact, the vertical arrows on each end are induced by inclusion, and the middle one is an isomorphism.

Thus each cohomology class in $H^*(B, b^+)$ corresponds to an element of $H^*(a^+)/\mathrm{im}(i')$ or to a cohomology class in $H^*(S)$. Those in the range of δ must be of the former type and it follows that $K^+(S) \equiv H^*(a^+)/\mathrm{im}(i')$ is "stable" under continuation in that it cannot suddenly decrease. Namely, given an element in the range of δ', one can choose a block so this element is in the range of δ. The same will then be true of the continuation of the block to nearby flows and so the corresponding element will again be in the range of δ'. Even though $K^+(S)$ cannot jump down, it can jump up with a corresponding change in the relation between S and a^+. Points (in an appropriate space of isolated invariant sets) where this happens are called "algebraic bifurcation points". Stable cohomology classes in $H^*(S)$ can also be defined in terms of the algebraic index and similar "bifurcation" phenomena can be described. The framework for this study is developed in [3].

Given a flow on X, suppose g is a real valued function which is non-increasing on orbits. Let constants C_i; $i = 0, 1, \ldots, n$ be given such that $\max g = C_0 > C_1 > \cdots > C_n = \min g$ and such that g is strictly decreasing on orbits at each point of the level surfaces $\{g = C_i;\ i = 1, \ldots, n-1\}$. Then the n sets $B_i \equiv \{x \mid C_{i-1} \geq g(x) \geq C_i;\ i = 1, \ldots, n\}$ are isolating blocks with $b_i^+ = \{g(x) = C_{i-1}\}$ and $b_i^- = \{g(x) = C_i\}$ excepting respectively the cases $i = 1$ and $i = n$ where $b_1^+ = b_n^- = \emptyset$.

Assume now that $\beta_i^r = \text{rank } H^r(B_i, b_i^+)$ is finite for all r and i. Then $\beta_i^r = \alpha_i^{r-1} + \alpha_i^r$ where $\alpha_i^{r-1} = \text{rank (im } \delta')$ and $\sigma_i^r = \text{rank (im j')}$. It is easy to prove inductively that $\sum_{i=1}^{n} \beta_i^r \geq \beta^r \equiv \text{rank } H^*(X)$. The first step goes as follows:

Let B_* denote the block $B_i \cup B_{i+1}$ and let $\beta_*^r = \text{rank}$ $H^*(B_*, b_*^+) = \text{rank } H^*(B_*, b_i^+)$.

In the sequence for the triple B_*, B_i, b_i^+ one uses excision to replace $H^*(B_*, B_i)$ by $H^*(B_{i+1}, b_{i+1}^+)$ and so obtains

$$\xrightarrow{\delta} H^*(B_{i+1}, b_{i+1}) \longrightarrow H^*(B_*, b_1^+) \longrightarrow H^*(B_1, b_1^+) \xrightarrow{\delta} .$$

Thus $\beta_*^r \leq \beta_i^r + \beta_{i+1}^r$. The inductive step is clear.

In this way we prove a general version of the Morse-Smale inequalities for flows on arbitrary compact metric spaces:

$$\sum_{i=1}^{n} (\alpha_i^{r-1} + \sigma_i^r) \geq \beta^r .$$

J. Auslander has shown in [4] that for any flow on X there is a continuous function g which is non-increasing on orbits and which is universal in that if any such function is strictly decreasing at x, the g is also. The set of points where g is not decreasing is a closed invariant set of the flow defined by Auslander to be the generalized recurrent set. This set is characterized by him in terms of prolongations; for gradient flows or Morse-Smale flows it is just the non-wandering set. Using this theorem and calculating α^{r-1} and σ_i^r for elementary singular points and periodic orbits (which can be done using any block, in particular the ones constructed above for critical points) one obtains the usual Morse-Smale inequalities. The sharper versions, in the general case, also follow from similar arguments.

There exist isolated invariant sets which cannot be continued to the empty set in a small neighborhood of the given flow but which have index $\bar{0}$. One way to see this is to make a block with $[B / b^+] = [B / b^-] = \bar{0}$, but such that $\pi_1(B, b^+) \neq 0$. (A cube with a knotted hole will do; b^+ is the top, b^- the bottom and the sides are filled with orbits running from top to bottom - to prevent these from being in ∂B, cut down the space.) If S were empty, then b^+ would be a strong deformation retraction of B which cannot be true. Thus the homotopy index is not "sharp". On the other hand, we cannot use $\pi_*(B, b^+)$ as an index for S since it depends on the block chosen. This suggests that some limiting process (as the blocks shrink to S) might be used. Regardless of this possibility the index defined here has the advantage that once a block is known, the index is also; in the example of the Hamiltonian system above it was the blocks, not the invariant sets, which were easy to find.

Other papers on isolated invariant sets (not specifically concerned with the index) are listed in the bibliography.

REFERENCES

[1] C. CONLEY and R.W. EASTON, Isolated invariant sets and isolating blocks, to appear in Trans. A.M.S.

[2] R. CHURCHILL, Algebraic Relations Between Invariant Sets and Their Asymptotic Sets, Thesis, University of Wisconsin, 1971.

[3] J.T. MONTGOMERY, Perturbation Theorems for Flows on a Compact Metric Space, Thesis, University of Wisconsin, 1971.

[4] J. AUSLANDER, Generalized recurrence in dynamical systems, Contributions to Differential Equations 3 (1964), 65-74.

[5] C. CONLEY, Invariant sets which carry a one-form, J. Diff. Eq. 8 (1970), 587-594.

[6] C. CONLEY, On the continuation of invariant sets of a flow, to appear in Proc. Internat. Congr. Mathematicians, Vol. VIII, Nice, 1970.

[7] R.W. EASTON, On the existence of invariant sets inside a submanifold convex to a flow, J. Diff. Eq. 6 (1970), 54-68.

[8] R.W. EASTON, Locating invariant sets, to appear in the Proceedings of the A.M.S. Summer Institute on Global Analysis (S. Smale and S.S. Chern, Ed.).

[9] R.W. EASTON, A flow near a degenerate critical point, in Advances in Differential and Integral Equations (John Nohel, Ed.), W.A. Benjamin, Inc., New York (to appear).

[10] E. ROD, Invariant Sets in the Monkey Saddle, Thesis, University of Wisconsin, 1971.

[11] T. WAZEWSKI, Sur un principe topologique de l'éxamen de l'allure asymptotique des intégrales des équations differéntielles ordinaires, Ann. Soc. Polon. Math. 20 (1947), 297-313.

[12] T. WAZEWSKI, Sur une methode topologique de l'éxamen de l'allure asymptotique des intégrales des équations differéntielles, Proc. Internat. Congr. Mathematicians, Vol. III, Amsterdam, 1954.

University of Wisconsin, Madison, Wisconsin

EQUATIONS MODELLING POPULATION GROWTH, ECONOMIC GROWTH, AND GONORRHEA EPIDEMIOLOGY

Kenneth L. Cooke* and James A. Yorke**

1. A Population Model

In this paper we report on studies of certain classes of equations which can be interpreted as models of biological or economic processes. In this section we shall formulate these equations with an interpretation as models of growth of a single-species biological population. In the following two sections we describe an infectious disease model and a model of capital growth which give rise to the same equations. In Section 4 we state our principal mathematical results on these equations, with some biological interpretation. Proofs of some of the theorems are in Section 5. The other proofs and further discussion can be found in [1]. In Section 6 we discuss the special case in which our equations are linear.

Let $x(t)$ denote the number of individuals in a population at time t (or the biomass of the population), and assume that every individual has life span L, here assumed to be a fixed constant. Assume that the number of births per unit time is some function of $x(t)$, say $g(x(t))$. Inasmuch as every individual dies at age L, the number of deaths per unit time at time t is $g(x(t-L))$. Since the difference

---

* Partially supported by N.I.H. grant GM 16197.

**Partially supported by N.S.F. grant GP 27284.

$g(x(t)) - g(x(t-L))$ is the net change in population per unit time, the growth of the population is governed by the equation

$$x'(t) = g(x(t)) - g(x(t-L)) \ . \tag{1}$$

This is one of the equations to be considered here. In case $g(x) = ax$ where a is a constant, Eq. (1) is a linear model suggested by one of the authors in [2].

Our second population model allows a distribution of life spans. Let $P(a)$ be the proportion surviving (or probability of survival) to age (at least) a. As before, let $g(x(t))$ be the number of births per unit time. Then the number of deaths per unit time at time t is

$$-\int_0^L g(x(t-s))P'(s)ds$$

since $-P'(s)ds$ is the population dying at an age in $[s, s+ds]$. Here L is the maximum life span and $P(L) = 0$. It follows that*

$$\begin{aligned} x'(t) &= g(x(t)) + \int_0^L g(x(t-s))P'(s)ds \\ &= g(x(t)) - \int_0^L g(x(t-s))p(s)ds \end{aligned} \tag{2}$$

where

$$P(a) = \int_a^L p(s)ds \ , \qquad P(0) = 1 \ , \qquad p(s) \geqslant 0 \ . \tag{3}$$

In Eq. (2) we have assumed the existence of the density $p(s)$. More generally we could write

$$x'(t) = g(x(t)) + \int_0^L g(x(t-s))dP(s) \ , \tag{4}$$

using a Riemann-Stieltjes integral. For the sake of mathematical simplicity we shall treat Eqs. (1) and (2) but not (4) in this paper. However, it should be noted that Eq. (4) includes Eqs. (1) and (2) as

*If we imagine that the population growth process starts at a fixed time $t = 0$ with a given age distribution, equations (1) and (2) are valid for $t > L$ but not necessarily for $0 < t < L$. In any case our interest is in large t.

special cases, as well as the equation

$$x'(t) = g(x(t)) - \sum_{j=1}^{N} p_j g(x(t-L_j)) \; . \tag{5}$$

This equation represents the case in which there are a finite number of possible life spans and a proportion p_j of individuals die at precisely age L_j (and $\sum_{j=1}^{N} p_j = 1$).

Remark. Equations (1) and (2) have first integrals, namely

$$x(t) = \int_{t-L}^{t} g(x(s))ds + c \tag{6}$$

$$x(t) = \int_{t-L}^{t} P(t-s)g(x(s))ds + c \; . \tag{7}$$

By inspection, the integral term in Eq. (6) is the number born in the preceding generation and the integral term in Eq. (7) is the number of individuals born in the past life span who are still alive. To have a correct biological interpretation we must have $c = 0$, since $x(t)$ is the size of the population. However, any solution of (6) or (7) is a solution of (1) or (2), respectively, no matter what c is, since by differentiating both sides of (6) and (7) we get (1) and (2).

If we let $r(t) = g(x(t))$, the number of births per unit time, we can derive an integral equation for $r(t)$ from (6) or (7). In fact we obtain

$$r(t) = g\left(\int_{t-L}^{t} r(s)ds + c \right) \tag{8}$$

$$r(t) = g\left(\int_{t-L}^{t} P(t-s)r(s)ds + c \right) . \tag{9}$$

Bartlett [3] has briefly discussed a linear equation similar to (9) which takes into account also that the number of offspring produced may depend on the age of the parent.

Lotka in [4] had formulated essentially the same model as our Eq. (2). See also [5], [6], in the latter of which he solved a linear integral equation similar to Eq. (7). Lotka's main assumptions were as follows.

1. The individuals are either all of one class as regards their general properties, and especially as regards those which affect the character of the limitation of their "life period"; or, if they belong to a number of different classes (e.g., males and females of a community of living organisms, etc.) then the relative proportion of individuals of each class among those formed during any element of time is constant.

2. The "length of life" of each individual is independent of the total number of individuals in the aggregate, and of the distribution of ages among them.

3. The general conditions of the system, in so far as they affect the "length of life" of the individuals are, on the average, uniform and constant throughout.

We assume, in addition, that:

4. The number of births per unit time is a function only of the population size.

2. Infectious Disease Model for Gonorrhea

The objective in a study of models such as these is not so much to obtain precise numerical predictions as it is to discover what various hypotheses imply about the general qualitative nature of the physical system.

Suppose we consider a population (of constant size N) which is attacked by an infectious disease. Let $x(t)$ $(0 \leq x(t) \leq N)$ be the size of the infective population, those individuals who have the disease and are capable of infecting others. Assume we are discussing a disease for which having the disease instills negligible immunity. Then the remainder of the population, $u(t) = N - x(t)$, is susceptible. We assume that the rate at which the susceptibles are infected depends only on the present size of infective and susceptible populations. Since N is constant, this rate depends only on $x(t)$ and may be written $g(x(t))$. Assume for the moment that every individual who catches the disease has the disease for time L (after which he becomes cured but susceptible). Hence people are infected at the rate $g(x(t))$ and are cured at the rate at which they were catching the disease time L ago, $g(x(t-L))$, so x satisfies Eq. (1). If we instead assume a variable cure time we find that Eq. (2) is satisfied and $P(a)$ is the probability of having the disease continuously for at least time a after the initial infection. It may also be seen that x satisfies Eqs. (6) and (7), respectively, with constant $c = 0$. If $c > 0$, this is interpreted as meaning that there is a constant subpopulation of size c of incurable infectious carriers of the disease.

The assumption that there is essentially no immunity gained by having the disease once or more is reasonable for gonorrhea (see [7]). Some diseases give "negative immunities" because the individual is weakened by having the disease and is more susceptible after having had it. If we assume that all individuals are equally susceptible it is reasonable to assume the rate of contraction of the disease depends linearly on the size of susceptible and infectious populations; that is, $g(x) = \gamma x(1-x)$ for some constant γ. In the case of venereal disease, some are definitely more susceptible than others.

Actually in the study of gonorrhea there are two populations to consider, the infected males and the infected females. Clinical treatment is necessary for cure. For males, gonorrhea is easily detected and is painful even a few days after initial infection, whereas with females infection can go undetected. Females can be infectious and be able to transmit the disease without even knowing they have it. Females represent the main reservoir of the disease in the population. It then seems reasonable to hypothesize that the number of males with the disease is (approximately) directly proportional to the size of this reservoir, that is, the number of infectious females.

Let r_m and r_f be the (time-independent) proportions of the population which are respectively male and female. Let $P_m(s)$ and $P_f(s)$ be respectively the fractions of the infected population which take longer than time s to be cured after the infection begins (so $P_m(0) = 1 = P_f(0)$). We take L to be any very large time so that $P_m(L)$ and $P_f(L)$ can be taken to be 0. We get Eq. (7) by letting $x(t)$ be the size of the total infected population and letting

$$P(s) = r_m P_m(s) + r_f P_f(s) .$$

Unlike most other diseases, gonorrhea is primarily nonseasonal. Actually there is a small seasonal component. About 10% more cases are reported each year during the summer months [8] than during the other seasons of the year. Incidentally, during these months, <u>fewer</u> than average illegitimate conceptions are reported, so the seasonal variation would not seem to be purely a result of the seasonal nature of promiscuity. For a report on the recent increases in the incidence of gonorrhea, see [9].

3. An Economic Interpretation

Let $x(t)$ denote the value of a capital stock at time t. We assume that the production of new capital depends only on $x(t)$ and the rate of production is $g(x(t))$. We assume equipment depreciates over a time L (the lifetime of the equipment) to value 0. Assume depreciation to be independent of the type of equipment and at time a after production the value of a unit of capital equipment has decreased in value by the factor $P(a)$ (so $P(0) = 1$, $P(L) = 0$). At any time, $x(t)$ equals the sum of the capital produced over the period $[t-L, t]$ plus a constant c denoting the value of non-depreciating assets; that is,

$$x(t) = \int_0^L P(a)g(x(t-a))da + c$$

which is the same as Eq. (7) (letting $s = t - a$). Eq. (2) is the differential equation form where $p(a) = \frac{-d}{da} P(a)$ is the rate of decrease in the value at age a for a capital unit whose value at production is 1.

4. Principal Results

We shall now state our principal results and give brief biological interpretations. We will let $x(t)$ denote a solution of Eq. (1) or (2), defined on $-L \leq t < T$. It is easy to see that if x cannot be extended to an interval with larger T then $|x(t)| \to \infty$ as $t \to T$. This is also true for Eq. (2) but we will not prove this; see [10]. We assume g is such that $T = \infty$ from now on. Our methods do not require this, but the notation becomes simpler. This assumption only says that we are considering populations whose size is a real number for each (large) time (allowing the population to be 0). It is sufficient to have, for some K, $|g(x)| \leq K + K|x|$.

Theorems 1 and 2 and the lemma and corollaries are true for any solution x of either Eq. (1) or Eq. (2). We usually discuss the more complicated Eq. (2) and let the reader supply the similar statements and proofs for the simpler Eq. (1).

THEOREM 1. Let $g(x)$ be continuously differentiable and such that every solution is continuable to $t = +\infty$. Suppose $t_0 \geq 0$ and let $x(t)$ be a solution of Eq. (1) or (2) satisfying

$$g(x(t_0)) \geq g(x(t_0 - s)) \quad \text{for all} \quad s \in [0,L] \; . \tag{10}$$

Then $x(t) \geq x(t_0)$ for all $t \geq t_0$.

This result has a biological interpretation. For the model of Section 1 it says that if the number of births per unit time is as large or larger now (time t_0) than at every time during the previous life span (times $t_0 - s$ for $s \in [0,L]$), then in the future the population will never be smaller than it is now. For the model of Section 2 it says that if the rate of new cases of infection is as large or larger now (time t_0) than at every time during $[t_0 - L, t_0]$ then in the future the number of infectives will never be smaller than it is now. The corresponding result with signs reversed is also true, that is,

THEOREM 1'. Suppose $t_0 \geq 0$ and

$$g(x(t_0)) \leq g(x(t_0 - s)) \quad \text{for all} \quad s \in [0,L] \; .$$

Then $x(t) \leq x(t_0)$ for all $t \geq t_0$.

This says that if the number of births per unit time (rate of infection) is now at least as small as every time during the past lifespan (past time period L), then the population (number of infectives) will never be larger than it is now.

COROLLARY. Suppose the hypotheses of Theorem 1 are satisfied and also assume g is a monotonic non-decreasing function. Then $x(t)$ is a monotonic non-decreasing function for $t \geq t_0$ (where t_0 is the same as in (10)).

THEOREM 2. Let $g(x)$ be continuously differentiable and such that every solution x of Eq. (1) or (2) is continuable to $t = \infty$. Then every solution satisfies

$$\begin{array}{llll} & \text{(i)} & x(t) \to +\infty & \text{as } t \to \infty \\ \text{or} & \text{(ii)} & x(t) \to \text{constant} & \text{as } t \to \infty \\ \text{or} & \text{(iii)} & x(t) \to -\infty & \text{as } t \to \infty . \end{array}$$

In terms of the model of Section 2, the conclusion of Theorem 2 is that the size of the infectious population $x(t)$ tends to constant size as $t \to \infty$; that is, the disease is endemic rather than epidemic. Of course, this model ignores time dependent (for example, seasonal) changes in the susceptibility of the population. It would be interesting if it could be proved that if g is periodic in time and nice in other ways there is a unique non-zero periodic solution of Eq. (6) or (7) (for each c) to which all non-zero solutions tend asymptotically.

Case (iii) has no biological or economic interpretation and in fact it does not seem reasonable to allow x to be negative. Our objective here is to include as many reasonable cases as possible rather than to exclude unreasonable cases.

Various corollaries of Theorem 2 are stated in Section 5, and additional results when g is a linear function are given in Section 6.

A modified version of our first problem postulates a time lag τ between conception and birth. Then the number of births at time t is $g(x(t-\tau))$ rather than $g(x(t))$. Eq. (1) and (2) must then be replaced by

$$x'(t) = g(x(t-\tau)) - g(x(t-\tau-L)) \tag{11}$$

$$x'(t) = g(x(t-\tau)) - \int_0^L g(x(t-\tau-s))p(s)ds . \tag{12}$$

In the infectious disease model, one can similarly assume that the rate at which susceptibles catch the disease depends on the size of the infective and susceptible populations a time τ in the past. In other words, there is a latent time τ between the time of exposure and the time that a susceptible becomes infectious. If $x(t)$ denotes the size of the infectious population and $S(t)$ the size of the susceptible population, we then obtain the equations (in the case analogous to (11))

$$S'(t) = -g(x(t), S(t)) + g(x(t-\tau-L), S(t-\tau-L))$$

$$x(t) = \int_{t-\tau-L}^{t-\tau} g(x(u), S(u))du \ .$$

Here $g(x,S)$ denotes the rate of new infections or exposures when the infective and susceptible populations are x and S, respectively, and we continue to assume that after recovery all individuals become susceptible again. These equations are generalizations of the Wilson-Burke model of infection [11].

These equations are mathematically more difficult than (1) and (2). Our only results are for the linear case of (11) and are given in Section 6. The linear case of Eq. (12) was formulated and partially analyzed by Wangersky and Cunningham [12] under the assumption that

$$P(s) = [1 - e^{-k(L-s)}][1 - e^{-kL}]^{-1} \ ,$$

where k is a parameter.

5. Proof of Theorem 1

For the following result we use only the fact that g is locally Lipschitzean, rather than "g is continuously differentiable." We say g is <u>locally Lipschitzean</u> if for each bounded interval $[\alpha,\beta]$ there is a constant γ (which may depend on α and β) such that

$$|g(y_1) - g(y_2)| \leqslant \gamma|y_1 - y_2| \quad \text{for all } y_1, y_2 \in [\alpha,\beta] \ .$$

The lemma says that if Eq. (10) is true $x(t)$ and $g(x(t))$ must increase before $x(t)$ can decrease.

<u>LEMMA 1</u>. Suppose Eq. (10) is true. If there is a $t_2 > t_0$ with $x(t_2) < x(t_0)$, then there must exist some $t_1 \in (t_0,t_2)$ such that

(13) $$x(t_1) > x(t_0) \quad \text{and} \quad g(x(t_1)) > g(x(t_0)) .$$

Proof. Let $t_2 > t_0$ and let $x(t)$ be a solution satisfying (10) with $x(t)$ defined for $t \in [t_0-L, t_2]$. Assume there is no $t_1 \in [t_0, t_2]$ such that Eq. (13) is true. We will prove that then $x(t_2) \geq x(t_0)$, which will complete the proof of the lemma. Define $g_0(x) = g(x(t_0))$ if $x > x(t_0)$ and $g(x) > g(x(t_0))$. Let $g_0(x) = g(x)$ elsewhere. Then for $t_1 \in [t_0-L, t_2]$

$$g_0(x(t_1)) = g(x(t_1)) .$$

Otherwise there would be a point $t_1 \in (t_0, t_2)$ where (13) is satisfied, or a point $t_1 \in [t_0-L, t_0]$ where $g(x(t_1)) > g(x(t_0))$, and the latter contradicts Eq. (10). Hence $x(t)$ satisfies Eq. (2) replacing g by g_0 everywhere.

Let $x_n(t)$ be a solution of

(14) $$x_n'(t) = \frac{1}{n} + g_0(x(t)) - \int_0^L p(s) g_0(x_n(t-s))ds \quad \text{for} \quad t \geq t_0$$

$$x_n(t) = x(t) , \qquad t \in [t_0-L, t_0] .$$

Notice that the right-hand derivative $x_n'(t_0) \geq n^{-1}$ so that $x_n(t) > x(t_0)$ for $t \in (t_0, t_0 + \delta)$ for some $\delta > 0$.

Claim: $x_n(t) \geq x(t_0)$ for all $t > t_0$. Suppose not. Then there is an n and $\tau > t_0$ such that $x_n(\tau) = x(t_0)$. Since $x_n'(t_0) \geq n^{-1}$, we may let τ be the smallest number strictly greater than t_0 such that $x_n(\tau) = x(t_0)$. However, since $x_n(t) \geq x(t_0)$ for $t \in [t_0, \tau]$, it follows that $g_0(x_n(t)) \leq g_0(x(t_0))$ for $t \in [t_0-L, \tau]$ and so

(15) $$x_n'(\tau) \geq \frac{1}{n} + g_0(x(t_0)) - \int_0^L p(s) g_0(x(t_0))ds \geq \frac{1}{n}$$

contradicting the definition of τ. Hence no such τ exists and the claim is proved.

Using the standard techniques of the Ascoli-Arzela theorem as usually applied to differential equations (see for example [13]), it follows that there is a function $x_0(t)$ and a subsequence $\{x_{n_i}(t)\}$

such that

$$x_0(t) = x(t) \quad \text{for} \quad t \in [t_0-L, t_0]$$

and $x_0(t)$ satisfies Eq. (2) and for each $T > 0$

$$x_{n_i}(t) \to x_0(t)$$

uniformly for $t \in [t_0-L, T]$. The fact that this sequence converges for all $T > 0$ follows from the assumption that all solutions of Eq. (2) can be defined for all large t; that is, $\sup(\text{domain } x(\cdot)) = \infty$. As a result for any $t > t_0$, $x_{n_i}(t)$ is defined for all but finitely many n. Since $x_n(t) > x(t_0)$ for $t > t_0$, it follows that $x_0(t) \geq x(t_0)$ for $t > t_0$. Since solutions of (2) are assumed to be uniquely determined by initial data (g is assumed to be locally Lipschitzean), it follows that $x(t) \equiv x_0(t)$ for $t \geq t_0-L$, and this lemma is proved.

Proof of Theorem 1: Suppose the result is false and there is some $t_2 > t_0$ with $x(t_2) < x(t_0)$. Consider the case where

$$g(y) \leq g(x(t_0)) \quad \text{for all} \quad y \leq x(t_0) \; . \tag{16}$$

Let J be the (closed) set of $t \in [t_0,t_2]$ such that $x(t) \geq x(t_0)$. Let γ be $\max\{g(x(t)) : t \in [t_0,t_2]\}$. Choose $t_1 \in J$ so that $g(x(t_1)) = \gamma$. We now apply Lemma 1 at t_1 (rather than t_0). We have $x(t_2) < x(t_1)$ but there exists no intermediate $t \in (t_1,t_2)$ with $x(t) > x(t_1)$ and $g(x(t)) > g(x(t_1))$, which contradicts the result in Lemma 1. Hence there is no $t_2 > t_0$ with $x(t_2) < x(t_0)$ and the theorem is proved in this case.

Consider the case in which (16) is not true. Define $g_2(y) = g(y)$ if either $y \geq x(t_0)$ or $g(y) \leq g(x(t_0))$. For all other y, let $g_2(y) = g(x(t_0))$. Let $x_2(t) = x(t)$ for $t \in [t_0-L,t_0]$ and for $t \geq t_0$ let x_2 satisfy Eq. (2), substituting g_2 for g. Since (16) holds for g_2, we have proved $x_2(t) \geq x(t_0)$ for all $t \geq t_0$. Therefore (by definition of g_2) $g(x_2(t)) = g_2(x_2(t))$, and x_2 also satisfies Eq. (2) (using g rather than g_2). By uniqueness of solutions $x(t) \equiv x_2(t)$ so the theorem is now proved.

The proof of the corollary follows immediately from Theorem 1 since if this result were not true, there would be a local maximum for both $x(t)$ and $g(x(t))$. This corollary does not say that such an $x(\cdot)$ would be unbounded. It could be bounded and asymptotically approach a constant value without ever achieving that value at a finite time.

The proof of Theorem 2 is given in [1].

Remark. It is clear that $x(t) = k$ is a solution of Eqs. (1) and (2) for *any* k. However, if we restrict attention to the biologically meaningful case, which is described by (6) or (7) with $c = 0$, only certain constants k provide solutions. In fact, k must satisfy

$$(17) \qquad k = Lg(k) \quad \text{or} \quad k = g(k)\int_0^L P(u)du$$

for Eq. (6) or (7), respectively. Such k are also the biologically meaningful constant solutions of Eq. (1) or (2). Hence the possible limiting values of $x(t)$ are those k for which (17) is satisfied.

Remark on boundedness and unboundedness. We now discuss the special case of Eq. (7) for which $c = 0$; that is,

$$(¤) \qquad x(t) = \int_0^L P(s)g(x(t-s))ds \ .$$

COROLLARY. Let $\pi_L = \int_0^L P(s)ds$. If $\pi_L g(y) \leq y$ for all large values of y, then all solutions of Eq. (¤) will be bounded above. If g is bounded, then each solution of Eq. (2) will be bounded.

Proof. We now prove the second half of the result. The first half is similar and the proof is omitted. If for some B, we have $g(y) < B$ for all y, then

$$(18) \qquad x(t) \leq \int_{t-L}^{t} P(t-s)Bds + c \leq BL + c \ .$$

Similarly if $g(y) \geq 0$ for all y and if $c \geq 0$, then

$$x(t) = \int_{t-L}^{t} P(t-s)g(x(s))ds + c \geq 0$$

and the case (iii) could not occur.

It is not clear whether (i) of Theorem 2 implies that $x(t)$ is monotonic on some interval $[T,\infty)$.

COROLLARY. If there is some $y_0 > 0$ such that

$$(\ast\ast) \qquad \pi_L g(y) > y \quad \text{for all } y \geq y_0$$

then there is a solution $x(t)$ of ($\ast$) such that $x(t) \to \infty$ as $t \to \infty$.

Proof. Let $x(t) = y_0$ for $t \in [-L,0)$ and let $x(0) = \pi_L g(y_0)$. For $t > 0$ let $x(t)$ be the unique solution of ($\ast$). Notice that $x(0)$ is chosen so that ($\ast$) is satisfied at $t = 0$. Also notice from Eq. ($\ast\ast$) that $x(0) > y_0$.

For $s \in (0,L]$ we have by ($\ast\ast$)

$$g(x(-s)) = g(y_0) = \frac{x(0)}{\pi_L} < g(x(0)) .$$

Therefore Eq. (10) is satisfied with $t_0 = 0$. It follows from Theorem 1 that $x(t) \geqslant x(0) > y_0$ for $t \geqslant 0$. From Theorem 2 we now know that either $x(t) \to \infty$ or $x(t) \to x_0$ (for some $x_0 > y_0$) as $t \to \infty$. To see that the latter is impossible notice that

$$x_0 = \lim_{t\to\infty} x(t) = \lim_{t\to\infty} \int_0^L P(s)g(x(t-s))ds = \int_0^L P(s)g(x_0)ds = \pi_L g(x_0) ,$$

which contradicts Eq. ($\ast\ast$). From Theorem 2, the only remaining possibility is that $x(t) \to \infty$ as $t \to \infty$, proving the corollary.

Remark on when the population cannot vanish.

COROLLARY. Assume $g(0) = 0$ and $g(y) \geqslant 0$ for all $y \geqslant 0$. Assume further than for some $\delta > 0$

$$g(y) > 0 \quad \text{for } y \in (0,\delta) .$$

Let $x(t)$ be a solution of ($\ast$) such that $x(t_0) > 0$ and $x(t) \geqslant 0$ for $t \in [t_0-L,t_0]$. Then

$$x(t) > 0 \quad \text{for all } t \geqslant t_0 .$$

Proof. Suppose the result is false. Let t_1 be the first time after t_0 such that $x(t_1) = 0$. This immediately contradicts (¤) (letting $t = t_1$) since the right-hand side of (¤) would be positive.

6. Linear Equations

In this section we study constant solutions of equations (1), (2), (6), and (7). As is customary in the stability theory of nonlinear differential equations, we shall now examine solutions of (6) or (7) which are close to constant solutions. Let k satisfy Eq. (7) and define

$$y(t) = x(t) - k \; . \tag{19}$$

Substituting into Eq. (6) we get

$$k + y(t) = \int_{t-L}^{t} [g(k) + g'(k)y(s) + \cdots]ds \; .$$

Discarding all but linear terms and using (17) we obtain the "variational equation"

$$y(t) = a\int_{t-L}^{t} y(s)ds \tag{20}$$

where $a = g'(k)$. All solutions of (20) also satisfy

$$y'(t) = a[y(t) - y(t-L)] \; . \tag{21}$$

The same procedure applied to Eq. (7) yields

$$y(t) = a\int_{t-L}^{t} P(t-s)y(s)ds \; ; \tag{22}$$

all differentiable solutions of Eq. (22) also satisfy (since $P(0) = 1$, $P(L) = 0$)

$$y'(t) = ay(t) - a\int_{0}^{L} p(s)y(t-s)ds \; . \tag{23}$$

The solutions of the linear equations (20)-(23) can be studied in terms of associated characteristic roots. For example, let $y = e^{\lambda t}$ and substitute into Eqs. (22) and (23). We obtain

(24) $$1 = a\int_0^L P(s)e^{-\lambda s}ds$$

(25) $$\lambda = a - a\int_0^L p(s)e^{-\lambda s}ds$$

respectively. The distribution of the roots λ of these equations is as follows.

THEOREM 3. Equations (24) and (25) have identical roots except for $\lambda = 0$. All non-real roots have negative real parts. The real roots of Eq. (24) are as follows. Let $a_0 = [\int_0^L P(s)ds]^{-1} = \pi_L^{-1}$.

(i) If $a \leqq 0$, there are no real roots.
(ii) If $0 < a < a_0$, there is one simple negative root.
(iii) If $a = a_0$, there is a simple root at $\lambda = 0$.
(iv) If $a > a_0$, there is one simple positive root.

There are no other real roots of Eq. (24), but Eq. (25) has an additional simple root at $\lambda = 0$ (so in case (iii) the root at 0 is a double root).

The proof of Theorem 3 is given in [1].

COROLLARY. The zero solution of Eq. (22) or (23) is stable if $a \leqq a_0$. When $a \leq a_0$, all solutions approach constant limits as $t \to +\infty$. If $a < a_0$, all solutions of Eq. (22) approach zero. If $a > a_0$, there are solutions of Eqs. (22) and (23) which are unbounded as $t \to +\infty$.

Remark. The proof of this corollary is omitted. A proof can be given of the type in [14]. (See Exercise 6 on page 112.)

The characteristic equations for Eqs. (20) and (21) are

(26) $$1 = a\int_0^L e^{-\lambda u}du$$

and

(27) $$\lambda = a[1 - e^{-L\lambda}]$$

respectively. The same kind of analysis leads to the following theorem. The distribution of roots of an equation equivalent to Eq. (27) has been thoroughly studied by Wright [15]. His results are more detailed than in Theorem 4, but do not directly include Theorem 4.

THEOREM 4. All non-real roots of Eqs. (26) and (27) have negative real parts. The real roots of Eq. (26) are located as follows:

(i) If $a \leqq 0$, there are no real roots.
(ii) If $0 < aL < 1$, there is a simple negative root.
(iii) If $aL = 1$, there is a simple root at $\lambda = 0$.
(iv) If $aL > 1$, there is a simple positive root.

There are no other real roots of Eq. (26), but Eq. (27) has an additional simple root at $\lambda = 0$ (in case (iii), 0 is a double root).

COROLLARY. The zero solution of Eq. (20) or (21) is stable if $aL \leqq 1$; all solutions approach constant limits as $t \to +\infty$. If $aL < 1$, all solutions of Eq. (20) approach zero. If $aL > 1$, there are solutions of Eqs. (20) and (21) which are unbounded as $t \to +\infty$.

Proof. This corollary is a consequence of well-known results for differential-difference equations, see [14]. It can alternatively be proved as a corollary of Theorems 1 and 2.

Finally, let us consider Eqs. (11) and (12), and their integrated forms

$$x(t) = \int_{t-L-\tau}^{t-\tau} g(x(s))ds + c \tag{28}$$

$$x(t) = \int_{t-L-\tau}^{t-\tau} P(t - s - \tau)g(x(s))ds + c \, . \tag{29}$$

In terms of our first biological model, the integral in Eq. (28) is the number of individuals born in the past generation $[t-L, t]$ (conceived in $[t-L-\tau, t-\tau]$). The integral in Eq. (29) is the number born in $[t-L,t]$ (conceived in $[t-L-\tau, t-\tau]$) and still alive. To have a correct biological interpretation we must have $c = 0$, but any solution of (28) or (29) is a solution of (11) or (12), respectively.

Any constant $x(t) = k$ is a solution of Eqs. (11) and (12), but $x(t) = k$ is a solution of Eq. (28) or (29) with $c = 0$ if and only if Eq. (17) is satisfied. Linearizing around such constant solutions (if any) we obtain the variational equations

(30) $$y(t) = a\int_{t-L-\tau}^{t-\tau} y(s)ds$$

(31) $$y(t) = a\int_{t-L-\tau}^{t-\tau} P(t-s-\tau)y(s)ds$$

where $a = g'(k)$. The differentiated forms of these equations are respectively

(32) $$y'(t) = a[y(t-\tau) - y(t-L-\tau)]$$

and

(33) $$y'(t) = a[y(t-\tau) - \int_0^L p(s)y(t-\tau-s)ds] .$$

We shall now study the characteristic equations of Eqs. (30) and (31), which are

(34) $$1 = ae^{-\tau\lambda}\int_0^L e^{-\lambda u}du$$

(35) $$\lambda = a[e^{-\tau\lambda}-e^{-(L+\tau)\lambda}]$$

respectively. These equations have the same roots except for $\lambda = 0$, which is always a root of Eq. (35) but is a root of Eq. (34) if and only if $1 = aL$.

<u>THEOREM 5.</u> The real roots of Eq. (34) are as follows.

(i) If $a \leq 0$, there are no real roots.
(ii) If $0 < a < L^{-1}$, there is one simple negative root.
(iii) If $a = L^{-1}$, there is a simple root at $\lambda = 0$ and no other real roots.
(iv) If $a > L^{-1}$, there is one simple positive root.

All pure imaginary roots can be obtained as follows. Let r be any non-zero integer and let

$$d = \frac{2\pi r}{L+2\tau} .$$

Then $\lambda = id$ is a root of Eq. (34) for the value

$$a = -\frac{1}{2} \frac{d}{\sin \tau d}$$

provided $\sin \tau d \neq 0$, but if $\sin \tau d = 0$, there is no choice of a for which $\lambda = id$ is a root. Moreover, if $\lambda = id$ is a root of Eq. (34) for a positive value of a, aL is greater than one and the equation also has a positive real root.

The proof of Theorem 5 is given in [1].

Remark. It follows from this theorem that there are positive values of a for which Eqs. (30) and (32) have periodic solutions. However, these solutions cannot be stable since whenever they occur there is also a positive root of the characteristic equation.

The complex roots $\lambda = b + id$ of Eq. (35) satisfy the equations

$$\frac{b}{a} = e^{-\tau b} \cos \tau d - e^{-(L+\tau)b} \cos(L+\tau)d$$

$$\frac{d}{a} = -e^{-\tau b} \sin \tau d + e^{-(L+\tau)b} \sin(L+\tau)d .$$

We have not determined the conditions for stability of Eqs. (30) or (32).

Acknowledgements

We would like to thank Melvin Meer and Wayne London for discussions respectively on the economic growth model and the infectious disease model. We also wish to thank Sandy Grabiner for pointing out an error in our original proof of Lemma 1.

REFERENCES

[1] K.L. COOKE and J.A. YORKE, Some equations modelling growth processes and gonorrhea epidemiology, Math. Biosciences, forthcoming.

[2] K.L. COOKE, Functional-differential equations: some models and perturbation problems, Differential Equations and Dynamical Systems, New York, Academic Press, 1967, 167-183.

[3] M.S. BARTLETT, Stochastic Population Models. London, Methuen and Co., Ltd., 1960.

[4] A.J. LOTKA, Studies on the mode of growth of material aggregates, Amer. J. Science 24 (1907), 199-216.

[5] A.J. LOTKA, Relation between birth rates and death rates, Science 26 (1907), 21-22.

[6] A.J. LOTKA, A problem in age distribution, Philosophical Magazine, Ser. 6, Vol. 21 (1911), 435-438.

[7] A.S. BENSON, editor, Control of Communicable Diseases. American Public Health Association, New York, 1970, 97.

[8] C.E. CORNELIUS III, Seasonality of Gonorrhea in the United States, HSMHA Health Reports, 86 (1971), 157-160.

[9] "The gonorrhea epidemic," Newsweek, April 26, 1971, p. 54.

[10] J.A. YORKE, Noncontinuable solutions of differential-delay equations, Proc. Amer. Math. Soc. 21 (1969), 648-652.

[11] E.B. WILSON and M.H. BURKE, The epidemic curve, Proc. Nat. Acad. Sci. USA, 28 (1942), 361-367.

[12] P.J. WANGERSKY and W.J. CUNNINGHAM, Time lag in population models, Cold Spring Harbor Symposia on Quantitative Biology 22 (1957), 329-338.

[13] A. STRAUSS and J.A. YORKE, On the fundamental theory of differential equations, SIAM Rev. 11 (1969), 236-246.

[14] R. BELLMAN and K.L. COOKE, Differential-Difference Equations. Academic Press, New York, 1963.

[15] E.M. WRIGHT, A non-linear difference-differential equation, J. Reine u. angew. Math., Band 194 (1955), 66-87.

KLC: Pomona College, Claremont, California

JAY: University of Maryland, College Park, Maryland

ABSOLUTE STABILITY OF SOME INTEGRO-DIFFERENTIAL SYSTEMS

C. Corduneanu

1. The aim of this paper is to investigate the absolute stability of some classes of integro-differential systems of the form

$$(S) \qquad \begin{cases} \dot{x}(t) = Ax(t) + \int_0^t B(t-s)x(s)ds + b\phi(\sigma(t)) , \\ \sigma = \langle c,x \rangle , \end{cases}$$

where x, b, c are real n-vectors, A is a constant $n \times n$ matrix, $B(t)$ is a measurable $n \times n$ matrix kernel such that $\|B(t)\| \in L(R_+,R)$ and $\phi(\sigma)$ is a real-valued function defined on the real line R.

We shall also consider some systems related to (S) and we shall find appropriate conditions that assure the absolute stability.

Let us notice that in the particular case $B(t) \equiv 0$, the system (S) reduces to the well known differential system of automatic control theory

$$(\Sigma) \qquad \dot{x}(t) = Ax(t) + b\phi(\sigma(t)) , \quad \sigma = \langle c,x\rangle ,$$

which has been thoroughly investigated by many authors (see [1], [8], [11], [12]).

The method we shall use in this paper consists in finding a nonlinear Volterra integral equation for $\sigma = \sigma(t)$ and then to apply various stability results which are available (see, for instance, [2], [3], [6], [13], [16], [25]).

The key result we need in the subsequent sections is due to S. I. Grossman and R. K. Miller [7] and regards the integro-differential system

$$\dot{x}(t) = Ax(t) + \int_0^t B(t-s)x(s)ds \ . \tag{1.1}$$

The resolvent matrix $R(t)$ associated with (1.1) is defined by

$$\dot{R}(t) = AR(t) + \int_0^t B(t-s)R(s)ds \ , \qquad t \in R_+ \ , \tag{1.2}$$

$$R(0) = I = \text{unit matrix of order } n \ . \tag{1.3}$$

<u>THEOREM 1.1</u>. (Grossman, Miller). Let A be a constant n by n matrix and $B(t)$ a measurable n by n matrix such that $\|B(t)\| \in L(R_+,R)$. A necessary and sufficient condition in order that $\|R(t)\| \in L(R_+,R)$ is given by

$$\det(sI - A - \tilde{B}(s)) \neq 0 \qquad \text{for} \qquad \operatorname{Re} s \geq 0 \ , \tag{1.4}$$

where $\tilde{B}(s)$ is the Laplace transform of $B(t)$, i.e.

$$\tilde{B}(s) = \int_0^\infty e^{-st}B(t)dt \ . \tag{1.5}$$

Condition (1.4) will play a central role in the sequel. It is useful to be pointed out that in the case of ordinary differential systems, i.e., when $B(t) \equiv 0$, it reduces to the well known stability condition for the matrix A.

2. Before going into details with respect to the system (S) or related integro-differential systems, we find it necessary to make a few comments concerning the application of Popov's frequency method to the investigation of absolute stability for various kinds of systems in which the delays occur.

The first of Popov's papers concerned with frequency domain stability criteria [16] deals with feedback systems described by the nonlinear integral equation

$$\sigma(t) = h(t) + \int_0^t k(t-s)\phi(\sigma(s))ds \ , \qquad t \in R_+ \ . \tag{E}$$

His subsequent papers [17], [18] treat the absolute stability problem for differential systems of the form (Σ).

In their joint paper [9], A. Halanay and V. M. Popov are concerned with systems involving delays, namely

$$(\Sigma_1) \qquad \dot{x}(t) = Ax(t) + Bx(t-\tau) + b\,\phi(\sigma(t-\tau)) \ , \qquad \sigma = \langle c,x \rangle \ ,$$

where $\tau > 0$ is a constant. They have proved that the results obtained in [17] can be easily adapted to the system (Σ_1).

In the author's papers [2], [3], the integral equation (E) is considered and using both frequency methods and compactness arguments, it was shown that there exists at least one solution of equation (E) defined on the positive half-axis R_+ and the behavior of such a solution is that required by the property of absolute stability.

Various systems related to (Σ_1) have been investigated by several authors and Halanay's book [8] contains some absolute stability results concerning systems with delay. In particular, the method of reducing such systems to integral equations is emphasized. Quite recently, Vl. Răsvan [22] has obtained frequency conditions for the absolute stability of automatic control systems of the form

$$(\Sigma_2) \qquad \begin{aligned} \dot{x}(t) &= Ax(t) + \sum_{k=1}^{r} B_k x(t-\tau_k) + b\,\phi(\sigma(t)) \ , \\ \sigma(t) &= \langle c_0, x(t) \rangle + \sum_{k=1}^{r} \langle c_k, x(t-\tau_k) \rangle \ . \end{aligned}$$

His method of investigation is mainly based on Popov's hyper-stability theory [19], but the method of reducing such systems to integral equation is also used.

Another recent contribution is due to J. Kato [10]. He shows that any system of the form

$$(\Sigma_3) \qquad \dot{x}(t) = F(x_t) + b\phi(\sigma(t)) \ , \qquad \sigma(t) = G(x_t) \ ,$$

can be also investigated with respect to its properties of absolute stability by using some known results related to integral equations of the form (E). In (Σ_3), F and G stand for some linear mappings from

$C([-h,0],R^n)$ into R^n, R^m respectively. As usual, x_t denotes the restriction of the mapping x to the interval $[t-h,t]$.

Among the papers devoted to the theory of integral equations of the form (E), we shall mention those of N. Luca [13], Kh. Geleg [6], V. A. Yakubovitch [25], Vl. Răsvan [22-23]. Some results of these authors will be used in the sequel in order to formulate various stability properties concerning the integro-differential system (S) or related systems.

3. Let us consider now the system (S) under the main assumptions of Theorem 1.1. Then, the resolvent matrix $R(t)$, defined by conditions (1.2) and (1.3), satisfies the basic property

$$\|R(t)\| \in L(R_+,R) \;. \tag{3.1}$$

By using the formula of variation of constants, we get from the first equation of (S):

$$x(t) = R(t)x^0 + \int_0^t R(t-s)b\phi(\sigma(s))ds \;. \tag{3.2}$$

From (3.2) and the last equation of (S) we obtain

$$\sigma(t) = \langle c,R(t)x^0 \rangle + \int_0^t \langle c,R(t-s)b\rangle\phi(\sigma(s))ds \;. \tag{3.3}$$

Equation (3.3) is of the form (E), with

$$h(t) = \langle c,R(t)x^0 \rangle \;, \tag{3.4}$$

$$k(t) = \langle c,R(t)b \rangle \;. \tag{3.5}$$

From (3.5) and Theorem 1.1, there results that the kernel of equation (3.3) is integrable on the positive half-axis. This property will allow us to use the frequency techniques.

It is useful to find the Laplace transform of the resolvent kernel $R(t)$:

$$\tilde{R}(s) = \int_0^\infty e^{-st}R(t)dt \;, \qquad \text{Re}\, s \geq 0 \;. \tag{3.6}$$

First of all, let us remark that condition (3.1) and equation (1.2) imply $\|\dot{R}(t)\| \in L(R_+,R)$. Indeed, the convolution product of two functions from L belong also to L. Therefore, we can take the Fourier-Laplace transform of both sides in (1.2) and we find

$$s\tilde{R}(s) - I = A\tilde{R}(s) + \tilde{B}(s)\tilde{R}(s), \quad \mathrm{Re}\, s \geq 0,$$

from which we get

$$\tilde{R}(s) = [sI - A - \tilde{B}(s)]^{-1}, \quad \mathrm{Re}\, s \geq 0. \tag{3.7}$$

The existence of the inverse matrix occuring in the right side of (3.7) is guaranteed by (1.4).

Now, we can easily formulate the following result.

<u>THEOREM 3.1</u>. Assume that the following conditions hold for the system (S):

1) A is a constant n by n matrix and $B(t)$ is a measurable n by n matrix kernel, with $\|B(t)\| \in L(R_+,R)$;

2) the stability condition (1.4) is verified;

3) $\phi(\sigma)$ is a continuous real valued function on R, such that

$$\sigma\phi(\sigma) > 0 \quad \text{for} \quad \sigma \neq 0 \tag{3.8}$$

and there exists a positive number φ with the property

$$|\phi(\sigma)| \leq \varphi, \quad \sigma \in R; \tag{3.9}$$

4) there exists $q \geq 0$, such that

$$\mathrm{Re}\{(1+i\omega q)\langle c,[i\omega I - A - \tilde{B}(i\omega)]^{-1}b\rangle\} \leq 0, \tag{3.10}$$

for any real ω.

Then, the system (S) has at least one solution (defined on R_+) for any initial condition $x(0) = x^0 \in R^n$, and any such solution satisfies

$$\lim_{t\to\infty} \|x(t)\| = 0. \tag{3.11}$$

<u>Proof</u>. Taking into account Theorem 1.1 we find easily that the integral equation (3.3) satisfies all the assumptions of Theorem 3.2.2 in [5]. Therefore, $\sigma(t)$ is such that $\lim \sigma(t) = 0$ as $t \to \infty$. If we consider

now that $x(t)$ and $\sigma(t)$ are related by (3.2), the conclusion of Theorem 3.1 follows from the fact that the convolution product tends to zero as $t \to \infty$, if at least one of the factors does, and from the simple remark that $\|R(t)\| \in L(R_+,R)$, $\|\dot{R}(t)\| \in L(R_+,R)$ imply $\|R(t)\| \to 0$ as $t \to \infty$.

Remark 1. Instead of condition (3.8), one can consider a more restrictive condition, namely,

$$(3.12) \qquad 0 < \sigma\phi(\sigma) \leq \lambda\sigma^2 \quad \text{for} \quad \sigma \neq 0 ,$$

with λ a positive number. Then, the corresponding frequency stability condition (3.10) changes into

$$(3.13) \qquad \mathrm{Re}\{(1+i\omega q)\langle c,[i\omega I-A-\tilde{B}(i\omega)]^{-1}b\rangle\} \leq \lambda^{-1} , \quad \omega \in R .$$

For details, see G. P. Szegö [24].

Remark 2. The requirement of boundedness for $\phi(\sigma)$ can be also relaxed (see [2] or [5]). An interesting open problem is to decide whether the theorem is valid or not when the boundedness condition or any other requirement concerning the order of growth of $\phi(\sigma)$ is dropped.

4. We shall consider now the following integro-differential system:

$$(S_1) \qquad \begin{cases} \dot{x}(t) = Ax(t) + \displaystyle\int_0^t B(t-s)x(s)ds + b\phi(\sigma(t)) , \\ \dot{\xi}(t) = \phi(\sigma(t)) , \\ \sigma = \langle c,x\rangle - \rho\xi , \end{cases}$$

where the quantities occuring above have the same meaning as in the preceding section. Of course, $\xi(t)$ is a scalar function. With respect to ρ, we shall assume it to be a positive constant.

Using again the formula of variation of constants and integrating the second equation of (S_1) we obtain

$$\sigma(t) = \langle c, R(t)x^0 \rangle - \rho\xi_0 + \int_0^t \{\langle c, R(t-s)b \rangle - \rho\}\phi(\sigma(s))ds \; . \tag{4.1}$$

Equation (4.1) is also of the form (E) and the following formulas give the corresponding $h(t)$ and $k(t)$:

$$h(t) = \langle c, R(t)x^0 \rangle - \rho\xi_0 \; , \tag{4.2}$$

$$k(t) = \langle c, R(t)b \rangle - \rho \; . \tag{4.3}$$

If we assume again that condition (1.4) holds, then it is clear that the results of [3] (see also [5]) can be applied to equation (4.1).

<u>THEOREM 4.1</u>. Let us assume that the following conditions hold for (S_1):

1) A is a constant n by n matrix and $B(t)$ is a measurable n by n matrix kernel, such that $\|B(t)\|$, $\|\dot{B}(t)\| \in L(R_+, R)$;

2) the stability condition (1.4) is fulfilled;

3) $\phi(\sigma)$ is a continuous real valued function on R, such that condition (3.8) is satisfied;

4) $\rho > 0$ and there exists $q \geq 0$, such that

$$\mathrm{Re}\{(1 + i\omega q)G(i\omega)\} \leq 0 \quad \text{for} \quad \omega \neq 0 \; , \tag{4.4}$$

where

$$G(i\omega) = \langle c, [i\omega I - A - \tilde{B}(i\omega)]^{-1} b \rangle - \rho(i\omega)^{-1} \; . \tag{4.5}$$

Then there exists at least one solution $x = x(t)$, $\xi = \xi(t)$ of the system (S_1), corresponding to an arbitrary initial condition $x(0) = x^0 \in R^n$, $\xi(0) = \xi_0 \in R$, and any such solution is defined on R_+ and satisfies

$$\lim_{t \to \infty} (\|x(t)\| + |\xi(t)|) = 0 \; . \tag{4.6}$$

<u>Proof</u>. The equation (4.1) for $\sigma(t)$ satisfies the conditions of Theorem 3.3.1 in [5]. Indeed, from condition 1) above there results that the resolvent $R(t)$ has (almost everywhere) a second derivative

which is integrable on R_+. From equation (1.2) one obtains

$$\ddot{R}(t) = A\dot{R}(t) + B(0)R(t) + \int_0^t \dot{B}(t-s)R(s)ds ,$$

almost everywhere on the positive half-axis. This shows that $h(t)$ given by (4.2) is such that $\dot{h}(t), \ddot{h}(t) \in L(R_+,R)$, no matter how we choose the initial values. It follows also that $k(t)+\rho$, $\dot{k}(t) \in L(R_+,R)$. From Theorem 3.3.1 in [5], there results that $\lim \sigma(t) = 0$ as $t \to \infty$. In order to obtain (4.6) we shall notice first that $x(t)$ is given by (3.2). As shown in the preceding section, under our assumptions there results $\lim \|x(t)\| = 0$ as $t \to \infty$. Using now the last equation of (S_1) we obtain $\lim \xi(t) = 0$ as $t \to \infty$ and (4.6) is thereby proven.

Remark. The existence of at least one solution for arbitrary initial values is guaranteed by Theorem 3.3.1 in [5] for any continuous function $\phi(\sigma)$ which satisfies (3.8). Therefore, no requirement on the order of growth of $\phi(\sigma)$ is made, as it happened in the preceding section. This seems to be rather interesting if we take into account the fact that the behavior of the kernel is better in the first case.

Another system to whom the same procedure can be applied is the following one:

$$(S_2) \qquad \begin{cases} \dot{x}(t) = Ax(t) + \int_0^t B(t-s)x(s)ds + b\xi(t) , \\ \dot{\xi}(t) = \phi(\sigma(t)) , \\ \sigma = \langle c,x \rangle - \rho\xi . \end{cases}$$

The corresponding integral equation for $\sigma(t)$ can be written as

$$(4.7) \qquad \sigma(t) = \langle c,R(t)x^0 \rangle + \langle c,R_1(t)b \rangle \xi_0 - \rho_1\xi_0 + \int_0^t \{\langle c,R_1(t-s)b \rangle - \rho_1\}\phi(\sigma(s))ds ,$$

with

$$(4.8) \qquad R_1(t) = -\int_t^\infty R(s)ds ,$$

$$\rho_1 = \rho - \langle c, (\int_0^\infty R(s)ds)b \rangle . \tag{4.9}$$

In order to obtain a stability result for (S_2), using Theorem 3.3.1 in [5] for equation (4.7), the following assumptions are necessary:

$$\|R_1(t)\| \in L(R_+, R) , \tag{4.10}$$

$$\rho_1 > 0 . \tag{4.11}$$

We leave to the reader the task of formulating the theorem on the absolute stability of the system (S_2).

5. In this section we are going to consider the integro-differential system

$$(S_3) \quad \begin{cases} \dot{x}(t) = Ax(t) + \int_0^t B(t-s)x(s)ds + b\phi(\sigma(t)) , \\ \dot{\xi}(t) = \mu\eta(t) + \alpha_1\phi(\sigma(t)) , \\ \dot{\eta}(t) = -\mu\xi(t) + \alpha_2\phi(\sigma(t)) , \\ \sigma = \langle c,x \rangle + \beta\xi + \gamma\eta , \end{cases}$$

where $\mu > 0$, α_1, α_2, β and γ are constants. For the quantities x, A, $B(t)$, b etc., we are keeping the same meaning as before.

The method of reducing (S_3) to an integral equation in $\sigma(t)$ follows the same lines as above. First, $x(t)$ from the first equation can be expressed by the formula (3.2). The second and third equations in (S_3) constitute a linear system if we regard $\sigma(t)$ as a known quantity. One obtains

$$\xi(t) = \xi_0 \cos\mu t + \eta_0 \sin\mu t + \int_0^t [\alpha_1 \cos\mu(t-s) + \alpha_2 \sin\mu(t-s)]\phi(\sigma(s))ds ,$$

$$\eta(t) = \eta_0 \cos\mu t - \xi_0 \sin\mu t + \int_0^t [\alpha_2 \cos\mu(t-s) - \alpha_1 \sin\mu(t-s)]\phi(\sigma(s))ds .$$

From the last equation of (S_3), the formula (3.2) and the above expressions for $\xi(t)$ and $\eta(t)$ we obtain an integral equation of the form (E) in $\sigma(t)$. The corresponding $h(t)$ and $k(t)$ are given by the following formulas:

$$h(t) = \langle c, R(t)x^0 \rangle + (\beta\xi_0 + \gamma\eta_0)\cos\mu t + (\beta\eta_0 - \gamma\xi_0)\sin\mu t, \tag{5.1}$$

$$k(t) = \langle c, R(t)b \rangle + (\beta\alpha_1 + \gamma\alpha_2)\cos\mu t + (\beta\alpha_2 - \gamma\alpha_1)\sin\mu t. \tag{5.2}$$

In other words, the corresponding integral equation has the form that was already investigated by A. Kh. Geleg [6] (see also [5]). Of course, we are always assuming that condition (1.4) holds, so that $R(t)$ is integrable.

<u>THEOREM 5.1</u>. Let us consider the integro-differential system (S_3) under the following assumptions:

1) and 2) - the same as in Theorem 3.1;

3) $\phi(\sigma)$ is a continuous bounded mapping from the real line into itself, such that

$$0 < \sigma\phi(\sigma) < r\sigma^2, \quad \sigma \neq 0, \tag{5.3}$$

with $0 < r \leq \infty$;

4) the following inequalities hold

$$\beta\alpha_1 + \gamma\alpha_2 < 0, \quad \beta\alpha_2 - \gamma\alpha_1 \leq 0; \tag{5.4}$$

5) the frequency condition

$$-\frac{1}{r} + \frac{1}{\mu}(\beta\alpha_2 - \gamma\alpha_1) + \mathrm{Re}\left\{\left(1 + \frac{i\omega}{\mu} \cdot \frac{\beta\alpha_2 - \gamma\alpha_1}{\beta\alpha_1 + \gamma\alpha_2}\right) \cdot \langle c, [i\omega I - A - \tilde{B}(i\omega)]^{-1} b \rangle\right\} \leq 0 \tag{5.5}$$

holds true for all real ω.

Then, for any solution of (S_3),

$$(5.6) \qquad \lim_{t\to\infty} (\|x(t)\| + |\xi(t)| + |\eta(t)|) = 0 .$$

Remark 1. The result of A. Kh. Geleg we used in order to obtain the above theorem does not ensure the existence of the solution of the integral equation for $\sigma(t)$. Therefore, we cannot conclude anything about the existence of the solution for (S_3) if we use the results in [6]. At the present time, several results concerning the continuability of solution of integro-differential equations are available [14], [15].

Remark 2. If we compare the frequency condition (5.5) with the corresponding ones from Theorem 3.1 and Theorem 4.1, we see that there is no parameter "q" in (5.5). Any quantity occurring in (5.5) is known and there is nothing to be chosen in order to satisfy the required inequality. The system (S_3) corresponds to the critical case when there exists a pair of pure imaginary characteristic roots and, therefore, it is more complicated than the preceding ones.

6. Instead of going further on the way of getting more and more sophisticated kernels for the integral equations to whom the integro-differential systems under investigation can be reduced, we shall consider now a problem with several nonlinearities:

$$(S_4) \qquad \begin{cases} \dot{x}(t) = Ax(t) + \int_0^t B(t-s)x(s)ds + C\xi(t) , \\ \dot{\xi}(t) = \phi(\sigma(t)) , \\ \sigma = Dx - \Gamma\xi . \end{cases}$$

In the system (S_4), C, D, Γ are some n by n matrices, and ξ, ϕ, σ are n-vectors.

The following vector integral equation is easily obtainable from (S_4):

$$(6.1) \qquad \sigma(t) = DR(t)x^0 - \Gamma\xi^0 + \int_0^t \{DR(t-s)C - \Gamma\}\phi(\sigma(s))ds .$$

We shall use now a result of N. Luca [13] (see also [5]). It is concerned with vector integral equations of the form (E) and leads to the following

THEOREM 6.1. Assume that the following conditions hold for the system (S_4):

1) and 2) - the same as in Theorem 4.1;

3) $\phi(\sigma)$ is a continuous mapping from R^n into itself, such that

$$\langle \sigma, \phi(\sigma) \rangle > 0 \quad \text{for } \sigma \neq 0 \; ; \tag{6.2}$$

4) there exist a matrix Q and a positive definite (on R^n) function $F(\sigma)$, such that

$$dF(\sigma) = \langle \phi(\sigma), Q d\sigma \rangle \; ; \tag{6.3}$$

5) Γ is a positive definite matrix and the frequency condition

$$\mathrm{Re}\left\{(I + i\omega Q)\left[D[i\omega I - A - \tilde{B}(i\omega)]^{-1}C - \Gamma(i\omega)^{-1}\right]\right\} \leqslant 0 \quad \text{for } \omega \neq 0 \; , \tag{6.4}$$

is satisfied.

Then, any solution of the system (S_4) which exists on the positive half-axis is such that

$$\lim_{t\to\infty} (\|x(t)\| + \|\xi(t)\|) = 0 \; . \tag{6.5}$$

The real part of a matrix, as it appears in (6.4), should be understood as follows. If A is a square matrix with complex entries, then $2\mathrm{Re}A = A + A^*$, where A^* denotes the complex conjugate of the transpose. Finally, if H is a matrix such that $H = H^*$, then it is called positive if and only if the quadratic form generated by H is positive definite.

The proof of Theorem 6.1 follows easily from that given in Luca's paper [13], if we take into account that the integral equation corresponding to $\sigma(t)$ from (S_4), i.e., equation (6.1), satisfies all the requirements occurring in the statement of Theorem 2 of the paper quoted above.

Remark. A recent result due to Vl. Răsvan [23] can be also applied in order to find stability conditions for the system (S_4).

7. Another kind of system to whom the frequency techniques can be applied is

$$(S_5) \qquad \begin{cases} \dot{x}(t) = Ax(t) + \int_0^t B(t-s)x(s)ds + \int_0^t b(t-s)\phi(\sigma(s))ds \,, \\ \sigma = \langle c,x \rangle \,, \end{cases}$$

where the meaning of the quantities occurring above is the same as in the preceding sections, excepting $b(t)$ which now denotes a measurable vector function from the positive half-axis into R^n.

Let us remark that the case corresponding to $B(t) \equiv 0$ has been investigated by the author in his paper [4]. It is obvious from (S_5) that the delay occurs also in the nonlinear part of the system.

The first equation in (S_5) leads easily to the formula

$$x(t) = R(t)x^0 + \int_0^t R(t-s)ds \int_0^s b(s-u)\phi(\sigma(u))du \,,$$

which can be also written as

$$(7.1) \qquad x(t) = R(t)x^0 + \int_0^t \left\{ \int_0^{t-u} R(t-u-v)b(v)dv \right\} \phi(\sigma(u))du \,.$$

Of course, we assume that the function $b(t)$ is sufficiently smooth in order to be able to change the order of integration. If we substitute $x(t)$ given by (7.1) in the second equation of (S_5) we get an integral equation of the form (E) with

$$(7.2) \qquad h(t) = \langle c,R(t)x^0 \rangle \,,$$

$$(7.3) \qquad k(t) = \langle c, \int_0^t R(t-u)b(u)du \rangle \,.$$

We can state the following result.

THEOREM 7.1. Consider the system (S_5) and assume that the following conditions hold:

1), 2) and 3) - the same as in Theorem 3.1;

4) $b(t)$ is a measurable function from R_+ into R^n such that

$$(7.4) \qquad \|b(t)\| \in L(R_+,R) \,;$$

5) there exists $q \geq 0$, such that

$$(7.5) \qquad \mathrm{Re}\{(1+i\omega q)\langle c,[i\omega I - A - \tilde{B}(i\omega)]^{-1}b(i\omega)\rangle\} \leq 0$$

for any real ω.

Then, the system (S_5) has at least one solution $x(t)$ for any initial value $x(0) = x^0 \in R^n$ and

$$\lim_{t\to\infty} \|x(t)\| = 0 . \tag{7.6}$$

The proof follows from the same result as the proof of Theorem 3.1.

Remark. It is possible to replace the second equation in (S_5) by another, for instance,

$$\sigma(t) = f(t) + \int_0^t \langle c(t-s), x(s) \rangle ds .$$

The method used above still applies. For the case $B(t) \equiv 0$, such problems were investigated by A. N. V. Rao [20], [21].

8. In concluding this paper, we shall point out some features related to the use of frequency techniques in the investigation of absolute stability for various classes of feedback systems.

First, the method of reducing the study of such systems to an integral equation of the form (E) allows to find efficient conditions for the absolute stability of systems in which the delays occur.

Second, the systems described by Volterra integral equations, as well as other systems whose investigation can be reduced to integral equations provide the most salient (and perhaps, the only) examples of systems with unbounded delay, in the existing literature.

Third, the hyperstability theory of V. M. Popov [19] seems not to be powerful enough in order to include all the results obtainable by the direct investigation of the integral equations. The following example shows what kind of difficulties might arise when trying to obtain the behavior of solutions of integral equations from the theory of hyperstability. The integral equation (E) occurs as follows in the investigation of feedback systems. We have a linear subsystem described by the input-output equation

$$\sigma(t) = h(t) + \int_0^t k(t-s)u(s)ds . \tag{L}$$

The feedback equation is

$$u(t) = \phi(\sigma(t)) , \tag{N}$$

which leads immediately to (E). In order to be able to apply the hyperstability theory, it is necessary to write (L) in the form

$$\sigma(t) = \int_{-\infty}^{t} k(t-s)u(s)ds \,, \tag{8.1}$$

in other words, we have to continue $u(t)$ on the negative half-axis, such that

$$h(t) = \int_{-\infty}^{0} k(t-s)u(s)ds \,. \tag{8.2}$$

But (8.2) constitutes a Fredholm equation of the first kind with $u(t), t \in (-\infty, 0]$, as unknown function. Since such equations have solution only under rather restrictive assumptions, it appears clearly that (L) can be written in the form (8.1) only for certain functions $h(t)$. Therefore, a direct investigation of the integral equations seems to be quite fruitful.

Finally, we shall notice that the results given in the preceding sections regard feedback systems whose uncontrolled part - described by equation (1.1) - involves unbounded delay.

REFERENCES

[1] A.M. AIZERMAN and F.R. GANTMAHER, Absolute Stability of Regulator Systems. Holden-Day, San Francisco, 1964.

[2] C. CORDUNEANU, Sur une équation intégrale non-linéaire, Analele St. Univ. Iaşi, 9 (1963), 369-375.

[3] C. CORDUNEANU, Sur une équation intégrale de la théorie du réglage automatique, Comptes Rendus Acad. Sci. Paris, 256 (1963), 3564-3567.

[4] C. CORDUNEANU, Quelques problèmes qualitatifs de la théorie des équations integro-différentielles, Coll. Math. 18 (1967), 77-87.

[5] C. CORDUNEANU, Integral Equations and Stability of Feedback Systems. Academic Press, New York (in print).

[6] A. Kh. GELEG, The absolute stability of nonlinear control systems with distributed parameters in critical cases, Aut. and Remote Control, 27 (1966), 525-534.

[7] S.I. GROSSMAN and R.K. MILLER, Nonlinear Volterra interro-differential systems with L^1-kernels (to appear).

[8] A. HALANAY, Differential Equations: Stability, Oscillations, Time Lag. Academic Press, New York, 1966.

[9] A. HALANAY and V.M. POPOV, On the stability of nonlinear automatic control systems with lagging argument, Aut. and Remote Control, 23 (1963), 783-386.

[10] J. KATO, Absolute Stability of Control Systems with Multiple Feedback, Lecture Notes in Mathematics, 144, Springer-Verlag, Berlin, Heidelberg, New York, 1970.

[11] S. LEFSCHETZ, Stability of Nonlinear Control Systems. Academic Press, New York, 1965.

[12] A.M. LETOV, Stability in Nonlinear Control Systems. Princeton Univ. Press, Princeton, N.J., 1961.

[13] N. LUCA, Sur quelques systèmes d'équations intégrales à noyau transitoire qui s'appliquent aux problèmes de réglage automatique, Analele St. Univ. Iaşi, X (1964), 347-355.

[14] R.K. MILLER and G.R. SELL, Existence, uniqueness and continuity of solution of integral equations, Annali Mat. Pura Appl. (IV), LXXX (1968), 135-152.

[15] L.W. NEUSTADT, On the solution of certain integral-like operator equations: existence, uniqueness and dependence theorems. Arch. Rat. Mechanics and Analysis, 38 (1970), 131-160.

[16] V.M. POPOV, Criterii de stabilitate pentru sistemele neliniare de reglare automată bazate pe utilizarea transformatei Laplace, Studii Cerc. Energ. IX (1959), 119-135.

[17] V.M. POPOV, Absolute stability of nonlinear systems of automatic control, Aut. and Remote Control, 22 (1961), 857-875.

[18] V.M. POPOV, A critical case of absolute stability, Aut. and Remote Control, 23 (1962), 1-21.

[19] V.M. POPOV, Hiperstabilitatea sistemelor automate. Ed. Acad. R.S.R., Bucureşti, 1966.

[20] A.N.V. RAO, On some systems of automatic control theory, Analele St. Univ. Iaşi, XV (1969), 47-57.

[21] A.N.V. RAO, Stability of multivariable nonlinear systems containing distributed elements, Bul. Inst. Pol. Iaşi, XV (1969), fasc. 3-4, 49-57.

[22] Vl. RĂSVAN, Asupra stabilizării sistemelor automate cu intirziere prin reacţii neliniare (to appear).

[23] Vl. RĂSVAN, Condiţii suficiente de stabilitate absolută in cazul reglării indirecte cu mai multe organe de execuţie (to appear).

[24] G.P. SZEGO, Sul comportamento asintotico di una equazione integrale non lineare, Analele St. Univ., Iaşi, XV (1969), 387-394.

[25] V.A. YAKUBOVITCH, Frequency conditions for the stability of solutions of nonlinear integral equation of automatic control (Russian), Vestnik Leningradsk. Univ., 1967, No. 7, 109-125.

Seminarul Matematic "A. Myller" Universitate, Iaşi, Romania

A NEW TECHNIQUE FOR PROVING THE EXISTENCE OF ANALYTIC FUNCTIONS

Stephen P. Diliberto

1. The Iteration Procedure

We shall present here a description of the principal features of a new technique for establishing the existence of analytic functions. We shall sketch a proof of the standard existence theorem for systems of first order analytic ordinary differential equations. The proof if completed is considerably longer than any other for that existence theorem. However, a comparison of this proof with earlier results using this procedure in our reports on Siegel's Normal Form Theorem [1], Convergence of Schroeder's Series [2], the Arnold-Moser Theorem [4], and Generic Stability of Hamiltonian Systems [3], shows that the arguments are quite close.

The starting point of these investigations concerned local normalization for a system of first order analytic ordinary differential equations

$$\frac{dx}{dt} = f(x) = Ax + X(x) \tag{1}$$

where $x = (x_1,\dots,x_n)$; $x = 0$ is a singular point; and Ax are the linear terms. Closely related to this is the system

$$(1)_\theta \qquad \left\{ \begin{aligned} \frac{d\theta_i}{dt} &= \omega_i \quad (i = 1,\dots,m) \\ \frac{dx}{dt} &= f(\theta,x) = Ax + g(\theta,\varepsilon) + \varepsilon X(\theta,x,\varepsilon) \end{aligned} \right\},$$

where A is constant as before and g and X (which is again quadratic) have period 2π in each θ_i $(i = 1,\ldots,n)$ and $\theta = (\theta_1,\ldots,\theta_n)$. For (1) the problem is to find a transformation

$$x = y + P(y) \tag{2}$$

which is analytic in y at $y = 0$, P has quadratic and higher terms, and the transformation (2) applied to (1) produces the equation for y

$$\frac{dy}{dt} = Ay . \tag{3}$$

For equation $(1)_\theta$ one uses a transformation of the type

$$x = y + S(\theta,\varepsilon) \tag*{$(2)_\theta$}$$

where S is a power series in ε so as to eliminate the g term in $(1)_\theta$ - S being multiply periodic of period 2π in each θ_i. The resulting equation for y being

$$\left\{\begin{aligned} \frac{d\theta_i}{dt} &= \omega_i \\ \frac{dy}{dt} &= Ay + \varepsilon Y(\theta,y,\varepsilon) \end{aligned}\right\} . \tag*{$(3)_\theta$}$$

A is as before - constant - and Y is multiply periodic in θ and quadratic in y.

The problems we posed are those of getting from (1) to (3) one step at a time by a procedure that eliminates the lowest degree term in X at each stage and the closely related problem of getting from $(1)_\theta$ to $(3)_\theta$ one step at a time by a procedure that at each stage eliminates the lowest degree terms in ε.

Namely, for X^k a vector function of $x^k = (x_1^k,\ldots,x_n^k)$ which is analytic in x^k at $x^k = 0$, and whose lowest terms are of degree k we see a transformation

$$x^k = x^{k+1} + P^k\,(x^{k+1}) \tag*{$(4)_k$}$$

where P^k is analytic at $x^{k+1} = 0$ and has lowest degree terms of degree k such that $(4)_k$ carries

$$(5)_k \qquad \frac{dx^k}{dt} = Ax^k + X^k(x^k)$$

into

$$(5)_{k+1} \qquad \frac{dx^{k+1}}{dt} = Ax^{k+1} + X^{k+1}(x^{k+1})$$

where X^{k+1} has degree at least $k+1$. We shall not need the corresponding statement on the $(1)_\theta$ to $(3)_\theta$ reduction.

Notation. Equation (M) of section P will be referred to as equation (P.M).

2. A Comparison of Some Problems

The problem of reducing (1.1) to (1.3) has a long history [5]. C. L. Siegel's result is that it can be done analytically if there exist positive constants c and γ such that for any integer vector $k = (k_1,\ldots,k_n)$ one has

$$(1) \qquad \left| \sum_{S=1}^{n} \lambda_S k_S \right|^{-1} \leq c\,\|k\|^\gamma .$$

The reduction of $(1.1)_\theta$ to $(1.3)_\theta$ is in general not possible. This problem is a special case of one studied by Moser [6]. If there exist positive constants c and γ such that for any integer vector $\ell = (\ell_1,\ldots,\ell_m) \neq 0$ one has (here $i = \sqrt{-1}$)

$$(2) \qquad \left| i \sum_{S=1}^{m} \ell_S \omega_S + \lambda_p \right|^{-1} \leq c\,\|\ell\|^\gamma \quad (p = 1,2,\ldots,n) .$$

Then a modified reduction is possible. In the simplest case with A having non-zero distinct roots if (2) holds there exists a $B(\varepsilon)$, analytic in ε such that the system

$$(1)'_\theta \qquad \begin{cases} \dfrac{d\theta}{dt} = \omega_i \quad (i = 1,\ldots,m) \\[2mm] \dfrac{dx}{dt} = (A+B(\varepsilon))x + g(\theta,\varepsilon) + \varepsilon X(\theta,x,\varepsilon) \end{cases}$$

does have a reduction to $(1.3)_\theta$ by a transformation $(1.2)_\theta$.

For generic stability of Hamiltonians at singular points condition (1) above is violated because the roots λ occur in pairs $\pm i\omega_s$ $(s = 1,2,\ldots,N)$ and $2N = n$. What is required is that there exist positive constants c and γ so that for any integer vector $k = (k_1,\ldots,k_N)$

$$(2) \qquad \left| \sum_{s=1}^{N} \omega_1 k_1 \right|^{-1} \leq c \, \|k\|^{\gamma} \, .$$

As in the Moser Theorem reduction to a linear form is not possible. This time however the modification is not to equation (1.1) but to (1.3). Let $r_i^2 = y_i^2 + y_{i+N}^2$ $(i = 1,2,\ldots,N)$. Then if (2) holds there does exist a transformation (1.2) carrying (1.1) into

$$(3) \qquad \frac{dy}{dt} = (A + B(r_1,\ldots,r_N))y$$

where $A + B$ is of stable type. Our own result [3] on generic stability, and indeed earlier works relating to it are not stated in the form just given. We do this to point out that the Moser result (and those of Arnold) sets out to do something that is essentially different from what one is attempting to do in getting the Birkhoff normal form - which is what equation (3) is.

For Hamiltonian systems the attempt to achieve a normal form of one sort or another may have a large class, say $\mathscr{F}$, of formal (not necessarily convergent) canonical changes of variables which achieve the type of normal form sought. Which elements of $\mathscr{F}$ one chooses does make a difference in the Birkhoff normal form problem. C. L. Siegel [7] and we [3] use the same partial differential equation for relating the new and old Hamiltonian with the generating function. Siegel chose a solution by a recursive method, we by an iterative method. Here choice of solution involves both the generating function and the new Hamiltonian. For one common class of such problems Siegel's transformation almost always diverges - while ours converges.

For singular points and periodic solutions our results show that the Arnold-Moser gaps do not exist in the analytic case for generic Hamiltonian systems, and in addition provide the first non-trivial example of orbital stability in a Hamiltonian system of degree greater than or equal three.

If one attempts to compare the proof idea in our convergence proofs with the type of proof idea associated with the names of Nash-Arnold-Moser-Schwartz-Sternberg [8], it appears our methods are more natural for analytic functions. Presumably our techniques could by sharpened by the use of smoothing operators.

3. The Convergence Problem

By using equation $(1.4)_k$ for different values of k one may define by induction a transformation

$$x' = T^{[k]}(x^{k+1}) . \tag{1}$$

Rather than map from an x^{k+1} space to x' and let the space change with each map, we introduce new variables y and x and replace (1) by

$$x = T^{[k]}(y) . \tag{1)*}$$

It is easily shown [1, p.11 and p. 16]:

THEOREM. If for all y with $\|y\| \leqslant R$ one has

(i) $\Sigma \|P^k\| \rightrightarrows$

(ii) $\Sigma \|(\frac{\partial P_i^k}{\partial y_j})\| \rightrightarrows$

(iii) $\Sigma \frac{1}{R} \|P^k\| < d$.

Then $T^{[k]} \rightrightarrows$ for $\|y\| < R\exp(-d)$.

The problem therefore is to establish (i), (ii) and (iii) for any iterative process.

4. Bounded Dominants

For an analytic function $f = \Sigma f_n z^n$ we define $\lceil f \rceil$ by $\lceil f \rceil = \Sigma |f_n| \, |z|^n$. We define $A(k,M,R)$ as the class of analytic functions whose power series with center at the origin converges for an open set including $|z| \leqslant R$ and on that disc is such that $\lceil f \rceil \leqslant M|z|^k$. We use the same definition for vector valued vector functions by putting

$|z| = |z_1| + \cdots + |z_n|$ if $z = (z_1,\ldots,z_n)$ and $\lceil f \rceil = \lceil f_1 \rceil + \cdots + \lceil f_m \rceil$ if $f = (f_1,\ldots,f_2)$. If f is analytic we shall let $f^{(L)}$ be the lowest degree terms in the expansion (about the origin) for f. Let $f^{(H)} = f - f^{(L)}$.

There are two important lemmas concerning the classes $A(k,M,R)$ [1]:

LEMMA 1. Let $f(z) \in A(k,M,R)$, $z = (z_1,\ldots,z_n)$. Then

$$f^{(H)}(z) \in A(k+1,\ n\frac{M}{R},\ R)\ .$$

LEMMA 2. Let P be a homogeneous polynomial of degree k. Let $P \in A(k,M,\infty)$. Then

$$\frac{\partial P}{\partial y_s} \in A(k-1,\ ck^2M,\ \infty)$$

where c is independent of k.

The utility of the classes $A(k,M,R)$ as distinct from the majorant method is that it allows one to make estimates for bounds on the function which are independent of estimates for the circle of convergence.

5. The Proof Plan for Convergence

The determination of P^k in $(1.4)_k$ arises in attempting to modify part of the vector function $Ax + X^k(x)$ that occurs in $(1.5)_k$. X^k is the part to be modified. We assume it belongs to $A(k,M_k,R_k)$. P^k will belong to $A(k,M_k',R_k')$. It follows easily that one has $R_k \geqslant R_k' \geqslant R_{k+1} \geqslant R_{k+1}' \geqslant \cdots$. If one can establish the existence of an R so that

$$(1) \qquad R_k \geqslant R\ ,$$

then the conditions of Section 3 become

$$\Sigma\, M_k'(R)^k \to$$

$$\Sigma\, M_k'\ k^2(R)^k \to\ .$$

In all of the studies we made [1], [2], [3], [4], the severest estimate is of the type

$$M_k' \leqslant M_k k^{\nu}$$

so that convergence of the transformation leads to establishing something like

$$\Sigma M_k k^{2+\nu}(R)^k \quad . \tag{2}$$

Three types of situations arise

C.1: $\quad M_{k+1} \leqslant (1 + d_k)M_k \; ; \qquad \sum_{k=1}^{N} d_k = D_N < D$

C.2: $\quad M_{k+1} \leqslant E\, M_k \; ; \qquad E > 1$

C.3: $\quad M_{k+1} \leqslant (1 + d_k)M_k^{1+(1/k-1)} \; ; \qquad \sum_{S=2}^{N} \frac{d_s}{s-1} = \bar{D}_N < D \quad .$

In the first, one has $M_k \leqslant (\exp D)M_2$ all k, in the second, one has $M_k \leqslant C^k M_2$ and in the third, one has $M_k \leqslant [M_2 \exp D]^{k-1}$.

We prove the last:

$$M_{k+1} \leqslant (1 + d_k)M_k^{1+(1/k-1)}$$

implies by an induction that

$$M_{k+1} \leqslant \prod_{s=2}^{k-1} \left\{(1+d_s)^{\frac{1}{s-1}\prod^{k-1}(1+\frac{1}{t})}\right\} M_2^{\prod^{k-1}(1+\frac{1}{t})}$$

$$\leqslant \prod_{s=2}^{k-1} \left\{(1+d_s)^{\frac{k-1}{s-1}}\right\} M_2^{k-1}$$

by using

$$\prod_h^k (1+\frac{1}{t}) \leqslant \frac{k}{h} \quad .$$

$$M_{k+1} \leqslant \left\{M_2 \prod_{s=2}^{k-1}(1+d_s)^{(1/s-1)}\right\}^{k-1} \leqslant \left\{M_2 \exp z(d_s/s-1)\right\}^{k-1} \quad .$$

Thus in cases C.1, C.2 and C.3 one readily establishes the existence of constants E and M so that for all k

$$M_k \leqslant ME^k \tag{3}$$

and (2) will then converge if only R is made smaller - which is always permissible.

What actually occurs is that the d_k in C.1 and C.3 depends on M_k:

$$d_k \leqslant ck^{\alpha}M_kR^{k-\beta} . \tag{4}$$

The problem then of proving the existence of M and E for which (3) holds can be solved thusly. Take M and E in (3) as a pair of unknowns to be solved for. Use (3) to eliminate M_4 in (4) and then use the resulting expression to eliminate d_k from our earlier estimate for C.1 and C.3.

Consider the C.3 case. Define

$$f_k(x) = c \sum_{s=\beta+1}^{k} \frac{s^{\alpha}x^{s-\beta}}{(s-1)}$$

$$f(x) = \sum_{s=\beta+1}^{\infty} \frac{s^{\alpha}x^{s-\beta}}{(s-1)} .$$

Since we are in the C.3 case we shall try to find E so that

$$M_k \leqslant E^{k-2} . \tag{3)'$$

Putting (3)' in 4 and using our earlier estimates one has

$$\begin{aligned} M_{k+1} &\leqslant \left\{M_2 \exp E^{\beta-2} f_k(RE)\right\}^{k-1} \\ &\leqslant \left\{M_2 \exp E^{\beta-2} f(RE)\right\}^{k-1} . \end{aligned}$$

Our problem then is that of finding number E satisfying

$$M_2 \exp E^{\beta-2} f(RE) < E .$$

Using the fact that $f(x) \to 0$ as $x \to 0$ it follows that if E is any number larger than M_2 it may be used as a solution if R is made small enough.

These ideas extend with very little change to the cases where

$$M_k \leq \prod_{t=1}^{p} (1 + d_{k,t})M_k$$

or

$$M_k \leq \prod_{t=1}^{p} (1 + d_{k,t})M_k^{1+(1/k-1)} .$$

6. A Proof Outline

Let

(1) $$\frac{dx}{dt} = f(x) .$$

Let $x^o = 0$ and assume $f(0) = \mathcal{E}^1 = (1,0,\ldots,0)$. If f is analytic we want to show there exist a transform R and its inverse T of the form

(2) $$x = y + P(y) = R(y)$$
$$y = x + Q(x) = T(x)$$

so that the differential equation for y is

(3) $$\frac{dy}{dt} = \mathcal{E}^1 .$$

To find a function $u(t,t^o,x^o)$ which is a solution of (1) and such that $u(t^o,t^o,x^o) = x^o$ we may use R and T. Namely put $y^o = T(x^o)$. Then the function $v(t,t^o,y^o) = (t-t^o)\mathcal{E}^1 + y^o$ is the solution function for (3) and so $u(t,t^o,x^o) = R((t-t^o)\mathcal{E}^1 + T(x^o))$. With R And T analytic so is u in all its arguments.

Suppose the reduction has gotten to the stage where

(4) $$\frac{dx^k}{dt} = \mathcal{E}^1 + X^k(x^k) \quad \text{where } X^k \in A(k,M_k,R_k) .$$

Then put

(5) $$x^k = x^{k+1} + P^k(x^{k+1}) , \quad P^k \in A(k+1,M_k',R_k') ,$$

where P^k is chosen so that

$$\frac{dx^{k+1}}{dt} = \mathcal{E}^1 + X^{k+1}(x^{k+1}) \quad \text{and} \quad X^{k+1} \in A(k+1, M_{k+1}, R_{k+1}) \;. \tag{6}$$

Differentiate (5) with respect to t, eliminate the time derivatives using (4) and (6), cancel an $\mathcal{E}^1$ from the R and L.H.S. There results

$$X^k(x^{k+1} + P^{k+1}) = P_x^{k+1}\,\mathcal{E}^1 + (I + P_x^{k+1})X^{k+1}$$

where

$$P_x^{k+1} = \left(\frac{\partial P_i^{k+1}}{\partial x_j^{k+1}} \right) \;.$$

Using the fact that

$$X^k(x^{k+1})^{(L)} = (X^k(x^{k+1} + P^{k+1}))^{(L)} \tag{7}$$

X^{k+1} will have the right degree if

$$X^k(x^{k+1})^{(L)} = P_x^{k+1}\,\mathcal{E}^1 \;.$$

The resulting equation for X^{k+1} can then be written as

$$(I + P_x^{k+1})X^{k+1} = X^k(x^{k+1} + P^{k+1}) - X^k(x^{k+1})^{(L)} \;. \tag{8}$$

Now in (7) both the R.H.S. and the L.H.S. are homogeneous polynomials (obtained by chopping off the tail of a series - actually tail, body, but not the head). If we substitute $x^{k+1} + P^{k+1}$ for x^{k+1} on the L.H.S. of (7) we obtain a function $X^k(x^{k+1} + P^{k+1})^{(L)}$ that is no longer a homogeneous polynomial. Likewise $X^k(x^{k+1}) + P^{k+1})^{(H)}$ is obtained by substituting $x^{k+1} + P^{k+1}$ for x^{k+1} in $X^k(x^{k+1})^{(H)}$. Equation (8) can be written as

$$X^{k+1} = (I + P_x^{k+1})^{-1}\Big\{[X^k(x^{k+1} + P^{k+1})^{(L)} - X^k(x^{k+1})^L] + X^k(x^{k+1} + P^{k+1})^{(H)}\Big\} \;. \tag{9}$$

From this point on the proof is straightforward, the last critical step being the regrouping involved in going from (8) to (9). The estimate on the first term in the { } uses Lemma 2 of Section 4 and the second term in { } uses Lemma 1 of Section 4.

REFERENCES

[1] S.P. DILIBERTO, A New Method for Establishing the Existence of Analytic Functions I: Siegel's Normal Form Theorem, O.N.R. Report, Berkeley, 1970.

[2] __________, A New Method for Establishing the Existence of Analytic Functions II: The Schroder Series, O.N.R. Report, Berkeley, 1970.

[3] __________, A New Method for Establishing the Existence of Analytic Functions III: General Stability of Hamiltonian Systems, O.N.R. Report, Berkeley, 1970.

[4] __________, A New Method for Establishing the Existence of Analytic Functions IV: The Arnold-Moser Theorems (MS).

[5] A.F. KELLEY, Using Change of Variables to Find Invariant Manifolds of Systems of Ordinary Differential Equations in a Neighborhood of a Critical Point, Periodic Orbit, or Periodic Surface, Ph.D. Thesis, University of California, Berkeley, 1963.

[6] T. MOSER, Convergent series expansions for quasi-periodic motions, Math. Ann. 169 (1967), 136-176.

[7] C.L. SIEGEL, Uber die Existence einer Normalform analytischer Hamiltonische Differentialgleichungen in der Nahe einer Gleichgewichtlosung, Math. Ann. 28 (1954), 144-170.

[8] STERNBERG, SHLOMO, Celestial Mechanics, Part II. W. A. Benjamin, New York, 1969.

University of California, Berkeley, California

POTENTIALS WITH CLOSED TRAJECTORIES ON SURFACES OF REVOLUTION

A. Halanay

The purpose of this paper is to extend a result obtained almost a century ago by J. Betrand [1] in the central forces problem and considered also by G. Darboux [2] in the same setting as here. Bertrand proved that in the central forces problem essentially only for the two potentials $-\frac{1}{r}$ and $\frac{1}{2}r^2$ all the trajectories are closed. Darboux studied the same question for motions on a furface of revolution and gave the explicit form of the potentials for the case of the sphere. Our method is somehow different from the one of Darboux and it seems that the results are new at least partially.

1. Consider first the central forces problem. The manifold is $R^2\backslash\{0\}$, the kinetic energy $\frac{1}{2}(u^2 + r^2v^2)$; the coordinates in $R^2\backslash\{0\}$ are (r,θ) and the coordinates in the tangent space at (r,θ) are (u,v); suppose the potential is $V(r)$, independent on θ. The corresponding Lagrange function is $L(r,\theta,u,v) = \frac{1}{2}(u^2 + r^2v^2) - V(r)$ and the system of differential equations describing the motion has the two global first integrals

$$E(r,\theta,u,v) = \frac{1}{2}(u^2 + r^2v^2) + V(r), \quad p(r,\theta,u,v) = r^2v .$$

By using the second one we get the system

$$\dot{r} = u, \quad \dot{u} = \frac{p^2}{r^3} - V'(r), \quad \dot{\theta} = \frac{p}{r^2} \quad ,$$

where the first two equations may be studied independently; the corresponding system has the global first integral

$$F(r,u) = \frac{1}{2}\left(u^2 + \frac{p^2}{r^2}\right) + V(r) = \frac{1}{2}u^2 + V_p(r) ,$$

$$V_p(r) = \frac{p^2}{2r^2} + V(r) .$$

The integral curves in the plane (r,u) are given by $F(r,u) = h$; the critical points of F correspond to the singular points of the system and are given by $u = 0$, $V_p'(r) = 0$.

Let r_p be a nondegenerate critical point of V_p; if $V_p''(r_p) < 0$ the $(r_p,0)$ is a saddle-point and for $h \neq h_p = V_p(r_p)$ the invariant manifold $I_{h,p}$ (corresponding to the energy h and angular momentum p) is a cylinder. If $V_p''(r) > 0$ then $(r_p,0)$ is a center, the trajectories for $h > h_p$ are closed, and the invariant manifold $I_{h,p}$ is a torus. To study the system on this torus we put $r - r_p = \rho\cos\phi$, $u = -\sqrt{V_p''(r_p)}\,\rho\sin\phi$ and the corresponding system is equivalent to an equation

$$\frac{d\theta}{d\phi} = \frac{p}{(r_p+\rho\cos\phi)^2\left[\sqrt{V_p''(r_p)}\,\sin^2\phi + \frac{1}{\rho\sqrt{V_p''(r_p)}}\,V_p'(r_p+\rho\cos\phi)\cos\phi\right]}$$

where $\rho(\phi)$ is defined by

$$\frac{1}{2}V_p''(r_p)\rho^2\sin^2\phi + V_p(r_p+\rho\cos\phi) = h .$$

We get for the rotation number on the torus the formula

$$\mu(h,p) = \frac{1}{2\pi}\int_0^{2\pi}\frac{p\,d\phi}{(r_p+\rho\cos\phi)^2\left[\sqrt{V_p''(r_p)}\,\sin^2\phi + \frac{1}{\rho\sqrt{V_p''(r_p)}}\,V_p'(r_p+\rho\cos\phi)\cos\phi\right]}$$

$$= \frac{1}{2\pi}\oint\frac{p\,dr}{r^2u} = \frac{1}{\pi}\int_{r_{min}}^{r_{max}}\frac{p\,dr}{r^2\sqrt{2(h - V_p(r))}} ,$$

where r_{min} and r_{max} are the solutions of the equation $h - V_p(r) = 0$ in a neighborhood of r_p for h closed to h_p. We may now ask to find the potentials $V(r)$ of class C^2 for which there exist invariant tori and all trajectories on these tori are closed, hence the potentials V such that μ is rational for all h,p. Since μ is continuous with respect to (h,p) that implies that μ should be a constant. From

$$\mu(h,p) = \frac{1}{2\pi}\int_0^{2\pi} \frac{p\,d\phi}{[r_p + O(\sqrt{h-h_p})]^2[V_p''(r_p) + O(\sqrt{h-h_p})]}$$

we see that if $\mu(h,p) = \frac{1}{q}$ for all h close to h_p we must have

$$\frac{p}{r_p^2\sqrt{V_p''(r_p)}} = \frac{1}{q} .$$

Let (r_1,r_2) be an interval such that $V'(r) > 0$ for $r \in (r_1,r_2)$ and let $r_0 \in (r_1,r_2)$; choose $p = r_0^2\sqrt{\frac{V'(r_0)}{r_0}}$.

For such p we see that $r_p = r_0$, $V_p''(r_0) = \frac{3V'(r_0)}{r_0} + V''(r_0)$. We get the differential equation

$$V''(r) + (3-q^2)\frac{V'(r)}{r} = 0 ,$$

hence $V(r) = Cr^{q^2-2}$.

If in the formula for $\mu(h,p)$ we consider further approximations with respect to $h - h_p$ the condition $\mu(h,p) = \frac{1}{q}$ gives

$$\frac{V_p'''(r_p)}{4r_pV_p''(r_p)} + \frac{5}{24}\left[\frac{V_p'''(r_p)}{V_p''(r_p)}\right]^2 + \frac{3}{r_p^2} + \frac{3}{4r_p}\frac{V_p'''(r_p)}{V_p''(r_p)} - \frac{V_p^{IV}(r_p)}{8V_p''(r_p)} = 0$$

and for $V(r) = Cr^{q^2-2}$ we get the condition

$$q^2 - 7 + \frac{5}{24}(q^2-7)^2 + 3 - \frac{q^4 - 12q^2 + 47}{8} = 0 .$$

We deduce that $q^2 = 1$ or $q^2 = 4$ and the result of Bertrand is thus obtained.

2. Consider now the case of the motion on a surface of revolution with equations $x = \psi(\xi)\cos\alpha$, $y = \psi(\xi)\sin\alpha$, $z = \phi(\xi)$, $\phi(\xi) = \int_0^\xi \sqrt{1-\psi'^2(\xi)}\,d\xi$ and the mechanical problem described correspondingly by the Lagrange

function

$$L(\xi,\alpha,u,v) = \frac{1}{2}(u^2 + \psi^2(\xi)v^2) - V(\xi) .$$

By using the first integral $\psi^2(\xi)v = p$ we get

$$\dot{\xi} = u, \quad \dot{u} = \frac{p^2\psi'(\xi)}{\psi^3(\xi)} - V'(\xi), \quad \dot{\alpha} = \frac{p}{\psi^2(\xi)}$$

and with $V_p(\xi) = \frac{p}{2\psi^2(\xi)} + V(\xi)$ the system is written $\dot{\xi} = u$, $\dot{u} = -V_p'(\xi)$, $\dot{\alpha} = \frac{p}{\psi^2(\xi)}$. If $V_p'(\xi_p) = 0$, $V_p''(\xi_p) > 0$ then in a neighborhood of ξ_p the invariant manifold $I_{h,p}$ is a torus.

By using the change of variables

$$\xi = \xi_p + \rho\cos\theta, \quad u = -\sqrt{V_p''(\xi_p)}\,\rho\sin\theta$$

we get for the rotation number $\mu(h,p)$

$$\mu(h,p) = \frac{1}{2\pi}\int_0^{2\pi} \frac{p\,d\theta}{\psi^2(\xi_p+\rho\cos\theta)[\sqrt{V_p''(\xi_p)}\sin^2\theta + \frac{1}{\rho\sqrt{V_p''(\xi_p)}}V_p'(\xi_p+\rho\cos\theta)\cos\theta]}$$

where ρ is defined by $\frac{1}{2}V_p''(\xi_p)\rho^2\sin^2\theta + V_p(\xi_p+\rho\cos\theta) = h$; we may write also

$$\mu(h,p) = \frac{1}{2\pi}\oint \frac{p\,d\xi}{\psi^2(\xi)u}, \quad \frac{1}{2}u^2 + V_p(\xi) = h$$

or

$$\mu(h,p) = \frac{1}{\pi}\int_{\xi_{min}}^{\xi_{max}} \frac{p\,d\xi}{\psi^2(\xi)\sqrt{2[h - V_p(\xi)]}} .$$

We deduce that if $\mu(h,p)$ has to be constant and equal to $\frac{1}{q}$ then

$$\frac{p}{\psi^2(\xi_p)\sqrt{V_p''(\xi_p)}} = \frac{1}{q},$$

$$\frac{5}{24}\left(\frac{V_p'''(\xi_p)}{V_p''(\xi_p)}\right)^2 - \frac{1}{8}\frac{V_p^{iv}(\xi_p)}{V_p''(\xi_p)} + \frac{V_p'''(\xi_p)}{V_p''(\xi_p)}\frac{\psi'(\xi_p)}{\psi(\xi_p)} + \frac{3\psi'^2(\xi_p)}{\psi^2(\xi_p)} - \frac{\psi''(\xi_p)}{\psi(\xi_p)} = 0.$$

The condition $V_p'(\xi_p) = 0$ gives $V'(\xi_p) = \frac{p^2\psi'(\xi_p)}{\psi^3(\xi_p)}$; if we give a ξ

such that $\frac{V'(\xi)\psi^3(\xi)}{\psi'(\xi)} > 0$ we may choose p such that $\xi_p = \xi$; since $V_p''(\xi) = V''(\xi) + \frac{p^2}{\psi^2(\xi)}\left(\frac{3\psi'^2(\xi)}{\psi^2(\xi)} - \frac{\psi''(\xi)}{\psi(\xi)}\right)$ the condition $p^2q^2 = \psi^4(\xi_p)V_p''(\xi_p)$ gives the differential equation for V:

$$\frac{V''}{V'} = \frac{q^2}{\psi\psi'} + \frac{\psi''}{\psi'} - \frac{3\psi'}{\psi} ;$$

hence the potentials with $\mu(h,p) = \frac{1}{q}$ are of the form

$$V'(\xi) = \frac{C\psi'(\xi)}{\psi^3(\xi)} e^{q^2 \int \frac{d\xi}{\psi(\xi)\psi'(\xi)}} .$$

We are not able to continue the discussion in this general situation and decide under what conditions there exists a q for which the condition given by considering further approximations is satisfied.

3. Complete results may be obtained in the particular case of surfaces of constant positive curvature $K = \mu^2$; for this case we have $\psi(\xi) = a \cos\mu\xi$, and the above formula gives

$$V'(\xi) = -\frac{C\ \mu\ \sin\mu\xi}{a^2\cos^3\mu\xi}\left(\frac{\cos\mu\xi}{\sin\mu\xi}\right)^{\frac{q^2}{a^2\mu^2}}$$

$$V_p(\xi) = -\frac{p^2}{2a^2}\,\mathrm{tg}^2\mu\xi - \frac{C}{2a^2 - \frac{q^2}{\mu^2}}(\mathrm{tg}\mu\xi)^{2-\frac{q^2}{a^2\mu^2}} + \frac{p^2}{2a^2} .$$

We have then

$$\mu(h,p) = \frac{1}{\pi}\int_{\xi_{min}}^{\xi_{max}} \frac{p\,d\xi}{a^2\cos^2\mu\xi\ \sqrt{2(h - V_p(\xi))}}$$

$$= \frac{1}{\pi a\mu}\int_{\rho_{min}}^{\rho_{max}} \frac{\frac{p}{a}\,d\rho}{\sqrt{2(h_1 - \frac{p^2}{2a^2}\rho^2 - C_1\rho^{2-q^2/a^2\mu^2})}} ,$$

$$h_1 = h - \frac{p^2}{2a^2} .$$

Recall that in the central forces problem we had

$$\mu(h,p) = \frac{1}{\pi}\int_{r_{min}}^{r_{max}} \frac{p\,dr}{r^2\sqrt{2(h - \frac{p^2}{2r^2} - V(r))}}$$

$$= \frac{1}{\pi}\int_{r_{min}}^{r_{max}} \frac{p\,dr}{r^2\sqrt{2(h - \frac{p^2}{2r^2} - Cr^{q^2-2})}}$$

$$= \frac{1}{\pi}\int_{\rho_{min}}^{\rho_{max}} \frac{p\,d\rho}{\sqrt{2(h - \frac{p^2}{2}\rho^2 - C\rho^{2-q^2}}} .$$

We see that the situation for surfaces of positive constant curvature is the same as in the central forces problem and hence we may deduce that the integral is independent on (h,p) iff $(\frac{q}{a\mu})^2 = 1$ or $(\frac{q}{a\mu})^2 = 4$.

The trajectories are closed iff $a\mu$ is rational. We may state

THEOREM. Consider the surface

$$x = \psi(\xi)\cos\alpha, \qquad y = \psi(\xi)\sin\alpha, \qquad z = \phi(\xi)$$

$$\psi(\xi) = a\cos\mu\xi \ , \qquad \phi(\xi) = \int_0^\xi \sqrt{1 - \psi'^2(\xi)}\, d\xi$$

and the mechanical problem described by the Lagrange function

$$L(\xi,\alpha,u,v) = \frac{1}{2}(u^2 + \psi^2(\xi)v^2) - V(\xi) .$$

Then there are essentially three potentials $V(\xi) \equiv 0$, $V(\xi) = tg^2\mu\xi$, $V(\xi) = -\frac{1}{tg\mu\xi}$ for which the rotation number on invariant tori $I_{h,p}$ is constant: the trajectories on invariant tori are closed iff $a\mu$ is rational.

4. For the pseudosphere $\psi(\xi) = e^{-\xi}$, $V_p'(\xi) = (p^2 - Ce^{-\frac{q^2}{2}e^{2\xi}})e^{2\xi}$ and the condition obtained from the second approximation gives the only solution $q = 0$ and that means that on the pseudosphere there are no potentials for which $\mu(h,p)$ is constant.

A similar calculation shows that the result is the same for surfaces of constant negative curvature $\psi(\xi) = a\,ch\,\mu\xi$. For a rotation conus $\psi(\xi) = \xi\sin\alpha$ the curvature is zero,

$$V(\xi) = \frac{C}{\left(\frac{q^2}{\sin^2\alpha} - 2\right)\sin^2\alpha}\,\xi^{\frac{q^2}{\sin^2\alpha} - 2},$$

$$\mu(h,p) = \frac{p}{\pi\sin^2\alpha}\int_{\rho_{min}}^{\rho_{max}} \frac{d\rho}{\sqrt{2\left(h - \frac{p}{2\sin^2\alpha}\rho^2 - \frac{C}{q^2-2\sin^2\alpha}\rho^{2-q^2/\sin^2\alpha}\right)}}$$

and we are again in the same situation as for the central forces problem. We deduce that $\mu(h,p)$ is independent on (h,p) iff $\frac{q^2}{\sin^2\alpha} = 1$ or $\frac{q^2}{\sin^2\alpha} = 4$; in these cases the rotation number is $\frac{1}{\sin\alpha}$ and $-\frac{1}{2\sin\alpha}$ respectively and it is rational iff $\sin\alpha$ is rational. The corresponding potentials are as in the central forces problem, $-\frac{1}{\xi}$ and $\frac{1}{2}\xi^2$.

Let us remark that essentially the results described above depend on the form of the metric on the surface; they are true for surfaces that may be continuously deformed preserving the length into a surface of revolution of constant positive curvature. From the viewpoint of the interior geometry all thsee surfaces are spheres, but we have seen that the solution depends essentially on a and μ, where $\mu = \sqrt{K}$, (K is the constant curvature of the surface). For the sphere $a = 1$, $\mu = 1$ the potentials are the ones obtained by Darboux.

5. There are also other mechanical problems for which similar questions may be considered.

Consider for example the problem described by the Lagrange function

$$L = \frac{I_1 + \mu\ell^2}{2}(\dot\theta^2 + \dot\phi^2\sin^2\theta) + \frac{I_3}{2}(\dot\psi + \dot\phi\cos\theta)^2 - V(\theta)$$

where ϕ, ψ, θ are the Euler angles.

We have two global integrals of angular momentum

$$p_1 = (I_1 + \mu\ell^2)u\sin^2\theta + I_3(v + u\cos\theta)\cos\theta$$

$$p_2 = I_3(v + u\cos\theta)$$

and the equation

$$(I_1 + \mu\ell^2)\dot{w} = (I_1 + \mu\ell^2)u^2 \sin\theta\,\cos\theta - I_3(v + u\cos\theta)u\sin\theta - V'(\theta)\ .$$

From the two first integrals we obtain u and v as functions of θ and p_1, p_2 ; denote $\tilde{p}_1 = \frac{p}{I_1 + \mu\ell^2}$, $p_2 = \frac{p}{I_1 + \mu\ell^2}$. We get the system

$$\dot{\theta} = w, \qquad \dot{w} = \frac{(\tilde{p}_1 - \tilde{p}_2\cos\theta)(\tilde{p}_1\cos\theta - \tilde{p}_2)}{\sin^3\theta} - \frac{1}{I_1 + \mu\ell^2} V'(\theta)$$

$$\dot{\phi} = \frac{\tilde{p}_1 - \tilde{p}_2\cos\theta}{\sin^2\theta}$$

$$\dot{\psi} = \frac{\tilde{p}_2(\cos^2\theta + a^2\sin^2\theta) - \tilde{p}_1\cos\theta}{\sin^2\theta}\ , \qquad a^2 = \frac{I_1 + \mu\ell^2}{I_3}$$

Considering also the energy integral

$$\frac{I_1 + \mu\ell^2}{2}(w^2 + u^2\sin^2\theta) + \frac{I_3}{2}(v + u\cos\theta)^2 + V(\theta) = h$$

and denoting

$$V_p(\theta) = \frac{1}{I_1 + \mu\ell^2} V(\theta) + \frac{1}{2}a^2\tilde{p}_2^2 + \frac{1}{2}\left(\frac{\tilde{p}_1 - \tilde{p}_2\cos\theta}{\sin\theta}\right)^2 ,$$

the system for θ, w is

$$\dot{\theta} = w\ , \qquad \dot{w} = -V_p'(\theta)\ .$$

If θ_p is a critical point for V_p and $V_p''(\theta_p) > 0$ the trajectories for $\frac{h}{I_1 + \mu\ell^2} > V_p(\theta_p)$ are closed; for

$$\theta - \theta_p = \rho\cos\tau\ , \qquad w = -\sqrt{V_p''(\theta_p)}\,\rho\sin\tau$$

we get

$$\frac{d\phi}{d\tau} = \frac{\tilde{p}_1 - \tilde{p}_2\cos(\theta_p + \rho\cos\tau)}{\sin^2(\theta_p + \rho\cos\tau)\left[\sqrt{V_p''(\theta_p)}\,\sin^2\tau + \frac{1}{\rho\sqrt{V_p''(\theta_p)}} V_p'(\theta_p + \rho\cos\tau)\cos\tau\right]}$$

$$\frac{d\psi}{d\tau} = \frac{\tilde{p}_2[\cos^2(\theta_p + \rho\cos\tau) + a^2\sin^2(\theta_p + \rho\cos\tau)] - \tilde{p}_1\cos(\tau_p + \rho\cos\tau)}{\sin^2(\theta_p + \rho\cos\tau)\left[\sqrt{V_p''(\theta_p)}\,\sin^2\tau + \frac{1}{\rho\sqrt{V_p''(\theta_p)}} V_p'(\theta_p + \rho\cos\tau)\cos\tau\right]}$$

where $\rho(\tau)$ is given by

$$\frac{1}{2} V_p''(\theta_p)\rho^2\sin^2\tau + V_p(\theta_p+\rho\cos\tau) = \frac{h}{I_1+\mu\ell^2} .$$

The motion is described by the rotation numbers

$$\mu_1(h,p) = \frac{1}{2\pi}\int_0^{2\pi} \frac{[\tilde{p}_1 - \tilde{p}_2 \cos(\theta_p+\rho\cos\tau)]d\tau}{\sin^2(\theta_p+\rho\cos\tau)[\sqrt{V_p''(\theta_p)}\ \sin^2\tau + \frac{1}{\rho\sqrt{V_p''(\theta_p)}} V_p'(\theta_p+\rho\cos\tau)\cos\tau]}$$

$$\mu_2(h,p) = \frac{1}{2\pi}\int_0^{2\pi} \frac{\tilde{p}_2[\cos^2(\theta_p+\rho\cos\tau)+a^2\sin^2(\theta_p+\rho\cos\tau)]-\tilde{p}_1\cos(\theta_p+\rho\cos\tau)}{\sin^2(\theta_p+\rho\cos\tau)[\sqrt{V_p''(\theta_p)}\sin^2\tau+\frac{1}{\rho\sqrt{V_p''(\theta_p)}} V_p'(\theta_p+\rho\cos\tau)\cos\tau]}d\tau.$$

These numbers may be written

$$\mu_1(h,p) = \frac{1}{2\pi}\oint \frac{\tilde{p}_1 - \tilde{p}_2 \cos\theta}{w \sin^2\theta} d\theta$$

$$\mu_2(h,p) = \frac{1}{2\pi}\oint \frac{\tilde{p}_2(\cos^2\theta+a^2\sin^2\theta)-\tilde{p}_1\cos\theta}{w \sin^2\theta} d\theta$$

$$\frac{1}{2} w^2 + V_p(\theta) = \frac{h}{I_1+\mu\ell^2}$$

or

$$\mu_1(h,p) = \frac{1}{\pi}\int_{\theta_{min}}^{\theta_{max}} \frac{\tilde{p}_1 - \tilde{p}_2\cos\theta}{\sqrt{2(\frac{h}{I_1+\mu\ell^2} - V_p(\theta))}\ \sin^2\theta} d\theta$$

$$\mu_2(h,p) = \frac{1}{\pi}\int_{\theta_{min}}^{\theta_{max}} \frac{\tilde{p}_2(\cos^2\theta+a^2\sin^2\theta) - \tilde{p}_1\cos\theta}{\sqrt{2(\frac{h}{I_1+\mu\ell^2} - V_p(\theta))}\ \sin^2\theta} d\theta .$$

Consider the particular case $a^2 = 1$ (i.e. $I_1 + \mu\ell^2 = I_3$) and denote $\frac{h}{I_1+\mu\ell^2} = h_1$. The notation numbers are in this case

$$\mu_1(h,p) = \frac{1}{\pi}\int_{\theta_{min}}^{\theta_{max}} \frac{\tilde{p}_1 - \tilde{p}_2 \cos\theta}{\sqrt{2(h_1 - V_p(\theta))}\ \sin^2\theta} d\theta$$

$$\mu_2(h,p) = \frac{1}{\pi}\int_{\theta_{min}}^{\theta_{max}} \frac{\tilde{p}_2 - \tilde{p}_1 \cos\theta}{\sqrt{2(h_1 - V_p(\theta))}\ \sin^2\theta}\, d\theta .$$

If we restrict ourselves to the family of trajectories with $u(0) = v(0)$ we deduce for this family from

$$u + v\cos\theta = \tilde{p}_1 , \quad v + u\cos\theta = \tilde{p}_2$$

that $\tilde{p}_1 = \tilde{p}_2$ and for the trajectories considered we will have $u(t) = v(t)$,

$$\mu_1(h,p) = \mu_2(h,p) = \frac{1}{2\pi}\int_{\theta_{min}}^{\theta_{max}} \frac{p\,d\theta}{\cos^2\frac{\theta}{2}\sqrt{2(h-V_p(\theta))}} ,$$

$$V_p(\theta) = \frac{1}{I_1+\mu\ell^2} V(\theta) + \frac{1}{2}p^2 + \frac{1}{2}p^2 \operatorname{tg}^2\frac{\theta}{2} .$$

We may write also

$$\mu_1(h,p) = \mu_2(h,p) = \frac{1}{\pi}\int_{\rho_{min}}^{\rho_{max}} \frac{p\,d\phi}{\sqrt{2h - p^2 - p^2\rho^2 - \tilde{V}(\rho)}}$$

and we are in the same situation as for the central forces problem; we deduce that for $V = 0$ and for essentially two other potentials $V(\theta) = -\operatorname{tg}\frac{\theta}{2}$, $V(\theta) = \frac{1}{2}\cot^2\frac{\theta}{2}$, the rotation numbers are rational, hence the trajectories considered are closed.

For $V \equiv 0$ (geodesic flow) a more detailed analysis shows that all trajectories are closed (μ_1 and μ_2 are rational even if $p_1 \neq p_2$), but this is no more true for the other two potentials.

Acknowledgement

Very stimulating in this research were the discussions with Andrei Iacob and Prof. Th. Hangan.

REFERENCES

[1] J. BETRAND, Théorème relatif au mouvement d'un point attiré vers un centre fixe, C.R. Acad. Sci. Paris, 77 (1873), 849-853.

[2] G. DARBOUX, Etude d'une question relative au mouvement d'un point sur une surface de révolution, Bull. Soc. Math. de France, T.V., (1876-1877), 100-113.

Academie, Republique Socialiste de Roumanie, Bucharest, Romania

LOCAL BEHAVIOR OF AUTONOMOUS NEUTRAL FUNCTIONAL DIFFERENTIAL EQUATIONS*

Jack K. Hale

A neutral functional differential equation is a model for an hereditary system which depends upon the present and past values of the state and the derivative of the state of the system as a function of time. In this generality, it does not seem possible to develop a general qualitative theory. In the past few years, my colleagues and I have introduced a special class of these systems which is simple enough to have many interesting mathematical properties and yet sufficiently general to include retarded functional differential equations, difference equations and several important applications. Considered separately, many of the results which have been obtained appear at first glance to be special. In this paper, we formulate some basic problems for a more restricted class of neutral functional differential equations in an abstract manner and then indicate the manner in which the alluded to special results contribute to a general qualitative theory in the neighborhood of an equilibrium point.

---

*This research was supported in part by the National Aeronautics and Space Administration under Grant No. NGL 40-002-015, in part by the Air Force Office of Scientific Research under Grant No. AF-AFOSR 67-0693D, and in part by the United States Army - Durham under Grant No. DA-31-124-ARO-D-270.

Suppose $r \geq 0$ is a given real number, E^n is an n-dimension linear vector space with norm $|\cdot|$ and $C = C([-r,0], E^n)$ is the space of continuous functions mapping the interval $[-r,0]$ into E^n with $|\phi| = \sup_{-r\leq\theta\leq 0} |\phi(\theta)|$ for $\phi \in C$. Let $\mathcal{B}(C,E^n)$ be the space of bounded linear operators with $\|B\| = \inf\{k: |B\phi| \leq k|\phi|\}$ for any $B \in \mathcal{B}(C,E^n)$. Let $\mathcal{G} \subset \mathcal{B}(C,E^n)$ be defined by

$$\mathcal{G} = \{G \subset \mathcal{B}(C,E^n): G(\phi) = \phi(0) - g(\phi),\ \ g(\phi) = \int_{-r}^{0} [d\mu(\theta)]\phi(\theta),$$
$$\mu \text{ an } n \times n \text{ matrix function of bounded variation and nonatomic at zero}\}$$

where we say an $n \times n$ matrix function μ of bounded variation on $[-r,0]$ is nonatomic at zero if there is a continuous nonnegative scalar function $\gamma(s)$, $s \geq 0$, $\gamma(0) = 0$ such that $|\int_{-s}^{0} [d\mu(\theta)]\phi(\theta)| \leq \gamma(s)|\phi|$ for all $s \geq 0$, $\phi \in C$. Let $\bar{\Omega}$ be the closure of an open bounded subset Ω of C,

$$\mathcal{F} = \{f: \Omega \to E^n,\ \ f, f' \text{ continuous and bounded in } \bar{\Omega}\}$$

and let $\|f\| = \sup_{\phi\in\bar{\Omega}}\{|f(\phi)| + \|f'(\phi)\|\}$ for $f \in \mathcal{F}$. If $A \geq 0$ and $x: [\sigma-r,\sigma+A] \to E^n$, then for any $t \in [\sigma,\sigma+A]$ we let $x_t: [-r,0] \to E^n$ be defined by $x_t(\theta) = x(t+\theta)$, $-r \leq \theta \leq 0$.

For any pair $(G,f) \in \mathcal{G} \times \mathcal{F}$, we define a neutral functional differential equation and designate it as NFDE(G,f) by a relation

$$\frac{d}{dt} G(x_t) = f(x_t) \ . \tag{1}$$

For any ϕ in $\bar{\Omega}$, a solution $x(\phi)$ of (1) through ϕ is a continuous function $x: [-r,A) \to E^n$ for some $A > 0$ such that $x_t(\phi) \in \bar{\Omega}$, $t \in [0,A)$, $G(x_t)$ is continuously differentiable on $(0,A)$ and satisfies (1) on $(0,A)$. For the purpose of this paper, we assume that all solutions remain in $\bar{\Omega}$ for all $t \geq 0$ for each NFDE(G,f) in $\mathcal{G} \times \mathcal{F}$. Of course, this implies that in our definition of $\mathcal{G}$ and $\mathcal{F}$ we have implicitly put limitations which will ensure this latter property.

For any given NFDE(G,f) and a solution $x(\phi,G,f)$ through ϕ, we can define a mapping

(2) $$T_{G,f}(t)\colon \bar{\Omega} \to \bar{\Omega}, \quad T_{G,f}(t)\phi = x_t(\phi,G,f)$$

for $t \geqslant 0$. The following properties are consequences of the same type of arguments used in [1, 2]:

(3) $$T_{G,f}(t)\phi \text{ is continuous in } (G,f,t,\phi) \in \mathcal{G} \times \mathcal{F} \times [0,\infty) \times \Omega.$$

(4) $$T_{G,f}(0) = I, \text{ the identity}$$

(5) $$T_{G,f}(t+s)\phi = T_{G,f}(t)T_{G,f}(s)\phi, \quad t \geqslant 0, \quad s \geqslant 0 .$$

In particular, $T_{G,f}$ is a dynamical system.

By a linear neutral functional differential equation, we mean a NFDE(G,L) with $(G,L) \in \mathcal{G} \times \mathcal{B}(C,E^n)$. In this case, the family of operators $\{T_{G,L}(t)\}$ form a strongly continuous semigroup of linear operators on C.

Our fundamental goal is to obtain as much information as possible about the properties of solutions of a NFDE(G,f) and to discover in what sense these properties are insensitive to small changes in (G,f) in the topology of $\mathcal{G} \times \mathcal{F}$. One can hope to eventually develop a qualitative theory in the spirit of the recent generic theory for ordinary differential equations. Unfortunately, at the present time we must limit our considerations to the neighborhood of an equilibrium point. Even for this theory, more restrictions are imposed on the equation. We indicate below the nature of the theory and occasionally give the implications of the hypotheses for some nonlocal problems in NFDE's.

Before embarking on details, it is worthwhile to mention that there is nothing magic about the space C as the underlying space for the definition of a NFDE. Any of the Sobolev spaces $W_p^{(k)}$ are perfectly legitimate and, in fact, as the global theory evolves, it certainly will be advantageous to have a norm which is differentiable.

For any $G \in \mathcal{G}$, let

$$(6)\qquad a_G = \inf\{a: \text{there is a constant } K = K(a) \text{ with the solution operator } T_G^0 \text{ of } G(x_t) = 0 \text{ satisfying } \|T_G^0(t)\| \leq Ke^{at},\ t \geq 0\},$$

that is, a_G is the type of the semigroup of linear operators $T_G^0(t)$ acting on the Banach space $C_G = \{\phi \in C: G(\phi) = 0\}$. Following [3], the operator G is said to be <u>stable</u> if $a_G < 0$.

In terms of a_G, a fundamental property which is a consequence of the proof in [12] is that, for any $(G,f) \in \mathcal{G} \times \mathcal{F}$, there is a continuous linear operator $\psi: C \to C_G$ such that

$$(7)\qquad T_{G,f}(t) = T_G^0(t)\psi + T_{G,f}^1(t)$$

where $T_G^0(t)$ is the semigroup of operators given in the definition (6) and $T_{G,f}^1(t)$ is completely continuous for $t \geq r$. In particular, if G is stable, then for any α, $a_D < -\alpha < 0$, there is a constant $K = K(\alpha)$ such that

$$(8)\qquad \|T_G^0(t)\| \leq Ke^{-\alpha t}, \quad t \geq 0 .$$

Consequently, $T_{G,f}(t)$ is a contraction operator plus a completely continuous operator for t sufficiently large and, in particular, has the fixed point property. Also, the representation (7) and relation (8) show that $T_{G,f}(t)$ has a smoothing property in the sense that the image of a bounded set under $T_{G,f}(t)$ gets closer and closer to a compact set as t gets large. This property has recently been exploited in a global theory of dissipative processes (see [4]).

Another basic implication of the hypothesis that G is stable is that the union of orbits with initial values in a compact set being bounded implies this union is precompact. See the above mentioned paper [4] for an application to dissipative processes and see [3] for a rather complete theory of stability using Liapunov functionals.

If $(G,L) \in \mathcal{G} \times \mathcal{B}(C,E^n)$, then the characteristic values of NFDE(G,L) are defined to be the roots λ of the characteristic equation

$$(9)\qquad \det \Delta(\lambda,G,L) \overset{\text{def}}{=} \det[\lambda G(e^{\lambda\cdot}I) - L(e^{\lambda\cdot}I)] = 0 .$$

It is shown in [5] that there are only a finite number of roots of (9) with real parts $\geq a$ if $a > a_G$. If $a > a_G$ and $\Lambda(a,G,L) = \{\lambda\colon \det \Delta(\lambda,G,L) = 0,\ \mathrm{Re}\,\lambda \geq a$, then it is a consequence of [6] that there are only a finite number of solutions of NFDE(G,L) of the form $p(t)e^{\lambda t}$, $\lambda \in \Lambda(a,G,L)$, $p(t)$ a polynomial. Furthermore, the initial values of these solutions form a finite dimensional subspace of C which is invariant under $T_{G,L}(t)$ and there is a complementary subspace of C also invariant under $T_{G,L}(t)$. If we designate the corresponding projection operators by $\pi(G,L,a)$ and $\pi^{\perp}(G,L,a) \overset{\text{def}}{=} I-\pi(G,L,a)$, then it is shown in [5] that there is a constant $K = K(a,G,L)$ such that

$$\|T_{G,L}(t)\pi^{\perp}(G,L,a)\| \leq Ke^{at}, \quad t \geq 0 . \tag{10}$$

Furthermore, $T_{G,L}(t)\pi(G,L,a)$ is equivalent to a linear ordinary differential equation with constant coefficients with the spectrum of the coefficient matrix being $\Lambda(a,G,L)$. We shall refer to $\pi(G,L,a)$, $\pi^{\perp}(G,L,a)$ as the projections associated with the decomposition of C by $\Lambda(a,G,L)$.

The estimate (10) was obtained using rather complicated arguments in Laplace transforms. Using the representation (7) for $T_{G,L}(t)$, one can also obtain this estimate from some general results in functional analysis dealing with completely continuous perturbations of bounded operators. It is instructive to investigate this approach for two reasons: 1) The ideas lead to an extension of Floquet theory for periodic systems which will not be exploited here; 2) Later remarks will indicate the need for a more detailed functional analytic investigation of completely continuous perturbations of bounded operators.

Suppose A is a bounded linear operator and B is a completely continuous linear operator mapping some Banach space X into itself. A complex number λ is called a normal eigenvalue of A if its generalized eigenspace X_λ is finite dimensional and $X = X_\lambda \oplus Y_\lambda$ where Y_λ is also invariant under A. A point λ is called a normal point of A if λ is either in the resolvent set of A or λ is a normal eigenvalue. Let $\tilde{\rho}(A)$ be the set of normal points of A. If C is any connected component of $\tilde{\rho}(A)$ and C contains at least one point of the resolvent set of $A+B$, then C is a connected component of $\tilde{\rho}(A+B)$ (see [7, p. 22]). To apply this to our problem, note that the spectral radius of $T_G^0(r)$ is $\exp(ra_G)$ and, therefore, $\exp \lambda r$ is in $\tilde{\rho}(T_G^0(r))$ for any λ with $\operatorname{Re}\lambda \geq a_1 > a_G$. Since $T_{G,L}(t)$ is a strongly continuous semigroup, there is certainly a λ with $\operatorname{Re}\lambda \geq a_1$ such that $\exp \lambda r$ is in the resolvent set of $T_{G,L}(r)$. Since $T_{G,L}^1(r)$ is completely continuous, any λ with $\operatorname{Re}\lambda \geq a_1$ is in $\tilde{\rho}(T_{G,L}(r))$. Consequently, the spectral radius of $T_{G,L}(r)\pi^{\perp}(G,L,a_1)$ is $\leq \exp a_1 r$ and estimate (10) follows by a standard procedure.

For a more detailed and complete discussion of the relationship between the spectrum of $T(t)$ and $\{\exp \lambda t, \lambda$ satisfying (9)$\}$, see [13].

From the preceding remarks, it looks as if we should give up Laplace transforms and use only functional analysis. However, recall that we wish to study properties of the equations which are insensitive to small changes in G,L. In particular, we need the following result: For any fixed $a > a_D$, there is a neighborhood $U_{G,L}$ of (G,L) in $\mathcal{G} \times \mathcal{B}(C,E^n)$ and a positive number $K = K(a,U_{G,L})$ such that

$$\|T_{\bar{G},\bar{L}}(t)\pi^{\perp}(\bar{G},\bar{L},a)\| \leq Ke^{at}, \quad t \geq 0, \quad \text{for all} \quad (\bar{G},\bar{L}) \in U_{G,L}. \tag{11}$$

The functional analytic approach does not seem to suffice for such general estimates and one must get more involved with the details of the spectrum of each operator. Estimate (11) could be obtained as a consequence of the method of integral manifolds to be mentioned later, but we prefer to proceed in a more detailed manner to obtain some results of independent interest. These propositions are stated as lemmas with only brief indications of the proofs.*

<u>LEMMA 1</u>. For any $\varepsilon > 0$, there exists an open neighborhood $U_G(\varepsilon)$ in $\mathcal{G}$ such that $a_{\bar{G}} < a_G+\varepsilon$ for every $\bar{G} \in U_G(\varepsilon)$. Furthermore, for any $a > a_G+\varepsilon$, there is a constant $K = K(a,U_G(\varepsilon),\varepsilon)$ such that

---

*At this conference, D. Henry has informed the author by personal communication that one can obtain the desired estimates from the integral representation of the projection operator $\pi^{\perp}(\bar{G},\bar{L},a)$.

$$\|T^{o}_{\overline{G}}(t)\| \leq Ke^{at}, \quad t \geq 0 ,$$

for every $\overline{G} \in U_G(\mathcal{E})$.

<u>Proof</u>. To find $a_{\overline{G}}$, we must estimate the solutions of $\overline{G}(x_t) = 0$ for $t \geq 0$. Writing this as $G(x_t) = (G-\overline{G})(x_t)$ and using the same type of argument as used in the proof of Lemma 3.4 of [3], one obtains the result.

<u>LEMMA 2</u>. For any $a > a_G$, there is an open neighborhood U_G of G in $\mathcal{G}$ and a $\delta = \delta(a,U_G) > 0$ such that $|\det \overline{G}(e^{\lambda \cdot}I)| \geq \delta$ on $\operatorname{Re} \lambda = a$ for all $\overline{G} \in U_G$.

<u>Proof</u>. Choose $\mathcal{E} > 0$ such that $a_G + \mathcal{E} < a$ and a neighborhood $U_G(\mathcal{E})$ as in Lemma 1. Then Lemma 3.4 of [5] implies that all roots of $\det \overline{G}(e^{\lambda \cdot}I) = 0$ are to the left of the line $\operatorname{Re} \lambda = a_{\overline{G}} < a + \mathcal{E}$ and there is a $\delta(a,\overline{G}) > 0$ such that $|\det \overline{G}(e^{\lambda \cdot}I)| \geq \delta(a,\overline{G})$ on $\operatorname{Re} \lambda = a$ for each $\overline{G} \in U_G(\mathcal{E})$. If no neighborhood and δ exists as stated in the lemma, then there are sequences $G_k \in U_G(\mathcal{E})$, $G_k \to G$ as $k \to \infty$ and $\lambda_k = \lambda_k(G_k)$, $\operatorname{Re} \lambda_k = a$, such that $|\det G_k(e^{\lambda_k \cdot}I)| \leq 1/k$, $k = 1,2,\ldots$. The proof now proceeds as in the proof of Lemma 3.4 of [5] using the appropriate generalization of Lemma 3.2 of [5].

<u>LEMMA 3</u>. For any $a > a_D$, there is an open neighborhood $U_{G,L}$ of (G,L) in $\mathcal{G} \times \mathcal{B}(C,E^n)$ such that for each $(\overline{G},\overline{L}) \in U_{G,L}$ the characteristic equation

$$\det \Delta(\lambda,\overline{G},\overline{L}) = 0$$

has the same finite number of characteristic roots with $\operatorname{Re} \lambda \geq a$. Furthermore, if $\pi(\overline{G},\overline{L},a)$, $\pi^{\perp}(\overline{G},\overline{L},a)$ are the projection operators associated with the decomposition of C by $\Lambda(a,\overline{G},\overline{L})$, then there is a constant $K = K(a,U_{G,L})$ such that (11) holds for every $(\overline{G},\overline{L}) \in U_{G,L}$. Furthermore, $\pi(\overline{G},\overline{L},a)$ is continuously differentiable in $\overline{G},\overline{L}$.

<u>Proof</u>. The remark about the number of roots is proved in the same manner as in the proof of Lemma 3.5 of [5]. The existence of the projection operator is a consequence of the general theory in [6]. The continuity and differentiability of the projection follows from the contour integral representation of the projection (see, e.g. [7, pp.

14-15]). To obtain the estimate (11), repeat the proof of Theorem 4.1 in [5].

As a first application of the previous results, let us consider as in [5] the saddle point property. Suppose G is stable, no roots of the characteristic equation (9) lie on the imaginary axis and $\pi(G,L,0)$, $\pi^{\perp}(G,L,0)$ are the projection operators associated with the decomposition of C by $\Lambda(0,G,L)$. Since G is stable, there is an $a_0 < 0$ such that $\pi(G,L,a) = \pi(G,L,0)$ for $a_0 \leq a \leq 0$. We know that $T_{G,L}(t)\pi(G,L,0)$ is equivalent to a linear ordinary differential equation whose coefficient matrix has spectrum with positive real parts and $\|T_{G,L}(t)\pi^{\perp}(G,L,0)\| \leq K \exp at$ for $t \geq 0$ and some $K > 0$, $a < 0$. Therefore, the solutions of the NFDE(G,L) in C is similar to a saddle point as shown in the accompanying Figure 1. From Lemma 3, there is a neighborhood U_G of G in $\mathcal{G}$ such that the same behavior is preserved for $(\overline{G},L)$ with $G \in U_G$ and $\pi(G,L,0)$ replaced by $\pi(\overline{G},L,0)$.

In what sense is this qualitative picture preserved for $(\overline{G},f)$ in a neighborhood of (G,L)? We say the <u>saddle point property of (G,L) is preserved</u> if (G,L) is as described above and there are an open neighborhood $U_{G,L}$ of (G,L) in $\mathcal{G} \times \mathcal{F}$ and a neighborhood W of $\phi = 0$ such that the following properties hold:

(i) In W, there is a unique equilibrium point $e(\overline{G},f)$ which is continuously differentiable in $(\overline{G},f)$ for each $(\overline{G},f) \in U_{G,L}$.

(ii) The stable and unstable manifolds of $e(\overline{G},f)$ in W are continuously differentiable in $(\overline{G},f) \in U_{G,L}$ and are diffeomorphic to those of (G,L) in W.

(iii) Any solution of the NFDE$(\overline{G},f)$ with initial value in W which is not on the stable manifold must leave W in a finite time for each $(\overline{G},f) \in U_{G,L}$.

<u>THEOREM 1</u>. The saddle point property of (G,L) is preserved.

<u>Proof</u>. Suppose $\overline{G} \in U_G$, the neighborhood given by Lemma 3. For any $f \in \mathcal{F}$, consider the NFDE$(\overline{G},f)$ written as

$$(12)\qquad \begin{aligned} \frac{d}{dt}\overline{G}(x_t) &= L(x_t) + h(x_t) \\ h(\phi) &\overset{\text{def}}{=} f(\phi) - L(\phi) \end{aligned} .$$

The variation of constants formula implies the initial value problem for (12) is equivalent to

$$x_t = T_{\overline{D},L}(t)\phi + \int_0^t T_{\overline{D},L}(t-s)X_0 h(x_s)ds \tag{13}$$

where X_0 is the $n \times n$ matrix function on $[-r,0]$ with $X_0(\theta) = 0$ for $-r \leqslant \theta < 0$, $X_0(0) = I$ (see [6, p. 13]). With the known estimates of $T_{\overline{D},L}(t)$ on $\pi(\overline{D},L,0)$, $\pi^{\perp}(\overline{D},L,0)$ given by Lemma 3, one can repeat the proof of a similar theorem for retarded equations given in [8] (see, also [9]).

In general, the solution operator $T_{G,f}(t)$ of a NFDE(D,f) need not be one-to-one. On the other hand, if the function μ of bounded variation used in the definition of G has a jump at $-r$ given by a nonsingular matrix B, then it is shown in [2] that $T_{G,f}(t)$ is a homeomorphism. Consequently, on an open dense set V in $\mathcal{G} \times \mathcal{F}$, one can assume $T_{G,f}(t)$ is a homeomorphism. We can thus introduce on V the concept of trajectory equivalence of NFDE's as in ordinary differential equations. The details have not been worked out, but there should be no essential difficulty in proving the following result: If $(G,L) \in V$ and (G,L) has the saddle point property, then there are a neighborhood W of $\phi = 0$ and a neighborhood $U \subset V$ of (G,L) such that $(\overline{G},f)$ is equivalent to (G,L) in W for each $(\overline{G},f) \in U$.

An important subclass of NFDE's are the retarded functional differential equations; namely, those NFDE(G,f) for which $G(\phi) = \phi(0)$. In this case, one can state conditions on (G,f) to ensure that $T_{G,f}(t)$ is one-to-one (see [2]), but this map will never be a homeomorphism if $r > 0$. Therefore, even the definition of trajectory equivalence seems to be a challenging problem.

Now let us turn to a more difficult question. Suppose we are given a NFDE(G,f) with G stable, $f(0) = 0$, $f = L+f_1$, $L = f'(0)$, p roots of the characteristic equation (9) have zero real parts and the remaining roots have negative real parts. Knowing the behavior of solutions of the NFDE(G,f), is it possible to discuss the behavior of solutions of the NFDE($\overline{G},\overline{f}$) for all $(\overline{G},\overline{f})$ in a neighborhood of (G,f)? A partial answer to this question is given below.

Since G is stable, there is a $d < 0$ such that no characteristic roots of (9) lie in the strip $d \leq \operatorname{Re} \lambda < 0$. Choose a real number a such that $d < a < 0$. Then Lemma 3 implies there are a neighborhood U_G of G in $\mathcal{G}$, projection operators $\pi(\overline{G},L,a)$, $\pi^{\perp}(\overline{G},L,a)$ associated with the decomposition of C by $\Lambda(a,\overline{G},L)$ and a constant $K = K(a,U_G)$ such that (11) is satisfied for each $(\overline{G},L) \in U_G \times \{L\}$. Furthermore, the dimension of the projection operators $\pi(\overline{G},L,a)$ is p for every $G \in U_G$. For any $\overline{f} \in F$, write the NFDE$(\overline{G},\overline{f})$ as

$$\text{(14)} \qquad \begin{aligned} \frac{d}{dt} G(x_t) &= L(x_t) + h_1(x_t) , \\ h_1(\phi) &= [\overline{f}(\phi)-f(\phi)] + [f(\phi)-L(\phi)] . \end{aligned}$$

As before, the variation of constants formula for (14) is

$$\text{(15)} \qquad x_t(\phi) = T_{\overline{G},L}(t)\phi + \int_0^t T_{\overline{G},L}(t-s)X_0 h_1(x_s)ds .$$

Using this relation and the known estimates on $T_{\overline{G},L}(t)$, one can apply the same reasoning as in [10] for retarded functional differential equations to obtain the following results.

THEOREM 2. Under the above hypotheses on (G,f), there are an open neighborhood $U_{G,f}$ of (G,f) in $\mathcal{G} \times F$, open neighborhoods V, W of $\phi = 0$, positive constants K, α and a diffeomorphism $h_{\overline{G},\overline{f}}: \pi(\overline{G},L,a)C \cap V \to W$ such that the set

$$M_{\overline{G},\overline{f}} = \{\phi \in C: \phi = h_{\overline{G},\overline{f}}\psi, \ \psi \in \pi(\overline{G},L,a)C\}$$

is a local integral manifold of the NFDE$(\overline{G},\overline{f})$ for $(\overline{G},\overline{f}) \in U_{G,f}$. The solutions of the NFDE$(\overline{D},\overline{f})$ on $M_{\overline{G},\overline{f}}$ is equivalent to an ordinary differential equation and $h_{\overline{G},\overline{f}}$ is differentiable in $(\overline{G},\overline{f}) \in U_{G,f}$. Finally, any solution $x(\phi)$ of the NFDE$(\overline{G},\overline{f})$ satisfies

$$|x_t(\phi) - h_{\overline{G},\overline{f}}\, \pi(\overline{G},L,a)x_t(\phi)| \leq Ke^{-\alpha t}|\phi|$$

as long as $x_t(\phi) \in W$, $\pi(\overline{G},L,a)x_t(\phi) \in V$.

COROLLARY 1. With $h_{\overline{G},\overline{f}}$ as in Theorem 2, any solution $x(\phi)$ satisfying $x_t(\phi) \in W$, $\pi(\overline{G},L,a)x_t(\phi) \in V$ for $t \leqslant 0$ must lie on $M_{\overline{G},\overline{f}}$.

Following the ideas used in the proof of Theorem 3.3 in [11] and Theorems 4.3 and 7.1 of [10] one can obtain

COROLLARY 2. Suppose $U_{G,f}$ is the neighborhood of Theorem 2, $(\overline{G},\overline{f}) \in U_{G,f}$ and there is a unique equilibrium point (say $\phi = 0$) of the NFDE$(\overline{G},\overline{f})$. If the zero solution of the ordinary differential equation on $M_{\overline{G},\overline{f}}$ is uniformly asymptotically stable (unstable), then the zero solution of the NFDE$(\overline{G},\overline{f})$ is uniformly asymptotically stable (unstable).

COROLLARY 3. Let $U_{G,f}, V, W$ be the neighborhoods of Theorem 2 and suppose $(\overline{G},\overline{f}) \in U_{G,f}$. If $x(\phi)$ is a solution of the NFDE$(\overline{G},\overline{f})$ with $\pi(G,L,a)x_t(\phi) \in V$, $x_t(\phi) \in W$ for $t \geqslant 0$, then the ω-limit set of ϕ belongs to $M_{\overline{G},\overline{f}}$ and is an invariant set. Furthermore, if there is a neighborhood on $M_{\overline{G},\overline{f}}$ which is positively invariant under the corresponding ordinary differential equation, then there is a neighborhood W' of $\phi = 0$ such that the ω-limit set of every solution of the NFDE$(\overline{G},\overline{f})$ with initial value $\phi \in W'$ belongs to $M_{\overline{G},\overline{f}}$.

As an application of Corollary 3 to a birfurcation problem, consider the system

$$\frac{d}{dt} G(x_t,\varepsilon) = f(x_t,\varepsilon)$$

where ε is a small real parameter and $G(\phi,\varepsilon)$, $f(\phi,\varepsilon)$ are very smooth functions of ϕ, ε for $\phi \in C$, $|\varepsilon| \leqslant \varepsilon_0$, $f(0,\varepsilon) = 0$. Suppose $L(\cdot,\varepsilon)$ is the derivative of $f(\phi,\varepsilon)$ with respect to ϕ evaluated at $\phi = 0$. Suppose the characteristic equation $\det \Delta(\lambda,G(\cdot,\varepsilon), L(\cdot,\varepsilon)) = 0$ has a pair of simple roots $\nu(\varepsilon) \pm i\mu(\varepsilon)$, $\varepsilon\nu(\varepsilon) > 0$ for $\varepsilon \neq 0$, $\nu(0) = 0$, $\mu(\varepsilon) > 0$, $|\varepsilon| \leqslant \varepsilon_0$, the remaining roots have negative real parts and there is a $\delta > 0$ such that $a_{G(\cdot,\varepsilon)} \leqslant -\delta < 0$, $|\varepsilon| \leqslant \varepsilon_0$. If we further assume that the zero solution of the NFDE$(G(\cdot,0), f(\cdot,0))$ is uniformly asymptotically stable, then Corollary 3 and a Poincaré-Bendixson argument imply there is at least one nontrivial periodic solution of the NFDE$(G(\cdot,\varepsilon), f(\cdot,\varepsilon))$ for $0 < \varepsilon \leq \varepsilon_1$, ε_1 sufficiently small.

Of course, to determine when the zero solution of a system with roots on the imaginary axis is uniformly asymptotically stable is an extremely difficult problem even for ordinary differential equations. Some success for NFDE's has recently been obtained in [11] for all roots on the imaginary axis being zero and for the general case by A. Hausrath in a Ph.D. thesis at Brown. Basic to this investigation is Corollary 2. This corollary implies that it is sufficient to discuss the ordinary differential equation on $M_{\overline{G},\overline{f}}$. Therefore, one can hope to approximate the manifold $M_{\overline{G},\overline{f}}$ to a sufficient accuracy so that a decision can be made. This is the approach that was taken in the above papers and the final sufficient conditions are expressed in terms of computable quantities.

REFERENCES

[1] J.K. HALE and M.A. CRUZ, Existence, uniqueness and continuous dependence in hereditary systems, Annali Mat. Pura Appl., 85 (1970), 63-82.

[2] J.K. HALE, Forward and backward continuation for neutral functional differential equations, J. Differential Equations, 9 (1971), 168-181.

[3] M.Z. CRUZ and J.K. HALE, Stability of functional differential equations of neutral type, J. Differential Equations, 7 (1970), 334-355.

[4] J.K. HALE, J.P. LASALLE and M. SLEMROD, Theory of a general class of dissipative processes, J. Math. Ana. Appl., to appear.

[5] M.A. CRUZ and J.K. HALE, Exponential estimates and the saddle point property for neutral functional differential equations, J. Math. Ana. Appl., 34 (1971).

[6] J.K. HALE and K.R. MEYER, A class of functional equations of neutral type, Mem. Amer. Math. Soc., 76 (1967), 65 pages.

[7] I.C. GOHBERG and M.G. KREIN, Introduction to the theory of linear nonself-adjoint operators, Trans. Math. Monographs Am. Math. Soc., 18 (1969).

[8] J.K. HALE and C. PERELLÓ, The neighborhood of a singular point for functional differential equations, Contr. to Differential Eqns., 2 (1964), 351-375.

[9] J.K. HALE, Functional Differential Equations, Applied Math. Sciences. Vol. 3, Springer-Verlag, 1971.

[10] N. CHAFEE, A birfurcation problem for a functional differential equation of finitely retarded type, J. Math. Ana. Appl., 34 (1971).

[11] J.K. HALE, Critical cases for neutral functional differential equations, J. Differential Eqns, 1971.

[12] J.K. HALE, A class of neutral equations with the fixed point property, Proc. Nat. Acad. Sci. USA, 67 (1970), 136-137.
[13] D. HENRY, Linear autonomous neutral functional differential equations, to appear.

Brown University, Providence, Rhode Island

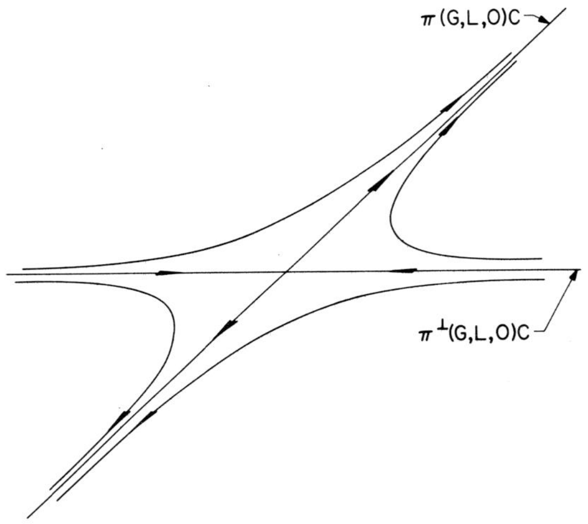

Fig. 1

EXTREMAL PROPERTIES OF BICONVEX CONTINGENT EQUATIONS

Hubert Halkin*

1. Introduction

Let F be a mapping from R^{n+1} into $\mathcal{P}(R^n)$, the set of all subsets of R^n. In other words for every $x \in R^n$, $t \in R^1$ we are given a subset $F(x,t)$ of R^n. To this mapping F we associate the contingent equation

$$\dot{x} \in F(x,t) .$$

We shall denote by I the set of all closed bounded intervals in R^1 with nonempty interior. If $a \in I$ then a^- will denote $\inf a$ and a^+ will denote $\sup a$, i.e. we shall have $a = [a^-,a^+]$. If $a \in I$ we shall denote by A_a the set of all absolutely continuous functions from a into R^n. We shall denote by A the set $\{(a,\psi): a \in I, \psi \in A_a\}$. A trajectory of $\dot{x} \in F(x,t)$ is a pair $(a,\psi) \in A$ such that $\dot{\psi}(t) \in F(\psi(t),t)$ for almost every $t \in a$. We shall denote by A^F the set of all trajectories of $\dot{x} \in F(x,t)$ and if $a \in I$ we shall denote by A_a^F the set of ψ such that $(a,\psi) \in A^F$. For $a \in I$, $t \in R^1$, $x \in R^n$ we shall also use the notation $A(t,x) = \{(a,\psi): (a,\psi) \in A, t \in a, \psi(t) = x\}$, $A^F(t,x) = A(t,x) \cap A^F$, $A_a(t,x) = \{\psi: (a,\psi) \in A(t,x)\}$, $A_a^F(t,x) = A_a(t,x) \cap A_a^F$. The scalar product of two elements x and y

---

*This research was sponsored by the Air Force Office of Scientific Research, Office of Aerospace Research, USAF, under Grant No. AFOSR-68-1529C.

in R^n will be denoted by $x \cdot y$, the Euclidean length of $x \in R^n$ will be denoted by $|x|$ and the Lebesgue measure of a set $T \subset R^1$ will be denoted by $\mu(T)$.

For the contingent equation $\dot{x} \in F(x,t)$ we shall define a simple optimization problem: given $x^-, x^+ \in R^n$, $a \in I$, find an element $\phi \in A_a^F(a^-, x^-)$ such that $x^+ \cdot \phi(a^+) \geq x^+ \cdot \psi(a^+)$ for all $\psi \in A_a^F(a^-, x^-)$. Whenever such ϕ exists we shall say that ϕ is an optimal solution for the problem (F, x^-, x^+, a).

The results given in this paper hold also if for the contingent equation $\dot{x} \in F(x,t)$ we define an optimization problem with the usual variety of more complex boundary conditions and objective functions. The novel feature of this paper lies in the particular structure which we assume for the contingent equation $\dot{x} \in F(x,t)$; this novel feature is most clearly understood in the context of the simple optimization problem defined here.

Before stating Theorem I, the central result of this paper, we must give a few definitions.

D.1.1. If $a \in I$, $\phi \in A_a$ and $\varepsilon > 0$ then the set $\{(x,t): t \in a, |x-\phi(t)| < \varepsilon\}$ is called the ε-tube around (a,ϕ) and is denoted by $\Theta(a,\phi,\varepsilon)$.

D.1.2. A mapping F from R^{n+1} into $\mathcal{P}(R^n)$ is measurable in t over R^{n+1} if for each $x \in R^n$ and each closed set $C \subset R^n$ the set $\{t: F(x,t) \cap C \neq \Phi\}$ is measurable.

D.1.3. A mapping G from R^n into $\mathcal{P}(R^n)$ is biconvex if $G(x)$ is convex whenever $x \in R^n$ and if $\mu G(x_1) + (1-\mu)G(x_2) \subset G(\mu x_1 + (1-\mu)x_2)$ whenever $x_1, x_2 \in R^n$ and $\mu \in [0,1]$.

D.1.4. A mapping F from R^{n+1} into $\mathcal{P}(R^n)$ is biconvex in x over R^{n+1} if for each $t \in R^1$ the mapping $F(\cdot,t)$ from R^n into $\mathcal{P}(R^n)$ is biconvex.

D.1.5. A mapping F from R^{n+1} into $\mathcal{P}(R^n)$ is integrally tamed on a tube $\Theta(a,\phi,\varepsilon)$ if there exists an integrable function γ from a into R^1 such that for all $(x,t) \in \Theta(a,\phi,\varepsilon)$ there is a $y \in F(x,t)$ with $|y| \leq \gamma(t)$.

THEOREM I. If $\phi \in A_a^F(a^-,x^-)$, if F is closed valued, measurable in t and biconvex in x over R^{n+1}, integrally tamed on $\Theta(a,\phi,\varepsilon)$ for some $\varepsilon > 0$ then ϕ is optimal for the problem (F, x^-, x^+, a) if and only if there exists a $p \in A_a(a^+,x^+)$ and a measurable set $T \subset a$ with $\mu(T) = \mu(a)$ such that $\dot{p}(t)\cdot\phi(t) + p(t)\cdot\dot{\phi}(t) \geq \dot{p}(t)\cdot x + p(t)\cdot y$ for all $t \in T$, all $x \in R^n$ and all $y \in F(x,t)$.

Remarks on Theorem I.

1) The sufficiency part of Theorem I is trivial since for all $\psi \in A_a^F(a^-,x^-)$ we have

$$p(a^+)\cdot\phi(a^+) - p(a^-)\cdot\phi(a^-) \geq p(a^+)\cdot\psi(a^+) - p(a^-)\cdot\psi(a^-) ,$$

i.e. $x^+\cdot\phi(a^+) \geq x^+\cdot\psi(a^+)$.

2) If in the inequality of Theorem I we let $x = \phi(t)$ we obtain $p(t)\cdot\dot{\phi}(t) \geq p(t)\cdot y$ for all $y \in F(\phi(t),t)$ which is the usual maximization of the Hamiltonian found in Pontryagin's Maximum Principle. We must remark that, besides this maximization of the Hamiltonian, the inequality of Theorem I contains the usual adjoint differential equation for p in the case of control problems $\dot{x} = f(x,u,t)$, $u \in \Omega$ such that $F(x,t) = \{f(x,u,t): u \in \Omega\}$ satisfies the assumptions stated here. Indeed if $\dot{\phi}(t) = f(\phi(t),u(t),t)$ we have then $\dot{p}(t)\cdot\phi(t) + p(t)\cdot f(\phi(t),u(t),t) \geq \dot{p}(t)\cdot x + p(t)\cdot f(x,u(t),t)$ for all $x \in R^n$, which if $f(x,u,t)$ is differentiable in x, leads to the usual adjoint equation: $\dot{p}(t) + \frac{\partial}{\partial x}(p(t)\cdot f(x,u(t),t)) = 0$.

3) For the convex problems of Bolza, which have been studied by R. T. Rockafellar [1, 2], the necessary conditions given here are equivalent to the necessary conditions obtained from Rockafellar's duality theory.

4) The following Theorem I*, which is a weaker form of Theorem I, is perhaps more intuitively pleasing.

D.1.6. If $b \in I$, then b^o is the interior of b, i.e. b^o is the open interval (b^-,b^+).

THEOREM I*. If $\phi \in A_a^F(a^-,x^-)$, if F is closed valued, measurable in t and biconvex in x over R^{n+1}, integrally tamed on $\Theta(a,\phi,\varepsilon)$ for some $\varepsilon > 0$ then ϕ is optimal for the problem (F, x^-, x^+, a) if and only if there exists a $p \in A_a(a^+, x^+)$ with $\frac{d}{dt}(p(t)\cdot\phi(t)) \geq \frac{d}{dt}(p(t)\cdot\psi(t))$ for all $(\psi, b) \in A^F$ and all $t \in b^0$ such that $\dot{p}(t)$, $\dot{\phi}(t)$ and $\dot{\psi}(t)$ exist.

We shall prove Theorem I* as a first step in the proof of Theorem I.

2. Reduction of Theorem I to the Integrally Bounded Case

We shall first give a few definitions.

D.2.1. For a bounded set $A \subset R^n$, the quantity $\sup_{x \in A}|x|$ will be denoted by $|A|$.

D.2.2. A mapping F from R^{n+1} into $\mathcal{P}(R^n)$ is integrally bounded on $a \in I$ if there exists an integrable function β from a into R^1 such that $|F(x,t)| \leq \beta(t)$ for all $(x,t) \in R^n \times a$.

D.2.3. A mapping F from R^{n+1} into $\mathcal{P}(R^n)$ is active on a tube $\Theta(a,\phi,\varepsilon)$ if for all $(x,t) \in \Theta(a,\phi,\varepsilon)$ we have $F(x,t) \neq \Phi$.

We now remark that in the proof of Theorem I we may without loss of generality replace the assumption: "F is integrally tamed on the tube $\Theta(a,\phi,\varepsilon)$" by the assumption: "$F$ is integrally bounded on a and active on $\Theta(a,\phi,\varepsilon)$". Indeed let γ be the integrable function from a into R^1 such that for all $(x,t) \in \Theta(a,\phi,\varepsilon)$ there is a $y \in F(x,t)$ with $|y| \leq \gamma(t)$, and let β be an integrable function from a into R^1 such that $\beta(t) = 1 + \gamma(t) + |\dot{\phi}(t)|$ for all $t \in a$ where $\dot{\phi}(t)$ exists and such that $\beta(t) = 1 + \gamma(t)$ for all $t \in a$ where $\dot{\phi}(t)$ does not exist. We then define a mapping $\tilde{F}$ from R^{n+1} into $\mathcal{P}(R^n)$ by the relation $\tilde{F}(x,t) = \{y\colon y \in F(x,t),\ |y| \leq \beta(t)$ whenever $t \in a\}$. It is an easy matter to prove that $\tilde{F}$ is closed valued, measurable in t, biconvex in x over R^{n+1}, integrally bounded on a and active on $\Theta(a,\phi,\varepsilon)$. If for some $t \in T$ we have $\dot{p}(t)\cdot\phi(t) + p(t)\cdot\dot{\phi}(t) \geq \dot{p}(t)\cdot x + p(t)\cdot y$ for all $x \in R^n$ and all $y \in \tilde{F}(x,t)$ then we shall have the same inequality for all $x \in R^n$ and all $y \in F(x,t)$ since $F(\cdot,t)$ is biconvex.

3. Some Properties of Contingent Equations

In this section we give two results, Theorem II and Theorem III, which will be used in the proof of Theorem I. The proof of those two theorems will appear in a forthcoming paper. If $b \in I$ then b^o will denote the interior of b, i.e. the open interval (b^-, b^+).

THEOREM II. If F is a mapping from R^{n+1} into $\mathcal{P}(R^n)$ which is closed valued, measurable in t, biconvex in x over R^{n+1}, integrally bounded on some $a \in I$ by a by some integrable function β from a to R^1, and active on a tube $\Theta(a,\phi,\varepsilon)$ for some $\phi \in A_a$ and some $\varepsilon > 0$ then for all $b \in I$ with $b \subset a$ and $\int_b(\beta(\tau) + |\dot{\phi}(\tau)|)d\tau \leqslant \frac{\varepsilon}{4}$, and for all $(x,t) \in \Theta(a,\phi,\frac{\varepsilon}{4})$ with $t \in b$ we have $A_b^F(t,x) \neq \Phi$.

THEOREM III. If F is a mapping from R^{n+1} into $\mathcal{P}(R^n)$ which is closed valued, measurable in t, biconvex in x over R^{n+1}, integrally bounded on some $a \in I$, active on the tube $\Theta(a,\phi,\varepsilon)$ for some $\phi \in A_a$ and some $\varepsilon > 0$ then there is a measurable set $T \subset a$ with $\mu(T) = \mu(a)$ and such that for all $t \in T$ the set $\{(x,y): |x-\phi(t)| < \frac{\varepsilon}{2}$, $y \in F(x,t)$, $y = \dot{\psi}(t)$ for some $(b,\psi) \in A^F(t,x)$ with $t \in b^o\}$ is dense in the set $\{(x,y): |x-\phi(t)| < \frac{\varepsilon}{2}$, $y \in F(x,t)\}$.

Remarks. The proof of Theorem II is a rather easy application of a result of Castaing [3] (based on the fact that a bounded biconvex mapping is Hausdorff continuous and hence uppersemicontinuous in the interior of the set where it is nonempty, that fact itself follows from Theorem IV proved in the following section). The proof of Theorem III is not used in proving Theroem I* but only in going from Theorem I* to Theorem I.

4. Continuity of Biconvex Mappings

We give below an inequality for biconvex mapping. This inequality will play an important role in the proof of Theorem I. We first state a few definitions.

D.4.1. If G is a biconvex mapping from R^n into $\mathcal{P}(R^n)$ then the set $\{x: x \in R^n, G(x) \neq \Phi\}$ will be denoted $A(G)$.

D.4.2. If $A \subset R^n$ and $\varepsilon > 0$ then $N_\varepsilon(A) = \{x: |x-y| < \varepsilon$ for some $y \in A\}$.

THEOREM IV. If G is a biconvex mapping from R^n into $\mathcal{P}(R^n)$, K is a compact subset of $\operatorname{int} A(F)$, $\rho > 0$ such that $N_{2\rho}(K) \subset A(G)$, $\gamma > 0$ such that for all $x \in \overline{N_\rho(K)}$ there is a $y \in G(x)$ with $|y| \leq \gamma$, then for all $x_1, x_2 \in K$, all $y_1 \in G(x_1)$ there exists a $y_2 \in G(x_2)$ such that $|y_1 - y_2| \leq |x_1 - x_2| \frac{|y_1| + \gamma}{|x_1 - x_2| + \rho}$.

Proof of Theorem IV. Let $\theta = \sup\{t: x_1 + t(x_2 - x_1) \in \overline{N_\rho(K)}\}$, we have then $\theta \geq 1 + \frac{\rho}{|x_1 - x_2|}$. Let $x^* = x_1 + \theta(x_2 - x_1) \in \overline{N_\rho(K)}$. There is some $y^* \in G(x^*)$ with $|y^*| \leq \gamma$. We have $x^* = (1-\theta)x_1 + \theta x_2$, $x_2 = \frac{\theta - 1}{\theta} x_1 + \frac{1}{\theta} x^*$, $y_2 = \frac{\theta - 1}{\theta} y_1 + \frac{1}{\theta} y^* \in G(x_2)$ and hence $y_2 - y_1 = \frac{1}{\theta}(y^* - y_1)$ which implies $|y_2 - y_1| \leq \frac{|y^*| + |y_1|}{\theta} \leq |x_1 - x_2| \frac{\gamma + |y_1|}{|x_1 - x_2| + \rho}$. This concludes the proof of Theorem IV.

5. Discrete Problem

In this section we study a discrete version of the optimization problem stated in Section 1. For this discrete problem we prove a necessary and sufficient condition, Theorem V, which will be used in the proof of Theorem I.

We are given a finite set $\mathcal{G} = \{G_1, \ldots, G_k\}$ where each G_i is a mapping from R^n into $\mathcal{P}(R^n)$. Let $\mathcal{D} = \{\phi = (\phi_1, \ldots, \phi_{k+1}): \phi_i \in R^n\}$, let $\mathcal{D}^{\mathcal{G}} = \{\phi: \phi \in \mathcal{D},\ \phi_{i+1} - \phi_i \in G_i(\phi_i)$ for $i = 1, \ldots, k\}$ and if $j \in \{1, \ldots, k\}$, $x \in R^n$ let $\mathcal{D}^{\mathcal{G}}(j,x) = \{\phi: \phi \in \mathcal{D}^{\mathcal{G}},\ \phi_j = x\}$. Given x^- and x^+ we then define the following optimization problem: find a $\phi \in \mathcal{D}^{\mathcal{G}}(1,x^-)$ such that $x^+ \cdot \phi_{k+1} \geq x^+ \cdot \psi_{k+1}$ for all $\psi \in \mathcal{D}^{\mathcal{G}}(1,x^-)$. If such ϕ exists we shall say that ϕ is an optimal solution for the problem $(\mathcal{G}, x^-, x^+)$.

D.5.1. We shall say that $\mathcal{G}$ is biconvex if for each $i \in \{1, \ldots, k\}$ the mapping G_i is biconvex.

THEOREM V. If $\mathcal{G}$ is biconvex, if $\phi \in \mathcal{D}^{\mathcal{G}}(1,x^-)$ then ϕ is an optimal solution of the problem $(\mathcal{G}, x^-, x^+)$ if and only if there exists a

$p \in \mathcal{D}(k+1, x^+)$ such that $(p_{i+1} - p_i)\cdot\phi_i + p_{i+1}\cdot(\phi_{i+1} - \phi_i) \geq (p_{i+1} - p_i)\cdot x + p_{i+1}\cdot y$ for all $i \in \{1,\ldots,k\}$, all $x \in R^n$ and all $y \in G_i(x)$.

Remark. The inequality of Theorem V may be written under many forms. The form stated above has the advantage of being a natural discretization of the inequality of Theorem I. The following simpler form: $p_{i+1}\cdot\phi_{i+1} - p_i\cdot\phi_i \geq p_{i+1}\cdot(x+y) - p_i\cdot x$ for all $i \in \{1,\ldots,k\}$, all $x \in R^n$ and all $y \in G_i(x)$ has obvious advantages.

Proof of Theorem V. The sufficiency part of Theorem V is trivial. Indeed if $\phi \in \mathcal{D}^{\mathcal{G}}(1,x^-)$, $\psi \in \mathcal{D}^{\mathcal{G}}(1,x^-)$, $p \in \mathcal{D}(k+1, x^+)$ and $p_{i+1}\cdot\phi_{i+1} - p_i\cdot\phi_i \geq p_{i+1}\cdot(x+y) - p_i\cdot x$ for all $i \in \{1,\ldots,k\}$, all $x \in R^n$ and all $y \in G_i(x)$ we shall have $p_{i+1}\cdot\phi_{i+1} - p_i\cdot\phi_i \geq p_{i+1}\cdot\psi_{i+1} - p_i\cdot\psi_i$ for all $i \in \{1,\ldots,k\}$ and hence $x^+\cdot\phi_{k+1} - p_1\cdot x^- \geq x^+\cdot\phi_{k+1} - p_1\cdot x^-$, i.e. $x^+\cdot\phi_{k+1} \geq x^+\cdot\psi_{k+1}$. We shall now proceed to the proof of the necessary condition.

For each $i \in \{1,\ldots,k+1\}$ we define a set $X_i = \{\psi_i\colon \psi \in \mathcal{D}^{\mathcal{G}}(1,x^-)\}$. We have then the relation $X_{i+1} = \{x+y\colon\ x \in X_i,\ y \in G_i(x)\}$ for each $i \in \{1,\ldots,k\}$. From the fact that $\mathcal{G}$ is biconvex we see immediately that each of the sets X_i will be convex. We define $p_{k+1} = x^+$. If ϕ is an optimal solution of the problem $(\mathcal{G}, x^-, x^+)$ we must have $p_{k+1}\cdot\phi_{k+1} \geq p_{k+1}\cdot x$ for all $x \in X_{k+1}$. The proof of Theorem V will then be completed by repeated application of the following intermediary result:

If for some $i \in \{1,\ldots,k\}$ we have $p_{i+1}\cdot\phi_{i+1} \geq p_{i+1}\cdot x$ for all $x \in X_{i+1}$ then there is an element $p_i \in R^n$ such that

(α) $p_i\cdot\phi_i \geq p_i\cdot x$ for all $x \in X_i$

(β) $p_{i+1}\cdot\phi_{i+1} - p_i\cdot\phi_i \geq p_{i+1}\cdot(x+y) - p_i\cdot x$ for all $x \in R^n$, all $y \in G_i(x)$.

If $p_{i+1} = 0$ we see that the intermediary result is indeed true by letting $p_i = 0$. If $p_{i+1} \neq 0$ we note that the two sets $K = \{(x,x+y)\colon x \in R^n,\ y \in G_i(x)\}$ and $L = \{(x,z)\colon x \in X_i,\ p_{i+1}\cdot z > p_{i+1}\cdot\phi_{i+1}\}$ are disjoint nonempty convex subsets of R^{2n}. Since $(\phi_i,\phi_{i+1}) \in K \cap \overline{L}$ we know that there is a hyperplane passing through (ϕ_i,ϕ_{i+1}) and separating K and $\overline{L}$, i.e. there is a nonzero vector $(\alpha,\beta) \in R^{2n}$ such that

(a) $\alpha\cdot\phi_i + \beta\cdot\phi_{i+1} \leq \alpha\cdot x + \beta\cdot z$ for all $x \in X_i$ and for all z with $p_{i+1}\cdot z \geq p_{i+1}\cdot\phi_{i+1}$.

(b) $\alpha\cdot x + \beta\cdot(x+y) \leq \alpha\cdot\phi_i + \beta\cdot\phi_{i+1}$ for all $x \in R^n$ and all $y \in G_i(x)$.

Since $\phi_i \in X_i$ we obtain from Relation (a) that $\beta\cdot\phi_{i+1} \leq \beta\cdot z$ for all z with $p_{i+1}\cdot z \geq p_{i+1}\cdot\phi_{i+1}$, i.e. we obtain that $\beta = \lambda p_{i+1}$ for some $\lambda \geq 0$. We have $\lambda > 0$ since $\lambda = 0$ would imply $\beta = 0$, which from Relation (b) would imply $\alpha = 0$, which cannot be since we have assumed that the vector (α,β) is not zero. If we now define $p_i = -\frac{\alpha}{\lambda}$ then Relation (a) gives Relation (α) and Relation (b) becomes Relation (β). This concludes the proof of the intermediary result and therefore concludes the proof of Theorem V.

6. Proof of Theorem I

For every $x \in R^n$ and $b \in I$ we define a set $R(x,b) = \{\psi(b^+)-x: \psi \in A_b^F(b^-,x)\}$. It is an easy matter to prove that if F is biconvex in x over R^{n+1} then for every $b \in I$ the mapping $R(\cdot,b)$ will be biconvex. If we are given a finite set $a^- = t_1 < t_2 < \cdots < t_{k+1} = a^+$ and if for each $i \in \{1,\ldots,k\}$ we let $a_i = [t_i,t_{i+1}]$, i.e. $a_i^- = t_i$ and $a_i^+ = t_{i+1}$, then $(\phi(t_1),\ldots,\phi(t_{k+1}))$ will be an optimal solution of the discrete optimization problem: maximize $x^+\cdot\phi(t_{k+1})$ subject to the constraints

(i) $\phi(t_1) = x^-$

(ii) $\phi(t_{i+1}) - \phi(t_i) \in R(\phi(t_i),a_i)$ for $i = 1,\ldots,k$.

According to Theorem V there thus exists a sequence $(p_1,\ldots,p_{k+1})$ such that

(α) $p_{k+1} = x^+$

(β) $(p_{i+1}-p_i)\cdot\phi(t_i) + p_{i+1}\cdot(\phi(t_{i+1})-\phi(t_i)) \geq (p_{i+1}-p_i)\cdot x + p_{i+1}\cdot y$ for all $x \in R^n$, all $i \in \{1,\ldots,k\}$ and all $y \in R(x,a_i)$.

For every $i \in \{1,\ldots,k\}$ and every $\delta > 0$ let $x_i(\delta) = \phi(t_i) + \delta(p_{i+1}-p_i)$. From ($\beta$) we then obtain $\delta|p_{i+1}-p_i|^2 \leq |p_{i+1}|\cdot|(\phi(t_{i+1})-\phi(t_i)) - y|$ for all $\delta > 0$, all $i \in \{1,\ldots,k\}$ and all $y \in R(x_i(\delta),a_i)$. We have already assumed that F is integrally bounded by some function β on a and that F is active on a tube $\Theta(a,\phi,\varepsilon)$ for some $\varepsilon > 0$. We shall now assume that the partition of a into $a_1,\dots,a_k$ is such that $\int_{a_i}\beta(\tau)d\tau \leqslant \frac{\varepsilon}{8}$ for each $i \in \{1,\dots,k\}$. It then follows that for each $i \in \{1,\dots,k\}$ and each $x \in R^n$ with $|x-\phi(t_i)| \leqslant \frac{\varepsilon}{4}$ we have

(a) $R(x,a_i)$ is nonempty (by Theorem II)

(b) $|R(x,a_i)| \leqslant \int_{a_i}\beta(\tau)d\tau \leqslant \frac{\varepsilon}{8}$.

Hence, by Theorem IV, it follows that for each $i \in \{1,\dots,k\}$ and each $x \in R^n$ with $|x-\phi(t_i)| \leqslant \frac{\varepsilon}{8}$ there is a $y \in R(x,a_i)$ with

$$|(\phi(t_{i+1})-\phi(t_i)) - y| \leqslant |x-\phi(t_i)| \frac{2\int_{a_i}\beta(\tau)d\tau}{|x-\phi(t_i)|+\frac{\varepsilon}{8}} \leqslant |x-\phi(t_i)|\frac{16\int_{a_i}\beta(\tau)d\tau}{\varepsilon}.$$

Hence if for each $i \in \{1,\dots,k\}$ we choose a $\delta_i > 0$ such that $\delta_i|p_{i+1}-p_i| < \frac{\varepsilon}{8}$ we shall obtain $\delta_i|p_{i+1}-p_i|^2 \leqslant |p_{i+1}|\,\delta_i|p_{i+1}-p_i|\frac{16\int_{a_i}\beta(\tau)d\tau}{\varepsilon}$, i.e. $|p_{i+1}-p_i| \leqslant |p_{i+1}|\frac{16}{\varepsilon}\int_{a_i}\beta(\tau)d\tau$. For every $i \in \{1,\dots,k\}$ we have $|p_i| \leqslant |p_{i+1}|(1+\frac{16}{\varepsilon}\int_{a_i}\beta(\tau)d\tau) \leqslant |x^+|\, e^{\frac{16}{\varepsilon}\int_a\beta(\tau)d\tau}$ and hence $|p_{i+1}-p_i| \leqslant \int_{a_i}\beta^*(\tau)d\tau$ if we let $\beta^*(t) = \frac{16}{\varepsilon}\beta(t)|x^+|\, e^{\frac{16}{\varepsilon}\int_a\beta(\tau)d\tau}$.

For each $m = 1,2,\dots$ we let $k = 2^m$ and we shall now consider a partition of the interval a into 2^m subintervals $a_{m,1}, a_{m,2},\dots, a_{m,2^m}$ of same length $\frac{a^+-a^-}{2^m}$. For each $m = 1,2,\dots$ we shall use the notation $t_{m,1} = a^-_{m,1} = a^-$, $t_{m,2^m+1} = a^+_{m,2^m} = a^+$ and $t_{m,i} = a^-_{m,i} = a^+_{m,i-1}$ whenever $i \in \{2,3,\dots,2^m\}$. Since β is integrable over the interval a there exists an $M < +\infty$ such that for all $m \geqslant M$ and all $j \in \{1,\dots,2^m\}$ we shall have $\int_{a_{m,j}}\beta(\tau)d\tau \leqslant \frac{\varepsilon}{8}$ and hence for all $m \geqslant M$ the preceeding results are applicable to the partition of the interval a into the subintervals $a_{m,1}, a_{m,2}, \dots, a_{m,2^m}$. Let $p_{m,1}, p_{m,2}, \dots, p_{m,2^m+1}$ be the sequence of p's corresponding to the partition of a into the 2^m subintervals $a_{m,1}, a_{m,2}, \dots, a_{m,2^m}$. Let p_m be the continuous function from the interval a into R^n such that for every $j \in \{1,2,\dots,2^m\}$ the function p_m is affine on each of the subintervals $a_{m,j}$ and such that for every $j \in \{1,2,\dots,2^m+1\}$ we have $p_m(t_{m,j}) = p_{m,j}$. For every $\eta > 0$ we know that there is a $\delta > 0$ such that for all measurable subset S of the interval a with $\mu(S) \leqslant \delta$ we have $\int_S\beta^*(\tau)d\tau \leqslant \frac{1}{2}\eta$; hence if S and S^* are measurable subsets of the interval a with $\mu(S) \leqslant \delta$, $0 < \mu(S) \leqslant \mu(S^*)$ we shall have $\frac{\mu(S)}{\mu(S^*)} \int_{S^*} \beta^*(\tau)d\tau \leqslant \eta$. We have then $|p_m(t_2) - p_m(t_1)| \leqslant \eta$ whenever $|t_2 - t_1| \leqslant \delta$, t_1 and $t_2 \in a$ and $m \geqslant M$. From the theorem of Arzela-Ascoli we know that there exists a subsequence p_M, p_{M+1}, ... which converges uniformly to some function p. From that uniform convergence it follows easily that $|p(t_2)-p(t_1)| \leqslant \int_{t_1}^{t_2} \beta^*(\tau)d\tau$ for all t_1 and $t_2 \in a$ (hence p is absolutely continuous) and that for every $m = 1,2,...$ and every $j \in \{1,2,...2^m\}$ we have

$$p(t_{m,j+1}) \cdot \phi(t_{m,j+1}) - p(t_{m,j}) \cdot \phi(t_{m,j}) \geqslant p(t_{m,j+1}) \cdot (x+y) - p(t_{m,j}) \cdot x$$

for all $x \in R^n$ and all $y \in R(x, a_{m,j})$. Let $t \in a^0$ such that $\dot{p}(t)$ and $\dot{\phi}(t)$ exist and let $(\psi,b) \in A^F$ such that $t \in b^0$ and $\dot{\psi}(t)$ exists; we shall prove by contradiction that $\frac{d}{dt}(p(t)\cdot\phi(t)) \geqslant \frac{d}{dt}(p(t)\cdot\psi(t))$. If $\frac{d}{dt}(p(t)\cdot\phi(t)) < \frac{d}{dt}(p(t)\cdot\psi(t))$ then there is an $\eta > 0$ such that $[t-\eta, t+\eta] \subset b$ and such that for all τ_1, τ_2 with $\tau-\eta \leqslant \tau_1 \leqslant t \leqslant \tau_2 \leqslant t+\eta$ and $\tau_1 < \tau_2$ we shall have $p(\tau_2)\cdot\phi(\tau_2) - p(\tau_1)\cdot\phi(\tau_1) < p(\tau_2)\cdot\psi(\tau_2) - p(\tau_1)\cdot\psi(\tau_1)$. Given that $t \in a^0$ and that $\eta > 0$ we can find an $m \geqslant M$ and a $j \in \{1,2,...,2^m\}$ such that $t-\eta \leqslant t_{m,j} \leqslant t \leqslant t_{m,j+1} \leqslant t+\eta$ and hence $p(t_{m,j+1})\cdot\phi(t_{m,j+1}) - p(t_{m,j})\cdot\phi(t_{m,j}) < p(t_{m,j+1})\cdot\psi(t_{m,j+1}) - p(t_{m,j})\cdot\psi(t_{m,j})$ which is a contradittion. We have thus completed the proof of Theorem I*. In order to prove Theorem I, which is stronger than Theorem I*, we shall now apply Theorem II which we have not used so far.

Let T be a measurable subset of a with $\mu(T) = \mu(a)$ and such that for all $t \in T$ the derivatives $\dot{p}(t)$ and $\dot{\phi}(t)$ exist and the set $\{(x,y): |x-\phi(t)| < \frac{\varepsilon}{2},\ y \in F(x,t),\ y = \dot{\psi}(t)$ for some $(b,\psi) \in A^F(t,x)$ with $t \in b^0\}$ is dense in the set $\{(x,y): |x-\phi(t)| < \frac{\varepsilon}{2},\ y \in F(x,t)\}$. We now prove Theorem I by contradiction. If for some $t \in T$, some $x \in R^n$, some $y \in F(x,t)$ we have $\dot{p}(t)\cdot\phi(t) + p(t)\cdot\dot{\phi}(t) < \dot{p}(t)\cdot x + p(t)\cdot y$, then, by the biconvexity of $F(\cdot,t)$, there will be some x^* with $|x^*-\phi(t)| < \frac{\varepsilon}{2}$ and some $y^* \in F(x^*,t)$ such that $\dot{p}(t)\cdot\phi(t) + p(t)\cdot\dot{\phi}(t) < \dot{p}(t)\cdot x^* + p(t)\cdot y^*$, then, by the density statement given above, there will be some $(b,\psi) \in A^F$ with $t \in b^0$ such that $\dot{p}(t)\cdot\phi(t) + p(t)\cdot\dot{\phi}(t) < \dot{p}(t)\cdot\psi(t) + p(t)\cdot\dot{\psi}(t)$. We have already seen earlier that this last inequality leads to a contradiction. This concludes the proof of Theorem I.

REFERENCES

[1] R.T. ROCKAFELLAR, Conjugate convex functions in optimal control and the calculus of variations, J. of Math. Anal. and Appl. 32 (1970), 174-222.
[2] R.T. ROCKAFELLAR, Existence and duality theorems for convex problems of Bolza, to appear in Trans. Amer. Math. Soc., 1971.
[3] C. CASTAING, Sur les équations différentielles multivoques, C. R. Acad. Sci. Paris 263 (1966), 63-66.

University of California at San Diego, La Jolla, California

ASYMPTOTIC DISTRIBUTION OF EIGENVALUES

W. A. Harris, Jr.* and Y. Sibuya**

We are concerned with the eigenvalue problem

(1) $$y''(x) - \lambda^2 p(x)y(x) = 0$$

(2) $$y \in L_2(-\infty,\infty)$$

where $p(x)$ is a sufficiently regular real valued function of the real variable x and $y \in L_2(a,b)$ means $\int_a^b |y(x)|^2 dx < \infty$. In particular we are concerned with the asymptotic distribution of large positive eigenvalues for this problem.

Problems of this nature arise naturally in investigations of wave phenomena, especially in quantum mechanics, see for example Kemble [4], Budden [1], and Jeffreys [3].

The boundary condition, $y \in L_2(-\infty,\infty)$ imposes certain restrictions on $p(x)$ as illustrated by the cases $p(x) = 1$, $p(x) = x$, and $p(x) = x^2 - 1$.

If $p(x) = 1$, the differential equation

$$y''(x) - \lambda^2 y(x) = 0$$

* Supported in part by the United States Army under Contract DA-ARO-D-31-124-71-G14.

**Supported in part by the National Science Foundation under Grant GP-7041X.

has two linearly independent solutions

$$y_{\pm}(x) = e^{\pm\lambda x}$$

and clearly no solution is square integrable. This is also the case if $p(x)$ has one sign and hence a necessary condition for a square integrable solution of the differential equation (1) is that $p(x)$ have at least one zero.

If $p(x) = x$, the differential equation

$$y''(x) - \lambda^2 xy(x) = 0$$

can be transformed into the Airy equation

$$y''(s) - sy(s) = 0$$

by the change of variable $s = \lambda^{2/3}x$ the asymptotic forms of the Airy functions show that no solution is square integrable in this case.

If $p(x) = x^2 - 1$, the differential equation

$$y''(x) - \lambda^2(x^2 - 1)\, y(x) = 0 \tag{3}$$

can be studied by means of the Weber functions. Indeed, for

$$\lambda = 2k + 1, \qquad k = 0,1, \ldots ,$$

the differential equation (3) admits the square integrable solutions

$$y(x) = e^{-\frac{\lambda x^2}{2}} P_k(\lambda^{\frac{1}{2}}x) , \qquad k = 0,1,\ldots ,$$

where P_k are the Hermite polynomials.

The zeros of $p(x)$ are referred to as turning points on transition points, and the problem of representation of solutions at such points is very different.

In intervals for which $p(x)$ does not change sign, e.g. $p(x) > 0$, the differential equation (1) possesses two solutions $y_{\pm}(x)$ which have the WKBJ representations

(4) $$y_{\pm}(x) = [A(x)^{-\frac{1}{2}} \exp(\pm\lambda \int^{x} A(\tau)d\tau)]\,[1 + O(\lambda^{-1})]$$

where $A(x) = +\sqrt{p(x)}$, under appropriate assumptions. If $p(x)$ has polynomial-like behavior as $(x) \to \infty$ and $p(x) > 0$ for $|x|$ sufficiently large, the representation (4) shows the existence of one solution $y_-(x)$ which approaches zero exponentially as $x \to +\infty$ and one solution $y_+(x)$ which approaches zero exponentially as $x \to -\infty$. Thus $y_-(x) \in L_2(a,+\infty)$ and $y_+(x) \in L_2(-\infty,a)$ for any finite a. The question of eigenvalues is thus the linear dependence of these two solutions. Unfortunately, the WKBJ representation (4) of these solutions, $y_{\pm}(x)$, is valid on disjoint intervals and we must provide connecting solutions.

We will assume that $p(x)$ satisfies the following hypotheses.

<u>H.1</u> $p(x)$ possesses an even, finite number of simple real zeros a_i, $i = 1,\ldots,2m$ with $a_1 > a_2 > \cdots > a_{2m}$.

<u>H.2</u> $p(x) = c_0 x\,[1 + q(x)]$ with $c_0 > 0$, $\ell \geq 0$ an integer and

$$\lim_{|x|\to\infty} q(x) = 0 .$$

Then subdominant solutions as indicated above will exist and appropriate connecting solutions may be constructed under the additional hypothesis

<u>H.3</u> $p(x)$ is a real analytic function for $x \in (-\infty,\infty)$; or

<u>H.3'</u> $p(x)$ is of class $C^n(-\infty,\infty)$, $n \geq 2$.

<u>THEOREM 1</u>. Let $p(x)$ satisfy hypothesis H.1, H.2, and H.3. For large λ the eigenvalues of the problem (1), (2) can be denoted by

$$\lambda_{k,\ell} = \frac{(\ell - \frac{1}{2})\pi}{\int_{a_{2k}}^{a_{2k-1}} \sqrt{|p(x)|}\,dx} + \sum_{i=1}^{N} \xi_{ki}\,\ell^{-i} + O(\ell^{-N-1})$$

as $\ell \to +\infty$, where N is any positive integer.

When $p(x)$ is a polynomial which possesses $2m$ simple real zeros and no other zeros, the problem (1), (2) has been considered by Evgrafov and Fedoryuk [2], Sibuya [7] and Olver $(m=1)$ [6]. These authors obtain the appropriate connecting solutions by using the theory of irregular singular points at $x = \infty$. In particular they are able to study the Stokes' curves in the whole complex plane and utilize them to construct such solutionn. Weinberg [8] has shown that $p(x)$ a real analytic function is sufficient to obtain these connection solutions.

If $p(x) = (x - a_1)(x - a_2)p(x)$, $x \in (\alpha,\beta)$, $-\infty < \alpha < a_2 < a_1 < \beta < +\infty$, $p(x) > 0$, Langer [5] has shown the existence of a change of variable (of class $C^3(\alpha,\beta)$)

$$x = x(s), \quad y(x) = [x'(s)]^{\frac{1}{2}} u(s) ,$$

which transforms the differential equation

$$\frac{d^2y}{dx^2} - \lambda^2 p(x)y = 0$$

into

$$\frac{d^2u}{ds^2} - \lambda^2(s^2 - 1)u = g(s)u$$

where

$$g(s) = \frac{x'''(s)}{2x'(s)} - \frac{3}{4}\left(\frac{x''(s)}{x'(s)}\right)^2 .$$

By considering $g(s)u$ as a perturbation term, and explicit representation of solutions of the differential equation

$$\frac{d^2u}{ds^2} - \lambda^2(s^2 - 1)u = 0 ,$$

in terms of Weber functions, he is able to treat the case $m = 1$, finite interval, and $p(x)$ of class C^2.

We can utilize these solutions of Langer as connecting solutions to obtain the corresponding result to Theorem 1 in case $p(x)$ is of class C^n, $n \geqslant 2$.

THEOREM 2. Let $p(x)$ satisfy hypothesis H.1, H.2, and H.3' $(n = 2)$. For large λ the eigenvalues of the problem (1), (2) can be denoted by

$$\lambda_{k,\ell} = \frac{(\ell - \frac{1}{2})}{\int_{a_{2k}}^{a_{ak-1}} \sqrt{|p(x)|}\,dx} + O(\ell^{-1})$$

as $\ell \to +\infty$.

REFERENCES

[1] K.G. BUDDEN, Radio Waves in the Ionosphere. University Press, Cambridge, England, 1961.

[2] M.A. EVGRAFOV and M.V. FEDORYUK, Asymptotic behavior as $\lambda \to \infty$ of the solution of the equation $w''z - p(z,\lambda)w(z) = 0$ in the complex z-plane, Uspehi Mat. Nauk. 21 (1966), 3-50.

[3] H. JEFFREYS, The free oscillations of water in an elliptical lake, Proc. Lond. Math. Soc. 23(2), (1923), 455-476.

[4] E.C. KEMBLE, The Fundamental Principles of Quantum Mechanics. Dover, New York, 1958.

[5] R.E. LANGER, The asymptotic solutions of a linear differential equation of the second order with two turning points, Trans. Amer. Math. Soc. 90 (1959), 113-142.

[6] F.W.J. OLVER, Error analysis of phase-integral methods, I, II, J. Res. NBS 69B(4), (1965), 271-290, 291-300.

[7] Y. SIBUYA, Subdominant solutions of the differential equation $y'' - \lambda^2(x - a_1)(x - a_2) \cdots (x - a_m)y = 0$, Acta Math. 119 (1967), 235-272.

[8] L. WEINBERG, Asymptotic distribution of eigenvalues for the boundary value problem $y'' - \lambda^2 p(x)y = 0$, $y \in L(-\infty,\infty)$, SIAM J. Math. Anal. (to appear).

WAH: University of Southern California, Los Angeles, California

YS: University of Minnesota, Minneapolis, Minnesota

ON LOGARITHMIC DERIVATIVES OF SOLUTIONS OF DISCONJUGATE LINEAR n^{TH} ORDER DIFFERENTIAL EQUATIONS*

Philip Hartman

1. The Main Theorem

This talk concerns disconjugate, n^{th} order, linear, homogeneous differential equations,

$$(\texttt{¤}) \qquad P_n(t,D)[x] \equiv x^{(n)} + a_{n-1}(t)x^{(n-1)} + \cdots + a_0(t)x = 0,$$

where $D[x] = dx/dt = x'$ and the coefficients are real-valued and continuous on a t-interval I. The differential equation (¤) is said to be disconjugate on I if every solution $x = x(t) \not\equiv 0$ has at most $n-1$ zeros (counting multiplicities) on I. Proofs will only be indicated, as the more novel results discussed here will appear shortly in [7] and related results have appeared in [5], [6] and [9]. It should be mentioned that a set of similar problems was considered by Levin [9] and by me [5], [6]. Although some of our arguments are similar, our main methods are different. My general method (of first finding a particular solution of (¤) by replacing (¤) by a suitable first order system and then using an induction on n) will be illustrated below. For a complete discussion of "disconjugacy", see Coppel [1].

---

*The study was supported by the Air Force Office of Scientific Research Contract No. F44620-67-C-0098.

A corollary and illustration of the main result below is the following:

COROLLARY. Let $I = [\gamma,\beta)$, $-\infty < \gamma < \beta \leq \infty$, be a half-open interval and let there exist constants $\alpha_0 < \alpha_1 < \cdots < \alpha_n$ satisfying $(-1)^{n+m} P_n(t,\alpha_m) \geq 0$ on I for $m = 0,\ldots,n$ (so that the characteristic polynomial

$$P_n(t,\lambda) \equiv \lambda^n + a_{n-1}(t)\lambda^{n-1} + \cdots + a_0(t) = 0$$

has n real zeros $\lambda_1(t),\ldots,\lambda_n(t)$ satisfying

$$\alpha_0 \leq \lambda_1(t) \leq \alpha_1 \leq \cdots \leq \alpha_{n-1} \leq \lambda_n(t) \leq \alpha_n$$

on I). Then (¤) is disconjugate on I and has positive, linearly independent solutions $x_1,\ldots,x_n$ such that

$$\alpha_0 \leq x_1'/x_1 \leq \alpha_1 < x_2'/x_2 < \cdots < \alpha_{n-1} < x_n'/x_n < \alpha_n .$$

This result is valid if there is no assumption or assertion involving α_0 and/or α_n.

For $n = 2$, this result is contained in Olech [10]. For $n = 3$ (where no $\alpha_0,\alpha_n = \alpha_3$ occur), the existence of x_2 satisfying $x_2 > 0$ and $\alpha_1 \leq x_2'/x_2 \leq \alpha_2$ was proved by Schuur [12]. Another proof of Schuur's result was given by Jackson [8] by reducing the problem to one for a non-linear second order equation for $r = x'/x$ (and, after this reduction, the result follows from Hartman [3]; cf. [4], Theorem 5.2, p. 434).

In order to state the main theorem leading to this COROLLARY, we introduce some terminology. The Wronskian determinant of k functions $u_1,\ldots,u_k$ will be denoted by

$$W(u_1,\ldots,u_k) \equiv \det(D^{i-1}[u_j])_{i,j=1,\ldots,k}; \quad \text{so that } W(u_1) \equiv u_1.$$

<u>Definition</u>. An ordered set of $n-1$ functions $u_1,\ldots,u_{n-1}$ of class C^n on a t-interval I is said to be a <u>$w_n(I)$-system</u> if $W(u_1,\ldots,u_k) > 0$ on I for $1 \leq k \leq n-1$.

The relationship between disconjugacy on I and the existence of a $W_n(I)$-system of solutions $x_1,\dots,x_{n-1}$ goes back at least to Frobenius; see Pólya [11] (and [6] or [9] for a completion of Pólya's results).

<u>Definition</u>. An ordered set of $n-1$ functions $u_1,\dots,u_{n-1}$ of class $C^n(I)$ is said to be a <u>$W_n(I)$-system</u> if

$$W(u_{i(1)},\dots,u_{i(k)}) > 0 \quad \text{on } I \quad \text{for} \quad 1 \leqslant i(1) < \cdots < i(k) \leqslant n$$

or equivalently [2]

$$W(u_j,\dots,u_m) > 0 \quad \text{on } I \quad \text{for} \quad 1 \leqslant j \leqslant m \leqslant n-1 .$$

An example of a set of functions which is a $W_n(I)$-system on any interval I is

(1.0) $\quad u_k(t) = \exp \alpha_k t \quad \text{for} \quad k = 1,\dots,n-1, \quad \text{where} \quad \alpha_1 < \cdots < \alpha_{n-1}.$

For, in this case, the Wronskian determinants above are positive Vandermonde determinants, up to exponential factors. COROLLARY results by applying the following theorem in the case (1.0).

<u>THEOREM</u>. Let $I = [\gamma,\beta)$, $-\infty < \gamma < \beta \leqslant \infty$. Let $u_0,\dots,u_{n-1} \in C^n(I)$ satisfy

(1.1) $\quad (-1)^{n+k} P_n(D)[u_k] \geqslant 0 \quad \text{for} \quad 0 \leqslant k \leqslant n-1 ,$

(1.2) $\quad u_1,\dots,u_{n-1}$ is a $W_n(I)$-system,

(1.3) $\quad W(u_0,\dots,u_k) \geqslant 0 \quad \text{for} \quad 0 \leqslant k \leqslant n-1 .$

Then (¤) has positive linearly independent solutions $x_1,\dots,x_n$ satisfying

(1.4) $\quad x_1'/x_1 \leqslant u_1'/u_1 < x_2'/x_2 < \cdots < u_{n-1}'/u_{n-1} < x_n'/x_n ,$

(1.5) $\quad W(x_1,u_0) \leqslant 0, \quad \text{so that} \quad u_0'/u_0 \leqslant x_1'/x_1 \quad \text{where} \quad u_0 > 0 ,$

and the inequalities

$$
(1.6)\quad \begin{cases} W(x_1,\ldots,x_k) > 0 \quad \text{for } 1 \leqslant k \leqslant n\,, \\ W(x_1,\ldots,x_k,u_j,\ldots,u_m) > 0 \;[\text{or} \geqslant 0] \quad \text{for } 1 \leqslant k < [\text{or} \leqslant] j \leqslant m < n\,, \\ W(x_1,\ldots,x_k,u_{k-1}) \leqslant 0 \quad \text{for } 1 \leqslant k \leqslant n, \\ W(x_1,\ldots,x_k,u_{k-1},u_j,\ldots,u_m) \leqslant 0 \quad \text{for } 1 \leqslant k < j \leqslant m < n. \end{cases}
$$

If, in addition, $u_n \in C^n(I)$ satisfies

$$
(1.7)\qquad P_n(D)[u_n) \geqslant 0\,,
$$

$$
(1.8)\qquad W(u_k,\ldots,u_n) > 0 \quad \text{for } 1 \leqslant k \leqslant n\,,
$$

then the solution $x_n(t)$ can be chosen to satisfy

$$
(1.9)\qquad W(x_n,u_n) > 0, \quad \text{i.e.,} \quad x_n'/x_n < u_n'/u_n\,,
$$

and $m = n$ is permitted in (1.6).

The conditions (1.1)-(1.2) imply that (¤) is disconjugate on I; [6] or [9]. THEOREM remains correct if there is no assumption or assertion concerning u_0, but if u_0 is not otherwise specified, we can always make the trivial choice $u_0 \equiv 0$. For $n = 2$, see Olech [10]; for partial results, $n = 3$, see Jackson [8].

Professor Coppel has called my attention to the fact that an incorrect version of this theorem was stated in [5], [6]. There $I = (\alpha,\beta)$ was an open interval $-\infty \leqslant \alpha < \beta \leqslant \infty$, the inequalities (1.6) did not occur, and the strict inequalities ">" in the assumption (1.8) and "<" in the conclusions (1.4), (1.9) were replaced by "⩾" and "⩽". It turns out that this incorrect statement becomes valid if the assertion of the "linear independence of $x_1,\ldots,x_n$" is omitted and, in fact, this corrected statement is a corollary of THEOREM above. In [9], Levin states an analogous form of THEOREM. He weakens $W_n(I)$ to $w_n(I)$ in (1.2), makes the additional assumption $u_{k-1} = o(u_k)$ as $t \to \beta$ but, in place of the inequalities (1.4), (1.5), (1.6), (1.9) for all t on I, he merely asserts $x_k = O(u_k)$, $u_{k-1} = O(x_k)$ as $t \to \beta$.

The proof of THEOREM depends on a refinement of the techniques of [6]. First, we prove the existence of a suitable x_1 and, second, we emply an induction on n (using the known solution x_1 and a variation of constants to reduce (¤) to an equation of order $n-1$). The existence of x_1 is established by replacing the nth order equation (¤) by suitable first order systems.

2. Wronskian Identities

Computations below depend on some identities for Wronskians which we state here for easy reference. Here $\sigma_0 = 1$, $\sigma_k = W(v_1,\dots,v_k)$, and $\hat{v}_j$ indicates the omission of v_j.

(2.1) $$[W(v_1,\dots,v_{k-1},x)/\sigma_k]' = W(v_1,\dots,v_k,x)\sigma_{k-1}/\sigma_k^2 \;;$$

(2.2) $$v_0^{k-1}\, W(v_0,\dots,v_k) = W[W(v_0,v_1),\dots,W(v_0,v_k)] \;;$$

(2.3) $$\begin{aligned}&\sigma_{m-1}W(x,v_1,\dots,\hat{v}_j,\dots,v_m)\\ &= \sigma_m W(x,v_1,\dots,\hat{v}_j,\dots,v_{m-1}) + W(x,v_1,\dots,v_{m-1})W(v_1,\dots,\hat{v}_j,\dots,v_m) \;;\end{aligned}$$

(2.4) $$\begin{aligned}&W(x,v_1,\dots,\hat{v}_j,\dots,v_m)\\ &= \sigma_m \sum_{k=j-1}^{m-1} W(x,v_1,\dots,v_k)W(v_1,\dots,\hat{v}_j,\dots,v_{k+1})/\sigma_k\sigma_{k+1} \;.\end{aligned}$$

(For these identities, see, e.g., (4.9), (2.3) and Corollary 2.2 in [6] and Proposition 15.1 in [7].)

3. The Solution x_1

The existence of x_1 is given by

LEMMA. Let $n > 1$, $I = [\gamma,\beta)$, $P_n(t,D)$, and $u_0,\dots,u_{n-1}$ as in THEOREM. Then (¤) has a solution $x = x_1(t)$ on I satisfying

(3.1) $$x_1 > 0 \text{ and } W(x_1,u_1,\dots,u_k) \geq 0 \quad \text{for } 1 \leq k < n \;,$$

(3.2) $$W(x_1,u_0) \leq 0 \text{ and } W(x,u_0,u_j,\dots,u_k) \leq 0 \quad \text{for } 1 < j \leq k < n;$$

(3.3) $$x_1,u_2,\dots,u_{n-1} \text{ is a } W_n(I)\text{-system.}$$

If, in addition, $u_n \in C^n(I)$ satisfies (1.7) and

(3.8) $$W(u_k,\dots,u_n) \geq 0 \quad \text{for } 1 \leq k \leq n \;,$$

then $k = n$ is permitted in (3.1) and (3.2) and, if (3.8) is strengthened to (1.8), then

(3.9) $$W(x_1, u_k, \ldots, u_n) > 0 \quad \text{for} \quad 2 \leq k \leq n .$$

Proof. We indicate the proof assuming, without loss of generality, that $u_0, \ldots, u_n$ exist and satisfy (1.1)-(1.3), (1.7), and (3.8). Let $\gamma < \beta_1 < \beta_2 < \cdots$, $\beta_N \to \beta$ as $N \to \infty$, and $I_N = [\gamma, \beta_N]$.

Assumption (A_N). For a moment, let N be fixed and assume the strict form of all inequalities in (1.3) and (3.8), so that $W(u_j, \ldots, u_m) > 0$ on I_N for $0 \leq j \leq m \leq n$.

Let $\omega_0 = 1$ and $\omega_k = W(u_1, \ldots, u_k)$ for $1 \leq k \leq n$. Let $y = (y_1, \ldots, y_n)$ be the vector with the components

(3.10) $$y_k = W(x, u_1, \ldots, u_{k-1})/\omega_k \quad \text{for} \quad 1 \leq k \leq n \quad (\text{so that } y_1 = x/u_1).$$

Then, by (2.1)

(3.11) $$y_k' = -y_{k+1}\omega_{k-1}\omega_{k+1}/\omega_k^2 \quad \text{for} \quad 1 \leq k \leq n-1 ,$$

(3.12) $$y_n' = -W(x, u_1, \ldots, u_n)\omega_{n-1}/\omega_n^2 .$$

To the last row of $W(x, u_1, \ldots, u_n)$ in (3.12), add $a_0(t)$ times the first row, $a_1(t)$ times the second, etc. When x is a solution of (¤), the last row becomes $(0, P_n(D)[u_1], \ldots, P_n(D)[u_n])$. Expanding the determinant along the last row, we see that

(3.13) $$W(x, u_1, \ldots, u_n) = \sum_{j=1}^{n} (-1)^{n+j} P_n(D)[u_j] \cdot W(x, u_1, \ldots, \hat{u}_j, \ldots, u_n) .$$

Hence (2.3) and Assumption (A_N) imply the existence of continuous $b^m(t)$ on I_N such that (3.12) becomes

(3.14) $$y_n' = -\sum_{m=1}^{n} b^m(t) y_m, \quad \text{where} \quad b^m(t) \geq 0 \quad \text{on} \quad I_N ;$$

in fact, we have

(3.15) $$b^m(t) = (\omega_{n-1}/\omega_m\omega_n) \sum_{j=1}^{m} (-1)^{n+j} P_n(D)[u_j] \cdot W(u_1, \ldots, \hat{u}_j, \ldots, u_m) .$$

Thus (¤) is equivalent to the first order system (3.11), (3.14) which is a system of the form $y' = -A(t)y$, where the entries in the matrix $A(t)$ are nonnegative on I_N.

Let $x = x(t) = x_{(N)}(t)$ be the solution of (¤) determined by initial conditions

$$(3.16)\qquad (x,x',\ldots,x^{(n-1)}) = (u_1,u_1',\ldots,u_1^{(n-1)}) \quad \text{at} \quad t = \beta_N,$$

so that $y_1 = x_1/u_1 = 1$, $y_k \geq 0$ at $t = \beta_N$. Hence $y_k \geq 0$, $y_k' \leq 0$ on I_N; that is, $x = x_{(N)}$ satisfies

$$(3.17)\qquad x > 0,\ W(x,u_1,\ldots,u_k) \geq 0 \ \text{on}\ I_N \ \text{for}\ 1 \leq k \leq n.$$

We now verify that $x = x_{(N)}(t)$ also satisfies

$$(3.18)\qquad W(x,u_0) \leq 0,\ W(x,u_0,u_2,\ldots,u_k) \geq 0 \ \text{on}\ I_N \ \text{for}\ 2 \leq k \leq n.$$

To this end, we replace $(u_1,\ldots,u_n)$ in the arguments above by $(u_0,u_2,\ldots,u_n)$. Let $\omega_1^0 = u_0$ and $\omega_k^0 = W(u_0,u_2,\ldots,u_k)$ for $2 \leq k \leq n$. Define the $(n-1)$-vector $z = (z_1,\ldots,z_{n-1})$ by

$$(3.19)\qquad z_1 = W(x,u_0)/\omega_2^0,\ z_k = W(x,u_0,u_2,\ldots,u_k)/\omega_{k+1}^0 \ \text{for}\ 2 \leq k \leq n-1.$$

Then, by (2.1),

$$(3.20)\qquad z_k' = -z_{k+1}\omega_k^0\omega_{k+2}^0/(\omega_{k+1}^0)^2 \ \text{for}\ 1 \leq k \leq n-2,$$

$$(3.21)\qquad z_{n-1}' = -W(x,u_0,u_2,\ldots,u_n)\omega_{n-1}^0/(\omega_n^0)^2.$$

Arguing as above,

$$W(x,u_0,\ldots,u_n) = -(-1)^n P_n(D)[u_0]W(x,u_2,\ldots,u_n) + \sum_{j=2}^{n}(-1)^{n+j}P_n(D)[u_j]\cdot W(x,u_0,u_2,\ldots,\hat{u}_j,\ldots,u_n).$$

Hence, by (2.4), (3.21) can be written as

$$(3.22)\qquad z_{n-1}' = -\sum_{m=1}^{n-1} c^m(t)z_m + q(t), \quad \text{where}\ c^m, q \geq 0 \ \text{on}\ I_N;$$

in fact, we have

$$c^m(t) = (\omega_{n-1}^0/\omega_m^0\omega_n^0)\sum_{j=2}^{m+1}(-1)^{n+j}P_n(D)[u_j]\cdot W(u_0,u_2,\ldots,\hat{u}_j,\ldots,u_{m+1}) \geq 0$$

and $q(t) = (-1)^n P_n(D)[u_0] \cdot W(x,u_2,\ldots,u_n)\omega^0_{n-1}/(\omega^0_n)^2$ is

$$q(t) = (-1)^n P_n(D)[u_0][\omega_n \omega^0_{n-1}/(\omega^0_n)^2] \sum_{m=1}^{n} y_m W(u_2,\ldots,u_m)/\omega_{m-1} \geq 0$$

if $x = x_{(N)}$. The initial conditions (3.16) imply that $z_k \leq 0$ at $t = \beta_N$. Hence, by (3.20) and (3.22), $z_k \leq 0$, $z_k' \geq 0$ on I_N. This gives (3.18).

From Assumption (A_N) and (2.4), we can deduce that $x = x_{(N)}$ satisfies

(3.23) $\quad W(x,u_0) \leq 0, \quad W(x,u_0,u_j,\ldots,u_m) \leq 0 \quad$ for $2 \leq j \leq m \leq n$.

For the solution $x = x_{(N)}$, we have (3.17) and (3.23) on I_N even without the Assumption (A_N). To see this, apply the result obtained in the case that u_0 and u_n are replaced by $u_0 + \varepsilon e^{-\alpha t}$ and $u_n + \varepsilon e^{\alpha t}$, where $\alpha = \alpha(N) > 0$ is large and fixed and $\varepsilon > 0$ is arbitrary, and let $\varepsilon \to 0$.

Let $c_{(1)}, c_{(2)}, \ldots$ be positive constants such that

$$c_{(N)} \sum_{k=0}^{n-1} \left| x^{(k)}_{(N)}(\gamma) \right| = 1 .$$

Then the sequence $c_{(1)}x_{(1)}(t)$, $c_{(2)}x_{(2)}(t), \ldots$ has a subsequence which is C^n-convergent on compact intervals of $[\gamma,\beta)$ and the limit function $x = x_1(t)$ is a solution of (¤) satisfying (3.1)-(3.2) with $k = n$ permitted, but "$x_1 > 0$" is replaced by "$0 \not\equiv x_1 \geq 0$". Actually, $x_1 > 0$ on I for $W(x_1,u_1) \geq 0$ implies that x_1/u_1 is non-increasing, so that if $x_1(t_0) = 0$ for some $t = t_0$, then $x_1(t) \equiv 0$ on $[t_0,\beta)$, hence on I.

The assertions (3.3) and (3.9) can easily be deduced using (2.4). This argument will be omitted here; cf. the proof of (v) in [6], Theorem 7.1_n, p. 331 and p. 338.

4. Proof of THEOREM

We indicate the proof under the assumption of the existence of u_n satisfying (1.7), (1.8).

The case $n = 1$ is trivial. Assume $n > 1$ and the validity of THEOREM when n is replaced by $n-1$. Let $x = x_1(t)$ be the solution of (¤) furnished by the LEMMA. Let

$$(4.1) \qquad Q_{n-1}(t,D)\,[v] = 0$$

be the $(n-1)^{st}$ order linear, homogeneous differential equation such that

$$(4.2) \qquad Q_{n-1}(t,D)\,[v] \equiv P_n(t,D)[x] \quad \text{if} \quad v = W(x_1,x)\,,$$

so that (4.1) has a positive leading coefficient and $v = v(t)$ is a solution if and only if (¤) has a solution $x = x(t)$ such that $v = W(x_1,x)$; cf. [6], p. 329. Put

$$(4.3) \qquad w_k = W(x_1,u_{k+1}) \quad \text{for} \quad 0 \leqslant k \leqslant n-1\,.$$

Then, we have by (2.2) and LEMMA,

$$(-1)^{n-1+k}Q_{n-1}(t,D)[w_k] \geqslant 0 \quad \text{for} \quad 0 \leqslant k \leqslant n-1\,,$$

$$w_1,\ldots,w_{n-2} \quad \text{is a} \quad W_{n-1}(I)\text{-system},$$

$$W(w_0,\ldots,w_k) \geqslant 0 \quad \text{for} \quad 0 \leqslant k \leqslant n-2,$$

$$W(w_k,\ldots,w_{n-1}) > 0 \quad \text{for} \quad 1 \leqslant k \leqslant n-1.$$

Thus, by the induction hypothesis, (4.1) has solutions $v_1,\ldots,v_{n-1}$ satisfying the analogues of (1.4)-(1.6) and (1.9).

There are solutions $x_2,\ldots,x_n$ of (¤) such that $v_{k-1} = W(x_1,x_k)$ for $2 \leqslant k \leqslant n$. In fact,

$$x_k(t) = x_1(t)\int_{\gamma}^{t}(v_{k-1}/x_1^2)ds + c_kx_1 \equiv x_{k0} + c_kx_1\,,$$

where c_k is an arbitrary integration constant. By (2.2), we can deduce (1.6) from its analogues for the $v_1,\ldots,v_{n-1}$ for all choices of the integration constants $c_2,\ldots,c_n$. We have to verify $x_k > 0$, (1.4), (1.5) and (1.9) for certain choices of $c_2,\ldots,c_n$.

The analogues of these inequalities for $v_1,\ldots,v_{n-1}$ and (2.2) imply that

$$(4.4) \qquad W(x_1,x_k,u_{k-1}) \leqslant 0 \leqslant W(x_1,x_k,u_k) \quad \text{for} \quad 2 \leqslant k \leqslant n.$$

We show that we can choose the integration constant c_k so that $c_k > 0$ and

$$W(x_k,u_{k-1}) < 0 < W(x_k,u_k) \quad \text{at} \quad t = \gamma \quad \text{for} \quad 2 \leq k \leq n.$$

In this case, $x_k > 0$ follows from $c_k > 0$, and (1.4), (1.5) and (1.9) from (2.1) and (4.4). (In these arguments, the case $k = 2$ requires special attention because of the possibility that $W(x_1,u_1)$ vanishes from some t, whereas $W(x_1,u_{k-1}) > 0$ for $k > 2$).

REFERENCES

[1] W.A. COPPEL, Disconjugacy, Australian National University, mimeographed notes 1970; to appear as Lecture Notes, Springer.

[2] M. FEKETE, Ueber ein Problem von Laguerre, Rend. Circ. Mat. Palermo, 34 (1912), 89-100.

[3] P. HARTMAN, On boundary value problems for systems of ordinary, nonlinear, second order differential equations, Trans. Amer. Math. Soc., 96 (1960), 493-509.

[4] P. HARTMAN, Ordinary Differential Equations. John Wiley & Sons, Inc., New York, 1964.

[5] P. HARTMAN, Disconjugate n^{th} order equations and principal solutions, Bull. Amer. Math. Soc., 74 (1968), 125-129.

[6] P. HARTMAN, Principal solutions of disconjugate n^{th} order linear differential equations, Amer. J. Math., 91 (1969), 306-362.

[7] P. HARTMAN, Corrigendum and addendum: Principal solutions of disconjugate n^{th} order linear differential equations, Amer. J. Math., 93 (1971), 439-451.

[8] L.K. JACKSON, Disconjugacy conditions for linear third order equations, J. Diff. Equations, 4 (1968), 369-372.

[9] A. YU. LEVIN, Non-oscillation of solutions of the equation $x^{(n)} + p_1(t)x^{(n-1)} + \cdots + p_n(t)x = 0$, Uspehi Mat. Nauk., 24 (1969), No. 2 (146), 43-96; Russian Math. Surveys, 24 (1969), No. 2, 43-99.

[10] C. OLECH, Asymptotic behavior of the solutions of second order differential equations, Bull. Acad. Polon. Sci. (Ser. Sci. Math. Astro. Phys.), 7 (1959), 319-326.

[11] G. PÓLYA, On the mean value theorem corresponding to a given linear homogeneous differential equation, Trans. Amer. Math. Soc., 24 (1922), 312-324.

[12] J.D. SCHUUR, The asymptotic behavior of solutions of third order linear differential equations, Proc. Amer. Math. Soc., 18 (1967), 391-393.

Johns Hopkins University, Baltimore, Maryland

UNIQUENESS AND EXISTENCE OF SOLUTIONS OF BOUNDARY VALUE PROBLEMS FOR ORDINARY DIFFERENTIAL EQUATIONS*

Lloyd K. Jackson

1. Introduction

An important and very well known property of linear ordinary differential equations is that the uniqueness of a solution of a boundary value problem implies the existence of a solution of the problem. The question of whether or not nonlinear equations could have this property has been studied only very recently. Our examination of this question will be in terms of considering the possible validity of the following Proposition.

PROPOSITION 1. Assume that for the differential equation

$$y^{(n)} = f(x,y,y',\dots,y^{(n-1)}) \tag{1}$$

the following conditions are satisfied:

(A) $f(x,y_1,y_2,\dots,y_n)$ is continuous on $(a,b) \times R^n$,

(B) Solutions of all initial value problems for (1) are unique and extend throughout (a,b),

and (C) For any $a < x_1 < x_2 < \cdots < x_n < b$ and any solutions $y_1(x)$ and $y_2(x)$ of (1), $y_1(x_i) = y_2(x_i)$ for $1 \leq i \leq n$ implies that $y_1(x) \equiv y_2(x)$ on (a,b).

---

*Partially supported by N.S.F. Grant G.P.-17321.

Then it follows that for any $a < x_1 < x_2 < \cdots < x_n < b$ and any real numbers c_i, $1 \leq i \leq n$, there is a solution $y(x)$ of (1) such that $y(x_i) = c_i$ for $1 \leq i \leq n$.

Hartman [1] has proven that, if solutions of initial value problems for (1) are unique and if all n-point boundary value problems for (1) have unique solutions which extend throughout (a,b), then all k-point problems,

$$y^{(n)} = f(x,y,y',\ldots,y^{(n-1)})$$

$$y^{(i)}(x_j) = c_{ij}, \quad 0 \leq i \leq n_j - 1, \quad 1 \leq j \leq k,$$

$$a < x_1 < x_2 < \cdots < x_k < b, \quad \sum_{j=1}^{k} n_j = n, \quad 2 \leq k \leq n-1,$$

also have solutions and the solutions are unique. Thus, if the validity of Proposition 1 is established, it follows that all k-point problems are also uniquely solvable.

The validity of Proposition 1 has been established for differential equations of order 2 and of order 3. In these cases in Hypothesis (B) one need not assume that solutions of initial value problems are unique. For equations of order greater than 3 the validity of Proposition 1, as it is stated above, is an open question. However, the conclusion of Proposition 1 has been established for equations of arbitrary order after the addition of a fourth Hypothesis (D) which is a compactness condition for bounded collections of solutions. This result will be stated in Section 2. In Section 3 we will review the results for equations of orders 2 and 3 in order to observe how it was possible to avoid the addition of Hypothesis (D) in these cases. Finally, in Section 4 we shall obtain a sufficient condition that an equation which satisfies Hypotheses (A), (B), and (C) will satisfy (D). However, the question of the validity of Proposition 1 for equations of order greater than 3 remains open.

2. A Strengthened Form of Proposition 1

Hartman [2] has proven the following strengthened form of Proposition 1.

THEOREM 1. Assume that $f(x,y,y',\ldots,y^{(n-1)})$ in equation (1) is such that Hypotheses (A), (B), and (C) are satisfied. In addition assume the condition:

(D) If $\{y_k(x)\}_{k=1}^{\infty}$ is a sequence of solutions of (1) which is uniformly bounded on a compact interval $[c,d] \subset (a,b)$, then, there is a subsequence $\{y_{k_j}(x)\}$ such that $\{y_{k_j}^{(i)}(x)\}$ converges uniformly on $[c,d]$ for each $0 \leq i \leq n-1$.

Then it follows that for any $a < x_1 < x_2 < \cdots < x_n < b$ and any real numbers c_i, $1 \leq i \leq n$, there is a solution $y(x)$ of (1) satisfying $y(x_i) = c_i$ for $1 \leq i \leq n$.

In [2] Theorem 1 is proven for vector systems $y' = g(x,y)$, $y = (y_0,y_1,\ldots,y_{n-1})$, $g = (g_0,g_1,\ldots,g_{n-1})$ which are such that in a solution vector $y(x) = (y_0(x),y_1(x),\ldots,y_{n-1}(x))$ the functions $y_1(x)$, $y_2(x)$, $\ldots$, $y_{n-1}(x)$ are successive pseudoderivatives of $y_0(x)$. The vector form of equation (1) is a special case of such a system. If (1) satisfies Hypotheses (A), (B), and (C), and if n-point boundary value problems are locally solvable, that is, if (1) also satisfies the Hypothesis:

(D*) For each $x_0 \in (a,b)$ there is an open interval $J(X_0) \subset (a,b)$ containing x_0 such that on $J(x_0)$ all n-point boundary value problems have solutions,

then all n-point boundary value problems for (1) on (a,b) have solutions, that is, the conclusion of Theorem 1 holds. Obviously, by Theorem 1, if (1) satisfies (A), (B), (C), and (D), then (D*) is satisfied. Conversely, it can be shown that, if (1) satisfies (A), (B), (C), and (D*), then (D) is also satisfied. In [2] these results are proven for systems $y' = g(x,y)$. It is also proven in [2] that the conclusion of Theorem 1 is valid for the half-open intervals $[a,b)$ and $(a,b]$ provided that it is assumed that Hypotheses (A) and (B) are satisfied on the half-open interval and the Hypotheses (C) and (D*) hold on (a,b).

Klaasen [3] has proven Theorem 1 by methods quite different from those used in [2].

3. Proposition 1 for Equations of Orders 2 and 3

The first proofs of Proposition 1 for equations of order 2 were given by Lasota and Opial [4] and Jackson [5]. Subsequently, a number of generalizations were given for the second order scalar equation. For example, the assumption of uniqueness of solutions of initial value problems was dropped, more general boundary conditions were considered, and the assumption of uniqueness of solutions of boundary value problems was weakened. For examples of these generalizations see Schrader and Waltman [6], Bailey and Shampine [7], Schrader [8] and Shampine [9]. Generalizations to 2-dimensional vector systems have been obtained by Hartman [2] and Waltman [10].

In [11] Jackson and Schrader proved that Proposition 1 is valid for equations of order 3. The proof does not require the assumption of uniqueness of solutions of initial value problems in Hypothesis (B). It would appear that in the case of equations of arbitrary order the extendability of all solutions to (a,b) is the essential point of Hypothesis (B), not the uniqueness of solutions of initial value problems.

In the proofs that have been given that Proposition 1 is valid for equations of orders 2 and 3 an essential part of the argument makes use of the fact that a bounded set of solutions of equation (1) is compact in the sense of Hypothesis (D) of Theorem 1. Consequently, a key part of the demonstration consists of showing that, in the case of equations of orders 2 and 3, Hypotheses (A), (B), and (C) imply that (D) is satisfied. In both of these cases the argument that this is so is based on the following well known Theorem [12, p. 14].

THEOREM 2. Assume that (1) satisfies Hypotheses (A) and (B). Then, if $\{y_k(x)\}$ is a sequence of solutions of (1) and $[c,d]$ is a compact subinterval of (a,b), either there is a subsequence $\{y_{k_j}(x)\}$ such that $\{y_{k_j}^{(i)}(x)\}$ converges uniformly on $[c,d]$ for each $0 \leqslant i \leqslant n-1$, or $\sum_{i=0}^{n-1} |y_k^{(i)}(x)| \to +\infty$ uniformly on $[c,d]$ as $k \to +\infty$.

It follows immediately that, if (1) is of order 2 and satisfies Hypotheses (A) and (B), then (1) also satisfies Hypothesis (D). For in this case, if $\{y_k(x)\}_{k=1}^{\infty}$ is a sequence of solutions of (1) with

$|y_k(x)| \leqslant M$ on $[c,d] \subset (a,b)$ for each $k \geqslant 1$, then for each $k \geqslant 1$ there is an $x_k \in [c,d]$ such that $|y_k(x_k)| + |y_k'(x_k)| \leqslant M + \frac{2M}{d-c}$. It then follows from Theorem 2 that Hypothesis (D) is satisfied.

For equations of order 3 such an immediate appeal to Theorem 2 does not seem to be possible. In [11] the appeal to Theorem 2 was preceded by the establishment of the following Lemmas.

LEMMA 3. If (1) is of order 3 and satisfies Hypotheses (A), (B), and (C), then solutions of 2-point boundary value problems for (1), when they exist, are unique.

LEMMA 4. Assume that (1) is of order 3 and satisfies Hypothesis (A). Then, given any compact interval $[c,d] \subset (a,b)$ and any fixed $M > 0$, there is a $\delta > 0$ such that $[x_1,x_2] \subset [c,d]$, $x_2 - x_1 \leqslant \delta$, and $|\alpha| \leqslant M$ implies that (1) has solutions $y_1(x)$ and $y_2(x)$ satisfying $y_i(x_1) = y_i(x_2) = \alpha$ for $i = 1,2$, $y_1'(x_1) = y_2'(x_2) = 0$, and $|y_i'(x)| \leqslant 1$, $|y_i''(x)| \leqslant 1$ on $[x_1,x_2]$ for $i = 1,2$.

LEMMA 5. Given a compact interval $[\alpha,\beta]$ of the reals and a positive number M, there is an $N > 0$, depending on $\beta - \alpha$ and M, such that $y(x) \in C^{(2)}[\alpha,\beta]$, $|y(x)| \leqslant M$ on $[\alpha,\beta]$, and $|y'(x)| + |y''(x)| > N$ on $[\alpha,\beta]$ imply that $y'(x_0) = 0$ for some $x_0 \in (\alpha,\beta)$.

The proofs of these Lemmas may be found in [11] and their proofs will not be discussed here. If one assumes that there is an equation (1) of order 3 that satisfies Hypotheses (A), (B), and (C), but not (D), then by Theorem 2 there is a compact interval $[c,d] \subset (a,b)$, an $M > 0$, and a sequence of solutions $\{y_k(x)\}_{k=1}^{\infty}$ such that $|y_k(x)| \leqslant M$ on $[c,d]$ for all $k \geqslant 1$ and such that $\sum_{i=0}^{2} |y_k^{(i)}(x)| \to +\infty$ uniformly on $[c,d]$. It follows from Lemma 5 that there is a $z(x) \in \{y_k(x)\}$ which is a solution of a boundary value problem of the type specified in Lemma 4 on an interval $[x_1,x_2] \subset [c,d]$ with $x_2 - x_1 \leqslant \delta$, δ as in Lemma 4, but not satisfying $|z'(x)| \leqslant 1$ and $|z''(x)| \leqslant 1$ on $[x_1,x_2]$. However, by Lemma 3 solutions of such problems are unique and the conclusion of Lemma 4 is contradicted. From this contradiction we conclude that an equation (1) of order 3 which satisfies (A), (B), and (C) must also satisfy (D).

In this approach to settling the question of Hypothesis (D) for equations of order 3 Lemma 3 plays an important role. The question answered in Lemma 3 could be phrased more generally for equations (1) of arbitrary order in the following way: If (1) satisfies (A), (B), and (C), does it follow that the solutions of all k-point boundary value problems, $2 \leq k \leq n-1$, for (1), when they exist, are unique? Some results of this type for equations of higher order may be found in [13]. Of course, if it is established that (A), (B), and (C) imply (D), then the validity of Proposition 1 is proven and the above question has then been answered affirmatively by Hartman in [1]. The point is that such results proven beforehand might be useful in proving that (A), (B), and (C) do imply (D) as was the case for equations of order 3. However, it seems very difficult, if not impossible, to extend the above method used for equations of order 3 to equations of higher orders. In the next section we consider another possible approach.

4. Generalized Solutions

Hartman [2] and Klaasen [3] in their proofs of Theorem 1 use the following formulation of the compactness condition: If $\{y_k(x)\}$ is a bounded monotone sequence of solutions of (1) on a compact interval $[c,d] \subset (a,b)$, then $\{y_k^{(i)}(x)\}$ converges uniformly on $[c,d]$ for each $0 \leq i \leq n-1$. In the presence of Hypotheses (A), (B), and (C) it can be shown that this formulation of the compactness condition is equivalent to that given in (D) of Theorem 1. Consequently, a possible method of proving Proposition 1 is to show that, if (1) satisfies (A), (B), and (C), then the limit of a bounded monotone sequence of solutions is a solution. In this section we shall examine some of the properties of the limit of a pointwise convergent sequence of solutions.

A family F of real valued continuous functions defined on a set I of the reals is said to be an n-parameter family of functions on I in case for any set x_i, $1 \leq i \leq n$, of n distinct points in I and any set of n real numbers c_i, $1 \leq i \leq n$, there is a unique $\phi(x)$ in F such that $\phi(x_i) = c_i$ for $1 \leq i \leq n$, [1].

If F is an n-parameter family of functions on an interval (a,b), a real valued function $g(x)$ on (a,b) is said to be F-convex on

(a,b) in case, for any $a < x_1 < x_2 < \cdots < x_n < b$, $\phi(x_i) = g(x_i)$ for $1 \leq i \leq n$ where $\phi(x) \in F$ implies $(-1)^{n+i}[\phi(x) - g(x)] \leq 0$ on (x_i, x_{i+1}) for $0 \leq i \leq n$ where $x_0 = a$ and $x_{n+1} = b$. Similarly, $g(x)$ is said to be F-concave in case $\phi(x_i) = g(x_i)$ for $1 \leq i \leq n$ implies $(-1)^{n+i}[\phi(x) - g(x)] \geq 0$ on (x_i, x_{i+1}) for $0 \leq i \leq n$.

If a particular equation (1) satisfies Hypotheses (A), (B), and (C), and, if for that equation all n-point boundary value problems on (a,b) do have solutions, then the set of all solutions of the equation would constitute an n-parameter family on (a,b). In this case F-convex and F-concave functions with respect to the family of solutions could be defined as above.

If it assumed that a given equation (1) satisfies only Hypotheses (A), (B), and (C), a concept of convexity or concavity with respect to the solution set can still be defined. However in this case it is not desirable to express the definitions in terms of solutions satisfying $y(x_i) = g(x_i)$ for $1 \leq i \leq n$ since such solutions may not exist. It appears that the following definitions given by Klassen [3] are suitable for our purposes. In these definitions we assume that we are dealing with a fixed equation (1) satisfying (A), (B), and (C).

A real valued function $g(x)$ defined on a subinterval $I \subset (a,b)$ is said to be a subfunction on I with respect to solutions of (1), or to be convex with respect to the solution set of (1), in case, for any set of points $x_1 < x_2 < \cdots < x_n$ contained in I and any solution $y(x)$ of (1), the inequalities $(-1)^{n+i}[y(x_i) - g(x_i)] < 0$, $1 \leq i \leq n$, imply $y(x) < g(x)$ on $I \cap [x_n, b)$, and the inequalities $(-1)^{n+i}[y(x_i) - g(x_i)] > 0$, $1 \leq i \leq n$, imply $(-1)^{n+1}[y(x) - g(x)] > 0$ on $(a, x_1) \cap I$. A superfunction with respect to solutions of (1) is similarly defined with all of the above inequalities between function values reversed.

A real valued function $g(x)$ on an interval $I \subset (a,b)$ is said to be a generalized solution of (1) on I in case it is simultaneously a subfunction and a superfunction on I with respect to solutions of (1). This definition can be reformulated in the following way: $g(x)$ is a generalized solution on $I \subset (a,b)$ in case, for any set of points $x_1 < x_2 < \cdots < x_n$ contained in I and any solution $y(x)$ of (1), the inequalities $(-1)^{n+i}[y(x_i) - g(x_i)] < 0$, $1 \leq i \leq n$, imply $y(x) < g(x)$ on $I \cap [x_n,b)$ and $(-1)^{n+1}[y(x) - g(x)] < 0$ on $I \cap (a,x_1]$, and the inequalities $(-1)^{n+i}[y(x_i) - g(x_i)] > 0$, $1 \leq i \leq n$, imply $y(x) > g(x)$ on $I \cap [x_n,b)$ and $(-1)^{n+1}[y(x) - g(x)] > 0$ on $I \cap (a,x_1]$.

THEOREM 6. Assume that (1) satisfies (A), (B), and (C) and that $\lim_{k\to\infty} y_k(x) = g(x)$ on $I \subset (a,b)$ where $\{y_k(x)\}$ is a sequence of solutions of (1). Then $g(x)$ is a generalized solution of (1) on I.

Proof. Assume that for $x_1 < x_2 < \cdots < x_n$ contained in I there is a solution $y(x)$ of (1) such that $(-1)^{n+i}[y(x_i) - g(x_i)] < 0$ for $1 \leq i \leq n$ but that also $y(x_0) > g(x_0)$ for some $x_0 > x_n$ in I. Then, since $\lim y_k(x) = g(x)$, there is a solution $y_m(z)$ of (1) such that $(-1)^{n+i}[y(x_i) - y_m(x_i)] < 0$ for $1 \leq i \leq n$ and $y(x_0) > y_m(x_0)$. This contradicts Hypothesis (C). The remainder of the proof is established in a similar way.

Having established that the limit of a bounded monotone sequence of solutions $\{y_k(x)\}$ of an equation (1) satisfying (A), (B), and (C) is a generalized solution, we would like to be able to show that the limit function is in fact a solution and that $\{y_k^{(i)}(x)\}$ converges uniformly for each $0 \leq i \leq n-1$. In approaching the problem from this direction one can take advantage of the fact that fixed point theorems can be used to prove local existence theorems for solutions of boundary value problems for (1). Using these local existence theorems and Hypotheses (A), (B), and (C), we can prove certain smoothness properties of a generalized solution of (1). Success for this approach depends on being able to prove a sufficient degree of smoothness of a generalized solution. We shall give some results in this direction and observe the difficulties involved.

THEOREM 7. Assume that (1) satisfies Hypothesis (A) and that $\phi(x) \in C^{(n-1)}[c,d]$ where $[c,d]$ is a compact subinterval of (a,b). Assume that $M > 0$ is such that $|\phi^{(i)}(x)| \leq M$ on $[c,d]$ for $0 \leq i \leq n-1$. Then there exists a $\delta > 0$ such that, for any $c \leq x_1 < x_2 < \cdots < x_n \leq d$ with $x_n - x_1 \leq \delta$, (1) has a solution

$y(x)$ with $y(x_i) = \phi(x_i)$ for $1 \leqslant i \leqslant n$ and $|y^{(j)}(x)| \leqslant 2M$ on $[x_1, x_n]$ for $0 \leqslant j \leqslant n-1$. Furthermore, δ can be chosen in such a way that, for each fixed set $x_1 < x_2 < \cdots < x_n$ satisfying the above conditions, there is an $\varepsilon > 0$ such that for any y_i, $1 \leqslant i \leqslant n$, with $|y_i - \phi(x_i)| < \varepsilon$, $1 \leqslant i \leqslant n$, (1) has a solution $y(x)$ satisfying $y(x_i) = y_i$, $1 \leqslant i \leqslant n$, and $|y^{(j)}(x)| \leqslant 3M$ on $[x_1, x_n]$ for $0 \leqslant j \leqslant n-1$.

Proof. The proof uses the Green's function for the boundary value problem $y^{(n)} = 0$, $y(x_i) = 0$, $1 \leqslant i \leqslant n$, and is a standard application of the Schauder-Tychonoff Fixed Point Theorem.

With the aid of this theorem we can prove convergence properties of sequences of solutions and smoothness properties of generalized solutions.

THEOREM 8. Assume that (1) satisfies (A) and (C) and that $\lim_{k\to\infty} y_k(x) = g(x)$ on $[c,d] \subset (a,b)$ where $\{y_k(x)\}$ is a sequence of solutions of (1). Then, if $g(x) \in C^{(n-1)}[c,d]$, $g(x)$ is a solution of (1) on $[c,d]$ and $\lim y_k^{(j)}(x) = g^{(j)}(x)$ uniformly on $[c,d]$ for each $0 \leqslant j \leqslant n-1$.

Proof. Let $M > 0$ be such that $|g^{(j)}(x)| \leqslant M$ on $[c,d]$ for $0 \leqslant j \leqslant n-1$. By Theorem 7 there is a $\delta > 0$ such that, if $c \leqslant x_1 < x_2 < \cdots < x_n \leqslant d$ is a fixed set of points with $x_n - x_1 \leqslant \delta$, there is an $\varepsilon > 0$ with the property that $|y_i - g(x_i)| < \varepsilon$ for $1 \leqslant i \leqslant n$ implies that (1) has a solution $y(x)$ satisfying $y(x_i) = y_i$ for $1 \leqslant i \leqslant n$ and $|y^{(j)}(x)| \leqslant 3M$ on $[x_1, x_n]$ for $0 \leqslant j \leqslant n-1$. It follows that there is an $N > 0$ such that $k \geqslant N$ implies $|y_k(x_i) - g(x_i)| < \varepsilon$ for $1 \leqslant i \leqslant n$. Hence by Hypothesis (C) and the choice of ε, $|y_k^{(j)}(x)| \leqslant 3M$ on $[x_1, x_n]$ for $0 \leqslant j \leqslant n-1$ and all $k \geqslant N$. The conclusion of the Theorem follows immediately.

Let $g(x)$ be a real valued function defined on (c,d). At a point $x_0 \in (c,d)$ where $g(x)$ has a finite right hand limit $g(x_0 + 0)$ we define

$$D^1 g(x_0 + 0) = \lim_{x \to x_0^+} \frac{g(x) - g(x_0 + 0)}{x - x_0}$$

provided the limit exists. The left derivative $D^1g(x_0 - 0)$ is similarly defined. Likewise, if $g(x_0 + 0)$ and $D^1g(x_0 + 0)$ exist and are finite, we define

$$D^2g(x_0+0) = \lim_{x \to x_0^+} \left\{ \frac{2}{(x-x_0)^2} [g(x) - g(x_0+0) - D^1g(x_0+0)(x - x_0)] \right\}$$

provided the limit exists. In general, if the limits defining $g(x_0+0)$ and $D^jg(x_0+0)$, $1 \leq j \leq k-1$, exist and are finite, we define

$$D^kg(x_0+0) = \lim_{x \to x_0^+} \left\{ \frac{k!}{(x-x_0)^k} [g(x) - g(x_0+0) - \sum_{j=1}^{k-1} \frac{D^jg(x_0+0)(x-x_0)^j}{j!}] \right\}$$

provided the limit exists. The left derivatives $D^jg(x_0-0)$ are defined in a corresponding way.

THEOREM 9. Assume that (1) satisfies (A) and (C) and that $g(x)$ is a bounded generalized solution of (1) on $(c,d) \subset (a,b)$. Then $g(x)$ has right and left limits at each point of (c,d) and $D^1g(x_0-0)$ and $D^1g(x_0+0)$ exist in the extended reals for all $x_0 \in (c,d)$. Furthermore, if at a point $x_0 \in (c,d)$ $D^jg(x_0+0)$ exists and is finite for each $1 \leq j \leq k-1 \leq n-2$, then the limit defining $D^kg(x_0+0)$ exists in the extended reals. The same assertion applies to the left derivative $D^kg(x_0-0)$.

Proof. Assume that for some $x_0 \in (c,d)$, $\liminf_{x \to x_0^+} g(x) < \limsup_{x \to x_0^+} g(x)$ and choose a real number r such that $\liminf_{x \to x_0^+} g(x) < r < \limsup_{x \to x_0^+} g(x)$. Then there exist sequences $\{s_k\}$ and $\{t_k\}$ in (c,d) such that $\lim s_k = \lim t_k = x_0$, $x_0 < s_{k+1} < t_k < s_k$ for each $k \geq 1$, $\lim g(s_k) = \limsup_{x \to x_0^+} g(x)$, and $\lim g(t_k) = \liminf_{x \to x_0^+} g(x)$. Let $y(x)$ be a solution of (1) satisfying the initial conditions $y(x_0) = r$ and $y^{(j)}(x_0) = 0$ for $1 \leq j \leq n-1$. This solution exists on $[x_0, x_0+\delta]$ for some $\delta > 0$, and, since $\lim_{x \to x_0} y(x) = r$, there is an $N > 0$ such that $k \geq N$ implies $x_0 < s_k < x_0 + \delta$ and $g(s_k) > y(s_k)$, $g(t_k) < y(t_k)$. This contradicts the fact that $g(x)$ is a generalized solution on (c,d). The existence of $g(x_0-0)$ is similarly proven.

Now assume that for some $x_0 \in (c,d)$ the limit defining $D^1g(x_0+0)$ does not exist in the extended reals. Then choose the real number r

such that

$$\liminf_{x\to x_0^+} \frac{g(x) - g(x_0+0)}{x - x_0} < r < \limsup_{x\to x_0^+} \frac{g(x) - g(x_0+0)}{x - x_0} .$$

If $y(x)$ is a solution of (1) satisfying initial conditions $y(x_0) = g(x_0+0)$, $y'(x_0) = r$, and $y^{(j)}(x_0) = 0$ for $2 \leq j \leq n-1$, again sequences $\{s_k\}$ and $\{t_k\}$ can be chosen so that $\lim s_k = \lim t_k = x_0$, $x_0 < s_{k+1} < t_k < s_k$ for each $k \geq 1$, and $g(s_k) > y(s_k)$, $g(t_k) < y(t_k)$ for all sufficiently large k. This again contradicts $g(x)$ being a generalized solution and we conclude that $D^1g(x_0+0)$ and $D^1g(x_0-0)$ exist in the extended reals for all $x_0 \in (c,d)$.

Finally, if we assume that for some $x_0 \in (c,d)$, $D^jg(x_0+0)$ exists and is finite for each $1 \leq j \leq k-1 \leq n-2$, then by considering a solution of (1) satisfying initial conditions $y(x_0) = g(x_0+0)$, $y^{(j)}(x_0) = D^jg(x_0+0)$ for $1 \leq j \leq k-1$, $y^{(k)}(x_0) = r$, and $y^{(j)}(x_0) = 0$ for $k+1 \leq j \leq n-1$, we can as above prove that the limit defining $D^kg(x_0+0)$ exists in the extended reals.

<u>COROLLARY 10</u>. If (1) satisfies (A) and (C) and $g(x)$ is a bounded generalized solution of (1) on $(c,d) \subset (a,b)$, then $g(x)$ has a finite derivative $g'(x)$ almost everywhere on (c,d).

<u>THEOREM 11</u>. Assume that (1) satisfies Hypotheses (A), (B), and (C). Let $\{y_k(x)\}$ be a sequence of solutions of (1) on $(c,d) \subset (a,b)$ such that $\{y_k(x)\}$ is uniformly bounded on (c,d) and $\lim y_k(x) = g(x)$ on (c,d). Then, if for some $x_0 \in (c,d)$ the derivatives $D^jg(x_0+0)$, $1 \leq j \leq n-1$, all exist and are finite or the derivatives $D^jg(x_0-0)$, $1 \leq j \leq n-1$, all exist and are finite, it follows that there is a subsequence $\{y_{k_j}(x)\}$ such that $\{y_{k_j}^{(i)}(x)\}$ converges uniformly on each compact subinterval of (a,b) for each $0 \leq i \leq n-1$.

<u>Proof</u>. Assume that for some $x_0 \in (c,d)$ the derivatives $D^jg(x_0+0)$, $1 \leq j \leq n-1$, exist and are finite. Let $p(x)$ be the polynomial

$$p(x) = g(x_0+0) + \sum_{j=1}^{n-1} \frac{D^jg(x_0+0)(x-x_0)^j}{j!} .$$

Then it follows from the definition of $D^{n-1}g(x_0+0)$ that given any

$\varepsilon > 0$ there is a $\delta > 0$ such that $x_0 + \delta < d$ and

$$|p(x) - g(x)| < \frac{\varepsilon(x-x_0)^{n-1}}{(n-1)!}$$

for $x_0 < x \leqslant x_0 + \delta$. Let d_0 be a fixed number satisfying $x_0 < d_0 < d$. By Theorem 7 there is a $\delta_0 < 0$, such that for $x_0 < x_1 < x_2 < \cdots < x_n \leqslant d_0$ with $x_i - x_{i-1} = \eta \leqslant \delta_0$ for each $1 \leqslant i \leqslant n$, (1) has a solution $y(x)$ with $y(x_i) = p(x_i)$ for $1 \leqslant i \leqslant n$ and $|y^{(j)}(x)| \leqslant 2M$ on $[x_1, x_n]$ for $0 \leqslant j \leqslant n-1$ where $|p^{(j)}(x)| \leqslant M$ on $[x_0, d_0]$ for $0 \leqslant j \leqslant n-1$. Furthermore, there is an $\varepsilon_0 > 0$ such that, if $|y_i - p(x_i)| < \varepsilon_0$ for $1 \leqslant i \leqslant n$, then (1) has a solution of $y(x)$ with $y(x_i) = y_i$ for $1 \leqslant i \leqslant n$ and $|y^{(j)}(x)| \leqslant 3M$ on $[x_1, x_n]$ for $0 \leqslant j \leqslant n-1$. It is not difficult to show that with equal spacing η between the x_i's a suitable ε_0 has the form $\varepsilon_0 = Mh_n\eta^{n-1}$ where h_n is a fixed constant depending on n. Now as noted above, if we choose $\varepsilon = \frac{Mh_n}{2n^{n-1}}$, there is an η, $0 < \eta \leqslant \delta_0$, such that $x_0 < x < x_0 + n\eta$ implies

$$|p(x) - g(x)| < \frac{\varepsilon(x-x_0)^{n-1}}{(n-1)!} \leqslant \frac{\varepsilon_0}{2(n-1)!} \leqslant \frac{\varepsilon_0}{2}.$$

For such a choice of $\eta > 0$ we have $|p(x_i) - g(x_i)| \leqslant \frac{\varepsilon_0}{2}$ for $1 \leqslant i \leqslant n$ where $x_i - x_{i-1} = \eta$ for $1 \leqslant i \leqslant n$. Consequently, if $N > 0$ is such that $k \geqslant N$ implies $|y_k(x_i) - g(x_i)| < \frac{\varepsilon_0}{2}$ for $1 \leqslant i \leqslant n$, then $|p(x_i) - y_k(x_i)| < \varepsilon_0$ for $k \geqslant N$ and $1 \leqslant i \leqslant n$. It follows from our construction and Hypothesis (C) that $|y_k^{(j)}(x)| \leqslant 3M$ on $[x_1, x_n]$ for $0 \leqslant j \leqslant n-1$ and all $k \geqslant N$. The conclusion of the Theorem follows.

Thus we see that, in order to prove that Hypotheses (A), (B), and (C) imply (D), it is sufficient to prove that, if $g(x)$ is the pointwise limit of a bounded sequence of solutions of (1) on $(c,d) \subset (a,b)$, then there is at least one $x_0 \in (c,d)$ at which either $D^j g(x_0+0)$, $1 \leqslant j \leqslant n-1$, or $D^j g(x_0-0)$, $1 \leqslant j \leqslant n-1$, are finite.

REFERENCES

[1] P. HARTMAN, Unrestricted n-parameter families, Rend. Circ. Mat. Palermo (2), 7 (1958), 123-142.

[2] P. HARTMAN, On n-parameter families and interpolation problems for nonlinear ordinary differential equations, Trans. Amer. Math. Soc., 154 (1971), 201-226.

[3] G. KLAASEN, Existence theorems for boundary value problems for n^{th} order ordinary differential equations, unpublished.

[4] A. LASOTA and Z. OPIAL, On the existence and uniqueness of solutions of a boundary value problem for an ordinary second order equation, Colloq. Math., 18 (1967), 1-5.

[5] L. JACKSON, Subfunctions and second order differential inequalities, Advances Math., 2 (1968), 307-363.

[6] K. SCHRADER and P. WALTMAN, An existence theorem for nonlinear boundary value problems, Proc. Amer. Math. Soc., 21 (1969), 653-656.

[7] P. BAILEY and L. SHAMPINE, Existence from uniqueness for two point boundary value problems, J. Math. Anal. Appl., 25 (1969), 569-574.

[8] K. SCHRADER, Existence theorems for second order boundary value problems, J. Diff. Eqns., 5 (1969), 572-584.

[9] L. SHAMPINE, Existence and uniqueness for nonlinear boundary value problems, J. Diff. Eqns., 5 (1969), 346-351.

[10] P. WALTMAN, Existence and uniqueness of solutions of boundary value problems for two dimensional systems of nonlinear differential equations, Trans. Amer. Math. Soc., 153 (1971), 223-234.

[11] L. JACKSON and K. SCHRADER, Existence and uniqueness of solutions of boundary value problems for third order differential equations, J. Diff. Eqns., 9 (1971), 46-54.

[12] P. HARTMAN, Ordinary Differential Equations. Wiley, New York, 1964.

[13] L. JACKSON and G. KLAASEN, Uniqueness of solutions of boundary value problems for ordinary differential equations, SIAM J. Appl. Math., 19 (1970), 542-546.

University of Nebraska, Lincoln, Nebraska

REALIZATION OF CONTINUOUS-TIME LINEAR DYNAMICAL SYSTEMS: RIGOROUS THEORY IN THE STYLE OF SCHWARTZ

R. E. Kalman* and M. L. J. Hautus**

1. Background and Intuitive Discussion

The problem of realization has become well-established in system theory during the last decade. It concerns the question of replacing the external description of a dynamical system (essentially a map from the space of input functions into the space of output functions) by an internal description (essentially a system of differential or difference equations formulated in terms of internal or state variables), using rigorous and mathematically natural constructions. The concept of realization is of major interest to "differential equationists" because it provides new reasons for studying important classes of differential equations as well as new viewpoints for attacking some of the classical problems.

We emphasize right away that dynamical systems are understood here in the modern sense: the input/output behavior of the system receives equal attention with the internal behavior. The classical mathematical theory of dynamical systems (e.g., topological dynamics) deals exclusively with the latter; this point of view is extremely restrictive for the modern applications.

*This work was supported in part by US Air Force Grant 71-1898.

**This work was supported in part by Netherlands Organization for the Advancement of Pure Research (Z.W.O.) and in part by the US Army under Contract DA 31-124-ARO-D-394.

Realization theory, as it exists today, is primarily algebraic. The results in the linear case (the theory hardly extends further at the moment) may be viewed as a chapter in the theory of modules over the ring of polynomials $k[z]$, k = field or just a commutative ring. The corresponding dynamical systems operate in discrete-time (that is, t = integer) and the internal dynamical equations constructed by realization theory are difference equations. A reasonably up-to-date account of the algebraic theory may be found in Kalman, Falb, and Arbib [5] and Kalman [3]. More recent results will appear in Kalman [4].

The continuous-time (t = real number) analog of the discrete-time realization problem is known to require the same (not merely the analogous) algebraic machinery of $k[z]$-modules. In fact, it is generally assumed that the continuous-time case can be reduced to the discrete-time case by a certain ad-hoc identification between the difference and differential operators. What has been missing until now are the precise conditions under which such a reduction will work. Since the functions on $\underline{\underline{R}}$ = reals are a much larger class than the functions on $\underline{\underline{Z}}$ = integers, it is of course obvious that an algebraic theory cannot cover all dynamical phenomena in the continuous-time case, even under severe restrictions such as linearity. It would be nice to have an axiomatic characterization of "continuous-time", to be erected over the well-developed algebraic setup in discrete time. But this seems hopelessly ambitious, at least for the present.

Since Heaviside, mathematicians have been continually tempted to try to identify the algebraic "quantity", "symbol", "operator" z (precisely, the indeterminate in the ring $k[z]$) with the differential operator d/dt. One of the most successful thrusts in that direction has been the theory of generalized functions ("distributions") developed by Sobolev and L. Schwartz. This theory does not employ any major algebraic ideas and consequently its application to system-theoretic questions until now has been rather superficial. Nevertheless, generalized functions constitute such a nice and well-developed machinery that we could expect it to serve as a kind of substitute for the unknown axiomatization of continuous time.

If the reader is willing to share the preceding judgments concerning the status of the continuous-time realization problem, then he would have to agree also to the following list of priorities concerning the questions which should be settled:

(1) How is a linear, constant input/output map to be defined via generalized functions?

(2) What classes of continuous-time systems can be analyzed completely in terms of an algebraic theory?

(3) Can all this be formulated in terms of generalized functions in the style of Schwartz or is it necessary to make nontrivial additions to Schwartz's theory?

To put it in more explicit terms, we wish to define a linear input/output map $f: u(\cdot) \mapsto y(\cdot)$ such that f can be "realized" via the system

$$(*) \qquad dx/dt = Fx + Gu(t) ,$$

$$(\dagger) \qquad y(t) = Hx(t) ,$$

where (F, G, H) are real constant matrices, $x \in \underline{\underline{R}}^n$, and $u(\cdot)$ is a suitably defined (possibly generalized) function with values in $\underline{\underline{R}}^m$, giving rise to a solution $x(\cdot)$ (which may also be a generalized function) of (*), which in turn defines $y(\cdot)$, with values in $\underline{\underline{R}}^p$, via (†).

A complete answer to even these relatively narrow questions would be rather difficult. We shall give one set of consistent answers which cover a very large part of, but possibly not all, the interesting applications of linear, constant, continuous-time systems.

Our answers to the three questions above will be, intuitively, the following:

(1) Using a category-theoretic characterization of f in the discrete-time case, we shall find a setup via generalized functions which has the same properties.

(2) From this setup for f, we can deduce all the main results of the algebraic theory of discrete-time systems. In particular, it will be true that the state space of the system defined by f is a (topological) module over the ring $\underline{\underline{R}}[z]$, with z being the operator d/dt taken in the sense of Schwartz.

(3) If the definition of Nerode equivalence classes can be extended to generalized functions, then the algebraic machinery will apply without further restrictions. This extension can be done in a way completely analogous to the standard extension of the operator d/dt from C^1 functions to generalized functions. So everything works "in the style of Schwartz" and there is no need for extraneous or restrictive technicalities.

We emphasize that the results presented here are by no means obvious consequences of the standard Schwartz theory, even though they happen to be (luckily!) compatible with it. Especially important is the fact that the setup defined below provides a merging of considerations involved "generalized functions" and "polynomial algebra", something that has been often claimed (but never really substantiated) for the so-called Laplace transform approach.

The present setup may be viewed as a successful theory of continuous-time systems; it gives a rigorous underpinning to results often assumed without proof (especially in the engineering literature). It is well to point out, nevertheless, that there may exist rival theories in which continuous-time systems possess different properties. Examples of such independent theories, which do not yet exist, would contribute substantially to clarifying the mystery of "continuous-time".

2. Review of the Algebraic Theory in the Discrete-Time Case

The parenthetical sentences give physical explanations which are not part of the mathematical definitions.

Let $R = k[z]$, the commutative ring with identity consisting of polynomials in z with coefficients in a field k and the usual definition of multiplication and addition. (R is the model of "signals" received by a dynamical system. The coefficient of z^{-k} is equal to the value of the signal at $t = k < 0$.)

Define the input space $\Omega = k^m[z]$, the (free) $k[z]$-module consisting of m-vector valued polynomials. (m is the number of input channels into the system.)

The <u>input/output map</u> f (of some underlying system) is defined as

(1) $f\colon \Omega \to k^p$ (f = k-homomorphism) .

(The value $f(\omega) \in k^p$ of f are the p numbers appearing at the p output channels of the system at $t=1$ in response to an input which lasts only until $t=0$.)

The basic idea of the algebraic theory is to extend the above definition of f to a $K[z]$-homomorphism, after which standard algebraic methods can be used.

Step 1. Define $k^p[[z^{-1}]]$, the commutative ring of formal power series in z^{-1} with coefficients in k; $\Gamma \subset k^p[[z^{-1}]]$, the subring of power series with first coefficient 0, is a $k[z]$ module, in which the scalar product is the usual product of a polynomial in z with a formal power series in z^{-1} followed by the deletion of all terms corresponding to nonnegative powers of z.

Step 2. Now extend f to

(2) $$\bar{f}\colon \Omega \to \Gamma\colon \omega \mapsto f(\omega) = \sum_{k \geqq 0} f(z^k \omega) z^{-k-1} .$$

The fact that this is a $k[z]$-homomorphism is not difficult to verify. Usually the bar over f is omitted after the initial construction has been explained.

In the usual applications $\ker f \neq 0$, and we call $X_f = \Omega/\ker f$ the state module of f. The existence of a $k[z]$-module structure on the state set X_f is the fundamental result of linear system theory; from this all else follows via pure algebra. The point is that the whole theory is rigidly deductive once the system-theoretic concept of an input-output map f has been identified with a module homomorphism.

Although the definition of X_f as a quotient space is natural from (2), from a system-theoretic viewpoint X_f is simply the set of Nerode equivalence classes. (The latter just happen to coincide with the quotient module of f.) The state $x(t)$ at time t is the Nerode equivalence class of all input functions ω which (i) are identical when restricted to (t,∞) and (ii) have the same outputs on (t,∞). See Kalman, Falb, and Arbib [5, p. 193 and 249]. The problem is to compute these equivalence classes as well as to study their dependence on t. This will be done in Section 6 for the linear, continuous-time case.

A (finite-dimensional), linear, constant, discrete-time dynamical system Σ is a triple of matrices (F, G, H) used in the relations

$$(**) \qquad x(t+1) = Fx(t) + Gu(t) ,$$

$$(\dagger\dagger) \qquad y(t) = Hx(t) ,$$

which are directly analogous to the continuous-time relations (*) and (†). Using (F, G, H) one can of course compute the map $f_\Sigma: (u(t_0), u(t_0 - 1), \ldots, u(t)) \mapsto y(1)$. If f_Σ coincides with an f given a priori, then Σ is a realization f. If n = minimum over the family of all realization of f, then Σ is canonical (or minimal). We call n = size F_Σ the dimension of Σ.

The canonical realization is unique (modulo a change of basis in k^n); it can be described directly in terms of the state module $\Omega/\ker f$. Denote the elements of the latter by $[\omega]_f$; then (e_k = unit vectors)

$$F:\ [\omega]_f \mapsto [z\omega]_f , \qquad G:\ u \mapsto \sum_{k=1}^{m} [e_k]_f u_k , \qquad H:\ [\omega]_f \mapsto f(\omega)(0).$$

Thus the solution of the realization problem, that is, the computation of (F, G, H) is essentially equivalent to the computation of the Nerode equivalence classes $[\omega]_f$.

3. Definition of Continuous-Time Input/Output Maps (Step 1)

From here on, we shall use, as much as possible, the standard concepts and notations of the Schwartz theory. See, for instance, Yoshida [8]. We shall give only the most essential explanations; for full details, consult Kalman and Hautus [6].

Since the theory of generalized functions has not yet been extended to include values over arbitrary fields, we shall assume for convenience that

$$(A_1) \qquad k := \underline{\underline{R}} = \text{reals} .$$

Let $\mathcal{E}'$ be the space of generalized functions with bounded support; that is, the dual space (with the Schwartz topology) of the space $\mathcal{E}$ of all C^∞ real-valued functions on the real line. Then it is natural to set

$$(A_2) \qquad \Omega := \mathcal{E}'^{m}_{(-\infty,0]} ,$$

where the subscript indicates that the supports of the generalized functions in $\mathcal{E}'$ belong to the half-line $(-\infty,0]$. It is well known that $\mathcal{E}'_{(-\infty,0]}$ is a (topological) commutative ring with identity: multiplication is by convolution and the identity is the Dirac delta function δ. Using this multiplication, it is clear that Ω may be viewed as a (topological) module over the ring

$$(A_3) \qquad R := \mathcal{E}'_{(-\infty,0]} .$$

We note further that, in analogy with $k[z]$, the ring R has no zero divisors (Titchmarsh's theorem, see Yoshida [8, p. 166]). R contains $\dot{\delta}$ and convolution with $\dot{\delta}$ may be identified with differentiation. Thus, if we write $\dot{\delta} = z$, then $z = d/dt$ and R will contain $z, z^2, \ldots,$ as $k[z]$. On the other hand, unlike $k[z]$, R is not generated by 1 -- but this fact is never used in the discrete-time theory.

In analogy with (1), we are now forced to assume also that

$$(A_4) \qquad f\colon \mathcal{E}'^{m}_{(-\infty,0]} \to \underline{\underline{R}} \qquad \text{(linear + continuous)} .$$

Although this assumption may appear completely natural, it is in fact much more restrictive (but therefore also easier to study) than some others which have occasionally been proposed for system-theoretic study. The reason for the restrictiveness becomes obvious from the following easy result.

<u>THEOREM 1</u>. Let $m = p = 1$. Then there exists a function $W_f \in \mathcal{E}_{[0,\infty)}$ such that $f(\omega) = \omega(W_f)$ for all $\omega \in \Omega$.

In other words, our definition (A_4) of f implies that the map f can be characterized via an <u>impulse response function</u> W_f which is of class C^∞.

Proof. Direct, via the standard duality theory of distributions. (The assumption $m = p = 1$ was stated only to simplify the notations.) ¤

The map f is automatically a continuous $\underline{\underline{R}}$-module homomorphism, according to (A_4). It remains to see how f can be extended canonically to a continuous $\mathcal{E}'_{(-\infty,0]}$-module homomorphism.

4. Free and Cofree Modules

The duality principle of linear system theory (see Kalman [2]) has been verified in so many cases that it is certainly desirable to satisfy it also in the present investigation. The principle states that the system behavior at the output must be the dual of that at the input (and vice versa). In elementary investigations, duality can be expressed via linear algebra (replace column vectors by row vectors, matrices by their transposes, etc.). But the only general and reliable way to define duality is via category theory. We wish to spare the reader here from the technicalities of the precise category-theoretic terminology of "free" and "cofree". We shall provide only some intuitive explanations. Full details, with some classical examples, may be found in Kalman and Hautus [6].

To say that $k^m[z]$ is a free $k[z]$-module over the k-module $k^m = U$ generated by the set $(e_1, \ldots, e_m)$ (e_k = unit vector) means essentially that $k^m[z]$ is generated by letting the ring $k[z]$ act on these quantities and that there are no relations between the elements so generated. In exactly the same way, we may view Ω defined by (A_2) as a free $\mathcal{E}'_{(-\infty,0]}$-module over $(e_1, \ldots, e_m)$. (In engineering terms, this is exactly equivalent to saying that every state in the system can be obtained by applying suitable inputs from $\mathcal{E}'_{(-\infty,0]}$ at each of the m terminals of the system and that there are no constraints between the states established by the different inputs in this way.)

The simplest definition of a cofree $k[z]$-module Γ over a k-module $k^p = Y$ is that every k-homomorphism $M \to Y$ ($M = k[z]$-module) can be extended to a unique $k[z]$-homomorphism $M \to \Gamma$. Since a cofree module over a fixed Y is unique (if it exists at all), it follows from Section 2 that $Y[[z^{-1}]]$ is exactly the cofree module Γ over Y.

The problem, then, is to find the cofree $\mathcal{E}'_{(-\infty,0]}$-module Γ over the $\underline{\underline{R}}$-module $\underline{\underline{R}}^p$. Γ is not difficult to determine and turns out to be $\mathcal{E}^p_{[0,\infty)}$. We cannot prove this fact without category theory, but we may easily verify that f extends to the $\mathcal{E}'_{(-\infty,0]}$ homomorphism

$$f: \ \mathcal{E}'^m_{(-\infty,0]} \to \mathcal{E}^p_{[0,\infty)} \ .$$

The verification, as before, is essentially equivalent to giving an explicit definition of the scalar multiplication in the module Γ.

Let $a \in \mathcal{E}'_{(-\infty,0]}$ and $\gamma \in \Gamma$. Define the scalar product

$$a \cdot \gamma: \ [0, \infty) \to \underline{\underline{R}}^p: \ t \mapsto a(\gamma(t - \cdot)) \ .$$

The last expression on the right represents the vector obtained by evaluating the generalized function a with respect to each of the components of the ordinary vector-valued function $\tau \mapsto \gamma(t - \tau)$. It can be shown that this evaluation is well defined. The module axioms are easily verified. So is the fact that this scalar multiplication is continuous.

5. Definition of Continuous-Time Input/Output Maps (Step 2)

We can now define the extension of the map f by

$$\text{(3)} \qquad \bar{f}: \ \Omega \to \Gamma: \ \omega \mapsto \bar{f}(\omega) \ = \ \gamma: \ t \mapsto f(\sigma_t \omega) \ ,$$

where σ_t is the left-shift operator by amount t. We omit the rather lengthy but straightforward technical verification of the fact that f is well-defined and is an $\mathcal{E}'_{(-\infty,0]}$-module homomorphism.

In the sequel we shall write simply f for $\bar{f}$ since there is no danger of confusion.

The formal definition we wanted to get is the following:

DEFINITION. A linear, constant, continuous-time input/output map f is an $\mathcal{E}'_{(-\infty,0]}$-module homomorphism between a free module Ω and a cofree module Γ.

Note that the duality principle is explicitly "built into" this definition.

6. Nerode Equivalence Classes for Generalized Functions

The concept of Nerode equivalence implies the following prescription for determining the state $x(t)$ of the canonical realization of f at time t:

(i) Apply an input $\omega \in \Gamma$ to the system initially $(t_0 < 0)$ at rest.

(ii) Shift ω to the left by amount t.

(iii) Truncate $\sigma_t\omega(\cdot)$ for $\tau > 0$.

(iv) Compute the cosets under f of the truncated $\sigma_t\omega$.

Let us denote the "truncation operator" by

$$\mathscr{S}: \omega \mapsto \mathscr{S}\omega: \tau \mapsto \begin{cases} 0, & \tau > 0, \\ \omega(\tau), & \tau \leqq 0. \end{cases}$$

Then the formula for the state is

$$x(t) = [\mathscr{S}\sigma_t\omega]_f . \tag{4}$$

The right-hand side is well defined for all ordinary functions. Moreover, if t is positive, then $\mathscr{S}$ acts trivially and we have simply $x(t) = [\sigma_t\omega]_f$. However, if $t \in$ support ω and ω is a generalized function, a serious technical difficulty arises: it is not possible to define $\mathscr{S}$ in a canonical way. (Various specific definitions of $\mathscr{S}$ are possible, but each of them would result in an ad-hoc definition of the Nerode equivalence classes.) The only possible way to avoid this dilemma seems to be the suggestion (originally due to the second author) to regard $x(\cdot)$ intrinsically as a generalized function, by extending the definition given by (4) to generalized functions.

We do this in two stages.

LEMMA 1. If $\omega \in \Omega$ is an ordinary continuous function, the map $\underline{\underline{R}} \to \mathscr{E}'^m_{(-\infty,0]}: t \mapsto \mathscr{S}\sigma_t\omega$ is continuous (with respect to the Schwartz topology of $\mathscr{E}'^m_{(-\infty,0]}$).

The proof is straightforward. The projection from Ω into the Nerode equivalence classes $\Omega/\ker f$ is a continuous map since f is continuous. This gives the second step.

LEMMA 2. Assume again that $\omega \in \Omega$ is an ordinary continuous function. Then for any $\phi \in \mathcal{D}$ (= C^∞ functions with bounded support on $\underline{\underline{R}}$), we have that

$$\int_{-\infty}^{\infty} [\mathcal{S}\,\sigma_t\omega]_f\,\phi(t)dt = [\mathcal{S}\,\omega*\check{\phi}]_f ,$$

where $\check{\phi}(t) = \phi(-t)$ and $*$ denotes convolution.

The proof is again a straightforward calculation, which we omit.

According to Lemma 2, if ω is an ordinary function then $x(\cdot)$, an ordinary function viewed as a generalized function, is explicitly given by

$$x(\phi) = [\mathcal{S}\,\omega*\check{\phi}]_f \quad . \tag{5}$$

Since $\omega*\check{\phi}$ is an ordinary (in fact, C^∞) function even when ω is a generalized function, the right-hand side of (5) is always defined. It is clearly a continuous function of ω. Since generalized functions can be approximated by ordinary functions, we obtain

THEOREM 1. Formula (5) gives the unique continuous extension of (4) to generalized functions.

Formula (5) is direct analog of the classical formula: if ω is a differentiable function, then

$$\dot{\omega}(\phi) = \omega(-\dot{\phi}) ;$$

using the right-hand side we can define the left-hand side also for generalized functions.

So (5) will serve as the canonical definition of Nerode equivalence classes and we have begun to substantiate our claim that the theory can be built up "in the style of Schwartz" if we adopt the definition of f given in Section 5.

7. The Internal Differential Equations

We wish to show that the vector-valued generalized function $\phi \mapsto x(\phi)$ $(\phi \in \mathcal{D})$ is a solution of a differential equation. This will lead to the construction of the constant matrices (maps) (F, G, H), and so to the solution of the realization problem.

The verification of these facts is trivial but instructive. Assume for convenience that $m = 1$ so that ω is a scalar generalized function. Then

$$\begin{aligned} \dot{x}(\phi) &= x(-\dot{\phi}) && \text{(definition of derivative);} \\ &= [-\mathcal{S}\,\omega * \dot{\check{\phi}}]_f && \text{(definition of Nerode classes);} \\ &= [\mathcal{S}\, z * \omega * \check{\phi}]_f && \text{(definition of } z = d/dt = \delta\text{);} \\ &= [z\,\mathcal{S}\,\omega * \check{\phi}]_f + [\delta]_f(\omega * \phi)(0) && \text{(direct computation);} \\ &= Fx(\phi) + G\,(\phi) && \text{(definition of convolution);} \end{aligned}$$

where

$$F\colon\ [\mathcal{S}\,\omega * \check{\phi}]_f \mapsto [\tfrac{d}{dt}\mathcal{S}\,\omega * \check{\phi}]_f \qquad \text{and} \qquad G = [\delta]_f$$

and (without having to appeal to the preceding constructions)

$$H\colon\ [\omega]_f \mapsto f(\omega)(0)\ .$$

Thus we have our

<u>MAIN THEOREM</u>. The definition (5) of Nerode equivalence classes given in Section 6 exhibits the state as a generalized function (of time) for any input $\omega \in \Omega$. The canonical realization (F, G, H) of f can be defined directly via the Nerode equivalence classes. Moreover, the state satisfies (*), a linear differential equation with constant coefficients, in the usual generalized-function sense.

<u>Remark</u>. Some authors, such as Balakrishnan [1], have attempted to show that the differential equation satisfied by $x(\phi)$ must be at least as general as

$$dx/dt \;=\; Fx + \sum_{k \geq 0} G_k d^k u/dt^k\ .$$

However, Balakrishnan's "proof" of this fact is defective because his constructions are not canonical. See also Kalman, Falb, and Arbib [5, p. 28, Theorem 1.1]. It is conceivable that these equations are the correct internal description of the system for some input/output map $\hat{f}$, but then the definition of $\hat{f}$ must of course be different from that of f given in Section 5. No such definition of $\hat{f}$ is known at present.

COROLLARY 1. If $X_f = \Omega/\ker f$ is finite-dimensional as an $\underline{\underline{R}}$-vector space, and if $m = 1$, then X_f is generated by $[1]_f, [z]_f, [z^2]_f, \ldots$.

Exactly the same result is true also in the discrete-time case. It is interesting that powers of z alone need be considered in generating X_f; other types of inputs, such as delayed impulses, need not be used in generating states. As a result of Corollary 1, the entire algebraic machinery of the discrete-time case is applicable to our specific continuous-time case.

Remark. The main theorem shows that it is possible to define the behavior of the output during the application of ω by the formula

$$y(\phi) = Hx(\phi) .$$

This information cannot be obtained directly from the input/output map f because of the impossibility of canonically truncating ω. But if $t \notin$ support ω, the output $y(t)$ is just $f(\omega)(t)$ and then $y(\cdot)$ is an ordinary function. This shows that the main theorem provides more insight into system behavior than is possible to obtain by direct use of the input/output map f.

8. Remarks Concerning the Literature of the Problem

An essential ingredient of the realization problem discussed here is the requirement that all constructions be canonical. We know of no other paper at this time (June 1971) which satisfies this desideratum. For instance, an early attempt by Balakrishnan [1], based on an extension of the Schwartz kernel theorem, requires ad hoc assumptions at many points and thus fails to provide a canonical setup for an analogue of our main theorem. A much more satisfactory result is due to Matsuo [7], who uses the standard form of the kernel theorem (see Yoshida [8, p. 156, Theorem 2]), but unfortunately his construction also fails to be completely canonical.

REFERENCES

[1] A.V. BALAKRISHNAN, Foundations of the state space theory of continuous systems, J. Comp. and Inf. Sci. 1 (1967), 91-116.

[2] R.E. KALMAN, On the general theory of control systems, in Proc. 1st IFAC Congress on Automatic Control, Moscow, 1960, Butterworths, London, Vol. 1, 481-492.

[3] R.E. KALMAN, Lectures on controllability and observability, in Proc. C.I.M.E. Summer School (Pontecchio Marconi, 1968), Edizioni Cremonese, Roma, 1969.

[4] R.E. KALMAN, Lectures on Algebraic System Theory, Springer Lecture Notes in Mathematics (to appear 1972).

[5] R.E. KALMAN, P.L. FALB and M.A. ARBIB, Topics in Mathematical System Theory. McGraw-Hill, New York, 1969.

[6] R.E. KALMAN and M.L.J. HAUTUS, Algebraic structure of linear dynamical systems. II. Continuous-time systems, to appear in Mathematical System Theory, 1972.

[7] T. MATSUO, Mathematical theory of linear continuous-time systems, Res. Reports of Automatic Control Laboratory, Nagoya University, 16 (1969), 11-17.

[8] K. YOSHIDA, Functional Analysis. Springer, 1965.

REK: *University of Florida, Gainesville, Florida, and Stanford University, Stanford, California*

MLJH: *Technical University, Einhoven, Netherlands*

DISSIPATIVE SYSTEMS*

J. P. LaSalle

1. Introduction

The research to be reported on here was completed recently by Hale, LaSalle, and Slemrod, and a complete paper [1] on the subject is to appear. The research developed in the following manner. Billotti under the direction of LaSalle completed a study [2] in 1969 of dissipative retarded functional differential equations, and he and LaSalle then developed these results in a more general fashion [3] which, however, assumed the strong "smoothing" of initial data as occurs for retarded functional differential equations. Discussions of this work with Hale and Slemrod interested them in further generalizations. Slemrod saw how to do this for parabolic and hyperbolic partial differential equations and Hale for a wide class of functional differential equations of neutral type. It was then decided to unite the disparate points of view by identifying the general hypotheses which lead to the principal results, and the culmination of this collaborative effort is the subject of this report.

*Research reported on here was supported in part by the Air Force Office of Scientific Research - AFOSR 693-67B, United States Army Research Office - DA-31-124-ARO-D-270, National Aeronautics and Space Administration - NGL 40-002-015, National Science Foundation - GP 15132 and Office of Naval Research - NONR N0014-67-A-0191-0009.

2. Processes

To present the background and the results succinctly it is convenient to begin with the concept of a process (think of this as a "generalized nonautonomous dynamical system").

Let $R = (-\infty,\infty)$, $R^+ = [0,\infty)$, X be a Banach space and $u\colon R \times X \times R^+ \to X$. Define

$$(\sigma,t)x = u(\sigma,x,t) \ , \qquad (\sigma,t)\colon X \to X \ .$$

X is the "state" space and interpret $(\sigma,t)x$ to be the state of the system at time $\sigma + t$ if initially the state at time σ was x.

The mapping u is said to define a process on X if u has the following properties:

(i) u is continuous,

(ii) $(\sigma,0) = I$, the identity mapping,

(iii) $(\sigma+s,t)(\sigma,s) = (\sigma,s+t)$, $\sigma \in R$ and $s,t \in R^+$.

Property (iii) corresponds to uniqueness in the forward direction of time. The (positive) motion or orbit through (σ,x) is $\bigcup_{t\geq 0} (\sigma,t)x$.

A dynamical system is an autonomous process: $(\sigma,t) = (0,t)$ for all $\sigma \in R$ and all $t \in R^+$. A motion is said to be periodic of period $\alpha > 0$ if $(\sigma,t+\alpha)x = (\sigma,t)x$ for all $t \in R^+$. A process is said to be periodic of period $\omega > 0$, if $(\sigma+\omega,t) = (\sigma,t)$ for all $\sigma \in R$ and all $t \in R^+$.

For a periodic process the Poincaré map $T\colon X \to X$ defined by

$$Tx = (\sigma,\omega)x$$

for some fixed σ defines a discrete dynamical system with a motion or orbit through x given by $\gamma^+(x) = \bigcup_{n=0}^{\infty} T^n x$. It follows easily that fixed points of T^k correspond to periodic motions of the periodic process of period $k\omega$. The limit set $L(x)$ of a discrete motion through x is

$$L(x) = \bigcap_{j=0}^{\infty} \mathrm{Cl} \bigcup_{n=j}^{\infty} T^n x \ .$$

A set M in X is said to be positively invariant if $TM \subset M$, negatively invariant if $M \subset TM$ and invariant if $M = TM$. It is easy to see that

LEMMA 1. If $\gamma^+(x)$ is precompact, then $L(x)$ is nonempty, compact and invariant.

Remark. It turns out later to be useful to note that the above result holds if x is replaced by an arbitrary compact set K.

In relation to applications one of the problems, as in the use of this lemma, in developing a general theory is to have results which depend upon determining boundedness of motions. One cannot, in general, give direct tests for compactness but can verify boundedness by use, for example, of Liapunov functions (see [1] for references to how this difficulty was overcome in developing a general stability theory).

3. The Principal Results To Be Generalized

Let $f\colon R \times R^n \to R^n$ be continuous and define a system of ordinary differential equations

$$\dot{x} = f(t,x) \quad (1)$$

Assume that the solution $\phi(t,\sigma,\xi)$, $\phi(\sigma,\sigma,\xi) = \xi$ for $\sigma \in R$ and $\xi \in R^n$ is unique, is defined for all $t \in [\sigma\ \infty)$, and depends continuously on (t,σ,ξ). Then $u(\sigma,\xi,t) = \phi(\sigma+t,\sigma,\xi)$ defines a process on R^n.

For a periodic system of ordinary differential equations (1) ($f(t,x) = f(t+\omega,x)$ for some $\omega > 0$, all $t \in R$ and all $x \in R^n$) let T be the corresponding Poincaré map defined above.

Then (1) (or T) is said to be dissipative if there is a bounded set B in R^n such that given $x \in R^n$ there is an integer $n(x)$ with the property that $T^n x \in B$ for all $n \geq n(x)$. (It is sufficient to assume only that $T^{n(x)} \in B$.)

This concept of dissipativeness for $n = 2$ was first studied by Levinson [4] in 1949 and more general results can be found in [5], [6], and [7]. The principal properties of dissipative systems, the objective of our generalization, are:

I. There is a maximal (nonempty) compact set J invariant under T.

II. J is globally asymptotically stable.

III. For some integer k_0, T^k has a fixed point for each $k \geqslant k_0$.

Levinson in [4] proved for $n = 2$, I and III (with $k_0 = 1$), and Pliss in [5] has the three results for general systems of ordinary differential equations. As will be pointed out later, it is now known for ordinary differential equations that $k_0 = 1$ (there is always a periodic solution of period ω).

4. Retarded Functional Differential Equations

With $r \geqslant 0$ given let $C = C([-r,0],R^n)$ be the space of continuous functions mapping $[-r,0]$ into R^n with the topology of uniform convergence. For any continuous x defined on $[\sigma-r,\sigma+A)$, $A > 0$ and any $t \in [\sigma,\sigma+A)$ define x_t in C by $x_t(\theta) = x(t+\theta)$, $\theta \in [-r,0]$. Let $f: R \times C \to R^n$ be continuous. A function $x = x(\sigma,\phi)$ defined on and continuous on $[\sigma-r,\sigma+A)$ is said to be a solution of the retarded functional differential equation

$$\dot{x}(t) = f(t,x_t) \tag{2}$$

on $[\sigma,\sigma+A)$ with initial value ϕ at σ if $x_\sigma = \phi$ and $x(t)$ satisfies (2) on $[\sigma,\sigma+A)$. Assume that each solution $x(\sigma,\phi)$ exists and is unique on $[\sigma,\infty)$, and $x(\sigma,\phi)t$ depends continuously on (σ,ϕ,t). Then $u(\sigma,\phi,t) = x_{\sigma+t}(\sigma,\phi)$ defines a process on C. Again assume (2) is periodic of period ω and T is the Poincaré map of C into itself defined by the periodic process u. Then the dissipative property is as before $(X = C)$.

H_1'') There is a bounded set B in X with the property that given $x \in X$ there is a positive integer $n(x)$ such that $T^n x \in B$ for all $n \geqslant n(x)$.

With the further, not unnatural assumption, on f that f takes bounded sets of $R \times C$ into bounded sets of R^n, then the solutions of (2) smooth initial data. In fact, expressed in terms of T, retarded functional differential equations have the following smoothing property ($X = C$):

H_2'') There is an integer n_0 such that given a bounded set B in X there is a compact set B^* in X such that $T^n x \in B$ for $n = 0,1,2,\dots,N$ $(N \geq n_0)$ implies $T^n(x) \in B^*$ for $n = n_0, n_0+1,\dots,N$.

The integer n_0 is the length of time it takes to smooth the initial data. For the retarded functional differential equation (2) n_0 is the smallest integer such that $n_0\omega \geq r$ and for the ordinary differential equation (1) $n_0 = 0$ (since $X = R^n$ is locally compact). With H_2'') it is sufficient to assume in H_1'') only that $T^n(x) \in B$ for $n(x) \leq n \leq n(x) + n_0$; that is, only long enough to smooth.

6. Functional Differential Equations of Neutral Type

For a more general definition of functional differential equations of neutral type and basic theorems concerning solutions and their properties see [8] and [9]. Here we consider a more special case.

Let C, f, and x_t be as before. Consider, in addition, the continuous map $D: R \times C \to R^n$ of the form

$$D(t,\phi) = \phi(0) + B_1(t)\phi(-r_1) + \cdots + B_k(t)\phi(-r_k)$$

where $0 \leq r_j \leq r$ and the B_j are uniformly continuous and bounded for $t \in R$. A function $x = x(\sigma,\phi)$ defined and continuous on $[\sigma-r,\sigma+A)$, $A > 0$, is said to be a solution of the neutral functional differential equation

$$\frac{d}{dt} D(t,x_t) = f(t,x_t) \tag{3}$$

on $(\sigma,\sigma+A)$ with initial value ϕ at σ if $x_\sigma = \phi$, $D(t,x_t)$ is continuously differentiable on $(\sigma,\sigma+A)$ and satisfies (3) on $(\sigma,\sigma+A)$. Assume for any (σ,ϕ) $R \times C$ that a solution of (3) exists on (σ,∞), is unique, and $x(\sigma,\phi)(t)$ is continuous in (σ,ϕ,t). The function $u(\sigma,\phi,t) = x_{\sigma+t}(\sigma,\phi)$ defines a process on C.

In this generality one cannot expect solutions to be smoother than the initial data and further restructions need to be placed on D (see [8]). More specifically, we shall assume that D is stable (see [8]); suffice it to say here that it is shown in [8] that D is stable if and only if the solutions of $D(t,x_t) = 0$ are uniformly asymptotically stable. If $D(t,\phi)$ and $f(t,\phi)$ are ω-periodic in t, D is stable, and f maps bounded sets of $R \times C$ into bounded sets of R^n, then from results in [8] and [10] it is not difficult to see that the Poincaré map T for this class of neutral functional differential equations has the following smoothing properties $(X = C)$:

H_2') To each bounded set B in X there corresponds a compact set B^* in X with the property that given $\varepsilon > 0$ there is an integer $n_0(\varepsilon,B)$ such that $T^n x \in B$ for $n \geq 0$ implies $T^n x \in B_\varepsilon^*$ for $n \geq n_0(\varepsilon,B)$, where B_ε^* is an ε-neighborhood of B^*.

H_4') For any compact set K in X, $\gamma^+(K) = \bigcup_{n=0}^{\infty} T^n K$ bounded implies $\gamma^+(K)$ is precompact.

The weaker smoothing here requires a stronger concept of dissipativeness:

H_1') There is a bounded set B in X such that for each x in X there is a neighborhood 0_x of x and an integer $N(x)$ such that $T^n 0_x \subset B$ for $n \geq N(x)$.

7. Partial Differential Equations

Certain types of parabolic and hyperbolic partial differential equations are known to define processes on appropriate Sobolev spaces. In the hyperbolic case there is some smoothing of initial data but this is not so for hyperbolic equations. However, when it is known that a hyperbolic equation defines a process on two Sobolev spaces X and Y with $X \subset Y$ algebraically and topologically and with the injection map completely continuous, then the smoothing effect is replaced by the fact that a bounded orbit in X is compact in Y.

8. General Hypotheses

As indicated by the brief discussion of the situation for partial differential equations one will, in general, want to consider a transformation T of two spaces X and Y with X embedded in Y as described above with the injection map assumed to be at least continuous. It then turns out that there are four hypotheses needed to obtain the generalizations of I, II, and III stated in Section 3. All of this requires considerable explanation and is more than we can enter into here (the reader is referred to [1]). The four hypotheses are of the following type:

H_1) A dissipative property

H_2) A smoothing property

H_3) A fixed point property

H_4) A smoothing property.

When $X = Y$ these hypotheses become

H_1') There is a bounded set $B \subset X$ such that for any $x \in X$, there is a neighborhood 0_x of x and an integer $N(x)$ such that $T^n 0_x \subset B$ for $n \geqslant N(x)$.

H_2') To each bounded set B in X there corresponds a compact set B^* in X with the property that given $\varepsilon > 0$ there is an integer $n_0(\varepsilon,B)$ such that $T^n x \in B$ for $n \geqslant 0$ implies $T^n x \in B_\varepsilon^*$ for $n \geqslant n_0(\varepsilon,B)$, where B_ε^* is an ε-neighborhood of B^*.

H_3') There is an integer k_0 such that for every closed bounded convex set $B \subset X$ and every integer $k \geqslant k_0$, if $T^n B$ is bounded for $0 \leqslant n \leqslant k$ and $T^k\colon B \to B$, then T^k has a fixed point in B.

H_4') For any compact set $B \subset X$, $\gamma^+(B)$ bounded implies $\gamma^+(B)$ precompact.

The results corresponding to I and II are implied by $H_1)$, $H_2)$, and $H_4)$, and the fixed point property III follows from $H_1)$, $H_2)$ and $H_3)$.

In the case of partial differential equations when the injection map is completely continuous it can be shown that $H_1)$ implies $H_1) - H_4)$, so that all that need be assumed is $H_1)$. When there is smoothing, as in the case of retarded or neutral functional, differential equations, then $H_1')$ implies $H_1') - H_4')$ and again only a dissipative assumption is required.

Let us examine at least the case of retarded functional differential equations in some detail. Here we have the smoothing property $H_2)''$ and need only assume the weaker form of dissipativeness $H_1)''$. It can then be shown that $H_1)''$ and $H_2'')$ imply $H_1') - H_4')$. In fact, one obtains

<u>THEOREM 1</u>. If T satisfies $H_1'')$ and $H_2'')$, then there is a compact set K in X with the property that given a compact set H in X there is an open neighborhood H_0 of H and an integer $N(H)$ such that $T^n(H_0) \subset K$ for all $n \geq N(H)$.

It can then be shown that $J = \bigcap_{n=0}^{\infty} T^n K$ is the maximal compact invariant set. This is done by showing first of all that J is well-defined (does not depend on the choice of K from Theorem 1) and that $J = L(K)$. Being a limit set J is nonempty, compact and invariant (see the remark below Lemma 1) and it is easy to see that J is the maximal compact invariant set and is a global attractor. Proving that J is stable is more difficult. The fixed point property III follows readily from Theorem 1 and Schauder's fixed point theorem.

If one assumes, in addition to $H_1'')$ and $H_2'')$ that T maps bounded sets into bounded sets, then it follows that T^k is compact for $k \geq n_0$ and one can show using Browder's extension of the Schauder fixed point theorem that T^k has a fixed point for each $k \geq n_0$ (in III, $k_0 = n_0$). This result has also been given by Horn in [11] as a consequence of his extension of Schauder's fixed point theorem that is slightly different from Browder's. A similar result for retarded functional differential equations which are uniformly bounded and uniformly ultimately bounded was given by Yoshizawa in [12] (see also [13]). In addition, Yoshizawa assumes that the f in (2) satisfies a Lipschitz condition.

Thus for ordinary differential equations we know for dissipative ordinary differential equations there is always a periodic solution of period ω and for dissipative retarded functional differential equations when the solution map maps bounded sets into bounded sets and $\omega \geq r$ there is a periodic solution of period ω.

One suspects that there should be better results and from conversations at this meeting with G. Stephen Jones and J. Hale, it seems clear that a dissipative retarded functional differential equation always has a periodic solution of period ω without any further assumptions.

From subsequent conversations with J. Hale it seems that a similar result is true for a restricted class of neutral functional differential equations (it appears necessary, for example, to assume that the operator D is autonomous).

9. Concluding Remarks

The rather abstract theory presented here shows how the theory of dissipative systems of ordinary differential equations can be extended to include a wide class of functional and partial differential equations. Since the basic hypotheses are all in terms of boundedness, finding sufficient conditions in terms of Liapunov functions is not too difficult and we are undertaking now to work out some nontrivial examples to illustrate how the theory can be applied.

REFERENCES

[1] J.K. HALE, J.P. LASALLE, and MARSHALL SLEMROD, Theory of a general class of dissipative process, to appear in J. Math. Anal. and Appl.

[2] J.E. BILLOTTI, Dissipative Functional Differential Equations, Ph.D. dissertation, Brown University, 1969.

[3] J.E. BILLOTTI and J.P. LASALLE, Periodic dissipative processes, to appear in the Bull. Amer. Math. Soc.

[4] N. LEVINSON, Transformation theory of non-linear differential equations of second order, Annals of Math., 45 (1944), 723-737.

[5] V.A. PLISS, Nonlocal Problems of the Theory of Oscillations. Academic Press, New York, 1966 (translation of 1964 Russian edition).

[6] J.P. LASALLE, A study of synchronous asymptotic stability, Annals of Math., 65 (1957), 571-581.

[7] R. REISSIG, G. SANSONE, and R. CONTI, Nonlineare Differentialgleichungen Höherer, Ordnung, Edizioni Cremonese, Roma, 1969.

[8] A.M. CRUZ and J.K. HALE, Stability of functional differential equations of neutral type, J. Diff. Eqns., 7 (1970), 334-355.
[9] J.K. HALE and A.M. CRUZ, Existence and continuous dependence for hereditary systems, Annali Mat. Pura Appl. (4), 85 (1970), 63-81.
[10] J.K. HALE, A class of neutral equations with the fixed point property, Proc. Nat. Acad. Sci. U.S.A., 67 (1970), 136-137.
[11] W.A. HORN, Some fixed point theorems for compact maps and flows in Banach spaces, Trans. Amer. Math. Soc., 149 (1970), 391-404.
[12] TARO YOSHIZAWA, Ultimate boundedness of solutions and periodic solution of functional differential equations, Colloques Internationaux sur les Vibrations Forcées dans les Systèmes Non-linéaires, Marseille, September 1964, 167-179.
[13] TARO YOSHIZAWA, Stability Theory by Liapunov's Second Method, Math. Soc. of Japan Publication, Tokyo, 1966.

Brown University, Providence, Rhode Island

RELAXATION OSCILLATIONS AND TURBULENCE

A. Lasota

1. Introduction

The term turbulence is due to O. Renolds. In his classical experiments fluids were allowed to flow through circular tubes. At the entry to the tube colouring matter was introduced and the velocity field was examined visually. When the mean velocity through the tube was sufficiently small, the flow was steady; that is, the velocity at any point was constant in time. When the mean velocity was sufficiently large, the flow was unsteady, irregular and very complicated. Such a flow is called turbulent. The physical phenomenon of turbulence has received a tremendous amount of study but the exact description of turbulent motion is rather hopeless.

In the case of turbulent flow we are only interested in the statistical properties of the motion. For example, if $v(t,x)$ denotes the velocity in time t at a given point x, we are interested in the mean velocity

$$v(x) = \lim_{t\to\infty} \frac{1}{t}\int_0^t v(s,x)ds .$$

Recently D. Ruelle and F. Takens noted that the statistical properties of turbulent motion can be explained (in some abstract situations) via the pointwise ergodic theorem [6].

Our purpose is to show that the phenomenon of turbulence occurs not only in the motion of fluids but also in the motion of a rigid body and even in the motion of a particle. In Section 2 we start from a definition of a turbulent dynamical system. In Section 3 a conjecture of S. M. Ulam concerning the existence of invariant measures is discussed. In Section 4 we consider the dynamical system given by the differential equation

$$y'' + f(y) = 0 \qquad \text{for} \quad y > p(x) \tag{1}$$

and the reflection law

$$y'(x + 0) = p'(x + 0) \qquad \text{for} \quad y = p(x) \; . \tag{2}$$

From the Ulam conjecture it follows that the dynamical system described by (1)-(2) is turbulent. Section 5 contains application of our theory to the mathematical theory of friction and some related problems.

2. Turbulent Dynamical Systems

Let X be a Hausdorff topological space and let μ be a given positive normalized measure defined on the σ-algebra of Borel subsets of X. In what follows, the term "measure" will be used for positive normalized Borel measures. A measure m will be called absolutely continuous if it is absolutely continuous with respect to μ. A measure m will be called discrete if it is supported at a single point.

Let T be a measurable transformation of X into itself. An invariant ergodic (with respect to T) measure m will be called <u>T-stable</u> if for any absolutely continuous measure m_0 the sequence of measures $\{m_0 T^{-k}\}$ converges weakly to m. Recall that a sequence of measures $\{m_k\}$ converges weakly to m if

$$\lim_{k\to\infty} \int_X f \, dm_k = \int_X f \, dm$$

for any real valued continuous bounded function f.

Observe that, in general, the ergodicity of the dynamical system (X,T,m) does not imply that the measure m is T-stable. As a simple example consider the unit interval $X = [0,1)$ with the usual Borel measure $d\mu = dx$. Define T on $[0,1)$ as

$$T(x) = x + c \pmod 1$$

where c is an irrational number. It is well known that the measure $m = \mu$ is invariant and ergodic but clearly m is not T-stable.

On the other hand mixing implies stability. Namely if m is equivalent to μ and a system (X,T,m) is mixing, then the measure m is T-stable.

Let T_λ $(\lambda > \lambda_0)$ be a family of measurable transformations from X into X.

Definition. The triple (X,T_λ,μ) will be called a turbulent dynamical system if the following conditions hold:

i) For every $\lambda > \lambda_0$ there exists a fixed point x_λ of T_λ, and the mapping $\lambda \to x_\lambda$ is continuous.

ii) For sufficiently small λ (say $\lambda_0 < \lambda < \lambda_1$) the discrete measure supported at x_λ is T_λ-stable.

iii) For sufficiently large λ (say $\lambda > \lambda_2$) there exists an absolutely continuous T_λ-stable measure m_λ.

Example. Consider the λ-adic transformations of the unit interval $[0,1)$ into itself given by

$$T_\lambda(x) = \lambda x \pmod 1, \quad \lambda > 0 .$$

Let μ be the usual measure on $[0,1)$ $(d\mu = dx)$. It is obvious that for any λ the point $x_\lambda = 0$ is a fixed point of the transformation T_λ. It is also easy to verify that for $\lambda < 1$ the measure supported at $x_\lambda = 0$ is T_λ-stable. In 1957 A. Rényi [4] proved that for any $\lambda > 1$ there exists an invariant (with respect to T_λ) absolutely continuous measure m_λ such that $dm_\lambda/dx > 0$. From the results of V. A. Rochlin [5] it follows that the system $([0,1),m_\lambda,T_\lambda)$ is mixing. This implies that for $\lambda > 1$ the measure m_λ is T_λ-stable.

3. Ulam's Conjecture

The problem of the existence of an absolutely continuous invariant measure for a given transformation T of the unit interval into itself is, in general, very difficult. For example the existence of such a measure for the "broken line"

$$T_a(x) = \begin{cases} 2x , & 0 \leqslant x \leqslant \frac{1}{2} \\ (2-a) - 2(1-a)x , & \frac{1}{2} \leqslant x \leqslant 1 \end{cases}$$

was stated by S. M. Ulam as an open question. According to the conjecture stated below the answer is affirmative for $0 \leqslant a \leqslant \frac{1}{2}$.

S. M. Ulam conjectured ([7], p. 74) that if $T: [0,1] \to [0,1]$ is sufficiently "simple" and its graph does not cross the line $y = x$ with a slope in absolute value less than 1, then there exists an absolutely continuous measure invariant with respect to T. Let us state this conjecture in detail (or rather a similar conjecture).

<u>CONJECTURE</u>. Let $0 = a_0 < \cdots < a_n = 1$ be a partition of the unit interval and let $\phi_i: [a_{i-1}, a_i] \to [0,1]$ be a sequence of C^2 functions such that

$$\phi_i'(x) > 1 \quad \text{for} \quad x \in [a_{i-1}, a_i] , \quad i = 1,\ldots,n . \tag{3}$$

Assume that the interior of the intersection

$$\bigcap_{i=1}^{n} \phi_i([a_{i-1}, a_i]) \tag{4}$$

is not empty and define the transformation T of the unit interval $[0,1]$ into itself by the condition

$$T(x) = \phi_i(x) \quad \text{for} \quad x \in (a_{i-1}, a_i) , \quad i = 1,\ldots,n$$

(the values $T(a_i)$ are arbitrary). Then there exists on $[0,1]$ a T-stable measure m which is absolutely continuous with respect to the Borel measure $d\mu = dx$.

This conjecture was formulated in terms of piecewise C^2 functions. Note that according to the results of W. Parry [3], in many cases, the problem may be reduced to piecewise linear functions.

Now let us mention shortly for what reason the existence of a T-stable measure m for a given transformation $T: X \to X$ is so important. First of all because m is invariant and ergodic. Thus according to the Birkhoff theorem for each integrable f we have

$$\lim_{n\to\infty} \frac{1}{n} \sum_{k=0}^{n-1} f(T^k(x)) = \int_X f \, dm \tag{5}$$

almost everywhere with respect to m. Consequently in order to find the time mean of f corresponding to the orbit

$$x, T(x), T^2(x), \ldots \tag{6}$$

it is enough to compute the space mean

$$\int_X f \, dm .$$

Moreover the T-stable measure m is unique and it can be found by the method of successive approximations as the limit of the sequence $\{m_0 T^{-k}\}$.

But the most important property of the T-stable measure m is related with the following probabilistic interpretation. Let m_0 denote the probability distribution of the initial value $x = x_0$ of the trajectory (6). As a matter of fact in any real physical situation the exact value of x_0 is not known; we do know its distribution m_0. The measure $m_1 = m_0 T^{-1}$ is the probability distribution of the value $x_1 = T(x_0)$ and, in general, $m_k = m_0 T^{-k}$ is the distribution of $x_k = T^k(x_0)$. Thus the T-stability of m means that the probability distribution of $\{x_k\}$ converges to m independently on the initial distribution m_0.

4. Relaxation Oscillations

Now we are going to describe in detail the dynamical system given by (1) and (2).

Suppose that the function $p: R \to R$ is continuous and satisfies

$$p(x) > 0 , \qquad p(-x) = p(x) , \qquad p(x+1) = p(x) \qquad \text{for } x \in R \tag{7}$$

$$p_1''(x) < 0 \qquad \text{for } 0 \leqslant x \leqslant 1 \tag{8}$$

where p_1 denotes the restriction of p to the closed unit interval $[0,1]$.

Suppose that the function $f: R \to R$ is C^2 and such that

$$f(y) > 0 \quad \text{for} \quad y > 0 \tag{9}$$

$$0 \leqslant f'(y) \leqslant \frac{\pi^2}{1 + y^3} \quad \text{for} \quad y > 0 \tag{10}$$

$$p_1''(x) + f(p_1(x)) = 0 \quad \text{for} \quad 0 \leqslant x \leqslant 1 \, . \tag{11}$$

Let x_0 be an arbitrary point. Given x_n consider the solution $y(x)$ of the differential equation

$$\lambda y''(x) + f(y(x)) = 0 \, , \qquad x \geqslant x_n$$

satisfying the initial condition

$$y(x_n) = p'(x_n + 0) \, .$$

Conditions (9), (10) and (11) imply that for any $\lambda > 1$ the solution is uniquely determined in a right hand side neighborhood of the point x_n, and that there exists exactly one point x_{n+1} such that

$$y(x) > p(x) \qquad \text{for} \quad x_n < x < x_{n+1}$$

and $y(x_{n+1}) = p(x_{n+1})$. It can be proved that

$$\lim_{n\to\infty} x_n = \infty \, .$$

Therefore for any x_0 and $\lambda > 1$ the trajectory $y(x)$ of the system (1)-(2) is uniquely determined for all $x \geqslant x_0$.

Denote by $C(R)$ the space of all continuous functions $y: R \to R$ with the supremum norm topology.

<u>THEOREM 1</u>. For every $\lambda > 1$ there exists exactly one periodic (with period 1) trajectory $y_\lambda(x)$ of (1)-(2). Moreover the mapping

$$(1,\infty) \ni \lambda \to y_\lambda \in C(R)$$

is continuous.

Of course, according to our description any trajectory of (1)-(2) is defined only to the right from its start point x_0. But without any loss of generality we may assume that periodic trajectories are defined on the whole real line R.

Proof (outlined). Given $\bar{x} \in [0,1)$ consider the trajectory $\bar{y}(x)$ which starts from $\bar{x}$. Denote by $\bar{\bar{x}}$ the first point such that $\bar{y}(\bar{\bar{x}}) = p(\bar{\bar{x}})$. We shall write $\bar{\bar{x}} = S_\lambda(\bar{x})$. It is easy to prove that

$$S_\lambda(\tfrac{1}{2}) < \frac{3}{2} \quad \text{and} \quad S_\lambda(x) > 1 \qquad \text{for} \quad x \in [0,1) .$$

From this it follows the existence of a point $x_\lambda \in (0,1)$ such that $S_\lambda(x_\lambda) = 1 + x_\lambda$. The point x_λ may be chosen as a start point of the periodic trajectory y_λ. The uniqueness of y_λ as well as the continuous dependence on λ is closely related with the fact that in the set

$$\{(x,\lambda) \; : \; 0 \leq x < 1, \;\; \lambda > 1, \;\; S_\lambda(x) < 2\}$$

the function $(x,\lambda) \to S_\lambda(x)$ is continuous, and strictly decreasing in x.

THEOREM 2. There exists a number $\lambda_1 > 1$ such that for each $1 < \lambda < \lambda_1$ the periodic trajectory y_λ is stable; that is

$$\lim_{x\to\infty} |y_\lambda(x) - y(x)| = 0$$

uniformly for all trajectories y starting from the interval $[0,1)$.

Proof (outlined). In accordance to (11) we may admit that for $\lambda = 1$ there exists exactly one trajectory, namely $y(x) = p(x)$. For this trajectory it is natural to admit $S_\lambda(x) \equiv 1$. Thus

$$\frac{dS_1(x)}{dx} = 0 \qquad \text{for} \quad x \in [0,1) .$$

From this, by the continuous dependence argument, we obtain

$$\left| \frac{dS_\lambda(x)}{dx} \right| < 1$$

for $x \in [0,1)$ and sufficiently small $\lambda - 1$. The last inequality is a sufficient condition for the stability of y_λ.

A close look at the proof of Theorems 1 and 2 teaches that in description of trajectories of (1)-(2) the important role plays only the function S_λ. As a matter of fact it is enough to consider the function

$$T_\lambda(x) = \langle S_\lambda^2(x)\rangle , \quad S_\lambda^2(x) = S_\lambda(S_\lambda(x))$$

where $\langle y\rangle$ denotes the fractional part of y. It can be proved (the proof is not really difficult but rather long and tedious) that the transformation T_λ satisfies for sufficiently large λ the hypotheses of our version of the Ulam conjecture. From this and Theorems 1 and 2 we may derive (if the conjecture is true) that the dynamical system

$$([0,1), \langle S_\lambda^2\rangle, \mu) , \qquad d\mu = dx$$

is turbulent.

5. Application

The phenomenon of turbulence occurs in many important technical systems which can be described by the one-dimensional oscillator (1) and the reflection law (2).

We may interpret the system (1)-(2) in the following way. A particle moves on the (x,y)-plane in the domain $y > p(x)$. In the direction of x-axis the speed v of the particle is constant. In the direction of y-axis the motion is governed by the equation

$$m \frac{d^2y}{dt^2} + f(y) = 0$$

where m is the mass of the particle and $-f$ is the force acting along the y-axis. Setting $x = vt$ we obtain

$$\lambda \frac{d^2y}{dx^2} + f(y) = 0 , \qquad \lambda = mv^2 ;$$

that is exactly equation (1). When the particle reaches the boundary $y = p(x)$ the total kinetic energy disappears. In consequence of the conditions $mv^2 > 1$ and

$$\frac{d^2p}{dx^2} + f(p(x)) = 0$$

the inertial force ejects the particle in the direction tangent to the boundary; that is

$$y'(x + 0) = p'(x + 0) \quad \text{for} \quad y = p(x) \ .$$

Such a physical situation occurs in the process of solid friction. Note that even for elastic bodies, like steel, the small (in the amplitude) collisions of rubbing surfaces are non-elastic [1]. The system (1)-(2) gives an adequate description of the motion when the wavy finish is periodic.

Another interesting example of system (1)-(2) occurs in the mathematical description of rotary drilling. During many years the dependence of the drilling efficiency on the rotating speed was theoretically unexplained. A reasonable agreement with experimental data was reached by a theory based on Theorems 1 and 2 (with the exact estimation of λ_1) [2].

REFERENCES

[1] F.P. BOWDEN and D. TABOR, Friction and Lubrication of Solids. 2nd Ed., Clarendon Press, Oxford, 1954.

[2] A. LASOTA and P. RUSEK, Problems of the stability of the motion in the process of rotary drilling with cogged bits, Archiwum Górnictwa, 15 (1970), 205-216 (Polish with Russian and German summary).

[3] W. PARRY, Symbolic dynamics and transformations of the unit interval, Trans. Amer. Math. Soc. 122 (1966), 368-378.

[4] A. RÉNYI, Representation for real numbers and their ergodic properties, Acta Math. Akad. Sci. Hungar. 8 (1957), 477-493.

[5] V.A. ROCHLIN, Exact endomorphisms of Lebesgue spaces, Izv. Akad. Nauk SSSR, 25 (1961), 499-530.

[6] D. RUELLE and F. TAKENS, On the nature of turbulence, IHES (1970).

[7] S.M. ULAM, A Collection of Mathematical Problems. Interscience Publishers, Inc., New York, 1960.

Jagellonian University, Krakow, Poland

GEOMETRIC DIFFERENTIAL EQUATIONS*

Solomon Lefschetz

My present topic refers to a scarcely recognized part of the vast topic of d.e. (= differential equations). The very title, however, presents an anomaly. While "d.e." is a clearly marked part of analysis, "geometry" offers considerable vagueness. However, while no question pertaining to d.e. can avoid a strong utilization of analysis, the "geometrical bent" will not always be as apparent.

Leaving aside the appropriateness of the title, my purpose is to present, however briefly, the profound contributions of a few outstanding authors, the first being Poincaré, the true founder of my topic.

Henri Poincaré. This work of foundation, dating almost a century, was accomplished by Poincaré in his great classic, Sur les courbes définies par une équation différentielle (Oervres, Vol. 2). For the first time the totality of the system of trajectories of a d.e. was studied. The particular first system selected was planar:

$$\dot{x} = f(x,y) \ , \qquad \dot{y} = g(x,y) \tag{1}$$

*Research reported on here was supported in part by the Air Force Office of Scientific Research - AFOSR 693-67B, United States Army Research Office - DA-31-124-ARO-D-270, and Office of Naval Research - NONR N0014-67-A-0191-0009.

where f and g are real polynomials. Poincaré pointed out the importance of critical points where both f and g vanish and realized, correctly enough, that as a first step he should avoid all highly special cases. Therefore, he limited the study to the elementary type. At such a point A, taken as the origin, (1) assumes the form

$$\dot{x} = ax + by + \cdots, \quad \dot{y} = cs + dy + \cdots \tag{2}$$

($\cdots$ terms of higher degree). Here a, b, c, d are real constants with $\delta = ad - bc \neq 0$ and characteristic roots defined by

$$\begin{vmatrix} a-r & b \\ c & d-r \end{vmatrix} = r^2 - (a+d)r + \delta = 0 .$$

Everything revolves around the values of the roots λ,μ. The characterization is this:

λ,μ real: node; both negative, stable; both positive, unstable.

λ,μ real and of opposite signs: saddle point.

λ,μ complex, $\lambda = \lambda' + i\lambda''$, $\mu = \lambda' - i\lambda''$, $\lambda' \neq 0$: focus $\lambda' < 0$, stable $\lambda' > 0$, unstable.

$\lambda' = 0$: center.

The center turns out to be the most troublesome. Basically, it begins with a succession of ovals and then turns into the preceding type, but no method has yet been found to say when this occurs.

Index. Let V be a vector system valid in the whole plane, for instance (f,g). Let A be an isolated critical point. In a small neighborhood of A the vector of V is only zero at A. Let γ be a small positive circle centered at A, and such that $V \neq 0$ on γ. Let P be a point of γ, V(P) its vector, θ its positive angle with some fixed direction. As P describes γ once, $\text{Var}\ \theta = m \cdot 2\pi$ and $m = \text{Ind}\ A$. Its basic properties are:

(a) Ind A is independent of γ (if small enough).

(b) Ind A may also be defined topologically as follows.

Take a large $\Gamma \supset A$. Let $V(P)$ cut Γ in P'. Then $P \to P'$ defines a map $\gamma \to \Gamma$ under which Γ is covered algebraically m times and $m = \text{Ind } A$.

(Via a classic theorem of L. E. J. Brouwer, this offers the natural extension of the index to higher dimensions.)

(c) When the Jordan curve J surrounds c.ps. (= critical points) A_h, $1 \leq h \leq n$, then

$$\text{Ind } J = \Sigma \text{ Ind } A_h .$$

(Note: Ind J defined like Ind A.) Hence, if J surrounds no c.p., Ind $A = 0$.

(d) Indices of elementary c.p. of $V(f,g)$ of (1) is $+1$, except for a saddle point when it is -1.

Some additional results of Poincaré.

A. Treatment of periodic systems (closed trajectories) - stability included.

B. Let Ω be a c.p. free plane region with piecewise analytic boundary Φ. If the trajectories along Φ all point into Ω or all point outside Ω then Ω contains a periodic solution (closed trajectory).

C. Let C be a circumference with topological transformation T in the positive direction. Let P be a point of C and let $P_1 = TP$, $P_2 = T^2P$, etc. Then there exists a σ such that if θ is an angular variable on C, and $\theta^n = T^nP$ then θ^n is such that

$$n\sigma - 2\pi \leq \theta^n \leq n\sigma + 2\pi .$$

Thus σ measures the mean rotation of T under indefinite affect of T.

$\frac{\sigma}{2\pi}$ rational $\Rightarrow$ there exist periodic points for every n.

$\frac{\sigma}{2\pi}$ irrational $\Rightarrow$ every point of C taken arbitrarily near every other under repeated iteration.

This has been applied to the complete description of the nature of solutions of a d.e. without c.p. on the torus (this under very general conditions).

D. Treatment of solutions of d.e. on a projective plane. Application to the behavior of polynomial systems at ∞.

E. On a smooth orientable surface Φ one may define d.e. and related V with its indices. Then Σ Ind $A_h = \chi(\Phi)$ = Euler-Poincaré characteristic.

F. Extension to 3-dimensions (somewhat incomplete).

G. Method of sections. This method was later extensively exploited by Birkhoff. Let γ be a closed analytic trajectory of a d.e. system in 3-space. At some point P of γ let Φ be a portion of an analytic surface not tangent to γ at P. Then the trajectories δ of the system very near γ intersect Φ in a first point Q very near P. Let Q_1 be the second intersection of δ with Φ very near P. The nature of the transformation $T: Q \to Q_1$ may and often does serve to tell a great deal about the neighborhood of γ. In particular, the fixed points of T, likewise those of the recurrent $T^1, T^2, \ldots,$ disclose the presence of new periodic trajectories of the system near γ. This is the famous method of sections.

H. Bifurcation theory for space analytic system which about a point split into several distinct systems (important in applications).

I. Poincaré's books on celestial mechanics (far ahead of their time) are replete with new theories and applications of d.e.

J. The Poincaré-Birkhoff theorem. Cosmology led Poincaré to surmise the following proposition: In the plane R, let Σ be the annular ring bounded by two circumferences C_1 and C_2, C_1 interior to C_2. Suppose that Σ has a topological area preserving transformation T (all important in dynamics) under which C_1 and C_2 rotate in opposite directions. Then T has generally two fixed points in Σ (the two may coincide).

A year or so before his death (aged 57), Poincaré published a long memoire describing his unfruitful endeavors to prove the full theorem, in the hope that a younger man might have more luck. The younger man G. D. Birkhoff published a remarkably short and ingenious proof of the theorem, about a year after Poincaré's death... .

General observation. It was characteristic of Poincaré that he constantly illustrated his results by concrete examples. In particular, he showed how his general 2-dimension theory could serve to provide a complete description of the full system of trajectories for comparatively simple systems, and in applications of various kinds his results turned out to be utterly useful.

G. D. Birkhoff (1884-1944). His work in our subject and related questions may be fairly described as a continuation of Poincaré's. However, in it the different directions pursued, analytical and geometric, are so mixed that the work of extricating the part belonging properly to my present topic is too difficult for me to separate. I shall, therefore, refer to the very competent article written by Marston Morse on Birkhoff's complete research (see beginning of Vol. I of Birkhoff's complete works). I shall limit my discourse to the proof of the Poincaré-Birkhoff theorem, its extension, and application and a few striking contributions on details made by Birkhoff.

A. Proof of the Poincaré-Birkhoff theorem. Birkhoff utilized in a fundamental way the property of invariance of area. If T is the initial transformation he introduced a new transformation T_ε: a radial ε shrinking of the ring (ε small) towards the center. This enabled him to define an invariant arc λ of $T_\varepsilon T$. Then considering the powers of $T_\varepsilon T$ and variation along λ of a vector joining a point to its predecessor, followed by a return along $-\lambda$, he showed that no fixed point implied a contradiction, from which the theorem follows.

B. Application to the billiard ball problem. The ball is assumed to roll on a plane table bounded by a convex curve. As it hits the boundary, the ball is reflected through an angle of "reflection" equal to the angle of incidence. Birkhoff shows (complete works, II, p. 333) that upon looking to the closed polygon paths one succeeds in proving that the associated dynamical systems has an infinity of periodic motions, some stable, others unstable.

In the course of the discussion Birkhoff introduced the highly interesting notion of minimax (intermediary between a maximum and a minimum).

Regarding the collection of periodic motions (discovered through the Poincaré surface of sections scheme) see also the very interesting and extensive paper of Birkhoff: Complete works, II, p. 111.

C. Extension of the Poincaré-Birkhoff theorem. Let r,θ be plane polar coordinates and let C be the circle $r = a > 0$. Let R be a ring bounded by C and a curve $\Gamma \supset C$. Let R_1, Γ_1 be a second similar system and let T be a topological map $R \to R_1$.

Theorem. If Γ and Γ_1 are met only once by any line θ = const., and T carries points of Γ and Γ_1 in opposite directions, then either (a) there are two distinct invariant points of R and R_1 under T or else (b) there is an annulus R_2 of R or R_1 (abutting on C) carried into part of itself by T (or T^{-1}).

D. Extension of the method of sections for dimension two. While Poincaré limited the method to a small open neighborhood Ω of a periodic solution Birkhoff extended the method to Ω's with boundary. This consisted of several disjoint periodic solutions of the basic system, or for that matter of the extension of the method to a surface of any genus.

E. This refers to a very interesting extension of the rotation character of a curve Γ in this direction: Let Γ be on an orientable surface Φ and thus let it have two distinct sides Γ_1 and Γ_2. Then Birkhoff showed that each of these two sides could have distinct rotation constants.

A. M. Liapunov. His name is indelibly attached to the concept of stability. His classical memoire, mainly known outside Russia by later French translation, Problème général de la stabilité du movement (Russian edition 1892, French translation, Annales de Toulouse, reproduced as Annals of Mathematics Studies), treats stability in a completely fundamental way. Much on the concept was known - and taken for granted - before Liapunov, but he not only organized the subject but also made many profound and new contributions to it. Evidently, Lagrange, Poincaré and a number of others could not and did not dispense with a knowledge of stability. For instance while the motion of the planets is "apparently" stable - in some practical sense say 10^4 years - how long did it last? (Who knows?) Besides, for example, the 3-bodies problem evoked seemingly different stability... .

Well at any rate since Liapunov the answer is unique. We recall a couple of the basic statements.

Let Σ be a dynamical system in n-space $x = (x_1, \ldots, x_n)$ and suppose that $x = 0$ is a solution. Then 0 is

(i) stable whenever given any open set $U \supset 0$ and time t_0, there exists another $V(U, t_0) \subset U$ such that if a motion starts in V at time t_0 it will remain in U for all $t > t_0$ (uniformly stable when V depends only on U); unstable whenever given U, t_0 as before, and whatever $V \subset U$, the motion at some $T > t_0$ reaches the boundary of U.

(ii) asymptotically stable wherever with the same data the motion $\to 0$.

(iii) conditionally stable wherever stability holds only for some subset $W \subset U$.

There are obvious applications to elementary critical points. The center is the only stable one not a. s.

Note that Liapunov (à la 1892) limited his study to analytical systems, holomorphic at the origin - an unimportant restriction.

Hope: Would someone take his memoire and shorten it appreciably, via suitable vector notations!

Major result. It is in a sense a generalization of the Lagrange stability theorem by extremal values of the potential. However, Liapunov's result does not require knowledge of the solutions. I will not state the actual theorem but rather a geometric interpretation for dimension two. Let $W(x_1, x_2)$ be positive definite in a region $\Omega(0)$, that is within a cylinder $x_1^2 + x_2^2 = r^2$. Let y be a third coordinate. The surface F: $y = W$ is a cup over Ω. The curves $W = \text{const.}$ are the projections of the horizontal sections of F. The paths Γ in $\Omega(0)$ are imaged into paths Δ on F. The projection $F \to \Omega(0)$ is topological.

Stability: $\dot{W} \leq 0$ implies that on F the path stays on the lower part of the cup, goes down sluggishly and need not reach 0.

A. st. $\dot{W} \leq -a < 0$. The path tends rapidly to the origin.

Instability. Along some Δ: $\dot{W} > \beta > 0$ - the path goes away from 0 and reaches the boundary of $\Omega(0)$.

The extension is easy.

For rapid applications see my book: D.E.: Geometric Theory, 2nd edition, p. 117.

Andronov and Pontryagin. These two authors have introduced in the late thirties a very novel and interesting concept: Structural Stability. Their announcing note in the Doklady gave no proofs. These were first provided by De Baggis and later considerably improved and expanded by Peixoto.

Briefly the idea is this. Given say an open bounded plane 2-cell region Ω with boundary B, which for a given d.e. has a finite set of elementary critical points and separatrices (lines emanating from saddle points). Suppose that (in some sense) the system undergoes an ε-deformation. Under what conditions does the phase - portrait of the system remain unchanged. This is known as structural stability. General conditions are: the system has only a finite number of elementary critical points, none a center; it has at most a finite number of closed paths in Ω, all limit - cycles, always stable or unstable on both sides; no separatrix joins two distinct saddle points. These are also sufficient conditions and (proved by Peixoto) they are also independent of ε.

It is understood throughout that B is analytical and crossed, not tangentially in the same direction by the solutions of the d.e.

Princeton University, Princeton, New Jersey, and
Brown University, Providence, Rhode Island

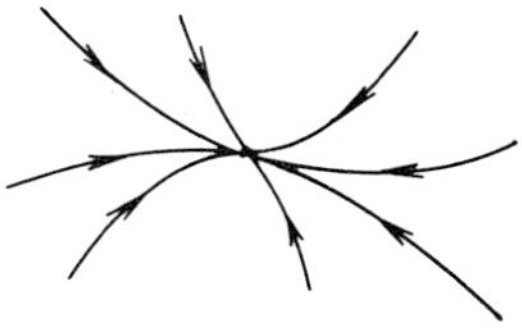

stable node
unstable; arrows
reversed

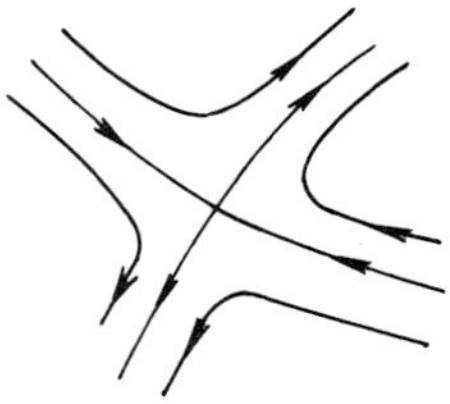

saddle point
1 stable } trajectory
1 unstable }
all the rest unstable

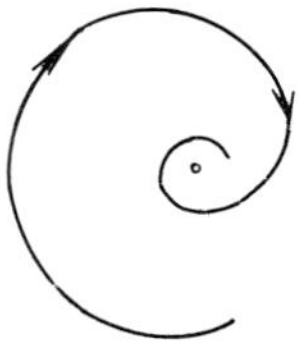

stable focus
unstable; arrows
reversed

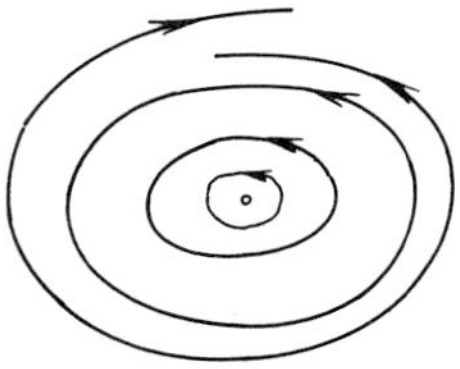

center stable
unstable; arrows
reversed

TAUBERIAN THEOREMS AND FUNCTIONAL EQUATIONS*

J. J. Levin and D. F. Shea

In this paper some of our results appearing in [11] are partially summarized. A number of extensions and refinements as well as a more abstract partial summary of [11] may be found in [12].

We study the asymptotic behavior as $t \to \infty$ of the bounded solutions of each of the equations

$$\text{(a)} \qquad x'(t) + \int_{-\infty}^{\infty} g(x(t-\xi))dA(\xi) = f(t) \qquad \left(' = \frac{d}{dt}, \quad -\infty < t < \infty\right)$$

$$\text{(b)} \qquad x(t) + \int_{-\infty}^{\infty} g(x(t-\xi))dA(\xi) = f(t) \qquad (-\infty < t < \infty)$$

$$\text{(c)} \qquad \int_{-\infty}^{\infty} x(t-\xi)dA(\xi) = f(t) \qquad (-\infty < t < \infty) ,$$

as well as of more general functional equations. Here g, A and f are prescribed functions and x is the unknown. It is assumed that

$$(1) \qquad \begin{cases} g \in C(\mathbf{C}), \quad A \in BV(-\infty,\infty) \quad (A(-\infty) = 0) , \\ f \in L^{\infty}(-\infty,\infty), \quad \lim_{t\to\infty} f(t) = f(\infty) \ \underline{\text{exists}}. \end{cases}$$

---

*This research was supported in part by the U. S. Army Research Office, Durham; and NSF Grant #GP-5728.

Both the scalar and vector cases (in the latter g, f, and x are vectors and A is a matrix) of (a), (b), and (c) are considered. Conditions (1) can be enormously generalized, e.g., instead of $f(\infty)$ existing one may assume $f = f(t,x)$ and $f(t,x) \to f^*(\infty)$ (a constant) $(t \to \infty)$ for each fixed x, or $f(t) = f_1(t) + f_2(t)$ with $f_2(\infty)$ existing but $f_1(t)$ periodic or almost periodic. The assumptions on g and A may also be generalized (see [11]).

The Pitt's form of Wiener's tauberian theorem is concerned with (c); Theorem 3c below is a generalization of that result. Because of its classical importance and the motivation that it supplies for the present investigation, (c) is carried along as a separate equation even though it is a very special case of (b). (To see this, in (b) set $g(x) = x$ and $A = -H + B$, where H is the unit step function and $B \in BV(-\infty,\infty)$, in order to obtain an equation of type (c).)

The analysis of the asymptotic behavior of the bounded solutions of Volterra equations, such as

$$x'(t) + \int_0^t g(x(t-\xi))dA(\xi) = f(t) \qquad (0 \leq t < \infty) , \tag{2}$$

is one of the motivations of this investigation. In (2) it is assumed that

$$\begin{cases} g \in C(\mathbf{C}), \quad A \in BV[0,\infty) \quad (A(0) = 0) , \\ f \in L^\infty(0,\infty), \quad \lim_{t\to\infty} f(t) = f(\infty) \ \underline{\text{exists}}. \end{cases} \tag{3}$$

If x is a bounded solution of (2) on $[0,\infty)$ it is easy to show that it as well as A and f may be extended to $(-\infty,0)$ in such a way that the extended functions satisfy an equation of type (a), with x bounded and (1) satisfied. Thus the study of the asymptotic behavior of the bounded solutions of (2) is contained in that of (a). Similar remarks apply to the Volterra equation analogues of (b) and (c).

A special case of (a) is the nonlinear system of ordinary differential equations

$$x'(t) + g(x(t)) = f(t) . \tag{4}$$

This follows on setting $A = \text{diag}(H,\ldots,H)$, where H is the unit step function, in (a). A similar choice of A in (2) also leads to

(4), but on $[0,\infty)$ instead of on $(-\infty,\infty)$. Thus the investigation of the asymptotic behavior of the bounded solutions of systems of ordinary differential equations, both on $(-\infty,\infty)$ and $[0,\infty)$ is contained in that of (a). Taking, for example, $A(\xi) = \rho_1 H(\xi) + \rho_2 H(\xi - 1)$ in (a) leads to

$$x'(t) + \rho_1 g(x(t)) + \rho_2 g(x(t-1)) = f(t) \ , \tag{5}$$

a differential-delay equation.

A special case of (b) is a form of the renewal equation. Other special cases of (b) occur in different applications. Thus, a rich variety of important and much studied equations are special cases of (a), (b), and (c).

It is emphasized that the existence and boundedness of any particular solution of (a), (b), or (c) being studied by the methods of this paper is taken as an hypothesis. (Other solutions of the same equation are allowed to be unbounded.) The splitting off of the questions of asymptotic behavior from those of existence and boundedness is not uncommon in the theory of ordinary differential equations. It is a most natural division here, where strong connections are established between classical tauberian problems and, for example, asymptotic questions for ordinary differential equations.

For simplicity we mainly confine ourselves to scalar equations here; see [11] for extensions to vector equations. The following definition is basic.

<u>Definition</u>. If $\{t_m\}$ is an increasing sequence of real numbers which satisfy $\lim_{m\to\infty} (t_m - t_{m-1}) = \infty$, then by a "$\psi$-sequence" associated with $\{t_m\}$ we mean a sequence of real valued functions $\{\psi_m(t)\}$ such that

$$(6) \quad \begin{cases} \psi_m \in C^\infty(-\infty,\infty), \quad \lim_{m\to\infty} \|\psi_m^{(j)}\|_\infty = 0 \quad (j = 1,2,\ldots), \quad \sum_{m=1}^{\infty} \psi_m(t) \equiv 1 \\ \psi_1(t) \equiv 1 \quad (t \leq t_1), \quad \psi_1(t) \equiv 0 \quad (t_2 \leq t), \quad \psi_1'(t) \leq 0 \quad (t_1 \leq t \leq t_2) \\ \psi_m(t) \equiv 0 \quad (t \leq t_{m-1} \text{ and } t \geq t_{m+1}), \quad \psi_m(t_m) = 1 \\ \psi_m'(t) \geq 0 \quad (t_{m-1} \leq t \leq t_m), \quad \psi_m'(t) \leq 0 \quad (t_m \leq t \leq t_{m+1}) \ . \end{cases}$$

Thus a ψ-sequence is a partition of unity in which the successive functions, the ψ_m, are varying slower and slower as $m \to \infty$.

The first result concerns equation (a), which we recall contains as very special cases the ordinary differential equation (4) and the differential-delay equation (5) (and corresponding systems in the vector formulation). The equation

$$\text{(a*)} \qquad y'(t) + \int_{-\infty}^{\infty} g(y(t-\xi))dA(\xi) = f(\infty) \qquad (-\infty < t < \infty)$$

is called the limit equation associated with (a) and plays a crucial role. (LAC below denotes absolute continuity on every compact interval and C_u^1 denotes a uniformly continuous derivative.)

<u>THEOREM 1a</u>. Let (1) hold and let $x(t) \in LAC \cap L^\infty(-\infty,\infty)$ satisfy (a) a.e. on $(-\infty,\infty)$. Define

$$\Gamma_a = \{y(t) \mid y \in C_u^1(-\infty,\infty),\ \|y\|_\infty \leq \|x\|_\infty,\ y \text{ satisfies (a*) on } (-\infty,\infty)\}.$$

Then there exist $y_m(t) \in \Gamma_a$ $(m = 1,2,\ldots)$ and sequences $\{t_m\}$, $\{\varepsilon_m\}$ of real numbers with $t_1 < t_2 < \cdots$, and $t_m - t_{m-1} \to \infty$, $\varepsilon_m \downarrow 0$ $(m \to \infty)$ such that

$$(7) \qquad \sup_{t_{m-2} \leq t \leq t_{m+2}} |x(t) - y_m(t)| < \varepsilon_m \qquad (m = 3,4,\ldots)$$

$$(8) \qquad \operatorname*{ess\,sup}_{t_{m-1} \leq t \leq t_{m+1}} |x'(t) - y_m'(t)| < \varepsilon_m \qquad (m = 3,4,\ldots) .$$

Moreover, there exists a ψ-sequence, $\{\psi_m\}$, associated with $\{t_m\}$ such that

$$(9) \qquad x(t) = \sum_{m=1}^{\infty} \psi_m(t) y_m(t) + \eta(t) \qquad (-\infty < t < \infty) ,$$

where

$$(10) \qquad \lim_{t\to\infty} \eta(t) = 0$$

$$(11) \qquad \lim_{t\to\infty} \left[\operatorname*{ess\,sup}_{t \leq \tau < \infty} |\eta'(\tau)| \right] = 0 .$$

Notice that $\Gamma_a \neq \emptyset$ is a conclusion and not an hypothesis of the theorem. Roughly speaking the result shows how a bounded solution of

(a) drifts, slower and slower as $t \to \infty$, among certain solutions of the limit equation (a*).

Theorem 1a is best possible in the sense of Theorem 2a below. It is related to but more precise than results usually derived from the analysis of the positive limit set of the solution of a functional equation. For ordinary differential equations the latter approach is well known; for applications of that approach to various special functional equations see, e.g., Hale [3] and Miller [14], [15]. (Note from (7) that (12) below is satisfied by the $y_m(t)$ of Theorem 1a.)

THEOREM 2a. Let g and A be as in (1) and let $\{t_m\}$ satisfy $\lim_{m\to\infty} (t_m - t_{m-1}) = \infty$. Let $\{\psi_m(t)\}$ be a ψ-sequence associated with $\{t_m\}$ and let $\{y_m(t)\}$ be a sequence satisfying

$$y_m \in C_u^1(-\infty,\infty), \qquad \|y_m\|_\infty \leq M < \infty$$

$$y_m'(t) + \int_{-\infty}^{\infty} g(y_m(t-\xi))dA(\xi) = \alpha \quad (-\infty < t < \infty) , \qquad (m = 1,2,\ldots)$$

where M and α are prescribed constants. Suppose further that

$$\lim_{m\to\infty} \left\{ \sup_{t_{m-1}\leq t \leq t_{m+2}} |y_m(t) - y_{m+1}(t)| \right\} = 0 . \tag{12}$$

Define

$$x(t) = \sum_{m=1}^{\infty} \psi_m(t)y_m(t) \qquad (-\infty < t < \infty) .$$

Then

$$\lim_{t\to\infty} \left\{ x'(t) + \int_{-\infty}^{\infty} g(x(t-\xi))dA(\xi) \right\} = \alpha .$$

Theorems 1a and 2a have analogues for equations (b) and (c). Only the analogues of Theorem 1a are stated here. The limit equations associated with (b) and (c) are respectively

$$y(t) + \int_{-\infty}^{\infty} g(y(t-\xi))dA(\xi) = f(\infty) \qquad (-\infty < t < \infty) \tag{b*}$$

$$\int_{-\infty}^{\infty} y(t-\xi)dA(\xi) = f(\infty) \qquad (-\infty < t < \infty) . \tag{c*}$$

($\mathcal{B}$ below denotes Borel measurability and C_u denotes uniform continuity.)

THEOREM 1b. Let (1) hold and let $x(t) \in \mathcal{B} \cap L^{\infty}(-\infty,\infty)$ satisfy (b) on $(-\infty,\infty)$ as well as the tauberian condition

$$\text{(T)} \qquad \lim_{t\to\infty,\tau\to 0} |x(t+\tau) - x(t)| = 0 .$$

Define

$$\Gamma_b = \{y(t) \,|\, y \in C_u(-\infty,\infty),\ \|y\|_\infty \leq \|x\|_\infty,\ y \text{ satisfies } (b^*) \text{ on } (-\infty,\infty)\}.$$

Then there exist $y_m(t) \in \Gamma_b$ $(m = 1,2,\ldots)$, sequences $\{t_m\}$, $\{\varepsilon_m\}$ as in Theorem 1a, and a ψ-sequence, $\{\psi_m(t)\}$, associated with $\{t_m\}$ which satisfy (7), (9), and (10).

THEOREM 1c. In Theorem 1b (b) may be replaced by (c) if also (b^*) and Γ_b are replaced by (c^*) and

$$\Gamma_c = \{y(t) \,|\, y \in C_u(-\infty,\infty),\ \|y\|_\infty \leq \|x\|_\infty,\ y \text{ satisfies } (c^*) \text{ on } (-\infty,\infty)\}$$

respectively.

Theorem 1c is, of course, a special case of Theorem 1b. As in Theorem 1a, one again has a bounded solution of a functional equation drifting among certain smooth solutions of the limit equation.

That (T) is a necessary condition for Theorem 1c is seen from the following observations. A computation shows that $x(t) = \sin(e^t)$ is a solution of (c) if A and f are given by

$$A(t) = \int_{-\infty}^{t} e^{-\xi} e^{-e^{-\xi}}\, d\xi, \qquad f(t) = (e^t + e^{-t})^{-1} .$$

Since the Fourier-Stieltjes transform

$$\text{(13)} \qquad \hat{A}(\lambda) = \int_{-\infty}^{\infty} e^{-i\lambda t} dA(t) \qquad (-\infty < \lambda < \infty)$$

of A is here given by $\hat{A}(\lambda) = \Gamma(1 + i\lambda) \neq 0$, it follows from known results of harmonic analysis (see [11]) that $\Gamma_c = \{0\}$. Thus, if Theorem 1c were applicable, (9) would imply $x(t) = \eta(t) \to 0$ $(t \to \infty)$, which is incompatible with $x(t) = \sin(e^t)$. Of course, $\sin(e^t)$ does not satisfy (T).

The proofs of Theorems 1a, 1b, 1c are too lengthy to be sketched here; however, a common key point is to establish that

(14) $$\lim_{s\to\infty}\left\{\inf_{y\in\Gamma}\left[\sup_{|t-s|\leq d}|x(t)-y(t)|\right]\right\} = 0 \quad \text{for each} \quad d > 0 ,$$

where $x(t)$ is the solution being studied and where Γ stands for Γ_a in the proof of Theorem 1a, etc. A little reflection shows that (14) is an intuitively reasonable statement; its importance lies in

<u>LEMMA 1</u>. Let $x(t)$, defined on $(-\infty,\infty)$, satisfy $\lim_{t\to\infty} \sup|x(t)| < \infty$ and let Γ be a collection of functions defined on $(-\infty,\infty)$ such that $\sup_{-\infty<t<\infty} |y(t)| < \infty$ for each $y \in \Gamma$. Let (14) hold.

Then there exist $y_m(t) \in \Gamma$ $(m = 1,2,\ldots)$, sequences $\{t_m\}$, $\{\varepsilon_m\}$ as in Theorem 1a, and a ψ-sequence, $\{\psi_m(t)\}$, associated with $\{t_m\}$ which satisfy (7), (9), and (10).

The tauberian condition (T) enters the proofs through

<u>LEMMA 2</u>. Let $x(t) \in L^\infty(-\infty,\infty)$ satisfy (T) and let $\{s_j\}$ be a sequence such that $\lim_{j\to\infty} s_j = \infty$. Then there exist a subsequence $\{s_{j_r}\}$ and a function $y(t)$ such that

$$\lim_{r\to\infty}\left\{\sup_{|t|\leq d} |x(t + s_{j_r}) - y(t)|\right\} = 0 \quad \text{for every} \quad d > 0$$

$$y(t) \in C_u(-\infty,\infty), \qquad \|y\|_\infty \leq \|x\|_\infty .$$

It is interesting to observe that if $x(t)$ satisfies the hypothesis of Theorem 1a, then it automatically satisfies (T). In the case of Theorem 1b it is easy to show that if A is also assumed to be absolutely continuous, then the hypothesis (T) follows as a consequence of the other assumptions. There are deeper conditions which guarantee the a priori satisfaction of (T).

The next set of results is concerned with the linear cases of (a) and (b), i.e., with

(ℓa) $$x'(t) + \int_{-\infty}^{\infty} x(t - \xi)dA(\xi) = f(t) \qquad (-\infty < t < \infty) ,$$

(ℓb) $$x(t) + \int_{-\infty}^{\infty} x(t - \xi)dA(\xi) = f(t) \qquad (-\infty < t < \infty) ,$$

and with (c). As noted above, (ℓb) and (c) are equivalent equations.

Note also (as is evident from Theorem 1b) that while Theorem 1c is concerned with (c) it makes no use of the linearity of that equation.

Define

$$S_a(A) = \{\lambda | \hat{A}(\lambda) = -i\lambda, \quad -\infty < \lambda < \infty\}$$

$$S_b(A) = \{\lambda | \hat{A}(\lambda) = -1, \quad -\infty < \lambda < \infty\}$$

$$S_c(A) = \{\lambda | \hat{A}(\lambda) = 0, \quad -\infty < \lambda < \infty\} ,$$

where $\hat{A}(\lambda)$ is given by (13). (In the vector cases of (ℓa), (ℓb), and (c) the corresponding spectral sets are defined by

$$S_a(A) = \{\lambda | \det[i\lambda E + \hat{A}(\lambda)] = 0, \quad -\infty < \lambda < \infty\}$$

$$S_b(A) = \{\lambda | \det[E + \hat{A}(\lambda)] = 0, \quad -\infty < \lambda < \infty\}$$

$$S_c(A) = \{\lambda | \det \hat{A}(\lambda) = 0, \quad -\infty < \lambda < \infty\} ,$$

where E is the identity matrix.)

<u>THEOREM 3a</u>. Let A and f be as in (1) and let $x(t) \in LAC \cap L^\infty(-\infty,\infty)$ satisfy (ℓa) a.e. on $(-\infty,\infty)$.

(i) If $S_a(A) = \emptyset$, then

$$x(t) = \frac{f(\infty)}{A(\infty)} + \eta(t) \qquad (-\infty < t < \infty) , \tag{15}$$

where $\eta(t)$ satisfies (10) and (11).

(ii) If $S_a(A) = \{\lambda_1,\ldots,\lambda_n\}$ and $\lambda_k \neq 0$ $(1 \leq k \leq n)$, then

$$x(t) = \frac{f(\infty)}{A(\infty)} + \sum_{k=1}^{n} c_k(t)e^{i\lambda_k t} + \eta(t) \qquad (-\infty < t < \infty) , \tag{16}$$

where $\eta(t)$ satisfies (10) and (11), and

$$c_k(t) \in C^\infty \cap L^\infty(-\infty,\infty) \qquad (1 \leq k \leq n) \tag{17}$$

$$\lim_{t\to\infty} c_k^{(j)}(t) = 0 \qquad (1 \leq k \leq n; \; j = 1,2,\ldots) \tag{18}$$

(iii) If $S_a(A) = \{\lambda_1,\ldots,\lambda_n\}$ where some $\lambda_k = 0$, then $f(\infty) = 0$ and

$$x(t) = \sum_{k=1}^{n} c_k(t)e^{i\lambda_k t} + \eta(t) \qquad (-\infty < t < \infty) , \tag{19}$$

where $\eta(t)$ and the $c_k(t)$ are as in (ii).

THEOREM 3b. Let A and f be as in (1) and let $x(t) \in \mathcal{B} \cap L^{\infty}(-\infty,\infty)$ satisfy (ℓb) on $(-\infty,\infty)$ as well as (T).

Then (i), (ii), and (iii) of Theorem 3a hold provided that

(α) $S_a(A)$ is replaced by $S_b(A)$,

(β) the term $A(\infty)$ in (15) and (16) is replaced by $1 + A(\infty)$,

(γ) the assertion that $\eta(t)$ satisfies (11) is not made.

THEOREM 3c. Let A and f be as in (1) and let $x(t) \in \mathcal{B} \cap L^{\infty}(-\infty,\infty)$ satisfy (c) on $(-\infty,\infty)$ as well as (T).

Then (i), (ii), and (iii) of Theorem 3a hold provided that

(α) $S_a(A)$ is replaced by $S_c(A)$,

(β) the assertion that $\eta(t)$ satisfies (11) is not made.

Pitt's form of Wiener's tauberian theorem is part (i) of Theorem 3c, i.e., the case in which $S_c(A) = \emptyset$ so that $\hat{A}(\lambda) \neq 0$ $(-\infty < \lambda < \infty)$; see Pitt [16]. Parts (ii) and (iii) generalize that well known result. In [11] we also extend Theorems 3a, 3b, 3c to allow for countable spectral sets $S_a(A)$, $S_b(A)$, $S_c(A)$. The proofs involve the application of Theorems 1a, 1b, 1c together with the concepts of narrow convergence, spectrum, and spectral synthesis formulated by Beurling [1], [2] in studying the harmonic analysis of bounded functions.

In all three theorems observe that (16) (or (19)) doesn't state that the solution, $x(t)$, is a trigonometric polynomial plus an error term which tends to zero as $t \to \infty$. Rather, (16) (or (19)) states that $x(t)$ "drifts" among a set of trigonometric polynomials whose frequencies are determined (by the $S(A)$ sets) but whose amplitudes, the $c_k(t)$, are changing slower and slower as $t \to \infty$. The function $\sin(\log(1 + t^2))$ exhibits the behavior of the $c_k(t)$.

These theorems are best possible; we illustrate this in the case of Theorem 3c(ii). Let $\alpha \in \mathbf{C}$, $c_k(t)$ $(1 \leqslant k \leqslant n)$ satisfy (17) and (18) (for $j = 1$), and $\eta \in \mathcal{B} \cap L^{\infty}(-\infty,\infty)$ satisfy (10). Define $x(t)$ by

$$(20) \qquad x(t) = \frac{\alpha}{A(\infty)} + \sum_{k=1}^{n} c_k(t) e^{i\lambda_k t} + \eta(t) \qquad (-\infty < t < \infty),$$

so that $x(t) \in \mathcal{B} \cap L^{\infty}(-\infty,\infty)$ obviously satisfies (T). Define

$$(21) \qquad f(t) = \int_{-\infty}^{\infty} x(t-\xi)\,dA(\xi) \qquad (-\infty < t < \infty) ,$$

where $x(t)$ is given by (20). It is not difficult to show that the f of (21) satisfies $f \in L^{\infty}(-\infty,\infty)$ and $\lim_{t\to\infty} f(t) = \alpha$, which completes the proof of the assertion.

As an application of the system formulation of Theorem 3a, consider the system of ordinary differential equations

$$(22) \qquad x'(t) + x(t)B = f(t) \qquad (0 \leq t < \infty) ,$$

where x and f are row vectors, $x = (x_1,\ldots,x_N)$, $y = (y_1,\ldots,y_N)$, and B is an N by N constant matrix. Suppose that $f \in L^{\infty}(0,\infty)$ and that $\lim_{t\to\infty} f(t) = 0$. Hence the limit system for (22) is

$$(22^*) \qquad y'(t) + y(t)B = 0 \qquad (-\infty < t < \infty) .$$

In this case $S_a(A)$ becomes

$$S(B) = \{\lambda \mid \det[i\lambda E + B] = 0, \quad -\infty < \lambda < \infty\} .$$

Suppose $S(B) = \{\lambda_1,\ldots,\lambda_n\}$ $(n \leq N)$. Let $\{\mu_1^{(k)},\ldots,\mu_{r_k}^{(k)}\}$ be a basis for the solutions of

$$\mu[i\lambda_k E + B] = 0 \qquad (1 \leq k \leq n) .$$

Define

$$y_s^{(k)}(t) = \mu_s^{(k)} e^{i\lambda_k t} \qquad (1 \leq k \leq n, \quad 1 \leq s \leq r_k) ,$$

so that the $y_s^{(k)}(t)$ span the space of solutions of (22^*) that are bounded on $(-\infty,\infty)$. Then from Theorem 3a it follows that

$$x(t) = \sum_{k=1}^{n}\left(\sum_{s=1}^{r_k} c_s^{(k)}(t)y_s^{(k)}(t)\right) + \eta(t) \qquad (-\infty < t < \infty) ,$$

where the $c_s^{(k)}(t)$ are scalar functions satisfying (17) and (18) and $\eta(t)$ is a vector function satisfying (10) and (11).

A natural and important problem is to find sufficient conditions on A and f which guarantee that the slowly varying functions, $c_k(t)$, of Theorems 3a, 3b, 3c are constants. One result in this direction is

THEOREM 4b. Let A and f be as in (1) and in addition let

$$S_b(A) = \{\lambda_1,\ldots,\lambda_n\}, \qquad \int_{-\infty}^{\infty} |t|^n |dA(t)| < \infty,$$

$$\int_{-\infty}^{\infty} e^{-i\lambda_k t} dA(t) \neq 0 \quad (1 \leqslant k \leqslant n), \qquad \int_0^{\infty} t^{n-1} |f(t) - f(\infty)| dt < \infty$$

hold. Let $x(t) \in \mathcal{B} \cap L^{\infty}(-\infty,\infty)$ satisfy (ℓb) on $(-\infty,\infty)$ as well as (T). Then

$$x(t) = \frac{f(\infty)}{1 + A(\infty)} + \sum_{k=1}^{n} \gamma_k e^{i\lambda_k t} + \eta(t) \qquad (-\infty < t < \infty), \tag{23}$$

where $\gamma_k \in \mathbf{C}$ $(1 \leqslant k \leqslant n)$ and $\eta(t)$ satisfies (10) (the term $f(\infty)/(1 + A(\infty))$ does not appear in (23) if one of the $\lambda_k = 0$).

The special case of Theorem 4b in which $n = 1$, $\lambda_1 = 0$, and certain other conditions are satisfied is a result of Karlin's [7] on the renewal equation. A similar result holds for (ℓa). In the Volterra equation analogues of (ℓa) and (ℓb), formulas for the constants γ_k are obtained which only involve the given functions A and f and not the solution $x(t)$.

Theorems 1a and 3a extend to systems with higher order derivatives, e.g., to systems of the form

$$x^{(r)}(t) + \sum_{j=0}^{r-1} \int_{-\infty}^{\infty} g_j(x^{(j)}(t - \xi)) dA_j(\xi) = f(t) \qquad (-\infty < t < \infty), \tag{24}$$

where $r \geqslant 2$ is an integer, $x = (x_1,\ldots,x_N)$, and each A_j is an N by N matrix. The linear case of (24) is, of course,

$$x^{(r)}(t) + \sum_{j=0}^{r-1} \int_{-\infty}^{\infty} x^{(j)}(t - \xi) dA_j(\xi) = f(t) \qquad (-\infty < t < \infty),$$

for which the appropriate spectral set is

$$S = \left\{\lambda \,\middle|\, \det\left[(i\lambda)^r E + \sum_{j=0}^{r-1} (i\lambda)^j \hat{A}_j(\lambda)\right] = 0, \quad -\infty < \lambda < \infty\right\}.$$

Some of our results concerning (24) extend classical ones of Volterra [17], Landau [8], Hardy and Littlewood [6].

In [11] and [12] various applications of Theorems 1a, 1b, 1c are made to specific nonlinear equations where, of course, g, A, and f are assumed to satisfy additional assumptions beyond those of (1). In each case the application involves studying the limit equation and, in particular, the set Γ_a, Γ_b, or Γ_c. Often, though not always, the object is to show that the Γ set contains a single element (which, because of the translation invariance of the limit equation, must be a constant). Since the limit equations are here assumed to be nonlinear, the methods of harmonic analysis aren't available and others are required, e.g., the method involving Liapounov functions.

One such application is the following result. It is related to earlier work of Levin [9], Levin and Nohel [10], Hale [3], Hannsgen [4], [5], Londen [13], and concerns the Volterra equation

$$x'(t) + \int_0^t g(x(t - \xi))a(\xi)d\xi = f(t) \qquad (0 \leq t < \infty) . \tag{25}$$

<u>THEOREM 5a</u>. Let

$$g \in C(-\infty,\infty), \quad xg(x) > 0 \quad (x \neq 0)$$

$$a \in C^{(2)}(0,\infty) \cap L^1(0,\infty), \qquad ta(t) \in L^1(0,\infty)$$

$$(-1)^k a^{(k)}(t) \geq 0 \quad (0 < t < \infty; \ k = 0,1,2), \qquad a \not\equiv 0$$

$$f \in L^\infty(0,\infty), \qquad \lim_{t\to\infty} f(t) = 0 ,$$

and let $x(t) \in LAC \cap L^\infty[0,\infty)$ satisfy (25) a.e. on $(0,\infty)$. Then

$$\lim_{t\to\infty} x(t) = 0 , \qquad \lim_{t\to\infty} \left[\operatorname*{ess\,sup}_{t\leq\tau<\infty} |x'(\tau)| \right] = 0 .$$

Applications are also made to various equations which are not of convolution type as well as to more general functional equations of the forms

$$x'(t) + Q(x)(t) = f(t) \quad \text{and} \quad Q(x)(t) = f(t) ,$$

where Q denotes an operator defined on L^∞. For these and other generalizations, some of which are mentioned above, see [11] and [12].

REFERENCES

[1] A. BEURLING, Un théorème sur les fonctions bornées at uniformément continues sur l'axe réel, Acta Math., 77 (1945), 127-136.
[2] _______, Sur une classe de fonctions presque-périodiques, C. R. Acad. Sci. (Paris), 225 (1947), 326-328.
[3] J.K.HALE, Sufficient conditions for stability and instability of autonomous functional-differential equations, J. Diff. Eqs., 1 (1965), 452-482.
[4] K.B. HANNSGEN, Indirect abelian theorems and a linear Volterra equation, Trans. A.M.S., 142 (1969), 539-555.
[5] _______, On a nonlinear Volterra equation, Mich. Math. J., 16 (1969), 365-376.
[6] G.H. HARDY and J.E. LITTLEWOOD, Contributions to the arithmetic theory of series, Proc. London Math. Soc. (2), 11 (1913), 411-478.
[7] S. KARLIN, On the renewal equation, Pacific J. Math., 5 (1955), 229-257.
[8] E. LANDAU, Über einen Satz von Herrn Esclangon, Math. Ann., 102 (1930), 177-188.
[9] J.J. LEVIN, The asymptotic behavior of the solution of a Volterra equation, Proc. A.M.S., 14 (1963), 534-541.
[10] J.J. LEVIN and J.A. NOHEL, Perturbations of a nonlinear Volterra equation, Mich. Math. J., 12 (1965), 431-447.
[11] J.J. LEVIN and D.F. SHEA, On the asymptotic behavior of the bounded solutions of some integral equations, I, II, III, J. Math. Anal. Appl., to appear.
[12] _______, Asymptotic behavior of the bounded solutions of some functional equations, to appear in a Symposium on Nonlinear Functional Analysis, Madison (1971).
[13] S.-O. LONDEN, The qualitative behavior of the solutions of a nonlinear Volterra equation, Mich. Math. J., to appear.
[14] R.K. MILLER, Asymptotic behavior of nonlinear delay-differential equations, J. Diff. Eqs., 1 (1965), 293-305.
[15] _______, Asymptotic behavior of solutions of nonlinear Volterra equations, Bull. A.M.S., 72 (1966), 153-156.
[16] H.R. PITT, Mercerian theorems, Proc. Cambridge Phil. Soc., 34 (1938), 510-520.
[17] V. VOLTERRA, Sur la théorie mathématique des phénomènes héréditaires, J. Math. Pures Appl., 7 (1928), 249-298.

University of Wisconsin, Madison, Wisconsin

PERTURBATIONS OF VOLTERRA EQUATIONS

John A. Nohel*

1. Introduction

We summarize a number of recent results concerning various asymptotic relationships between solutions of the perturbed system

$$x(t) = f(t) + \int_0^t a(t,s)[x(s) + g(s,x(s))]ds \tag{1}$$

and the unperturbed linear system

$$y(t) = f(t) + \int_0^t a(t,s)y(s)ds \tag{2}$$

on the interval $0 \leqslant t < \infty$. Here x, y, f, g are vector functions with n components with f, g given and a is a given n by n matrix. The perturbations g are to be thought of as being small. We shall also discuss the initial value problem on $0 \leqslant t < \infty$ for the integrodifferential system

$$\text{(1d)} \qquad \begin{cases} x'(t) = F(t) + A(t)x(t) + \int_0^t B(t,s)x(s)dx + G(t,x(t)) \\ x(0) = x_0 \end{cases}$$

in relation to the unperturbed linear system

---

*Supported by U.S. Army Research Office - Durham.

$$\text{(2d)} \qquad \begin{cases} y'(t) = F(t) + A(t)y + \int_0^t B(t,s)y(s)ds \\ y(0) = x_0 , \end{cases}$$

where F, G, x_0 are given vectors and A, B are given n by n matrices.

Systems (1) and (1d) arise in a variety of applications: control theory, nuclear reactor dynamics, superfluidity, heat transfer, visco-elasticity, biology, genetics and others. While (1d) can be reduced to (1) by an integration, doing so is undesirable for certain purposes.

The present results form an extension of the classical Perron, Poincaré, Weyl, Levinson stability theorems for perturbed systems of ordinary differential equations to (1) and (1d). We remark that perturbation problems in which the unperturbed system is a nonlinear Volterra integral equation also arise in applications. These cannot in general be treated by the methods discussed here. The interested reader is referred to [1], [2], [3], [4]. For the study of the asymptotic behavior of bounded solutions of general functional equations (which include Volterra equations), which may be regarded as perturbations of an entirely different sort (namely of certain limiting equations), we refer the reader to [5].*

2. Some "Variation of Constants" Formulae

With the kernel a(t,s) in (1) we associate the resolvent kernel r(t,s) which is defined to be the (unique) solution of the system (see [6], [7])

$$\text{(3)} \qquad r(t,s) = a(t,s) + \int_s^t a(t,u)r(u,s)du \qquad (0 \leq s \leq t < \infty) .$$

With the kernel A(t,s) in (1d), we associate the kernel R(t,s) which is defined to be the solution of the initial value problem (see also [8] where a different definition is taken)

---

*See also J. J. Levin, these proceedings.

$$(3d)\qquad \begin{cases} \frac{\partial R}{\partial t}(t,s) = A(t)R(t,s) + \int_s^t B(t,u)R(u,s)du \\ R(s,s) = I\,, \qquad t \geq s \geq 0\,, \end{cases}$$

where I is the identity matrix. If $s > t \geq 0$ we define $r(t,s) \equiv 0$, $R(t,s) \equiv 0$.

Let $LL^1(S)$ denote the set of all measurable functions on S such that the seminorms

$$\|f\|_\Sigma = \int_\Sigma |f(t)|dt$$

are finite for all Σ compact subsets Σ of S. We shall require throughout that at least $r(t,s) \in LL^1(R^+ \times R^+)$, where $R^+ = \{t\colon 0 \leq t < \infty\}$. It can be verified by direct substitution that under a variety of hypotheses (see remarks following Theorem 1)

$$(4)\qquad y(t) = f(t) + \int_0^t r(t,s)f(s)ds$$

is a solution of (2). As far as $R(t,s)$ is concerned, it is reasonable to assume (see remarks following Theorem 1d) that $R(t,s)$ is continuous; one also readily verifies that

$$(4d)\qquad Y(t) = R(t,0)x_0 + \int_0^t R(t,s)F(s)ds$$

is in a variety of senses a solution to the system (2d). Another straightforward calculation shows that the perturbed system (1) is equivalent to the system

$$(5)\qquad x(t) = y(t) + \int_0^t r(t,s)g(s,x(s))ds\,,$$

where $y(t)$ is given by (4), while the system (1d) is equivalent to the system

$$(5d)\qquad x(t) = Y(t) + \int_0^t R(t,s)G(s,x(s))ds$$

where $Y(t)$ is given by (4d).

Taking the special case $B(t) \equiv 0$ it follows from (3d) that $R(t,s) = \Phi(t)\Phi^{-1}(s)$, where Φ is the fundamental matrix of $y' = A(t)y$

for which $\Phi(0) = I$, and (4d) and (5d) are classical variations of constants formulae for ordinary differential equations.

3. Stability

We shall be interested in stability type results of continuous solutions of (1) and (1d) on R^+. Let $BC = \{\phi$ continuous on R^+: $\|\phi\| = \sup_{t\in R^+} |\phi(t)| < \infty\}$ where $|\ |$ is any convenient norm in $\mathbf{R}^n$. We make the following assumptions

(H_1) $r(t,s) \in LL^1(R^+ \times R^+)$

(H_2) $g(t,x)$ is continuous in (t,x) for $t \in R^+$, $|x| < \infty$ and $g(t,0) \equiv 0$.

(H_3) For every $\alpha > 0$ there exists a $\beta > 0$ such that $|g(t,x)| \leq \alpha|x|$, uniformly in t, whenever $|x| \leq \beta$; the following result is established in [7, Theorem 1]:

THEOREM 1. Suppose $y \in BC$ is a solution of (2). Let there exist a constant $K > 0$ such that

$$\int_0^t |r(t,s)|ds \leq K \qquad (t \in R^+) , \tag{6}$$

and let

$$\lim_{h\to 0} [\int_t^{t+h} |r(t+h,s)|ds + \int_0^t |r(t+h,s) - r(t,s)|dx] = 0 \tag{7}$$

for any t on R^+. Then given λ, $0 < \lambda < 1$, there exists a number $\varepsilon_0 > 0$ such that $0 < \varepsilon \leq \varepsilon_0$ and $\|y\| \leq \lambda\varepsilon$ together imply that (1) has at least one solution $x \in BC$ and $\|x\| \leq \varepsilon$. If in addition $\lim_{t\to\infty} y(t) = 0$ and if for each $T > 0$

$$\lim_{t\to\infty} \int_0^T |r(t,s)|ds = 0 , \tag{8}$$

then also $\lim_{t\to\infty} x(t) = 0$.

The proof of Theorem 1 makes use of the Schauder-Tychonoff fixed point theorem applied to the operator Ω defined by the right hand side of (5) acting on the subspace $S_\varepsilon = \{\phi \in BC: \|\phi\| \leq \varepsilon\}$.

We observe that if $f \in BC$ and if r satisfies (6), (7), then by (4) $y \in BC$ and $\|y\| \leq (1 + K)\|f\|$. Thus if $\|f\| \leq \frac{\lambda\varepsilon}{1+K}$, Theorem 1 implies that (1) has a solution $x \in BC$ such that $\|x\| \leq \varepsilon$. If in addition $f \to 0$ as $t \to \infty$ and if r satisfies (8), then from (4) $y \to 0$ and by Theorem 1 the solution $x \to 0$. Thus Theorem 1 is a stability type result with the forcing term f playing the role of initial conditions in ordinary differential equations. A somewhat different result and proof is obtained in [9, Theorem 1*].

If assumption (H_3) is strengthened as follows:

$(\tilde{H}_3)$ For every $\alpha > 0$ there exists a $\beta > 0$ such that $|g(t,u) - g(t,v)| \leq \alpha|u-v|$, uniformly in t, whenever $|u| \leq \beta$, $|v| \leq \beta$,

then the above mentioned mapping Ω is a contradiction on S_ε and the solution x of (1) is unique (see [6, Theorem 3 and Corollary 3.1]). For earlier versions of this latter result in the convolution case, see [10], [11]. This technique has also been used in [6] to establish the existence of asymptotically periodic and almost periodic solutions of (1) assuming that (2) has a solution with this property; see also [12].

The applicability of results like Theorem 1 depends mainly on satisfying assumptions (6) and (8). The following are criteria which do this in terms of the original kernel $a(t,s)$.

(i) $a(t,s) = a(t-s)$

<u>PALEY-WIENER THEOREM</u> [13]. Let $a(t) \in L^1(R^+)$; then the resolvent kernel $r(t) \in L^1(R^+)$ if and only if

$$\det(I - \hat{a}(s)) \neq 0 \qquad (\mathrm{Re}\ s \geq 0),$$

where $\hat{a}(s)$ is the Laplace transform of $a(t)$.

(ii) If $a(t) \notin L^1(R^+)$ (e.g. one may have $a(t) = t^{-\sigma}$ $(0 < \sigma < 1)$) one can often apply Laplace transforms and Tauberian theorems to (3) and satisfy conditions (6) and (8). Examples of this, though in a different context, may be found in [14], [15], [16].

(iii) If $n = 1$, $a(t,s) = A(t-s)B(s)$, then it is shown in [6, Theorem 6] (see also Miller [17]), that if $A(t)$ is completely monotonic on $0 < t < \infty$ and if $B(t) \in BC$ and $B(t) \geq 0$ then $r(t,s)$ satisfies (6) and (7). If, in addition, $\lim_{t\to\infty} A(t) = 0$, then $r(t,s)$ satisfies (8).

Finally, we remark that various generalizations of Theorem 1 have been obtained recently in an abstract setting using the concept of admissibility of a pair of subspaces with respect to an operator (in this case the resolvent operator). For this approach we refer the reader to [18], [19], [20]. Similar results are established in [8, Theorems 4 and 5].

We now turn to the differential system (1d).

THEOREM 1d. Let $A \in LL^1(R^+)$, $B \in LL^1(R^+ \times R^+)$. Let $G(t,x)$ satisfy (H_2) and (H_3) and let $R(t,s)$ satisfy (6). Let $Y(t) \in BC$ be a solution of (2d) satisfying the initial condition $Y(0) = x_0$. Then given $0 < \lambda < 1$ there exists an $\varepsilon_0 > 0$ such that $0 < \varepsilon \leq \varepsilon_0$ and $\|Y\| \leq \lambda\varepsilon$ together imply that the system (1d) has at least one solution $x \in BC$, $x(0) = x_0$, and $\|x\| \leq \varepsilon$. If in addition $\lim_{t\to\infty} Y(t) = 0$ and if $R(t,s)$ satisfies (8), then $\lim_{t\to\infty} x(t) = 0$.
(As was the case for (1), if $G(t,x)$ satisfies $(\tilde{H}_3)$, the solution x is unique.)

The proof of Theorem 1d is essentially the same as that of Theorem 1 making use of (5d) (equivalent to (1d)). Observe that the hypothesis concerning A, B implies from (3d) and (integrating (3d)) from

$$R(t,s) = I + \int_s^t [A(u) + \int_u^t B(\sigma,u)d\sigma]R(u,s)du$$

that $R(t,s)$ is continuous in (t,s) for $0 \leq s \leq t$. Thus (7) is automatically satisfied. Also observe that, $R(t,0) \in BC$, $F \in BC$ and (6) implies (from (4d)) that $Y \in BC$; moreover, if $\|F\|$ and $|x_0|$ are sufficiently small $\|Y\| \leq \lambda\varepsilon$. Thus Theorem 1d is a stability theorem which extends well known results in ordinary differential equations (see [21, p. 327]).

The applicability of Theorem 1d depends on $R(t,s)$ satisfying (6) and (8). This is a more difficult problem than in Theorem 1. An important result has recently been obtained by Grossman and Miller [22] for the special case $A(t) \equiv A$ a constant matrix, $B(t,s) \equiv B(t-s)$. Then (3d) becomes

$$(R) \qquad R'(t) = AR(t) + \int_0^t B(t-s)R(s)dx \ , \qquad R(0) = I \ .$$

(Evidently the classical ODE case has $B \equiv 0$, and all the eigenvalues of A having negative real parts, in which case (6) and (8) are obviously satisfied.)

<u>THEOREM</u>* (Generalized Paley-Wiener for (R)). Let $B(t) \in L^1(R^+)$. Then $R(t) \in L^1(R^+)$ if and only if

$$(PW) \qquad \det\,(sI - A - \hat{B}(s)) \neq 0 \qquad (\mathrm{Re}\ s \geqslant 0) \ ,$$

where $\hat{B}(s)$ is the Laplace transform of $B(t)$.

To obtain this result they make crucial use of recent work of the second author [23] concerning the initial value problem

$$(9) \qquad \begin{cases} y'(t) = Ay(t) + \int_0^t B(t-s)y(s)ds & (t \geqslant \tau \geqslant 0) \\ y(t) = h(t) & 0 \leqslant t \leqslant \tau, \end{cases}$$

where $B \in LL^1(R^+)$ and where h is a given continuous function on $0 \leqslant t \leqslant \tau$ (the usual initial value problem has $\tau = 0$); $y \equiv 0$ is a solution of (9) and one defines its <u>uniform stability</u> and <u>uniform asymptotic stability</u> (with respect to τ) in the same way it is done for delay equations. (E.g. $y \equiv 0$ is uniformly stable if given $\varepsilon > 0$ and $\tau \geqslant 0$ there exists a $\delta = \delta(\varepsilon)$, independent of τ, such that whenever h is a given continuous function for which $\sup_{0 \leqslant t \leqslant \tau} |h(t)| \leqslant \delta$, then (9) has a solution $y(t,\tau,h)$ for $t \geqslant \tau$ satisfying $y(\tau,\tau,h) = h(\tau)$ and $|y(t,\tau,h)| \leqslant \varepsilon$.) In rough outline these concepts are related to $R(t) \in L^1(0,\ )$ and the Paley-Wiener Theorem for (R) as follows:

*An independent proof of this result was recently obtained by D. S. Shea using a method similar to Paley-Wiener [13].

(i) If $B \in L^1(0,\infty)$ and if the solution $y \equiv 0$ of (9) is uniformly asymptotically stable, then condition (PW) is satisfied.

(ii) If $B,R \in L^1(0,\infty)$ then

(a) the solution $y \equiv 0$ of (9) is uniformly asymptotically stable, and

(b) for any τ, h the solution $y(t,\tau,h)$ of (9) is in $L^p(0,\infty)$ $(1 \leq p \leq \infty)$.

(iii) If $B \in L^1(0,\infty)$ and if the solution $y \equiv 0$ is uniformly asymptotically stable, then $R(t) \in L^1(0,\infty)$.

To prove the sufficiency of (PW) is suffices to show that condition (PW) implies that the solution $y(t) \equiv 0$ of (9) is uniformly asymptotically stable. The necessity follows from (iia) and (i). Statements (i), (ii), (iii) are established in [22] and [23]. The proof of (iii) involves proving a converse theorem, showing that (9) possesses a Liapunov functional of a certain type; this is done by adapting to (9) the ingenious construction of Massera for ordinary differential equations [24].

4. Asymptotic Equivalence

We say that systems (1) and (2) are asymptotically equivalent if given a bounded solution $y(t)$ of (2) on R^+ there exists a bounded solution $x(t)$ of (1) on R^+ such that $\lim_{t\to\infty} (x(t) - y(t)) = 0$ and conversely. The same definition holds for systems (1d) and (2d). It is clear that this is a more general concept than asymptotic stability. In place of assumptions (H_2), (H_3) in Section 3, we now assume

$$(H_4) \quad \begin{cases} g(t,x) \text{ is continuous in } (t,x) \text{ for } t \in R^+, \ |x| < \infty \\[2ex] |g(t,x)| \leq \begin{cases} \lambda(t)|x| & \text{if } |x| \geq 1 \\ \lambda(t) & \text{if } |x| < 1 \end{cases} \\[2ex] \text{where } \lambda(t) \geq 0, \ \lambda \in BC, \ \lim_{t\to\infty} \lambda(t) = 0 \text{ and} \\[1ex] \|\lambda\| K \leq \frac{1}{2} \text{ where } K \text{ is the a priori constant in (6).} \end{cases}$$

One can then establish the following result [7, Theorem 3]:

THEOREM 3. Let g satisfy (H_4), and let the resolvent r satisfy (6), (7), (8). Then the systems (1) and (2) are asymptotically equivalent.

A part of this result was established in [9, Theorem 2] under a different and more stringent hypothesis concerning the original kernel a and by different methods without, however, requiring the condition $\|\lambda\|K \leq \frac{1}{2}$. It is also shown in [9, Ex. 3] that the requirement $\lim_{t\to\infty} \lambda(t) = 0$ cannot be improved to $\lambda \in L^1(R^+)$. For the asymptotic equivalence of systems (1d) and (2d), one has:

THEOREM 3d. Let G satisfy (H_4) and let the differential resolvent R satisfy (6) and (8). Then the systems (1d) and (2d) are asymptotically equivalent.

In the proof of Theorem 3 (similarly 3d), one uses the Schauder-Tychonoff fixed point theorem to establish the existence of a bounded solution of (1) given a bounded solution of (2) (it is here where one needs the condition $\|\lambda\|K \leq \frac{1}{2}$), and then a direct limiting argument to show that the difference between $x(t)$ and $y(t)$ tends to zero. For the converse given a bounded solution $u(t)$ of (1), one defines

$$v(t) = u(t) - \int_0^t r(t,s)g(s,u(s))ds$$

and shows (using the resolvent equation) that v is a (bounded) solution of (2); then one verifies directly that $\lim_{t\to\infty} (u(t) - v(t)) = 0$.

The condition $\|\lambda\|K \leq \frac{1}{2}$ in (H_4) can be dropped if one strengthens the other condition on $g(t,x)$ slightly (see [7, Theorem 4]); we have:

THEOREM 4. Let $g(t,x)$ be continuous in (t,x) for $t \in R^+$, $|x| < \infty$, and let $|g(t,x)| \leq \lambda(t)|x|^{\sigma}$ $(0 \leq \sigma < 1)$, where $\lambda(t) \geq 0$, $\lambda \in BC$, $\lim_{t\to\infty} \lambda(t) = 0$ and let r satisfy (6), (7), (8). Then (1) and (2) are asymptotically equivalent.

A similar result holds for systems (1d) and (2d). Theorem 4 is proved by a different argument in that, instead of using the Schauder-Tychonoff theorem, one obtains an a priori estimate for the solution of

(1) from which one can deduce the existence of a bounded solution of (1) given a bounded solution of (2). The remainder of the argument is the same as in the proof of Theorem 3.

REFERENCES

[1] J.J. LEVIN and J.A. NOHEL, Perturbations of a nonlinear Volterra equation, Mich. Math. J. 12 (1965), 431-447.

[2] J.J. LEVIN, A nonlinear Volterra equation, not of convolution type, J. Diff. Eqns. 4 (1968), 176-186.

[3] K.B. HANNSGEN, On a nonlinear Volterra equation, Mich. Math. J. 16 (1969), 365-376.

[4] R.C. MacCAMY and J.S.W. WONG, Stability theorems for some functional differential equations, Trans. AMS (to appear).

[5] J.J. LEVIN and D.F. SHEA, On the asymptotic behavior of the bounded solutions of some integral equations, I, II, III, J. Math. Anal. Appl. (to appear).

[6] R.K. MILLER, J.A. NOHEL, and J.S.W. WONG, Perturbations of Volterra integral equations, J. Math. Anal. Appl. 25 (1969), 676-691.

[7] J.A. NOHEL, Asymptotic relationships between systems of Volterra equations, Annali di Mat. Pura ed Applic. (to appear).

[8] S.I. GROSSMAN and R.K. MILLER, Perturbation theory for Volterra integrodifferential systems, J. Diff. Eqns. 8 (1970), 457-474.

[9] A. STRAUSS, On a perturbed Volterra integral equation, J. Math. Anal. Appl. 30 (1970), 564-575.

[10] R.K. MILLER, On the linearization of Volterra interral equations, J. Math. Anal. Appl. 23 (1968), 198-208.

[11] J.A. NOHEL, Remarks on nonlinear Volterra equations, Proc. U.S.-Japan Seminar on Differential and Functional Equations. W.A. Benjamin, Inc., 1967, pp. 249-266.

[12] J.L. KAPLAN, On the asymptotic behavior of Volterra integral equations, J. Math. Anal. Appl. (to appear).

[13] R.E.A.C. PALEY and N. WIENER, Fourier Transforms in the Complex Domain. Amer. Math. Soc., Providence, 1934.

[14] J.J. LEVIN and J.A. NOHEL, A system of integrodifferential equations occurring in reactor dynamics, J. Math. Mich. 9 (1960), 347-368.

[15] T.A. BRONIKOWSKI, An integrodifferential system which occurs in reactor dynamics, Arch. Rational Mech. Anal. 37 (1970), 363-380.

[16] K.B. HANNSGEN, Indirect abelian theorems and a linear Volterra equation, Trans. Amer. Math. Soc. 142 (1969), 539-555.

[17] R.K. MILLER, On Volterra integral equations with nonnegative integrable resolvents, J. Math. Anal. Appl. 22 (1968), 319-340.

[18] C. CORDUNEANU, Problèmes globaux dans le théorie des équations intégrals de Volterra, Ann. Mat. Pura Appl. 67 (1965), 349-363.

[19] R.K. MILLER, Admissibility and nonlinear Volterra integral equations, Proc. Amer. Math. Soc. 25 (1970), 65-71.

[20] J.A. NOHEL, Perturbations of Volterra equations and admissibility, Proceedings 2nd Japan-U.S. Seminar on Ordinary Differential and Functional Equations, Springer-Verlag (to appear).

[21] E.A. CODDINGTON and N. LEVINSON, Theory of Ordinary Differential Equations. McGraw-Hill Book Co., Inc., 1955.

[22] S.I. GROSSMAN and R.K. MILLER, Nonlinear Volterra integrodifferential systems with L^1 kernels (to appear).

[23] R.K. MILLER, Asymptotic stability properties of linear Volterra integrodifferential equations, J. Diff. Eqns. (to appear).

[24] J.L. MASSERA, On Liapunov's conditions for stability, Ann. Math. 50 (1949), 705-721.

Ecole Polytechnique Fédérale, Lausanne, Switzerland, and University of Wisconsin, Madison, Wisconsin

INVOLUTORY MATRIX DIFFERENTIAL EQUATIONS*

William T. Reid

1. Introduction

The interrelations between linear matrix differential systems and associated Riccati matrix differential equations are well known, and of frequent use in a variety of instances of variational problems, control theory, and transmission line phenomena. In particular, when the linear system is Hamiltonian and hermitian in nature there is an intimate connection between such systems and the generalized Sturmian theory, emanating from the work of Marston Morse [5,6]. There is a corresponding case wherein the linear differential system is symmetric, but non-real, and in which the basic solvability theorems are strict duals of those in the hermitian case. The purpose of the present paper is to present a unified discussion of these two alternative cases, for which the term "involutory" has been adopted. The discussion is phrased in the context of a generalized differential system which is equivalent to a type of linear vector Riemann-Stieltjes integral equation, of the form considered previously by the author in papers [10],[11], [12].

*This research was supported by the Air Force Office of Scientific Research, Office of Aerospace Research, United States Air Force, under Grant AFOSR-71-2069. The United States Government is authorized to reproduce and distribute reprints for governmental purposes notwithstanding any copyright notation hereon.

Matrix notation is used throughout; in particular, one column matrices are called vectors, and the linear vector space of ordered n-tuples of complex numbers, with complex scalars, is denoted by $\mathbf{C}_n$. The $n \times n$ identity matrix is denoted by E_n, or merely by E when there is no ambiguity, and 0 is used indiscriminately for the zero matrix of any dimensions. For a matrix M the transpose is denoted by $\tilde{M}$, and the conjugate transpose by M^*. If $M = [M_{\alpha j}]$, $N = [N_{\alpha j}]$, $(\alpha = 1,\ldots,n;\ j = 1,\ldots,k)$ are $n \times k$ matrices, for typographical simplicity the symbol $(M;N)$ is used to denote the $2n \times k$ matrix whose j^{th} column has elements $M_{1j},\ldots,M_{nj},N_{1j},\ldots,N_{nj}$. If a matrix function $M(t)$ is a.c.,(absolutely continuous), on $[a,b]$, then $M'(t)$ denotes the matrix of derivatives at values where these derivatives exist, and zero elsewhere. If $M(t)$ is (Lebesgue) integrable on $[a,b]$, then $\int_a^b M(t)dt$ signifies the matrix of integrals of respective elements of $M(t)$. For a given interval $[a,b]$ on the real line the symbols $\mathfrak{C}_{mn}[a,b]$, $\mathcal{L}_{mn}[a,b]$, $\mathcal{L}^2_{mn}[a,b]$, $\mathcal{L}^\infty_{mn}[a,b]$, $\mathcal{BV}_{mn}[a,b]$ and $\mathcal{A}_{mn}[a,b]$ are used to denote the class of $m \times n$ matrix functions $[M_{\alpha\beta}(t)]$, $(\alpha = 1,\ldots,m;\ \beta = 1,\ldots,n)$, on $[a,b]$ which are respectively continuous,(Lebesgue) integrable, (Lebesgue) measurable and $|M_{\alpha\beta}(t)|^2$ integrable, measurable and essentially bounded, of b.v. (bounded variation), and a.c. on $[a,b]$. For brevity, the double subscript is reduced to merely m for the m-dimensional vector case specified by $n = 1$, and both subscripts are omitted in the scalar case $m = 1$, $n = 1$. If matrix functions $M(t)$ and $N(t)$ are equal a.e. (almost everywhere) on their interval of definition, we write simply $M(t) = N(t)$. An $m \times n$ matrix function $M(t)$ is said to be locally a.c., or locally of b.v., on I if $M \in \mathcal{A}_{mn}[a,b]$, or $M \in \mathcal{BV}_{mn}[a,b]$, for arbitrary compact subintervals $[a,b]$ of I.

If $P(t)$, $M(t)$, $Q(t)$ are matrix functions of respective dimensions $m \times p$, $p \times q$, $q \times n$ and $P \in \mathfrak{C}_{mp}[a,b]$, $M \in \mathcal{BV}_{pq}[a,b]$, $Q \in \mathfrak{C}_{qn}[a,b]$, then the symbol $\int_a^b P(t)[dM(t)]Q(t)$ denotes the $m \times n$ matrix with elements given by the Riemann-Stieltjes integrals

$$\sum_{\alpha=1}^{p} \sum_{\beta=1}^{q} \int_a^b P_{i\alpha}(t)Q_{\beta j}(t)dM_{\alpha\beta}(t), \qquad (i = 1,\ldots,m;\ j = 1,\ldots,n)\ ;$$

the symbols $\int_a^b [dM(t)]Q(t)$ and $\int_a^b P(t)[dM(t)]$ designate $\int_a^b E_p[dM(t)]Q(t)$ and $\int_a^b P(t)[dM(t)]E_q$, respectively.

2. Formulation of the Problem

In the class of $m \times n$ matrices with complex elements, let Θ denote either the operation of transpose, "$\sim$", or conjugate transpose, "$*$". We clearly have the following properties in the class of finite dimensional matrices, with the understanding that in each case the dimensions of the involved matrices are such that the operations are meaningful:

$$(M_1 + M_2)^\Theta = M_1^\Theta + M_2^\Theta, \quad (M_1 M_2)^\Theta = M_2^\Theta M_1^\Theta ;$$

$$(cM)^\Theta = \bar{c}M^\Theta, \text{ in case } \Theta = *; \quad (cM)^\Theta = cM^\Theta \text{ in case } \Theta = \sim.$$

Also, if a matrix function $M(t)$ is integrable on an interval $[a,b]$ on the real line, then $[\int_a^b M(t)dt]^\Theta = \int_a^b M^\Theta(t)dt$. Correspondingly, if $M(t)$ is locally a.c., then $M^\Theta(t)$ is locally a.c., and if M possesses a finite derivative $M'(t_0)$ at a value $t = t_0$ then M^Θ possesses a finite derivative $M^{\Theta\prime}(t_0)$ and $M^{\Theta\prime}(t_0) = [M'(t_0)]^\Theta$.

For the class of finite dimensional matrices with elements in the complex field, in each of the above cases the mapping $M \to M^\Theta$ is obviously the standard involution corresponding to the respective conjugations $a + ib \to a - ib$ and $a + ib \to a + ib$ in the complex base field. If $M = M^\Theta$ then M is termed Θ-<u>symmetric</u>, with the obvious substitutions of <u>hermitian</u> and <u>symmetric</u> in the respective cases with $\Theta = *$ and $\Theta = \sim$. Since "Θ" is the astronomical symbol for the sun, it is suggested that in the following text the symbol "Θ" be read as "sol".

We shall consider the generalized matrix differential system

$$\text{(2.1)} \quad \begin{aligned} \Delta[u,v](t) &\equiv -dv(t) + [C(t)u(t) - A^\Theta(t)v(t)]dt + [dM(t)]u(t) = 0 , \\ L_2[u,v](t) &\equiv u'(t) - A(t)u(t) - B(t)v(t) = 0 , \end{aligned}$$

where A, B, C and M are $n \times n$ matrix functions defined on an interval I on the real line which satisfy the following hypothesis.

($\mathcal{H}$) (1) $B(t) = B^{\circledast}(t)$, $C(t) = C^{\circledast}(t)$ and $M(t) = M^{\circledast}(t)$ for $t \in I$;

(2) for arbitrary compact subintervals $[a,b] \subset I$, the matrix functions A, B, C belong to $\mathcal{L}^{\infty}_{nn}[a,b]$, and $M \in \mathcal{BV}_{nn}[a,b]$.

By a solution of (2.1) is meant a pair of vector functions $u(t)$, $v(t)$ which belong to $\mathcal{A}_n[a,b] \times \mathcal{BV}_n[a,b]$ for arbitrary compact subintervals $[a,b]$ of I, with $L_2[u,v](t) = 0$ on I and

$$(2.2) \qquad v(t) = v(\tau) + \int_{\tau}^{t} [C(s)u(s) - A^{\circledast}(s)v(s)]ds + \int_{\tau}^{t} [dM(s)]u(s)$$

for $(\tau,t) \in I \times I$.

From the results of [11; Sec. 2] it follows that for given vectors u_0, v_0 of $\mathbf{C}_n$ and $\tau \in I$ there exists a unique solution of (2.1) satisfying the initial conditions $u(\tau) = u_0$, $v(\tau) = v_0$. In particular, if $M(t) \equiv 0$ then $(u(t);v(t))$ is a solution of (2.1) if and only if $u(t)$ and $v(t)$ are both locally a.c. on I and $y(t) = (u(t);v(t))$ is a solution in the Carathéodory sense of the ordinary differential system

$$(2.3) \qquad \mathcal{J}y'(t) + \mathcal{A}(t)y(t) = 0 ,$$

where

$$\mathcal{J} = \begin{bmatrix} 0 & -E_n \\ E_n & 0 \end{bmatrix}, \qquad \mathcal{A}(t) = \begin{bmatrix} C(t) & -A^{\circledast}(t) \\ -A(t) & -B(t) \end{bmatrix}.$$

Corresponding to the vector system (2.1), we have the matrix system

$$(2.1_M) \qquad \begin{aligned} \Delta[U,V](t) &\equiv -dV(t) + [C(t)U(t) - A^{\circledast}(t)V(t)]dt + [dM(t)]U(t) = 0 , \\ L_2[U,V](t) &\equiv U'(t) - A(t)U(t) - B(t)V(t) = 0 , \end{aligned}$$

where it is understood that $n \times k$ matrix functions $U(t)$, $V(t)$ are a solution of (2.1_M) if each column vector function of the $2n \times k$ matrix $(U(t);V(t))$ is a solution of (2.1).

For a given compact subinterval $[a,b]$ of I, let $\mathcal{D}[a,b]$ denote the class of n-dimensional vector functions η for which there exists an associated $\zeta \in \mathcal{L}^2_n[a,b]$ such that $L_2[\eta,\zeta](t) = 0$ on $[a,b]$. The subclass of $\mathcal{D}[a,b]$ on which $\eta(a) = 0 = \eta(b)$ will be denoted by

$\mathcal{D}_0[a,b]$, and the fact that η is a member of $\mathcal{D}[a,b]$ or $\mathcal{D}_0[a,b]$ with associated ζ will be denoted by $\eta \in \mathcal{D}[a,b]:\zeta$ or $\eta \in \mathcal{D}_0[a,b]:\zeta$.

If $(\eta_\alpha,\zeta_\alpha) \in \mathcal{L}_n^2[a,b] \times \mathcal{L}_n^2[a,b]$ for $\alpha = 1,2$, we shall denote by $J[\eta_1:\zeta_1,\eta_2:\zeta_2;a,b]$ the functional

$$(2.4) \qquad J[\eta_1:\zeta_1,\eta_2:\zeta_2;a,b] = \int_a^b \{\zeta_2^\Theta(t)B(t)\zeta_1(t) + \eta_2^\Theta(t)C(t)\eta_1(t)\}dt + \int_a^b \eta_2^\Theta(t)[dM(t)]\eta_1(t) .$$

Now for $\eta_\alpha \in \mathcal{D}[a,b]:\zeta_\alpha$, $(\alpha = 1,2)$, the vector functions ζ_α are in general not determined uniquely; the value of the corresponding functional (2.4) is independent of the particular choice of the ζ_α, however, and consequently the symbol for the integral (2.4) is reduced to $J[\eta_1,\eta_2;a,b]$. Also, for brevity we write $J[\eta;a,b]$ for $J[\eta,\eta;a,b]$ in case $\eta \in \mathcal{D}[a,b]$. That is, if $\eta \in \mathcal{D}[a,b]:\zeta$ then

$$(2.4') \qquad J[\eta;a,b] = \int_a^b \{\zeta^\Theta(t)B(t)\zeta(t) + \eta^\Theta(t)C(t)\eta(t)\}dt + \int_a^b \eta^\Theta(t)[dM(t)]\eta(t) .$$

Corresponding to the results for the case $\Theta = *$, (see [11; Sec. 2] and [15; Ch. VII, Sec. 4] for the case of ordinary differential equations), we have the following basic result, which in essence states that (2.1) is the "Euler-Lagrange" system for the functional (2.4') subject to the differential equation constraint $L_2[\eta,\zeta](t) = 0$.

<u>THEOREM 2.1</u>. If $[a,b] \subset I$, and $u \in \mathcal{A}_n[a,b]$, then there exists a v such that $(u;v)$ is a solution of (2.1) if and only if there exists a v_1 such that $u \in \mathcal{D}[a,b]:v_1$, and $J[u:v_1,\eta:\zeta;a,b] = 0$ for all $\eta \in \mathcal{D}_0[a,b]:\zeta$.

From Theorem 2.3 of [11] we have that $(u;v)$ is a solution of (2.1) if and only if $\hat{u}(t) = u(t)$, $\hat{v}(t) = v(t) - M(t)u(t)$ are individually locally a.c. on I, and $\hat{y} = (\hat{u};\hat{v})$ is a solution of the ordinary differential system

$$(2.3') \qquad \mathcal{J}\hat{y}'(t) + \hat{\mathcal{A}}(t)\hat{y}(t) = 0 ,$$

where

$$\hat{\mathcal{A}}(t) = \begin{bmatrix} \hat{C}(t) & -\hat{A}^{\odot}(t) \\ -\hat{A}(t) & -\hat{B}(t) \end{bmatrix}, \tag{2.5}$$

and

$$\hat{A} = A + BM, \quad \hat{B} = B, \quad \hat{C} = C - A^{\odot}M - MA - MBM . \tag{2.6}$$

Corresponding to (2.3'), we have the matrix differential equation

$$\mathcal{J}\hat{Y}'(t) + \hat{\mathcal{A}}(t)\hat{Y}(t) = 0 . \tag{2.3$'_M$}$$

In view of hypothesis $(\mathcal{H})$, it follows readily that the matrix functions $\hat{A}$, $\hat{B}$, and M of (2.6) satisfy corresponding conditions; that is

$$\hat{B}(t) = \hat{B}^{\odot}(t), \quad \hat{C}(t) = \hat{C}^{\odot}(t) , \tag{2.7}$$

and for arbitrary compact subintervals $[a,b]$ of I the matrix functions $\hat{A}$, $\hat{B}$, $\hat{C}$ belong to $\mathcal{L}^{\infty}_{nn}[a,b]$. In view of conditions (2.7), we have $\hat{\mathcal{A}}^{\odot}(t) = \hat{\mathcal{A}}(t)$.

Now if $\hat{Y}(t) = (\hat{U}(t);\hat{V}(t))$ is any solution of the matrix differential equation (2.3$'_M$) with $\hat{U}(t)$ non-singular on a subinterval I_0 of I, then $\hat{W}(t) = \hat{V}(t)\hat{U}^{-1}(t)$ is a solution of the Riccati matrix differential equation

$$\hat{\mathcal{K}}[\hat{W}](t) \equiv \hat{W}'(t) + \hat{W}(t)\hat{A}(t) + \hat{A}^{\odot}(t)\hat{W}(t) + \hat{W}(t)\hat{B}(t)\hat{W}(t) - \hat{C}(t) = 0 . \tag{2.8}$$

Moreover, in view of (2.7) it follows readily that if $\hat{W} = \hat{W}_0(t)$ is a solution of (2.8) on a subinterval I_0 of I then $\hat{W} = \hat{W}_0^{\odot}(t)$ is also a solution of this equation on I_0. Consequently, (2.8) is called $\odot$-involutory, or merely involutory, whenever hypothesis $(\mathcal{H})$ holds for either $\odot = *$ or $\odot = \sim$. When consideration is directed to a particular one of these instances, (2.8) is called hermitian in case $\odot = *$, and symmetric whenever $\odot = \sim$. In either case, $\hat{W}(t)$, $t \in I_0$, is a $\odot$-symmetric solution of (2.8), that is $\hat{W}(t) \equiv \hat{W}^{\odot}(t)$ on I_0, if and only if there exists a value $s \in I_0$ such that $\hat{W}(s) = \hat{W}^{\odot}(s)$.

As is well known (see, for example, [11; Section 2] or [15; Problem 8, p. 101]), a matrix function $\hat{W}(t)$, $t \in I_0$, is a solution of (2.8) if and only if there exists a solution $\hat{Y}(t) = (\hat{U}(t);\hat{V}(t))$ of

$(2.3'_M)$ with $\hat{U}(t)$ non-singular on I_0 and such that $\hat{W}(t) = \hat{V}(t)\hat{U}^{-1}(t)$ on this interval. Now if $Y(t) = (U(t);V(t))$ is the solution of (2.1_M) such that $\hat{U}(t) = U(t)$, $\hat{V}(t) = V(t) - M(t)U(t)$, then $\hat{V}(t)\hat{U}^{-1}(t) = V(t)U^{-1}(t) - M(t)$, so that $W(t) = V(t)U^{-1}(t)$ is of b.v. on I_0. Moreover, for W and $\hat{W}$ related by $W = \hat{W} + M$, and $\hat{A}$, $\hat{B}$, $\hat{C}$ defined by (2.6), we have

$$C(t) - WA(t) - A^{\ominus}(t)W - WB(t)W \equiv \hat{C}(t) - \hat{W}\hat{A}(t) - \hat{A}^{\ominus}(t)\hat{W} - \hat{W}\hat{B}(t)\hat{W} ,$$

and consequently $\hat{W}(t) = \hat{V}(t)\hat{U}^{-1}(t)$ is a solution of the Riccati matrix differential equation (2.8) on a subinterval I_0 if and only if there exists a value $s \in I_0$ and constant matrix Ψ such that $W(t) = \hat{W}(t) + M(t)$ is the solution on I_0 of the Riccati matrix integral equation

$$(2.9)\quad W(t) - \int_s^t [C(r) - W(r)A(r) - A^{\ominus}(r)W(r) - W(r)B(r)W(r)]dr = M(t) + \Psi .$$

<u>THEOREM 2.2</u>. If hypothesis $(\mathcal{H})$ holds, and $s \in [a,b] \subset I$, the following conditions are equivalent:

(i) the $n \times n$ matrix Ψ is such that the integral equation (2.9) has a solution $W(t)$ on $[a,b]$;

(ii) if $\hat{Y}(t) = (\hat{U}(t);\hat{V}(t))$ is the solution of $(2.3'_M)$ determined by the initial conditions $\hat{U}(s) = E$, $\hat{V}(s) = \Psi$, then $\hat{U}(t)$ is non-singular on $[a,b]$;

(iii) if $Y(t) = (U(t);V(t))$ is the solution of (2.1_M) determined by the initial conditions $U(s) = E$, $V(s) = M(s) + \Psi$, then $U(t)$ is non-singular on $[a,b]$.

Moreover, when these conditions hold the involved matrix functions $W(t)$, $\hat{U}(t)$, $\hat{V}(t)$, $U(t)$, $V(t)$ are related by the equations

$$(2.10)\quad W(t) = M(t) + \hat{V}(t)\hat{U}^{-1}(t) = V(t)U^{-1}(t) , \quad \text{for } t \in [a,b] .$$

3. <u>Examples of Involutory Systems</u>

As indicated in the introduction, important instances of differential systems satisfying the involutory condition occur as accessory differential systems for simple integral problems of the calculus of variations (see, for example, Bliss [2; Sec. 81], and Reid [7]). For a detailed formulation and discussion of such ordinary differential equation problems, the reader is referred to [15; Ch. VII, Sec. 2] and [16; Ch. III, Sec. 2]. Special examples of generalized differential systems include some differential systems with interface conditions, and self-adjoint systems of linear difference equations (see [10; Secs. 2,6]). A particularly simple system

$$(3.1) \qquad -dv(t) + [d\mu(t)]u(t) = 0, \quad u'(t) - [1/r(t)]v(t) = 0 ,$$

occurs in the solution of a maximum problem treated by the author in [14]. Simple modifications of the above examples provide instances of each of the cases $\odot = \sim$ and $\odot = *$.

A class of differential systems which satisfies the involutory condition with $\odot = \sim$, and for which the coefficient matrix functions are non-real, is presented by certain differential equations in the complex plane. Consider the second order linear homogeneous differential equation

$$(3.2) \qquad \frac{d}{dz}\left(r(z)\frac{df}{dz}\right) - p(z)f = 0 ,$$

where $r(z)$ and $p(z)$ are holomorphic functions of the complex variable z in a simply connected region R, and $r(z) \neq 0$ for $z \in R$. Under the substitution $f_1(z) = f(z)$, $f_2(z) = r(z)(df/dz)$ the equation (3.2) is equivalent to the first order system

$$(3.3) \qquad \frac{df_1}{dz} = \frac{1}{r(z)} f_2 , \qquad \frac{df_2}{dz} = p(z)f_1 .$$

If $(f_\alpha(z))$, $(\alpha = 1,2)$ is a solution of (3.3) in R, and Γ is a rectifiable curve in R with a.c. representation $z = z(t)$, $t_1 \leqslant t \leqslant t_2$, then $u(t) = f_1(z(t))$, $v(t) = f_2(z(t))$ is a solution of the differential system

$$(3.4) \qquad -v'(t) + [p(z(t))z'(t)]u(t) = 0, \quad u'(t) - \left[\frac{z'(t)}{r(z(t))}\right] v(t) = 0 ,$$

for $t_1 \leqslant t \leqslant t_2$. This system is of the form (2.1) with $n = 1$, $A(t) \equiv 0$, $C(t) = p(z(t))z'(t)$, $B(t) = z'(t)/r(z(t))$, and $M(t) \equiv 0$.

Clearly (3.4) is symmetric in general, but hermitian in only the very special instance of holomorphic functions $r(z)$, $p(z)$ and curve Γ such that the thus defined functions $C(t)$ and $B(t)$ are real-valued for $t \in [t_1,t_2]$.

If for an involutory system (2.1) the matrix functions A, B, C are real-valued and $M \equiv 0$, then for $(u;v)$ a solution of this system the corresponding vector functions $u_1 = v - iu$, $v_1 = v + iu$ satisfy a differential system

$$(3.5)\qquad \begin{aligned} -v_1'(t) + C_1(t)u_1(t) - \tilde{A}_1(t)v_1(t) &= 0 , \\ u_1'(t) - A_1(t)u_1(t) - B_1(t)v_1(t) &= 0 , \end{aligned}$$

with $A_1 = \frac{1}{2}\{i[C - B] + A - \tilde{A}\}$, $B_1 = \frac{1}{2}\{-i[C + B] - A - \tilde{A}\}$, $C_1 = \frac{1}{2}\{i[C + B] - A - \tilde{A}]$, which is a complex differential system of the form (2.1) that is involutory with $\circledcirc = \sim$, but in general not involutory with $\circledcirc = *$. In his study of oscillation phenomena for such an original system (2.1), Atkinson [1; Ch. 10] employed the matrix system corresponding to (3.5). In particular, if $U(t)$, $V(t)$ are $n \times n$ matrix functions such that $Y(t) = (U(t);V(t))$ is a conjugate basis for (2.1) in the sense to be defined in the next section, then $Y_1(t) = (U_1(t);V_1(t)) = (V(t) - iU(t);V(t) + iU(t))$ is such that $W_1(t) = V_1(t)U_1^{-1}(t)$ is unitary and the corresponding Riccati matrix differential equation satisfied by $W_1(t)$ is linear.

4. Properties of Solutions of Involutory Systems

If $\hat{y}_\alpha = (\hat{u}_\alpha;\hat{v}_\alpha)$, $(\alpha = 1,2)$, are solutions of (2.3'), then in view of the relations $\mathcal{J}^\circledcirc = -\mathcal{J}$ and $\hat{\mathcal{A}}^\circledcirc(t) \equiv \hat{\mathcal{A}}(t)$ we have $-[\hat{y}_2^\circledcirc(t)]'\mathcal{J} + \hat{y}_2^\circledcirc(t)\hat{\mathcal{A}}(t) = 0$ and $\{\hat{y}_2^\circledcirc(t)\mathcal{J}\hat{y}_1(t)\}' = 0$, so that $\hat{y}_2^\circledcirc(t)\mathcal{J}\hat{y}_1(t)$ is constant on I. As a consequence of the relation $\hat{u} = u$, $\hat{v} = v - Mu$ between solutions $(u;v)$ of (2.1) and solutions $(\hat{u};\hat{v})$ of (2.3'), we have the following result.

<u>LEMMA 4.1</u>. If $y_1(t) = (u_1(t);v_1(t))$ and $y_2(t) = (u_2(t);v_2(t))$ are solutions of (2.1), then the function

$$(4.1)\qquad \{y_1|\Theta|y_2\}(t) = y_2^{\Theta}(t)\mathcal{J}y_1(t) = v_2^{\Theta}(t)u_1(t) - u_2^{\Theta}(t)v_1(t)$$

is constant on I.

If $y_\alpha(t) = (u_\alpha(t);v_\alpha(t))$, $(\alpha = 1,2)$, are solutions of (2.1) such that the constant function (4.1) is zero, then they are said to be (mutually) Θ-conjugate solutions of (2.1). If the matrix coefficient functions A, B, C are real-valued, $M = 0$, and $\Theta = \sim$, then for real-valued solutions of this system the concept of Θ-conjugacy is that of conjugacy as introduced by von Escherich [3] for accessory differential systems associated with a variational problem; for this case, see also Bliss [2; especially, Secs. 11, 23, 36, 81] and Reid [7]. In case the coefficient matrix functions are complex-valued, and $\Theta = *$, the concept of Θ-conjugacy is the condition called conjoinedness by the author (see [8], [15; Ch. VII] for ordinary differential systems, and [10], [11], [12] for generalized differential systems). For differential systems (2.3) the involutory operation Θ has been introduced in [16; Ch. III].

If $Y(t) = (U(t);V(t))$ is a $2n \times r$ matrix satisfying (2.1_M) and whose column vectors are r linearly independent solutions of (2.1) which are mutually Θ-conjugate, these solutions form a basis for a Θ-conjugate family of solutions of dimension r, consisting of the set of all solutions of (2.1) which are linear combinations of these column vectors. The following result is basic for the study of such systems.

THEOREM 4.1. The maximal dimension of a Θ-conjugate family of solutions of (2.1) is n; moreover, a given Θ-conjugate family of solutions of dimension $r < n$ is contained in a Θ-conjugate family of dimension n.

In case $\Theta = *$, the result of this theorem may be established by exactly the same steps as used by the author [8; Lemma 2.3]; see also [15; Ch. VII, Theorem 2.1]. The case $\Theta = \sim$ is more elementary in detail, since in this instance each solution $y(t) = (u(t);v(t))$ of (2.1) is self-Θ-conjugate; that is, $\tilde{y}(t)\mathcal{J}y(t) \equiv 0$.

If $Y(t) = (U(t);V(t))$ is a solution of (2.1_M) whose column vectors are n linearly independent solutions of (2.1) which are mutually

Θ-conjugate, then for brevity $Y(t)$ is called a <u>Θ-conjugate basis</u> for (2.1). In view of the above presented relations between solutions of (2.1_M), ($2.3'_M$), (2.8), and (2.9), we have that $Y(t)$ is a Θ-conjugate basis for (2.1) with $U(t)$ non-singular on a subinterval $[a,b]$ of I if and only if the matrix function $W(t) = V(t)U^{-1}(t)$ is a Θ-symmetric solution of (2.9) on $[a,b]$.

<u>THEOREM 4.2</u>. Suppose that $Y_0(t) = (U_0(t);V_0(t))$ is a solution of (2.1_M) with $U_0(t)$ non-singular on a subinterval I_0 of I. For K the constant $n \times n$ matrix $K = -\{Y_0|\Theta|Y_0\}(t) = U_0^{\Theta}(t)V_0(t) - V_0^{\Theta}(t)U_0(t)$, and $s \in I_0$, let $T = T(t,s|Y_0)$ denote the solution of the differential system

$$(4.2) \qquad T'(t) = -U_0^{-1}(t)B(t)U_0^{\Theta-1}(t)KT(t) , \qquad T(s) = E .$$

Then $Y(t) = (U(t);V(t))$ is a solution of (2.1_M) on I_0 if and only if

$$(4.3) \qquad U(t) = U_0(t)N(t), \quad V(t) = V_0(t)N(t) + U_0^{\Theta-1}(t)[K_1 - KN(t)] ,$$

where K_1 is a constant matrix and

$$(4.4) \qquad N(t) = T(t,s|Y_0)[K_0 + S(t,s|Y_0)K_1] ,$$

$$(4.5) \qquad S(t,s|Y_0) = \int_s^t T^{-1}(r,s|Y_0)U_0^{-1}(r)B(r)U_0^{\Theta-1}(r)dr .$$

In particular, if $Y(t) = (U(t);V(t))$ and $Y_0(t) = (U_0(t);V_0(t))$ are solutions of (2.1_M) related by (4.3), then $\{Y|\Theta|Y_0\}(t) \equiv -K_1$.

By using the relations between solutions of (2.1_M) and ($2.3'_M$) one may show that the result of the above theorem are equivalent to the corresponding results for ($2.3'_M$). In turn, for this latter system the result may be established as in the particular case of $\Theta = *$, for which the reader is referred to [9; Sec. 3].

It is to be remarked that if $Y_0(t) = (U_0(t);V_0(t))$ is a Θ-conjugate basis for (2.1), then $K = -\{Y_0|\Theta|Y_0\}$ is 0, so that $T(t,s|Y_0) \equiv E$, and the matrix function $S(t,s|Y_0)$ defined by (4.5) becomes

$$(4.6) \qquad S(t,s|Y_0) = \int_s^t U_0^{-1}(r)B(r)U_0^{\Theta-1}(r)dr ,$$

which is clearly Θ-symmetric for $(t,s) \in I_0 \times I_0$.

As in the case for ordinary differential systems with $\Theta = *$ (see [9; Theorem 3.1], [15; Problem 5, p. 310]), one may establish for (2.1) the following result.

COROLLARY 1. Suppose that $Y_0(t) = (U_0(t);V_0(t))$ and $K = -\{Y_0|\Theta|Y_0\}$ are as in Theorem 4.2 on a subinterval I_0 of I, and that $T(t,s|Y_0)$ is the solution of the differential system (4.2). If $Y(t) = (U(t);V(t))$ is any solution of (2.1_M), then for $(s,t) \in I_0 \times I_0$ we have

$$(4.7) \quad \{Y|\Theta|Y_0\} + KU_0^{-1}(t)U(t) = T^{\Theta-1}(t,s|Y_0)[\{Y|\Theta|Y_0\} + KU_0^{-1}(s)U(s)] .$$

For a nondegenerate subinterval I_0 of I, let $\Lambda(I_0)$ denote the linear space of n-dimensional vector functions v which are solutions of $v'(t) + A^{\Theta}(t)v(t) = 0$ and satisfy $B(t)v(t) = 0$ on I_0; clearly $v \in \Lambda(I_0)$ if and only if $(u;v) = (0;v)$ is a solution of (2.1) on I_0. If $\Lambda(I_0)$ is zero-dimensional, (2.1) is said to be <u>normal</u> on I_0, or to have <u>abnormality of order zero on</u> I_0, whereas if $\Lambda(I_0)$ has dimension $d = d(I_0) > 0$ the system (2.1) is said to be <u>abnormal</u>, with <u>order of abnormality</u> d on I_0. Also, (2.1) is said to be <u>identically normal</u> on I if this system is normal on every nondegenerate subinterval of I. Values t_1 and t_2 are said to be (<u>mutually</u>) <u>conjugate</u> (with respect to (2.1)) if there exists a solution $y(t) = (u(t);v(t))$ of this system such that $u(t_1) = 0 = u(t_2)$, and $u(t) \not\equiv 0$ on the subinterval of I with endpoints t_1 and t_2. The system (2.1) is said to be <u>disconjugate</u> on a subinterval I_0 of I if I_0 contains no pair of points t_1, t_2 which are conjugate.

If $s \in I$ and $Y(t;s) = (U(t;s);V(t;s))$ is the solution of (2.1_M) determined by the initial conditions $U(s;s) = 0$, $V(s;s) = E$, then whenever (2.1) is normal on every subinterval of I that has s as an endpoint it follows that a value $t \neq s$ is conjugate to s if and only if $U(t;s)$ is singular. Now if $Y_0(t) = (U_0(t);V_0(t))$ is a solution of (2.1) with $U_0(t)$ non-singular on a subinterval I_0 of I, and $s \in I_0$, then upon choosing $Y(t) = Y(t;s)$ in Theorem 4.2 we have that $N(s) = 0$, so that $K_0 = 0$ in (4.4), and $U(t;s) = U_0(t)T(t,s|Y_0)S(t,s|Y_0)K_1$, where $K_1 = -\{Y(\cdot;s)|\Theta|Y_0\}$ is the non-singular matrix $U_0^{\Theta}(s)$.

Consequently, a value $t \in I_0$ is such that $U(t;s)$ is singular if and only if $S(t,s|Y_0)$ is singular, and we have the following result.

COROLLARY 2. Suppose that $Y_0(t) = (U_0(t);V_0(t))$ and $S(t,s|Y_0)$ are as in Theorem 4.2 on a subinterval I_0 of I. If $s \in I_0$, and (2.1) is normal on every subinterval of I_0 that has s as an endpoint, then a value $t \in I_0$ distinct from s is such that $S(t,s|Y_0)$ is singular if and only if t is conjugate to s, relative to (2.1).

COROLLARY 3. Suppose that I is an open interval (a_0,b_0), $(-\infty \leqslant a_0 < b_0 \leqslant \infty)$, on which (2.1) is identically normal, while this system is disconjugate on a subinterval $I_0 = (c_0,b_0)$ of I and $Y_0(t) = (U_0(t);V_0(t))$ is a solution of (2.1_M) with $U_0(t)$ non-singular on I_0. Then for $s \in I_0$ the matrix $S(t,s|Y_0)$ of (4.5) is non-singular for $t \in I_0$, $t \neq s$; moreover, if there exists a value $s \in I_0$ such that $S^{-1}(t,s|Y_0) \to 0$ as $t \to b_0$, then $S^{-1}(t,r|Y_0) \to 0$ as $t \to b_0$ for arbitrary $r \in I_0$.

The first conclusion of this corollary is an immediate consequence of the result of Corollary 2. Now if s and r are values on I_0 the fundamental matrix solution of (4.2) possesses the well known property $T(t,s|Y_0) = T(t,r|Y_0)T(r,s|Y_0)$, and one may verify directly that $S(t,s|Y_0) = T(s,r|Y_0)[S(t,r|Y_0) - S(s,r|Y_0)]$, from which the final conclusion is a ready consequence. In the case of $\Theta = *$ the result of Corollary 3 is of fundamental importance in the characterization of a principal solution of (2.1) at the endpoint b. In this connection the reader is referred to Hartman [4], and Reid [9; 15, Ch. VII, Secs. 3, 5].

5. Properties of Solutions of (2.9)

If $\hat{W} = \hat{W}_0(t)$ is a solution of the Riccati matrix differential equation (2.8) on a subinterval I_0 of I, and $s \in I_0$, let $\hat{G}(t) = \hat{G}(t,s|\hat{W}_0)$, $\hat{H}(t) = \hat{H}(t,s|\hat{W}_0)$ be specified as the solutions of the differential systems

$$(5.1) \qquad \hat{G}' + [\hat{A}^{\Theta} + \hat{W}_0\hat{B}]\hat{G} = 0, \quad \hat{G}(s) = E ,$$

$$(5.2)\qquad \hat{H}' + \hat{H}[\hat{A} + \hat{B}\hat{W}_0] = 0\ , \qquad \hat{H}(s) = E\ ,$$

and define the matrix function $\hat{F}(t,s|\hat{W}_0)$ as

$$(5.3)\qquad \hat{F}(t,s|\hat{W}_0) = \int_s^t \hat{H}(r,s|\hat{W}_0)\hat{B}(r)\hat{G}(r,s|\hat{W}_0)dr\ .$$

It then follows (see, for example, [11; Sec. 4], [15; Problem 9, p. 102]) that $\hat{W}(t)$ is a solution of (2.8) on I_0 if and only if the constant matrix $\hat{\Gamma} = \hat{W}(s) - \hat{W}_0(s)$ is such that the matrix function $E + \hat{F}(t,s|\hat{W}_0)\hat{\Gamma}$ is non-singular for $t \in I_0$, and in this case

$$(5.4)\qquad \hat{W}(t) = \hat{W}_0(t) + \hat{G}(t,s|\hat{W}_0)\hat{\Gamma}[E + \hat{F}(t,s|\hat{W}_0)\hat{\Gamma}]^{-1}\hat{H}(t,s|\hat{W}_0)\ .$$

Moreover, the non-singularity of $E + \hat{F}(t,s|\hat{W}_0)\hat{\Gamma}$ is equivalent to the non-singularity of $E + \hat{\Gamma}\hat{F}(t,s|\hat{W}_0)$, and hence $\hat{W}(t)$ is a solution of (2.8) on I_0 if and only if $E + \hat{\Gamma}\hat{F}(t,s|\hat{W}_0)$ is non-singular and

$$(5.4')\qquad \hat{W}(t) = \hat{W}_0(t) + G(t,s|\hat{W}_0)[E + \hat{\Gamma}\hat{F}(t,s|\hat{W}_0)]^{-1}\hat{\Gamma}\hat{H}(t,s|\hat{W}_0)\ .$$

Now as noted in Section 2, $\hat{W} = \hat{W}_0(t)$ is a solution of (2.8) on a subinterval I_0 of I if and only if $\hat{W} = \hat{W}_0^{\Theta}(t)$ is a solution of this equation on I_0. As $\hat{B}(t) \equiv \hat{B}^{\Theta}(t)$, we have the following result.

<u>LEMMA 5.1</u>. If $\hat{W}_0(t)$ is a solution of (2.8) on a subinterval I_0 of I, and $\hat{G}(t,s|\hat{W}_0)$, $\hat{H}(t,s|\hat{W}_0)$, $\hat{F}(t,s|\hat{W}_0)$ are defined by (5.1), (5.2), (5.3), then

$$(5.5)\qquad \begin{aligned} &(a)\quad \hat{G}^{\Theta}(t,s|\hat{W}_0) = \hat{H}(t,s|\hat{W}_0^{\Theta})\ ,\\ &(b)\quad \hat{H}^{\Theta}(t,s|\hat{W}_0) = \hat{G}(t,s|\hat{W}_0^{\Theta})\ ,\\ &(c)\quad \hat{F}^{\Theta}(t,s|\hat{W}_0) = \hat{F}(t,s|\hat{W}_0^{\Theta})\ .\end{aligned}$$

If $\hat{Y}_0(t) = (\hat{U}_0(t);\hat{V}_0(t))$ is a solution of $(2.3_M')$ such that $\hat{U}_0(t)$ is non-singular on I_0 and $\hat{W}_0(t) = \hat{V}_0(t)\hat{U}_0^{-1}(t)$, then in view of the definitive property of $\hat{H}(t,s|\hat{W}_0)$ as the solution of (5.2) we have that

$$(5.6)\qquad \hat{H}(t,s|\hat{W}_0) = \hat{U}_0(s)\hat{U}_0^{-1}(t)\ .$$

Moreover, if $\hat{W}_0(t)$ is a Θ-symmetric solution of (2.8) we have from (5.5-b) that $\hat{G}(t,s|\hat{W}_0) = \hat{H}^{\Theta}(t,s|\hat{W}_0) = \hat{U}_0^{\Theta-1}(t)\hat{U}_0^{\Theta}(s)$. If $Y_0(t) = (U_0(t);V_0(t))$

is a solution of (2.1_M) with $U_0(t)$ non-singular on a subinterval I_0 of I, and $\hat{Y}_0(t) = (\hat{U}_0(t);\hat{V}_0(t)) = (U_0(t);V_0(t) - M(t)U_0(t))$ is the corresponding solution of $(2.3'_M)$, then for $W_0(t) = V_0(t)U_0^{-1}(t)$, $\hat{W}_0(t) = \hat{V}_0(t)\hat{U}_0^{-1}(t)$, and matrix functions $\hat{A}$, $\hat{B}$, $\hat{C}$ defined by (2.6) we have $\hat{A} + \hat{B}\hat{W}_0 = A + BW_0$, $\hat{A}^\Theta + \hat{W}_0\hat{B} = A^\Theta + W_0B$. Consequently, if $s \in I$ and $G(t) = G(t,s|W_0)$, $H(t) = H(t,s|W_0)$, and $F(t,s|W_0)$ are defined by

$$G' + [A^\Theta + W_0B]G = 0\ , \qquad G(s) = E\ , \tag{5.7}$$

$$H' + H[A + BW_0] = 0\ , \qquad H(s) = E\ , \tag{5.8}$$

$$F(t,s\ W_0) = \int_s^t H(r,s|W_0)B(r)G(r,s|W_0)dr\ , \tag{5.9}$$

we have $G(t,s|W_0) = \hat{G}(t,s|\hat{W}_0)$, $H(t,s|W_0) = \hat{H}(t,s|\hat{W}_0)$, and $F(t,s|W_0) = \hat{F}(t,s|\hat{W}_0)$. In view of the above results, and Theorem 2.2, we have the following theorem.

<u>THEOREM 5.1</u>. Suppose that hypothesis $(\mathcal{H})$ holds, and $W = W_0(t)$ is a solution of the Riccati matrix integral equation (2.9) with $\Psi = \Psi_0$ on a subinterval I_0 of I containing $t = s$. If $G(t,s|W_0)$, $H(t,s|W_0)$, $F(t,s|W_0)$ are defined by (5.7), (5.8), (5.9), then a matrix function $W(t)$ is a solution of (2.9) on I_0 if and only if $\Gamma = \Psi - \Psi_0$ is such that $E + F(t,s|W_0)\Gamma$ is non-singular on I_0, and in this case

$$\begin{aligned} W(t) &= W_0(t) + G(t,s|W_0)\Gamma[E + F(t,s|W_0)\Gamma]^{-1}H(t,s|W_0)\ , \\ &= W_0(t) + G(t,s|W_0)[E + \Gamma F(t,s|W_0)]^{-1}\Gamma H(t,s|W_0)\ . \end{aligned} \tag{5.10}$$

If $W = W_0(t)$ is a Θ-symmetric solution of (2.9) with $\Psi = \Psi_0$ on a subinterval I_0 of I containing $t = s$, then $\Psi_0 = \Psi_0^\Theta$. Also, if $Y_0(t) = (U_0(t);V_0(t))$ is a Θ-conjugate basis for (2.1) for which $W_0(t) = V_0(t)U_0^{-1}(t)$ on I_0, then for $t \in I_0$ we have $H(t,s|W_0) = U_0(s)U_0^{-1}(t)$, $G(t,s|W_0) = U_0^{\Theta-1}(t)U_0^\Theta(s)$, and $F(t,s|W_0) = U_0(s)S(t,s|Y_0)U_0^\Theta(s)$. Therefore, as a direct consequence of the above theorem we have the following result.

THEOREM 5.2. Suppose that hypothesis $(\mathcal{H})$ holds, and $W = W_0(t)$ is a Θ-symmetric solution of (2.9) with $\Psi = \Psi_0$ on a subinterval I_0 of I containing $t = s$, and that $Y_0(t) = (U_0(t);V_0(t))$ is a Θ-conjugate basis for (2.1) determined by initial conditions $Y_0(s) = (M;W_0(s)M)$, with M non-singular. Then $U_0(t)$ is non-singular on I_0, and a matrix function $W = W(t)$, $t \in I_0$, is a solution of (2.9) on this interval if and only if the matrix $\Gamma_1 = U_0^{\Theta}(s)[\Psi - \Psi_0]U_0(s)$ is such that $E + S(t,s|Y_0)\Gamma_1$ is non-singular for $t \in I_0$, and in this case

$$
\begin{aligned}
W(t) &= W_0(t) + U_0^{\Theta-1}(t)\Gamma_1[E + S(t,s|Y_0)\Gamma_1]^{-1}U_0^{-1}(t) , \\
&= W_0(t) + U_0^{\Theta-1}(t)[E + \Gamma_1 S(t,s|Y_0)]^{-1}\Gamma_1 U_0^{-1}(t) .
\end{aligned}
\tag{5.11}
$$

6. Transformations for (2.1) and (2.9)

In this section we shall consider the transformation of (2.1) and (2.9) under a substitution

$$
u^0(t) = T^{-1}(t)u(t) , \qquad v^0(t) = T^{\Theta}(t)v(t) , \tag{6.1}
$$

where $T(t)$ is an $n \times n$ matrix function that is non-singular and locally a.c. on I. The following result is immediate formally; a precise derivation involving the differential expression $\Delta[u,v](t)$ may be attained by the integral method of [10; Lemma 2.1] and [11; Theorems 2.1, 2.2].

THEOREM 6.1. If hypothesis $(\mathcal{H})$ holds and $T(t)$ is an $n \times n$ matrix function which is non-singular and locally a.c. on I, then $y(t) = (u(t);v(t))$ is a solution of (2.1) on I if and only if $y^0(t) = (u^0(t);v^0(t))$ defined by (6.1) is a solution of the involutory system

$$
\begin{aligned}
\Delta^0[u^0,v^0](t) &\equiv -dv^0(t) + [C^0(t)u^0(t) - A^{0\Theta}(t)v^0(t)]dt \\
&\qquad + [dM^0(t)]u^0(t) = 0 , \\
L_2^0[u^0,v^0](t) &\equiv u^{0\prime}(t) - A^0(t)u^0(t) - B^0(t)v^0(t) = 0 ,
\end{aligned}
\tag{2.1^0}
$$

where

$$A^{o}(t) = T^{-1}(t)[A(t)T(t) - T'(t)], \qquad B^{o}(t) = T^{-1}(t)B(t)T^{\Theta-1}(t) ,$$

(6.2)

$$C^{o}(t) = T^{\Theta}(t)C(t)T(t) , \qquad M^{o}(t) = \int_{\tau}^{t} T^{\Theta}(s)[dM(s)]T(s) .$$

Without writing it specifically, the generalized matrix differential system corresponding to (2.1^{o}) will be denoted by (2.1^{o}_{M}).

If $y_{\alpha}(t) = (u_{\alpha}(t);v_{\alpha}(t))$, $(\alpha = 1,2)$, are solutions of (2.1), and $y^{o}_{\alpha}(t) = (u^{o}_{\alpha}(t);v^{o}_{\alpha}(t)) = (T^{-1}(t)u_{\alpha}(t);T^{\Theta}(t)v_{\alpha}(t))$, $(\alpha = 1,2)$, are the associated solutions of (2.1^{o}), then one may verify readily that $\{y_1|\Theta|y_2\} = \{y^{o}_1|\Theta|y^{o}_2\}$. In particular, $y_1(t)$ and $y_2(t)$ are Θ-conjugate solutions of (2.1) if and only if the corresponding $y^{o}_1(t)$ and $y^{o}_2(t)$ are Θ-conjugate solutions of (2.1^{o}). Also, $Y(t) = (U(t);V(t))$ is a Θ-conjugate basis for (2.1) if and only if the associated $Y^{o}(t) = (U^{o}(t);V^{o}(t)) = (T^{-1}(t)U(t);T^{\Theta}(t)V(t))$ is a Θ-conjugate basis for (2.1^{o}).

If $\Phi(t)$ is a fundamental matrix solution of $\Phi'(t) + A^{\Theta}(t)\Phi(t) = 0$, then $T(t) = \Phi^{\Theta-1}(t)$ is a fundamental matrix solution of $T'(t) - A(t)T(t) = 0$, and with this choice of $T(t)$ the matrix functions of (6.2) are

$$A^{o}(t) \equiv 0 , \qquad B^{o}(t) = \Phi^{\Theta}(t)B(t)\Phi(t) ,$$

(6.3)

$$C^{o}(t) = \Phi^{-1}(t)C(t)\Phi^{\Theta-1}(t), \qquad M^{o}(t) = \int_{\tau}^{t} \Phi^{-1}(s)[dM(s)]\Phi^{\Theta-1}(s) .$$

As in the case of ordinary differential equations (see Reid [13; Sec. 3], [16; Ch. III, Sec. 4]), such a $T(t)$ will be referred to as a <u>reducing transformation</u> for (2.1), and the resulting system (6.2) as a <u>reduced system</u>.

If (2.1) has order of abnormality equal to d on a subinterval I_0 of I, the fundamental matrix solution $\Phi(t)$ of $\Phi'(t) + A^{\Theta}(t)\Phi(t) = 0$ may be chosen so that the last d column vectors of $\Phi(t)$ form a basis for the linear vector space of n-dimensional vector functions $v(t)$ which are solutions of the differential equation $v'(t) + A^{\Theta}(t)v(t) = 0$, and satisfy $B(t)v(t) = 0$ on I_0. The transformation matrix $T(t) = \Phi^{\Theta-1}(t)$ is such that the matrix $B^{o}(t)$ of (6.3) is of the form $B^{o}(t) = \mathrm{diag}\{\hat{B}(t);0\}$ where $\hat{B}(t)$ is an $(n-d) \times (n-d)$ Θ-symmetric matrix function on I_0. Such a choice of $T(t)$ will be referred to as a <u>preferred reducing transformation for</u> (2.1).

For $T(t)$ a preferred reducing transformation, let the corresponding matrix functions $C^o(t)$ and $M^o(t)$ of (6.2) be written as

$$(6.4)\qquad C^o(t) = \begin{bmatrix} \hat{C}_{11}(t) & \hat{C}_{12}(t) \\ \hat{C}_{21}(t) & \hat{C}_{22}(t) \end{bmatrix}, \quad M^o(t) = \begin{bmatrix} \hat{M}_{11}(t) & \hat{M}_{12}(t) \\ \hat{M}_{21}(t) & \hat{M}_{22}(t) \end{bmatrix},$$

where $\hat{C}_{11}(t) = \hat{C}^{\Theta}_{11}(t)$ and $\hat{M}_{11}(t) = \hat{M}^{\Theta}_{11}(t)$ are $(n-d) \times (n-d)$, $\hat{C}_{12}(t) = \hat{C}^{\Theta}_{21}(t)$ and $\hat{M}_{12}(t) = \hat{M}^{\Theta}_{21}(t)$ are $(n-d) \times d$, and $\hat{C}_{22}(t) = \hat{C}^{\Theta}_{22}(t)$ and $\hat{M}_{22}(t) = \hat{M}^{\Theta}_{22}(t)$ are $d \times d$. Moreover, let $\eta(t)$, $\zeta(t)$ be $(n-d)$-dimensional vector functions defined as $\eta(t) = (u_\alpha(t))$, $\zeta(t) = (v_\alpha(t))$, $(\alpha = 1,\dots,n-d)$, and $\rho(t)$, $\sigma(t)$ be d-dimensional vector functions $\rho(t) = (u_{n-d+\beta}(t))$, $\sigma(t) = (v_{n-d+\beta}(t))$, $(\beta = 1,\dots,d)$. Then the generalized differential system (2.1) may be written as

$$(6.5)\qquad \begin{aligned} -d[\zeta(t) - \hat{M}_{12}(t)\rho(t_o)] + \hat{C}_{11}(t)\eta(t)dt + [d\hat{M}_{11}(t)]\eta(t) &= 0, \\ -d[\sigma(t) - \hat{M}_{22}(t)\rho(t_o)] + \hat{C}_{21}(t)\eta(t)dt + [d\hat{M}_{21}(t)]\eta(t) &= 0, \\ \eta'(t) - \hat{B}(t)\zeta(t) &= 0, \\ \rho'(t) &= 0. \end{aligned}$$

In particular, if $u(t) = (\eta(t);\rho(t))$, $v(t) = (\zeta(t);\sigma(t))$ are solutions of (2.1) defining t_1 and t_2 as conjugate points with respect to this system, then $\rho(t) \equiv 0$, and consequently such a pair of points are conjugate with respect to (2.1) if and only if they are conjugate with respect to the <u>truncated preferred reduced system</u>

$$(6.6)\qquad \begin{aligned} -d\zeta(t) + \hat{C}_{11}(t)\eta(t)dt + [d\hat{M}_{11}(t)]\eta(t) &= 0, \\ \eta'(t) - \hat{B}(t)\zeta(t) &= 0. \end{aligned}$$

If $W(t)$ is a solution of (2.9), and $Y(t) = (U(t);V(t))$ is a solution of (2.1_M) such that $W(t) = V(t)U^{-1}(t)$, then for $Y^o(t) = (U^o(t);V^o(t)) = (T^{-1}(t)U(t);T^{\Theta}(t)V(t))$ the corresponding solution of the matrix system (2.1^o_M) we have that $W^o(t) = V^o(t)U^{o-1}(t) = T^{\Theta}(t)W(t)T(t)$ is a solution of the involutory Riccati integral equation

$$(2.9^o)\qquad W^o(t) - \int_s^t [C^o(r) - W^o(r)A^o(r) - A^{o\Theta}(r)W^o(r) - W^o(r)B^o(r)W^o(r)]dr = M^o(t) + \Psi^o,$$

where $\Psi^o = T^\Theta(s)\Psi T(s)$. Also, $W(t)$ is a Θ-symmetric solution of (2.9) if and only if $W^o(t) = T^\Theta(t)W(t)T(t)$ is a Θ-symmetric solution of (2.9^o).

If $T(t)$ is a preferred reducing transformation for (2.1), so that $A^o(t) \equiv 0$, $B^o(t) = \mathrm{diag}\{\hat{B}(t);0\}$, while $C^o(t)$ and $M^o(t)$ are given by (6.4), upon writing $W^o(t)$ and Ψ^o as the corresponding partitioned matrices

$$(6.7) \qquad W^o(t) = \begin{bmatrix} W^o_{11}(t) & W^o_{12}(t) \\ W^o_{21}(t) & W^o_{22}(t) \end{bmatrix}, \qquad \Psi^o = \begin{bmatrix} \Psi^o_{11} & \Psi^o_{12} \\ \Psi^o_{21} & \Psi^o_{22} \end{bmatrix},$$

the Riccati matrix integral equation (2.9^o) may be written as the system

$$(6.8) \qquad W^o_{\alpha\beta}(t) - \int_s^t [\hat{C}_{\alpha\beta}(r) - W^o_{\alpha 1}(r)\hat{B}(r)W^o_{1\beta}(r)]dr = \hat{M}_{\alpha\beta}(t) + \Psi^o_{\alpha\beta} ,$$

for $\alpha,\beta = 1,2$. Clearly the interval of existence of $W^o(t)$ is that determined by the equation in $W^o_{11}(t)$ given by $\alpha = 1$, $\beta = 1$ in (6.8), which is the Riccati matrix integral equation associated with the truncated preferred reduced system (6.6).

7. Transformation of the Functional $J[\eta;a,b]$

For the case of $\Theta = *$ a basic result for the study of oscillation phenomena of solutions of (2.1), and associated boundary problems, is the so-called "Legendre" or "Clebsch" transformation of the functional $J[\eta;a,b]$ defined by (2.4'). The instance of ordinary differential equations is treated in [15; Ch. VII, Sec. 4]; for a generalized differential equation, see [11; Theorem 3.4]. For the general involution Θ a corresponding result is presented in the following theorem.

THEOREM 7.1. Suppose that $[a,b] \subset I$, and that $U(t)$, $V(t)$ are $n \times r$ matrix functions such that $V \in \mathcal{BV}_{nr}[a,b]$ and $U \in \mathcal{D}[a,b]:V$. If $\eta \in \mathcal{D}[a,b]:\zeta$ and there exists an a.c. r-dimensional vector function h such that $\eta(t) = U(t)h(t)$ on $[a,b]$, then

$$(7.1) \qquad J[\eta;a,b] = \int_a^b \{[\zeta - Vh]^\Theta B[\zeta - Vh] - h^\Theta[U^\Theta V - V^\Theta U]h'\}dt + \eta^\Theta Vh\Big|_a^b + \int_a^b \eta^\Theta(t)\{\Delta[U,V](t)\}h(t) .$$

In particular, if $Y(t) = (U(t);V(t))$ is a Θ-conjugate basis for (2.1) with $U(t)$ non-singular on $[a,b]$, then for $h(t) = U^{-1}(t)\eta(t)$ we have

$$J[\eta;a,b] = \eta^{\Theta}Vh\Big|_a^b + \int_a^b [\zeta - Vh]^{\Theta}B[\zeta - Vh]dt \; . \tag{7.2}$$

REFERENCES

[1] F.V. ATKINSON, Discrete and Continuous Boundary Problems. Academic Press, New York, 1960.

[2] G.A. BLISS, Lectures on the Calculus of Variations. Univ. of Chicago Press, Chicago, 1946.

[3] G. VON ESCHERICH, Die zweite Variation der einfachen Integrale, Wiener Sitzungsberichte (8), 107 (1898), 1191-1250.

[4] P. HARTMAN, Self-adjoint, nonoscillatory systems of ordinary, second order, linear differential equations, Duke Math. J., 24 (1957), 25-36.

[5] M. MORSE, A generalization of the Sturm separation and comparison theorems in n-space, Math. Annalen, 103 (1930), 72-91.

[6] M. MORSE, The Calculus of Variations in the Large. Amer. Math. Soc. Colloquium Publ., XVIII, (1934).

[7] W.T. REID, Boundary problems of the calculus of variations, Bull. Amer. Math. Soc., 42 (1937), 633-666.

[8] W.T. REID, Oscillation criteria for linear differential systems with complex coefficients, Pacific J. of Math., 6 (1956), 733-751.

[9] W.T. REID, Principal solutions of non-oscillatory self-adjoint linear differential systems, Pacific J. of Math., 8 (1958), 147-169.

[10] W.T. REID, Generalized linear differential systems, J. of Math. and Mech., 8 (1959), 705-726.

[11] W.T. REID, Generalized linear differential systems and related Riccati matrix integral equations, Illinois J. of Math., 10 (1966), 701-722.

[12] W.T. REID, Variational methods and boundary problems for ordinary linear differential systems, Proc. U.S.-Japan Sem. on Differential and Functional Equations, Univ. of Minnesota, Minneapolis, Minn., June 26-30, 1967; W.A. Benjamin, Inc., 267-299.

[13] W.T. REID, Monotoneity properties of solutions of hermitian Riccati matrix differential equations, SIAM J. of Math. Anal., 1 (1970), 195-213.

[14] W.T. REID, A maximum problem involving generalized linear differential equations of the second order, J. of Differential Equations, 8 (1970), 283-293.

[15] W.T. REID, Ordinary Differential Equations. John Wiley and Sons, New York, 1971.

[16 W.T. REID, Riccati Differential Equations. To be published by Academic Press.

University of Oklahoma, Norman, Oklahoma

CONTROL THEORY OF HYPERBOLIC EQUATIONS RELATED TO CERTAIN QUESTIONS IN HARMONIC ANALYSIS AND SPECTRAL THEORY*
(An Outline)

David L. Russell

1. Introduction

In a 1967 paper "Nonharmonic Fourier Series in the Control Theory of Distributed Parameter Systems" [14] we have shown that the classical results of Paley and Wiener [12], Levinson [10], Schwartz [17] and others man be used to advantage in studying the controllability of the wave equation in a single space dimension. The purpose of the present article goes beyond such a simple application of existing results in harmonic analysis to control problems. We wish to show in addition that the study of control problems for certain hyperbolic partial differential equations leads to some interesting, and perhaps unexpected, consequences in harmonic analysis. Thus there is a two-way interplay between these two subjects, only recently becoming apparent, and we may hope for deeper studies of this relationship in the future.

Because our purpose is to uncover this relationship, we will not attempt great generality in our presentations. Many of the results which we will obtain are valid for any second order linear hyperbolic partial differential equation in two independent variables x and t

*Supported in part by the Office of Naval Research under Contract NR-041-404.

whose coefficients depend only upon x. However, such a complete treatment would introduce complications which would obscure our main points. Hence we shall focus our attention in this paper on systems related to partial differential equations of the form

$$\rho(x) \frac{\partial^2 w}{\partial t^2} - p(x) \frac{\partial^2 w}{\partial x^2} + q(x) \frac{\partial w}{\partial t} + r(x) \frac{\partial w}{\partial x} = 0 , \tag{1.1}$$

$$0 \leq x \leq 1 , \quad 0 \leq t < \infty ,$$

where the coefficient functions ρ, p, q and r are twice continuously differentiable for $0 \leq x \leq 1$ and

$$\rho(x) \geq \rho_0 > 0 , \quad p(x) \geq p_0 > 0 , \quad 0 \leq x \leq 1 .$$

If (1.1) is thought of as a model for small vibrations of a flexible string, ρ is the linear mass density and p is the modulus of elasticity.

We shall impose boundary conditions of the form

$$A_0 \frac{\partial w}{\partial t}(0,t) + B_0 \frac{\partial w}{\partial x}(0,t) \equiv 0 , \quad 0 \leq t < \infty , \tag{1.2}$$

$$A_1 \frac{\partial w}{\partial t}(1,t) + B_1 \frac{\partial w}{\partial x}(1,t) \equiv f(t) , \quad 0 \leq t < \infty , \tag{1.3}$$

with the proviso that

$$\frac{A_0}{B_0} \neq \pm\left(\frac{\rho(0)}{p(0)}\right)^{\frac{1}{2}} , \quad \frac{A_1}{B_1} = \pm\left(\frac{\rho(1)}{p(1)}\right)^{\frac{1}{2}} . \tag{1.4}$$

If we again use the physical analogy of the flexible string, the boundary condition (1.2) corresponds to a fixed end $(B_0 = 0)$, an end free to move in the direction of the w axis $(A_0 = 0)$, or an end free to move but with positive or negative friction $(A_0 \neq 0$, $B_0 \neq 0)$. The reason for the restrictions (1.4) will become clear later. The boundary condition (1.3) at $x = 1$ can be interpreted similarly with $f(t)$ a "control" force at our disposal with which we attempt to influence the evolution of solutions of (1.1).

We will find it convenient to put our problem in a certain standard form. The change of independent variable

$$\xi = \int_0^x \left(\frac{\rho(s)}{p(s)}\right)^{\frac{1}{2}} ds$$

carries (1.1) into an equation of the form

$$\frac{\partial^2 w}{\partial t^2} - \frac{\partial^2 w}{\partial \xi^2} + a(\xi)\frac{\partial w}{\partial t} + b(\xi)\frac{\partial w}{\partial \xi} = 0 \tag{1.5}$$
$$0 \leqslant \xi \leqslant \ell \equiv \xi(1) , \quad 0 \leqslant t < \infty .$$

The coefficients $a(\xi)$, $b(\xi)$ are now continuously differentiable functions of ξ. This second order scalar equation can be replaced by the first order two dimensional system

$$\frac{\partial}{\partial t}\binom{u}{v} - \begin{pmatrix} 0 & 1 \\ 1 & 0 \end{pmatrix} \frac{\partial}{\partial \xi}\binom{u}{v} + \begin{pmatrix} a(\xi) & b(\xi) \\ 0 & 0 \end{pmatrix}\binom{u}{v} = 0 , \tag{1.6}$$

where $u = \frac{\partial w}{\partial t}$, $v = \frac{\partial w}{\partial \xi}$. Every solution of (1.5) in class C^m, $m \geqslant 2$, corresponds to a solution of (1.6) of class C^{m-1}. It should be noted however that two solutions of (1.5) differing by a non-zero constant are carried into the same solution of (1.6). Otherwise the correspondence is complete in both directions. The appropriate boundary conditions are now

$$\alpha_0 u(0,t) + \beta_0 v(0,t) \equiv 0 , \tag{1.7}$$

$$\alpha_1 u(\ell,t) + \beta_1 v(\ell,t) \equiv f(t) , \tag{1.8}$$

with the condition

$$\frac{\alpha_0}{\beta_0} \neq \pm 1 , \quad \frac{\alpha_1}{\beta_1} \neq \pm 1 . \tag{1.9}$$

We have arrived at the system (1.6) because we wished to introduce our topic by means of the familiar scalar equation (1.1). But all of the work which we do is done just as easily if we generalize (1.6) slightly to

$$\frac{\partial}{\partial t}\binom{u}{v} - \begin{pmatrix} 0 & 1 \\ 1 & 0 \end{pmatrix} \frac{\partial}{\partial \xi}\binom{u}{v} + \begin{pmatrix} a_{11}(\xi) & a_{12}(\xi) \\ a_{21}(\xi) & a_{22}(\xi) \end{pmatrix}\binom{u}{v} = 0 \tag{1.10}$$

where the real coefficients $a_{ij}(\xi)$ are continuously differentiable for $0 \leq \xi \leq \ell$. We retain the boundary conditions (1.7), (1.8).

By studying the controllability of the system (1.6), (1.7), (1.8), we are able to prove certain theorems about the operator

$$(1.11) \qquad L\begin{pmatrix} u \\ v \end{pmatrix} = \begin{pmatrix} 0 & 1 \\ 1 & 0 \end{pmatrix} \frac{d}{d\xi}\begin{pmatrix} u \\ v \end{pmatrix} - \begin{pmatrix} a_{11}(\xi) & a_{12}(\xi) \\ a_{21}(\xi) & a_{22}(\xi) \end{pmatrix}\begin{pmatrix} u \\ v \end{pmatrix}$$

with boundary conditions of the type (1.7), (1.8). In particular, if the (in general complex) eigenvalues of L are $\{\lambda_k\}$, we will be able to establish that $\{e^{\lambda_k t}\}$ form a Riesz basis for $L^2[0,2\ell]$ in a way very different from that pursued by Paley and Wiener, Levinson, Schwartz and others. Moreover, by showing that the controls which bring solutions of (1.10), (1.7), (1.8) to zero at time $t = 2\ell$ can be synthesized by means of a linear feedback control law, we prove a rather unusual characterization of the dual basis of $L^2[0,2\ell]$ relative to $\{e^{\lambda_k t}\}$ which has possible application to numerical computation of the functions $\{q_k(t)\}$ which are biorthogonal to $\{e^{\lambda_k t}\}$.

2. Principal Results

In this section we state our theorems for the system (1.10), (1.7), (1.8) and supply proofs where they are reasonably short. The proofs of Theorems 1 and 3 are long and are not given here.

The basis of our work is the question of finite time controllability. This topic has been studied earlier by the author [14], [15] and in a thesis by J. Grainger [7]. The present work begins with a statement of these results in terms of "finite energy" solutions, i. e., generalized solutions of (1.10), (1.7), (1.8) for which

$$\int_0^\ell [|u(\xi,t)|^2 + |v(\xi,t)|^2] d\xi < \infty, \quad t \geq 0 .$$

Appropriate existence, uniqueness and regularity theorems for such solutions may be found in [9] and [11]. Although we have taken all of the coefficients in our partial differential equation and boundary conditions to be real, we will find it convenient to consider complex solutions.

THEOREM 1. Let initial and terminal states (u_0,v_0) and (u_1,v_1) be given at the times $t = 0$ and $t = 2\ell$, respectively, with u_0, v_0, u_1, v_1 all in $L^2[0,\ell]$. Then there is exactly one function $f \in L^2[0,2\ell]$ such that the solution (u,v) of (1.10), (1.7), (1.8) which satisfies

$$(2.1) \qquad u(\xi,0) = u_0(\xi), \quad v(\xi,0) = v_0(\xi) \quad \text{a.e. in} \quad [0,\ell]$$

also satisfies

$$u(\xi,2\ell) = u_1(\xi), \quad v(\xi,2\ell) = v_1(\xi) \quad \text{a.e. in} \quad [0,\ell]$$

and there is a positive constant P, independent of u_0, v_0, u_1, v_1 such that

$$(2.3) \qquad \int_0^{2\ell} |f(t)|^2 dt \leq P\int_0^{\ell} (|u_0(\xi)|^2 + |v_0(\xi)|^2 + |u_1(\xi)|^2 + |v_1(\xi)|^2) d\xi \ .$$

Moreover, there is a second positive constant $\hat{P}$ such that when $u_1 = v_1 = 0$

$$(2.4) \qquad \int_0^{\ell} (|u_0(\xi)|^2 + |v_0(\xi)|^2) d\xi \leq \hat{P} \int_0^{2\ell} |f(t)|^2 dt \ .$$

Also, when $u_1 = v_1 = 0$ the condition (1.9) can be replaced by the weaker restriction

$$\frac{\alpha_0}{\beta_0} \neq 1 \ , \qquad \frac{\alpha_1}{\beta_1} \neq -1$$

and the existence of f satisfying (2.3) is still assured. However, (2.4) cannot be proved in this case.

The proofs of the main theorems stated in this outline are given in a more complete version of this paper [18].

We have shown in [14] and [15] that it is in general impossible to satisfy the given initial and terminal conditions if less time is allowed, while the control f is not unique if more time is allowed.

The system (1.10), (1.7), (1.8) with $f \equiv 0$ has the form

$$\frac{d}{dt} \binom{u}{v} = L \binom{u}{v}$$

where L is the differential operator defined by (1.11) with domain Δ

in $L^2[0,\ell] \otimes L^2[0,2\ell]$ consisting of pairs of functions (u,v) whose first derivatives, taken in the sense of the theory of distributions, lie in $L^2[0,\ell]$ and which satisfy

$$(2.5) \qquad \alpha_0 u(0) + \beta_0 v(0) = \alpha_1 u(\ell) + \beta_1 v(\ell) = 0 .$$

The adjoint of L is the operator

$$L^*\binom{w}{z} = -\begin{pmatrix} 0 & 1 \\ 1 & 0 \end{pmatrix} \frac{d}{d\xi}\binom{w}{z} - \begin{pmatrix} a_{11}(\xi) & a_{21}(\xi) \\ a_{12}(\xi) & a_{22}(\xi) \end{pmatrix} \binom{w}{z}$$

defined on the domain Δ^* which differs from Δ in that (w,z) belonging to it satisfy

$$(2.6) \qquad \alpha_0 w(0) - \beta_0 z(0) = \alpha_1 w(\ell) - \beta_1 z(\ell) = 0 .$$

Very general results due to Birkhoff [1], Schwartz [16], Kramer [8] and others show that L is a <u>spectral operator</u>, in particular it has a sequence of complex simple eigenvalues $\{\lambda_k\}$ such that the associated normalized eigenvectors (ϕ_k, ψ_k) form a Riesz basis in $L^2[0,\ell] \otimes L^2[0,\ell]$, i.e., each (u,v) in that space has a unique development

$$(2.7) \qquad \binom{u}{v} = \sum c_k \begin{pmatrix} \phi_k \\ \psi_k \end{pmatrix}$$

with

$$(2.8) \qquad m_1 \sum |c_k|^2 \leqslant \left\| \binom{u}{v} \right\| \leqslant m_2 \sum |c_k|^2$$

for fixed positive constants m_1, m_2. The adjoint operator L^* has eigenvalues $\{\bar{\lambda}_k\}$ which are the complex conjugates of the $\{\lambda_k\}$ and eigenvectors (ϕ_k^*, ψ_k^*) such that

$$(2.9) \qquad \left(\begin{pmatrix} \phi_\ell \\ \psi_\ell \end{pmatrix}, \begin{pmatrix} \phi_\ell^* \\ \psi_\ell^* \end{pmatrix} \right)_{L^2[0,\ell]\otimes L^2[0,\ell]} = \delta_{k\ell} = \begin{cases} 1, & k = \ell \\ 0, & k = \ell \end{cases} .$$

Now let (u,v) be a (possibly complex) solution of (1.10), (1.7), (1.8) and (w,z) a (possibly complex) solution of

$$(2.10) \qquad \frac{\partial}{\partial t}\binom{w}{z} = \begin{pmatrix} 0 & 1 \\ 1 & 0 \end{pmatrix} \frac{\partial}{\partial \xi}\binom{w}{z} + \begin{pmatrix} a_{11}(\xi) & a_{21}(\xi) \\ a_{12}(\xi) & a_{22}(\xi) \end{pmatrix} \binom{w}{z},$$

satisfying boundary conditions of the form (2.6). If u_0, v_0, u_1, v_1 all belong to $C^1[0,\ell]$ one easily justifies the following computation in the rectangle

$$D = \{(\xi,t) \,|\, 0 \leqslant \xi \leqslant \ell,\ \ 0 \leqslant t \leqslant 2\ell\}:$$

$$\begin{aligned}
0 &= \iint_D \Bigg\{ \Bigg[\binom{u}{v}, \Bigg(\frac{\partial}{\partial t}\binom{w}{z} - \begin{pmatrix} 0 & 1 \\ 1 & 0 \end{pmatrix} \frac{\partial}{\partial \xi}\binom{w}{z} - \begin{pmatrix} a_{11} & a_{21} \\ a_{12} & a_{22} \end{pmatrix} \binom{w}{z} \Bigg) \Bigg] \\
&\quad + \Bigg[\Bigg(\frac{\partial}{\partial t}\binom{u}{v} - \begin{pmatrix} 0 & 1 \\ 1 & 0 \end{pmatrix} \frac{\partial}{\partial \xi}\binom{u}{v} + \begin{pmatrix} a_{11} & a_{12} \\ a_{21} & a_{22} \end{pmatrix} \binom{u}{v} \Bigg), \binom{w}{z} \Bigg] \Bigg\} d\xi dt \\
(2.11) \quad & \\
&= \iint_D \left\{ \frac{\partial}{\partial t}(u\overline{w} + v\overline{z}) - \frac{\partial}{\partial \xi}(u\overline{z} + v\overline{w}) \right\} d\xi dt \\
&= \int_0^{\ell} [(u(\xi,2\ell)\overline{w}(\xi,2\ell) + v(\xi,2\ell)\overline{z}(\xi,2\ell)) \\
&\qquad - (u(\xi,0)\overline{w}(\xi,0) + v(\xi,0)\overline{z}(\xi,0))] d\xi \\
&\quad + \int_0^{2\ell} [(u(0,t)\overline{z}(0,t) + v(0,t)\overline{w}(0,t)) \\
&\qquad - (u(\ell,t)\overline{z}(\ell,t) + v(\ell,t)\overline{w}(\ell,t))] dt \,.
\end{aligned}$$

If we expand the solution $(u(\xi,t),v(\xi,t))$ as in (2.7):

$$\begin{pmatrix} u(\xi,t) \\ v(\xi,t) \end{pmatrix} = \sum c_k(t) \begin{pmatrix} \phi_k(\xi) \\ \psi_k(\xi) \end{pmatrix},$$

and note that

$$e^{-\overline{\lambda}_k(T-t)} \begin{pmatrix} \phi_k^*(\xi) \\ \psi_k^*(\xi) \end{pmatrix} = \begin{pmatrix} w(\xi,t) \\ z(\xi,t) \end{pmatrix}$$

is a solution of (2.10), we may substitute in (2.11) and use (2.9) and the boundary conditions satisfied by (u,v) and (w,z) at 0 and ℓ

to see that

$$(2.12)\qquad c_k(T) - c_k(0)e^{-\lambda_k T} = \begin{cases} \int_0^{2\ell} \dfrac{\overline{\phi_k^*(\ell)}}{\beta_1} e^{-\lambda_k(T-t)} f(t)dt, & \beta_1 \neq 0 \\[2ex] \int_0^{2\ell} \dfrac{\overline{\psi_k^*(\ell)}}{\alpha_1} e^{-\lambda_k(T-t)} f(t)dt, & \beta_1 = 0 . \end{cases}$$

Now let

$$(2.13)\qquad \begin{pmatrix} u(\xi,0) \\ v(\xi,0) \end{pmatrix} = \begin{pmatrix} u_0(\xi) \\ v_0(\xi) \end{pmatrix} = \sum c_k(0) \begin{pmatrix} \phi_k(\xi) \\ \psi_k(\xi) \end{pmatrix}$$

be steered by means of the control f to the zero terminal state $u_1(\xi) \equiv v_1(\xi) \equiv 0$. Then in (2.12) $c_k(T) = 0$ and thus

$$(2.14)\qquad -c_k(0) = \begin{cases} \int_0^{2\ell} \dfrac{\overline{\phi_k^*(\ell)}}{\beta_1} e^{\lambda_k t} f(t)dt , & \beta_1 \neq 0 \\[2ex] \int_0^{2\ell} \dfrac{\overline{\psi_k^*(\ell)}}{\alpha_1} e^{\lambda_k t} f(t)dt , & \beta_1 = 0 , \end{cases}$$

for all k.

In order to perform the calculations (2.11) we assumed u_0, v_0 belong to $C^1[0,\ell]$. One may verify easily, however, that (2.14) will also hold for u_0, v_0 in $L^2[0,\ell]$ through approximation of (2.13) by finite partial sums and use of the inequalities of Theorem 1. We leave this to the reader. Then it follows from (2.8) that if $\{c_k(0)\}$ is any sequence of complex numbers with

$$\sum |c_k(0)|^2 < \infty$$

the <u>moment problem</u> (2.14) has a solution $f \in L^2[0,2\ell]$. Moreover, using (2.3), (2.4) and (2.8) we see that there are positive numbers K_1 and K_2, independent of $\{c_k(0)\}$, such that

$$(2.15)\qquad K_1 \sum |c_k(0)|^2 \leqslant \int_0^{2\ell} |f(t)|^2 dt \leqslant K_2 \sum |c_k(0)|^2 .$$

We have used here the facts, provable with a minimum of difficulty, that when $\beta_1 \neq 0$, $|\phi_k^*(\ell)|$ is bounded away from 0 and ∞ and when $\beta_1 = 0$, $|\psi_k^*(\ell)|$ is bounded away from 0 and ∞. Then (2.14) and (2.15) together prove

THEOREM 2. Let the eigenvalues of L be $\{\lambda_k\}$ and let $\{c_k\}$ be any sequence of complex numbers with $\Sigma|c_k|^2 < \infty$. Then the moment problem:

$$c_k = \int_0^{2\ell} e^{\lambda_k t} f(t)dt \quad \text{for all } k, \tag{2.16}$$

has a unique solution of $f \in L^2[0,2\ell]$ such that

$$K_3 \sum |c_k|^2 \leq \int_0^{2\ell} |f(t)|^2 dt \leq K_4 \sum |c_k|^2$$

for certain positive constants K_3 and K_4 independent of $\{c_k\}$.

This theorem implies that the functions $\{e^{\lambda_k t}\}$ form a Riesz basis for the space $L^2[0,2\ell]$, i.e., every function $g \in L^2[0,2\ell]$ has an expansion

$$g(t) = \sum \gamma_k e^{\lambda_k t},$$

convergent in $L^2[0,2\ell]$, with the property

$$K_5 \sum |\gamma_k|^2 \leq \|g\|_{L^2[0,2\ell]} \leq K_6 \sum |\gamma_k|^2$$

for positive constants K_5 and K_6. The coefficients γ_k are given by

$$\gamma_k = \int_0^{2\ell} g(t)\overline{q}_k(t)dt$$

where $q_k(t)$ is the solution of (2.16) with $c_k = 1$, $c_\ell = 0$, $\ell \neq k$. The sequence $\{q_k\}$ is the biorthogonal sequence for $\{e^{\lambda_k t}\}$, or the dual basis for $L^2[0,2\ell]$ relative to the basis $\{e^{\lambda_k t}\}$. Another interpretation is that q_k is the unique control function steering the initial state

$$(2.17) \qquad \begin{pmatrix} u_0(\xi) \\ v_0(\xi) \end{pmatrix} = -r_k \begin{pmatrix} \phi_k(\xi) \\ \psi_k(\xi) \end{pmatrix}, \quad r_k = \begin{cases} \dfrac{\beta_1}{\overline{\phi_k^*(\ell)}}, & \beta_1 \neq 0 \\[2ex] \dfrac{\alpha_1}{\overline{\psi_k^*(\ell)}} & \beta_1 = 0 , \end{cases}$$

to zero in time $t = 2$.

Now one could also prove all of these results by the Fourier transform methods of Paley and Wiener [12], Levinson [10] and Schwartz [17], provided one had sufficiently good asymptotic estimates of the location of the eigenvalues $\{\lambda_k\}$. In this respect the interesting thing about Theorem 2 is that it has been proved without detailed reference to the location of these eigenvalues. Even the necessary information that L is a spectral operator can be proved rather easily with the partial differential equations methods we employ together with a general theorem in [16]. Of course our work is quite special since it applies only to sequences $\{\lambda_k\}$ consisting of eigenvalues of operators L defined above whereas the work of the authors cited applies to much more general sequences.

The familiar results to the effect that the functions $\{e^{\lambda_k t}\}$ are excessive in $L^2[0,T]$ if $T < 2\ell$ and deficient but linearly independent in $L^2[0,T]$ if $T > 2\ell$ can also be proved using methods like these. How this would be done should be clear from the work in [14] and [15] together with what we have already written here so we will not go into details.

While the proof of Theorem 1 [18] is constructive, the method used is not particularly well adapted to computation. Thus it is significant that this control f can be synthesized by means of a linear feedback control law, provided a linear relationship holds between the initial and terminal states. This, and other consequences, follow from

<u>THEOREM 3</u>. Let u_0 and v_0 lie in $L^2[0,\ell]$ and let γ be any real number. Let

$$(2.18) \qquad \sigma^+ = \exp\left(-\frac{1}{2}\int_0^\ell [a_{11}(\xi) + a_{12}(\xi) + a_{21}(\xi) + a_{22}(\xi)]d\xi \right) ,$$

$$(2.19) \qquad \sigma^- = \exp\left(-\frac{1}{2}\int_0^\ell [-a_{11}(\xi) + a_{12}(\xi) + a_{21}(\xi) - a_{22}(\xi)]d\xi\right) ,$$

where the $a_{ij}(\xi)$ are the coefficients appearing in (1.10). Let (u,v) be a solution of (1.10), (1.7) and (1.8). Then

$$(2.20) \qquad u(\xi,2\ell) \equiv \gamma u_0(\xi) , \qquad v(\xi,2\ell) \equiv \gamma v_0(\xi)$$

if and only if the solution (u,v) satisfies the boundary condition

$$(2.21) \qquad \left(\frac{\sigma^+}{\beta_0 - \alpha_0} - \frac{\gamma\sigma^-}{\beta_0 + \alpha_0}\right) u(\ell,t) + \left(\frac{\sigma^+}{\beta_0 - \alpha_0} + \frac{\gamma\sigma^-}{\beta_0 + \alpha_0}\right) v(\ell,t) = \int_0^\ell [h_1(\xi)u(\xi,t) + h_2(\xi)v(\xi,t)]d\xi$$

where h_1 and h_2 are certain continuous functions depending only upon the a_{ij}, α_0, β_0 and γ. When the a_{ij} are all zero, h_1 and h_2 vanish identically. When $\gamma = 0$ it is sufficient to assume $\frac{\alpha_0}{\beta_0} \neq 1$ and the term $\frac{\gamma\sigma^-}{\beta_0 + \alpha_0}$ disappears.

An immediate consequence of Theorem 3 is the feedback law for the control f. If we put

$$(2.22) \qquad \alpha_2 = \frac{\sigma^+}{\beta_0 - \alpha_0} - \frac{\gamma\sigma^-}{\beta_0 + \alpha_0} , \qquad \beta_2 = \frac{\sigma^+}{\beta_0 - \alpha_0} + \frac{\gamma\sigma^-}{\beta_0 + \alpha_0}$$

we verify readily that

$$(\alpha_2)^2 + (\beta_2)^2 = 2\left[\frac{(\sigma^+)^2}{(\beta_0 - \alpha_0)^2} + \frac{(\gamma\sigma^-)^2}{(\beta_0 + \alpha_0)^2}\right] > 0 .$$

If the vector (α_2,β_2) is a multiple of (α_1,β_1), say $(\alpha_2,\beta_2) = c(\alpha_1,\beta_1)$, $c \neq 0$, then (1.8) and (2.21) together yield

$$f(t) = \int_0^\ell \left[\frac{h_1(\xi)}{c} u(\xi,t) + \frac{h_2(\xi)}{c} v(\xi,t)\right]d\xi .$$

If (α_2,β_2) and (α_1,β_1) are linearly independent, then one can find a third vector (α_3,β_3) in R^2 such that (α_2,β_2) and (α_3,β_3) are linearly independent and

$$(\alpha_2,\beta_2) = c_1(\alpha_1,\beta_1) - c_2(\alpha_3,\beta_3)$$

with $c_1 \neq 0$. Then

$$f(t) = \frac{c_2\alpha_3}{c_1} u(\ell,t) + \frac{c_2\beta_3}{c_1} v(\ell,t) + \int_0^\ell \left[\frac{h_1(\xi)}{c_1} u(\xi,t) + \frac{h_2(\xi)}{c_1} v(\xi,t)\right] d\xi \ .$$

Thus we have proved

THEOREM 4. Let initial and terminal conditions (u_0,v_0) and (u_1,v_1) be given satisfying (2.20). Then the control f steering the solution (u,v) of (1.10), (1.7), (1.8) from (u_0,v_0) to $(u_1,v_1) = \gamma(u_0,v_0)$ satisfies a feedback law

$$f(t) = \mu u(\ell,t) + \nu v(\ell,t) + \int_0^\ell [k_1(\xi)u(\xi,t) + k_2(\xi)v(\xi,t)]d\xi$$

where k_1 and k_2 lie in $C^1[0,\ell]$ and (μ,ν) is either the zero vector or else (α_1,β_1) and (μ,ν) are linearly independent.

Whether or not this synthesis of the control function f could be useful in applications is not entirely clear. The first thing to be checked is whether its use would give rise to serious instabilities. Since $u(\ell,t)$ and $v(\ell,t)$ are measured at the precise point $\xi = \ell$ where the control f is applied, this is a defnnite possibility.

If we take $\gamma = 0$ in Theorem 3 we see that a solution of (1.10), (1.7) and

$$u(\ell,t) + v(\ell,t) = \frac{\beta_0 + \alpha_0}{\sigma^+} \int_0^\ell [h_1(\xi)u(\xi,t) + h_2(\xi)v(\xi,t)]d\xi \tag{2.23}$$

always satisfies $u(\xi,2\ell) \equiv v(\xi,2\ell) \equiv 0$. Then from (2.17) and the remarks accompanying it we see that the functions $\{q_k(t)\}$ biorthogonal to $\{e^{\lambda_k t}\}$ can be computed by solving (1.10), (1.7), (2.23) with the initial state (2.17) and then using (1.8). Since the computation of h_1 and h_2 can be carried out once and for all (see Section 4 in [18]) by solving a relatively simple partial differential equation, we have here a possible method for the numerical calculation of the functions $\{q_k(t)\}$. We remark that (1.10), (1.7) and (2.23) is a system whose solutions can be approximated rather easily using the method of characteristics [3].

Now we will make some comments about the implications of Theorem 3 in a general mathematical sense, not particularly relation to control problems. Fixing γ as in Theorem 3, we consider the unbounded operator L_1 defined in the Hilbert space $L^2[0,\ell] \oplus L^2[0,\ell]$ by (1.11) but with domain Δ_1 consisting of pairs of functions (u,v) satisfying

$$\alpha_0 u(0) + \beta_0 v(0) = 0 \tag{2.24}$$

and (cf. (2.21), (2.22))

$$\alpha_2 u(\ell) + \beta_2 v(\ell) = \int_0^\ell [h_1(\xi)u(\xi) + h_2(\xi)v(\xi)]d\xi \tag{2.25}$$

and having first derivatives in $L^2[0,\ell]$. It is not difficult to verify that Δ_1 is dense in $L^2[0,\ell] \oplus L^2[0,\ell]$.

Solutions (u,v) of (1.10), (2.24), (2.25) have the form

$$\binom{u}{v} = e^{L_1 t}\begin{pmatrix} u_0 \\ v_0 \end{pmatrix},$$

where $e^{L_1 t}$ is the strongly continuous semi-group (group if $\gamma \neq 0$) generated by the operator L_1. Theorem 3 gives us certain information about this semigroup (group) which in turn indicates some interesting properties of the operator L_1.

If $\gamma = 0$ the semigroup $e^{L_1 t}$ has the property

$$e^{L_1(2\ell)}\begin{pmatrix} u_0 \\ v_0 \end{pmatrix} = 0$$

for all (u_0,v_0) in $L^2[0,\ell] \oplus L^2[0,\ell]$. Thus

$$e^{L_1 t} = 0$$

for $t \geqslant 2\ell$. Thus we have a somewhat unusual example of a strongly continuous semigroup which vanishes identically after a certain time, in this case 2ℓ. From results in [5] we see that the spectrum of L_1 must be empty in this case.

This result can be proved more or less directly when the a_{ij} are all zero (so that h_1 and h_2 are also zero). In this case the boundary condition (2.25) becomes

$$u(\ell) + v(\ell) = 0 . \tag{2.26}$$

When the a_{ij} are not all identically zero the properties of the operator L with a right hand boundary condition of the form (2.26) are rather elusive. This is one of the singular cases encountered by Birkhoff [1] and others in their pioneering work on the spectral properties of such operators. The significance of our work lies in the fact that we have shown that if in this singular case we replace the boundary condition (2.26) by

$$u(\ell) + v(\ell) = \frac{\beta_0 - \alpha_0}{\sigma^+} \int_0^\ell [h_1(\xi)u(\xi) + h_2(\xi)v(\xi)]d\xi$$

$$\left(\alpha_2 = \beta_2 = \frac{\sigma^+}{\beta_0 - \alpha_0} \quad \text{if} \quad \gamma = 0\right)$$

then once again we have an operator whose spectrum is empty. We remark that one can give examples to show that this is not generally true for the boundary condition (2.26) when the a_{ij} are non-zero.

Now we take up the case $\gamma \neq 0$. Theorem 3 then shows that the group $e^{L_1 t}$ has the property

$$e^{L_1(2\ell)}\begin{pmatrix} u_0 \\ v_0 \end{pmatrix} \doteq \gamma\begin{pmatrix} u_0 \\ v_0 \end{pmatrix}$$

so that $e^{L_1(2\ell)} = \gamma I$. Letting

$$\rho = \frac{\log|\gamma|}{2\ell}$$

it is clear that the group $e^{(L_1 - \rho I)t}$ is <u>periodic</u> with period 2ℓ when $\gamma > 0$:

(2.27) $$e^{(L_1-\rho I)2\ell} = I , \quad \gamma > 0 ,$$

and anti-periodic when $\gamma < 0$, i.e.

$$e^{(L_1-\rho I)2\ell} = -I , \quad \gamma < 0 .$$

Consider the case $\gamma > 0$. We define a new inner product $\langle\ ,\ \rangle$ in $L^2[0,\ell] \oplus L^2[0,\ell]$ by

$$\left\langle \begin{pmatrix} u_0 \\ v_0 \end{pmatrix}, \begin{pmatrix} \hat{u}_0 \\ \hat{v}_0 \end{pmatrix} \right\rangle = \int_0^{2\ell} \left(e^{(L_1-\rho I)t} \begin{pmatrix} u_0 \\ v_0 \end{pmatrix}, e^{(L_1-\rho I)t} \begin{pmatrix} \hat{u}_0 \\ \hat{v}_0 \end{pmatrix} \right) dt ,$$

where $(\ ,\ )$ is the usual inner product in that space. Because the operators $e^{(L_1-\rho I)t}$ are uniformly bounded and have uniformly bounded inverses (the latter a consequence of (2.27)) we see that the norm $\langle\!\langle\ \rangle\!\rangle$ associated with the inner product $\langle\ ,\ \rangle$ is equivalent to the usual norm $\|\ \|$ associated with $(\ ,\ )$ in the sense that

$$r_1 \langle\!\langle \begin{pmatrix} u \\ v \end{pmatrix} \rangle\!\rangle \leqslant \left\| \begin{pmatrix} u \\ v \end{pmatrix} \right\| \leqslant r_2 \langle\!\langle \begin{pmatrix} u \\ v \end{pmatrix} \rangle\!\rangle$$

for certain fixed positive constants r_1, r_2. The periodicity of the group $e^{(L_1-\rho I)t}$ when $\gamma > 0$ shows that the inner product $\langle\ ,\ \rangle$ is invariant under the action of the group. Thus $e^{(L_1-\rho I)t}$ is a unitary group with respect to this inner product in $L^2[0,\ell] \oplus L^2[0,\ell]$. Stone's theorem [13] then shows that $L_1 - \rho I$ is anti-hermitian with respect to this inner product with a representation

$$L_1 - \rho I = \int_{-\infty}^{\infty} i\mu \, dE(\mu)$$

where $E(\mu)$ is the spectral measure associated with $L_1 - \rho I$. Since $e^{(L_1-\rho I)t}$ is periodic, however, we can show easily that the support of $E(\mu)$ must be a subset of the points

$$0, \ \pm \frac{k\pi}{\ell} , \quad k = 1,2,3,\ldots .$$

Thus, with respect to the usual inner product $(\, , \,)$, L_1 is a spectral operator with spectrum a subset of the points

$$(2.28) \qquad \rho, \ \rho \pm i \frac{k}{\ell}, \quad k = 1,2,3,\dots .$$

When $\gamma < 0$ we can argue in much the same way to show that L_1 is a spectral operator whose spectrum is a subset of the points

$$(2.29) \qquad \rho \pm i \frac{(k - \frac{1}{2})\pi}{\ell}, \quad k = 1,2,3,\dots .$$

When the a_{ij} are all zero, which implies $\sigma^+ = \sigma^- = 1$ and h_1 and h_2 are zero, one can verify directly that the spectrum of the operator L with boundary conditions

$$(2.30) \qquad \begin{aligned} &\alpha_0 u(0) + \beta_0 v(0) = 0 , \\ &\left(\frac{1}{\beta_0 - \alpha_0} - \frac{\gamma}{\beta_0 + \alpha_0} \right) u(\ell) + \left(\frac{1}{\beta_0 - \alpha_0} + \frac{\gamma}{\beta_0 + \alpha_0} \right) v(\ell) = 0 \end{aligned}$$

consists of precisely the points (2.28) or (2.29), depending upon whether $\gamma > 0$ or $\gamma < 0$, respectively, and that each such point is an eigenvalue of single multiplicity. If the a_{ij} are not zero and we consider the operator L with boundary conditions (2.30), the eigenvalues are again simple and approach the values (2.28) or (2.29) asymptotically. The perturbation in L brought about by introducing the non-zero a_{ij} gives rise to a perturbation in the eigenvalues. Thus it is of some interest to be able to prove that this perturbation of the eigenvalues can be "undone", not by removing the a_{ij}, but by changing the right hand boundary condition. Specifically, our result is

<u>THEOREM 5</u>. There exist continuous functions h_1 and h_2 such that the operator

$$L_1\begin{pmatrix} u \\ v \end{pmatrix} = \begin{pmatrix} 0 & 1 \\ 1 & 0 \end{pmatrix} \frac{d}{d\xi} \begin{pmatrix} u \\ v \end{pmatrix} - \begin{pmatrix} a_{11}(\xi) & a_{12}(\xi) \\ a_{21}(\xi) & a_{22}(\xi) \end{pmatrix} \begin{pmatrix} u \\ v \end{pmatrix}$$

with boundary conditions

$$\alpha_0 u(0) + \beta_0 v(0) = 0 , \tag{2.31}$$

$$\left(\frac{\sigma^+}{\beta_0 - \alpha_0} - \frac{\gamma\sigma^-}{\beta_0 + \alpha_0} \right) u(\ell) + \left(\frac{\sigma^+}{\beta_0 - \alpha_0} + \frac{\gamma\sigma^-}{\beta_0 + \alpha_0} \right) v(\ell)$$
$$= \int_0^\ell [h_1(\xi)u(\xi) + h_2(\xi)v(\xi)]d\xi , \quad \gamma \text{ real} , \tag{2.32}$$

is a spectral operator whose spectrum coincides (multiplicity included) with that of the operator

$$L_0\binom{u}{v} = \begin{pmatrix} 0 & 1 \\ 1 & 0 \end{pmatrix} \frac{d}{d\xi} \binom{u}{v}$$

with boundary conditions (2.31) and

$$\left(\frac{1}{\beta_0 - \alpha_0} - \frac{\gamma}{\beta_0 + \alpha_0} \right) u(\ell) + \left(\frac{1}{\beta_0 - \alpha_0} + \frac{\gamma}{\beta_0 + \alpha_0} \right) v(\ell) = 0 . \tag{2.33}$$

Remarks. When $a_{11}(\xi) + a_{22}(\xi) \equiv 0$, $\sigma^+ = \sigma^-$ and the boundary conditions (2.32) and (2.33) differ only by an integral term.

If we want the boundary condition (2.33) to have a given form

$$\alpha u(\ell) + \beta v(\ell) = 0 \tag{2.34}$$

we can do so by setting

$$\gamma = \frac{\beta - \alpha}{\beta + \alpha} \left(\frac{\beta_0 + \alpha_0}{\beta_0 - \alpha_0} \right) .$$

The only boundary condition (2.33) which cannot be realized in this way is

$$u(\ell) - v(\ell) = 0 .$$

Proof of Theorem 5. We have already established that the spectrum of L_1 is a subset of the spectrum of L_0. When $\gamma = 0$ the spectra of L_1 and L_0 have been shown to be empty in both cases so there is nothing to prove. Hence we need only show that when $\gamma \neq 0$ each point in (2.28) or (2.29) belongs to the spectrum of L_1 and that each of these points is a simple eigenvalue.

Let us consider a boundary value control system

$$(2.35)\qquad \frac{\partial}{\partial t}\begin{pmatrix} u \\ v \end{pmatrix} - \begin{pmatrix} 0 & 1 \\ 1 & 0 \end{pmatrix}\frac{\partial}{\partial \xi}\begin{pmatrix} u \\ v \end{pmatrix} + \begin{pmatrix} a_{11}(\xi) & a_{12}(\xi) \\ a_{21}(\xi) & a_{22}(\xi) \end{pmatrix}\begin{pmatrix} u \\ v \end{pmatrix} = 0 ,$$

$$(2.36)\qquad \alpha_0 u(0,t) + \beta_0 v(0,t) = 0 ,$$

$$(2.37)\qquad \alpha_2 u(\ell,t) + \beta_2 v(\ell,t) - \int_0^{\ell}[h_1(\xi)u(\xi,t) + h_2(\xi)v(\xi,t)]d\xi = g(t) ,$$

where α_2 and β_2 are given by (2.22) and $g \in L^2[0,2\ell]$.

Now consider the following adjoint system:

$$(2.38)\qquad \begin{aligned} &\frac{\partial}{\partial t}\begin{pmatrix} w \\ z \end{pmatrix} - \begin{pmatrix} 0 & 1 \\ 1 & 0 \end{pmatrix}\frac{\partial}{\partial \xi}\begin{pmatrix} w \\ z \end{pmatrix} - \begin{pmatrix} a_{11}(\xi) & a_{21}(\xi) \\ a_{12}(\xi) & a_{22}(\xi) \end{pmatrix}\begin{pmatrix} w \\ z \end{pmatrix} \\ &\qquad + \frac{w(\ell,t)}{\beta_2}\begin{pmatrix} h_1(\xi) \\ h_2(\xi) \end{pmatrix} = 0 \end{aligned}$$

$$(2.39)\qquad \alpha_0 w(0) - \beta_0 z(0) = 0 ,$$

$$(2.40)\qquad \alpha_2 w(\ell) - \beta_2 z(\ell) = 0 .$$

(If $\beta_2 = 0$ we replace $\frac{w(\ell,t)}{\beta_2}$ in (2.38) by $\frac{z(\ell,t)}{\alpha_2}$.) Then we compute, using (2.36) and (2.39)

$$\frac{d}{dt}\left(\begin{pmatrix} u(\cdot,t) \\ v(\cdot,t) \end{pmatrix}, \begin{pmatrix} w(\cdot,t) \\ z(\cdot,t) \end{pmatrix}\right)$$

$$= \int_0^{\ell}\left[\left(\begin{pmatrix} 0 & 1 \\ 1 & 0 \end{pmatrix}\frac{\partial}{\partial \xi}\begin{pmatrix} u \\ v \end{pmatrix} - \begin{pmatrix} a_{11}(\xi) & a_{12}(\xi) \\ a_{21}(\xi) & a_{22}(\xi) \end{pmatrix}\begin{pmatrix} u \\ v \end{pmatrix}, \begin{pmatrix} w \\ z \end{pmatrix}\right)\right.$$

$$\left. + \left(\begin{pmatrix} u \\ v \end{pmatrix}, \begin{pmatrix} 0 & 1 \\ 1 & 0 \end{pmatrix}\frac{\partial}{\partial \xi}\begin{pmatrix} w \\ z \end{pmatrix} + \begin{pmatrix} a_{11}(\xi) & a_{21}(\xi) \\ a_{12}(\xi) & a_{22}(\xi) \end{pmatrix} - \frac{w(\ell,t)}{\beta_2}\begin{pmatrix} h_1(\xi) \\ h_2(\xi) \end{pmatrix}\right)\right]d\xi$$

$$= \int_0^{\ell} \left[\left(\begin{pmatrix} 0 & 1 \\ 1 & 0 \end{pmatrix} \frac{\partial}{\partial\xi} \binom{u}{v}, \binom{w}{z} \right) + \left(\binom{u}{v}, \begin{pmatrix} 0 & 1 \\ 1 & 0 \end{pmatrix} \frac{\partial}{\partial\xi} \binom{w}{z} \right) \right] d\xi$$

$$- \frac{\overline{w(\ell,t)}}{\beta_2} \int_0^{\ell} [h_1(\xi)u(\xi,t) + h_2(\xi)v(\xi,t)]d\xi \;=\; \frac{\overline{w(\ell,t)}}{\beta_2}\, g(t) \;,$$

the last equality following immediately when we integrate the term

$$\left(\begin{pmatrix} 0 & 1 \\ 1 & 0 \end{pmatrix} \frac{\partial}{\partial\xi} \binom{u}{v}, \binom{w}{z} \right)$$

by parts and then use (2.37) and (2.40). (Again, if $\beta_2 = 0$ we replace $\frac{\overline{w(\ell,t)}}{\beta_2}$ by $\frac{\overline{z(\ell,t)}}{\alpha_2}$.) Thus, if $\binom{u}{v}$ and $\binom{w}{z}$ satisfy (2.35) and (2.38) and the given boundary conditions, we have

$$\left(\begin{pmatrix} u(\cdot,2\ell) \\ v(\cdot,2\ell) \end{pmatrix}, \begin{pmatrix} w(\cdot,2\ell) \\ z(\cdot,2\ell) \end{pmatrix} \right) - \left(\begin{pmatrix} u(\cdot,0) \\ v(\cdot,0) \end{pmatrix}, \begin{pmatrix} w(\cdot,0) \\ z(\cdot,0) \end{pmatrix} \right)$$

(2.41)

$$= \int_0^{2\ell} \frac{\overline{w(\ell,t)}}{\beta_2}\, g(t)dt \;.$$

Suppose now we set $u(\xi,0) \equiv v(\xi,0) \equiv 0$ and consider the following problems:

(a) Letting $\binom{u}{v}$ solve (2.35), (2.36), (2.37) for these zero initial data and for arbitrary $g \in L^2[0,2\ell]$, are terminal states $(u(\cdot,2\ell), v(\cdot,2\ell))$ dense in $L^2[0,T]$?

(b) Can the zero state $(u(\cdot,2\ell), v(\cdot,2\ell)) = (0,0)$ be reached using some $g \neq 0$ in $L^2[0,2\ell]$?

We will show that the answer to (a) is "yes" and the answer to (b) is "no". Assuming this for the moment, we can complete the proof of Theorem 2.

Suppose λ_j were an eigenvalue of L_1 with multiplicity greater than 1. Since L_1 has been shown to be similar to an anti-hermitian operator, there must then exist two linearly independent eigenvectors (w_j,z_j) and $(\hat{w}_j,\hat{z}_j)$ of L_1^* corresponding to the eigenvalue $\overline{\lambda}_j$ of L_1^*. Then both

$$(2.42) \qquad \begin{bmatrix} w(\xi,t) \\ z(\xi,t) \end{bmatrix} = e^{\overline{\lambda}_j(2\ell-t)} \begin{bmatrix} w_j(\xi) \\ z_j(\xi) \end{bmatrix}$$

and

$$(2.43) \qquad \begin{bmatrix} \hat{w}(\xi,t) \\ \hat{z}(\xi,t) \end{bmatrix} = e^{\overline{\lambda}_j(2\ell-t)} \begin{bmatrix} \hat{w}_j(\xi) \\ \hat{z}_j(\xi) \end{bmatrix}$$

solve $\frac{d}{dt}\begin{pmatrix} w \\ z \end{pmatrix} + L_1^*\begin{pmatrix} w \\ z \end{pmatrix} = 0$, which is the abstract form of (2.38), (2.39), (2.40). Indeed (see [2] for related material), L_1^* is the operator

$$L_1^*\begin{pmatrix} w \\ z \end{pmatrix} = -\begin{bmatrix} 0 & 1 \\ 1 & 0 \end{bmatrix} \frac{\partial}{\partial \xi}\begin{pmatrix} w \\ z \end{pmatrix} - \begin{bmatrix} a_{11}(\xi) & a_{21}(\xi) \\ a_{12}(\xi) & a_{22}(\xi) \end{bmatrix}\begin{pmatrix} w \\ z \end{pmatrix} + \frac{w(\ell)}{\beta_2}\begin{bmatrix} h_1(\xi) \\ h_2(\xi) \end{bmatrix}$$

with domain boundary defined by conditions of the form (2.39), (2.40). Substituting (2.42) and (2.43) into (2.41) and recalling that we are taking $u(\cdot,0) = v(\cdot,0) = 0$, we have

$$\left(\begin{bmatrix} u(\cdot,2\ell) \\ v(\cdot,2\ell) \end{bmatrix}, \begin{bmatrix} w(\cdot,2\ell) \\ z(\cdot,2\ell) \end{bmatrix}\right) = \int_0^{2\ell} \frac{\overline{w}_j(\ell)}{\beta_2} e^{\lambda_j(2\ell-t)} g(t)dt$$

$$\left(\begin{bmatrix} u(\cdot,2\ell) \\ v(\cdot,2\ell) \end{bmatrix}, \begin{bmatrix} \hat{w}(\cdot,2\ell) \\ \hat{z}(\cdot,2\ell) \end{bmatrix}\right) = \int_0^{2\ell} \frac{\overline{w}_j(\ell)}{\beta_2} e^{\lambda_j(2\ell-t)} g(t)dt .$$

Then for all states $\begin{bmatrix} u(\cdot,2\ell) \\ v(\cdot,2\ell) \end{bmatrix}$ reachable from zero via (2.35), (2.36), (2.37) with controls $g \in L^2[0,2\ell]$ we have

$$\left(\begin{bmatrix} u(\cdot,2\ell) \\ v(\cdot,2\ell) \end{bmatrix}, \frac{\beta_2}{\overline{\hat{w}}_j(\ell)}\begin{bmatrix} w(\cdot,2\ell) \\ z(\cdot,2\ell) \end{bmatrix} - \frac{\beta_2}{\overline{w}_j(\ell)}\begin{bmatrix} \hat{w}(\cdot,2\ell) \\ \hat{z}(\cdot,2\ell) \end{bmatrix}\right) = 0 .$$

But this cannot be so if, as we claim, the answer to (a) is "yes". Thus, assuming the positive answer to (a), L_1^*, and hence L_1, has simple eigenvalues.

If some number $\rho \pm i \frac{j\pi}{\ell}$ (or $\rho \pm \frac{(j - \frac{1}{2})\pi}{\ell}$) is missing from the spectrum of L_1, assume it is $\rho + i \frac{j\pi}{\ell}$ for definiteness, then we note that

$$g_j(t) = e^{(-\rho + i\frac{j\pi}{\ell})t}$$

has the property that

$$\int_0^{2\ell} e^{\overline{\lambda}_k t} g_j(t)dt = \int_0^{2\ell} e^{-i\frac{k\pi}{\ell}t} e^{i\frac{j\pi}{\ell}t} dt = 0$$

for all λ_k which are eigenvalues of L_1. Letting $\binom{w_k}{z_k}$ be the eigenvector of L_1^* corresponding to its eigenvalue $\overline{\lambda}_k$ and setting

$$\begin{pmatrix} w_k(\xi,t) \\ z_k(\xi,t) \end{pmatrix} = e^{\overline{\lambda}_k(2\ell - t)} \begin{pmatrix} w_k(\xi) \\ z_k(\xi) \end{pmatrix},$$

we find, after substitution in (2.41), again with $(u(\cdot,0),v(\cdot,0)) = (0,0)$, that

$$\left(\begin{pmatrix} u(\cdot,2\ell) \\ v(\cdot,2\ell) \end{pmatrix}, \begin{pmatrix} w_k \\ z_k \end{pmatrix} \right)$$

for all k. Since the eigenvectors of L_1^* span $L^2[0,\ell] \oplus L^2[0,\ell]$, we conclude that

$$\begin{pmatrix} u(\cdot,2\ell) \\ v(\cdot,2\ell) \end{pmatrix} = \binom{0}{0}$$

and thus $g_j(t)$ is a non-zero control taking $(0,0)$ into $(0,0)$. Hence if, as we will show, the answer to (b) is "no", we conclude that each of the numbers (2.28) an eigenvalue of L_1 when $\gamma > 0$ and each of the numbers (2.29) is an eigenvalue of L_1 when $\gamma < 0$.

Now to complete the proof of Theorem 5, we take up questions (a) and (b). Let initial and terminal states (u_0,v_0), (u_1,v_1) be given, u_0, v_0, u_1, v_1 all in $L^2[0,\ell]$. By Theorem 1 there is a unique f in $L^2[0,2\ell]$ such that if (u,v) solves (2.35), (2.36), (2.1) with

$$\alpha_2 u(\ell,t) + \beta_2 v(\ell,t) = f(t), \tag{2.44}$$

then $(u(\xi,2\ell),v(\xi,2\ell)) = (u_1(\xi),v_1(\xi))$, a.e. Then let

$$g(t) = f(t) - \int_0^{\ell} [h_1(\xi)u(\xi,t) + h_2(\xi)v(\xi,t)]d\xi$$

and we have

$$\alpha_2 u(\ell,t) + \beta_2 v(\ell,t) - \int_0^{\ell} [h_1(\xi)u(\xi,t) + h_2(\xi)v(\xi,t)]d\xi = g(t) ,$$

so g, which clearly lies in $L^2[0,2\ell]$, steers (2.35), (2.36), (2.37) from (u_0,v_0) to (u_1,v_1). Thus the answer to (a) is, indeed, "yes".

Passing to question (b), if g steers a solution of (1.10), (2.36), (2.37) from $(0,0)$ to $(0,0)$ then

$$f(t) = g(t) + \int_0^{\ell} [h_1(\xi)u(\xi,t) + h_2(\xi)v(\xi,t)]dt$$

steers a solution of (2.35), (2.36), (2.44) from $(0,0)$ to $(0,0)$. Then Theorem 1 shows that $f(t) = 0$ a.e. in $[0,2\ell]$ so that

$$g(t) = -\int_0^{\ell} [h_1(\xi)u(\xi,t) + h_2(\xi)v(\xi,t)]dt \tag{2.45}$$

a.e. in $L^2[0,\ell]$. Then the solution (u,v) satisfies

$$\alpha_2 u(\ell,t) + \beta_2 v(\ell,t) = 0 \quad \text{a.e. in } [0,2\ell]$$

which implies $(u(\xi,t),v(\xi,t)) = (0,0)$ a.e. and we have, from (2.45),

$$g(t) \equiv 0 \quad \text{a.e. in } L^2[0,2\ell] ,$$

showing that the answer to (b) is "no". With this the proof of Theorem 5 is complete.

REFERENCES

[1] G.D. BIRKHOFF, Boundary value and expansion problems of ordinary linear differential equations, Trans. Amer. Math. Soc. 9 (1908), 373-395.

[2] R.H. COLE, General boundary conditions for an ordinary linear differential system, Trans. Amer. Math. Soc. 111 (1964), No. 3, 521-550.

[3] L. COLLATZ, The Numerical Treatment of Differential Equations. John Wiley and Sons, New York, 1960.

[4] R. COURANT and D. HILBERT, Methods of Mathematical Physics, Vol. II, Partial Differential Equations. Interscience Publ. Co., New York, 1962, Chapter V.

[5] N. DUNFORD and J.T. SCHWARTZ, Linear Operators, Part I, General Theory. Interscience Publ. Co., New York, 1958.

[6] P.R. GARABEDIAN, Partial Differential Equations. John Wiley and Sons, New York, 1964.
[7] J.J. GRAINGER, Boundary-value control of distributed systems characterized by hyperbolic differential equations, Doctoral thesis, Electrical Engineering Dept., Univ. of Wisconsin, Madison, 1969.
[8] H.P. KRAMER, Perturbations of differential operators, Pacific J. Math. 7 (1957), 1405-1435.
[9] P.D. LAX, On Cauchy's problem for hyperbolic equations and the differentiability of solutions of elliptic equations, Comm. Pure Appl. Math. 8 (1955), 615-633.
[10] N. LEVINSON, Gap and Density Theorems. Amer. Math. Soc. Colloq. Publ. Vol. 26 (1940).
[11] J.L. LIONS and E. MAGENES, Problèmes aux limites non-homogènes et applications. Vol. I and II, Dunod, Paris, 1968.
[12] R.E.A.C. PALEY and N. WIENER, Fourier Transforms in the Complex Domain. Amer. Math. Soc. Colloq. Publ. Vol. 19 (1934).
[13] F. RIESZ and B. SZ.-NAGY, Functional Analysis. F. Ungar Publ. Co., New York, 1955. (See pp. 380 ff.)
[14] D.L. RUSSELL, On boundary-value controllability of linear symmetric hyperbolic systems, in Mathematical Theory of Control, Academic Press, New York, 1967.
[15] ____________, Nonharmonic Fourier series in the control theory of distributed parameter systems, J. Math. Anal. Appl. 18 (1967), No. 3, 542-560.
[16] J.T. SCHWARTZ, Perturbations of spectral operators and applications, Pacific J. Math. 4 (1954), 415-458.
[17] L. SCHWARTZ, Étude des sommes d'exponentielles. Deuxième edition, Hermann, Paris, 1959.
[18] D.L. RUSSELL, Control theory of hyperbolic equations related to certain questions in harmonic analysis and spectral theory, to appear in J. Math. Anal. Appl.

University of Wisconsin, Madison, Wisconsin

THE HOMOLOGY OF INVARIANT SETS OF FLOWS*

Robert J. Sacker

1. Introduction

In this paper we consider a flow $\Pi: W \times R \to W$ where W is a metric space and R is the reals. As usual Π is continuous, $\Pi(x,0) = x$ for all $x \in W$, and for all $s,t \in R$, $\Pi(\Pi(x,t),s) = \Pi(x,t+s)$. We denote $\Pi(x,t)$ by $x \cdot t$ and for $A \subset W$, $T \subset R$, $A \cdot T \equiv \{p \cdot t: p \in A, t \in T\}$. We will focus our attention on a compact subset $X \subset W$ and in particular on the class $I(X)$ of all closed subsets of X which are invariant under Π, i.e., $A \in I(X)$ if $A \cdot t = A$ for all $t \in R$.

Our aim is to assign to X certain groups which describe the manner in which the closed invariant subsets are situated in X and how they are tied together. The invariant sets are characterized as projective chains and are shown to form a chain complex. The abovementioned groups then turn out to be the homology groups of this complex. These groups are related to the ordinary homology groups of the space via exact sequences.

This paper represents a summary of partial results. Further details and proofs will appear elsewhere.

For motivation consider the two examples of flows in the plane $W = R^2$ given in Figs. 1 and 2 where $X = \{(x,y): x^2 + y^2 \leqslant 1\}$:

*This research was partially supported by U.S. Army Grant DA-ARO-D-31-124-71-G14.

Fig. 1

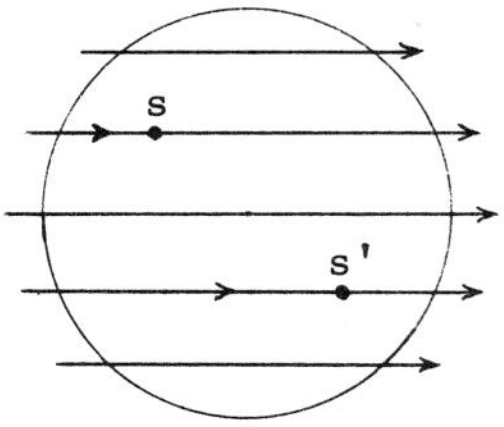

Fig. 2

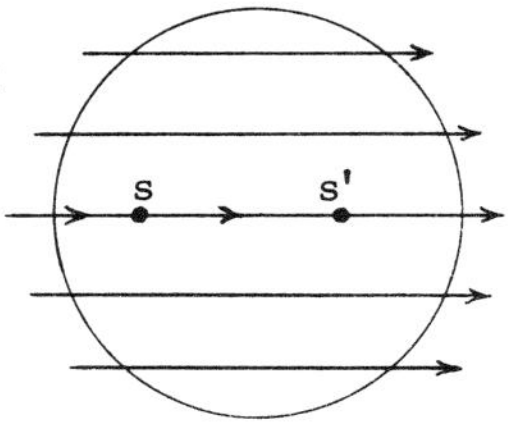

In each case we have parallel flow interrupted by two singularities, s and s', and in each case s and s' form the boundary of a 1-simplex. However only in Fig. 2 do s and s' form the boundary of an <u>invariant</u> 1-simplex in X. This will be borne out in the definition of the special groups $\hat{H}(X)$ when we see that in Fig. 1, $\hat{H}_0(X) = G \oplus G$ and in Fig. 2, $\hat{H}_0(X) = G$, i.e., s and s' generate distinct homology classes in the first case and the same class in the second. Here G is the underlying group of coefficients. In these examples, of course, singular homology would suffice. In general, however, it is inadequate, for considering the space $X \subset R^2$ to consist of two singularities s_1 and s_2, a periodic orbit p, and two orbits ℓ_1 and ℓ_2 with $\alpha(\ell_i) = s_i$ and $\omega(\ell_i) = p$, $i = 1,2$, where α and ω denote respectively the $(t \to -\infty)$ and $(t \to +\infty)$ limit sets of the indicated orbit.

2. <u>Projective Chains</u> [1]

An open set $U \subset X$ is called regular if $U = \text{int}\,\overline{U}$. For these sets, $U \subset U' \Leftrightarrow \overline{U} \subset \overline{U}'$. A finite open <u>grating</u> of X, $\alpha = \{U_1,\ldots,U_n\}$, is a collection of disjoint regular open subsets such that $\overline{\alpha} = \{\overline{U}_1,\ldots,\overline{U}_n\}$ is a covering of X. We call $\overline{\alpha}$ the corresponding <u>closed grating</u>. For gratings α and β, we say β refines $\alpha(\beta > \alpha)$ if for each $V \in \beta$ there exists $U \in \alpha$ such that $V \subset U$. From the disjointness assumption we see that U is necessarily unique and if we let $\mathcal{G}(X) = \{\alpha, >\}$ be the collection of all finite open gratings of X then $\mathcal{G}(X)$ is a partially ordered set.

The nerve N_α of a grating $\alpha = \{U\} = \{U_1,\ldots,U_\ell\}$ is an abstract simplicial complex consisting of all k-simplexes $\sigma^k = \langle U_{i_0},\ldots,U_{i_k}\rangle$,

$k = 0,\dots,\ell-1$, where $\{i_0,\dots,i_k\} \subset \{1,2,\dots,\ell\}$ is a subset of integers such that

$$\bar{U}_{i_0} \cap \cdots \cap \bar{U}_{i_k} \neq \emptyset .$$

If $\beta > \alpha$, $\beta = \{V\}$, there is a unique simplicial mapping $\Pi_\alpha^\beta : N_\beta \to N_\alpha$ defined by the property $\Pi_\alpha^\beta \langle V_i \rangle = \langle U_j \rangle$ if $V_i \subset U_j$, and extended to the simplexes by linearity. We define the q-dimensional chain group of the complex N_α (with coefficients in G an abelian group) to be

$$C_{q,\alpha}(X) = F_G[\sigma_1^q, \dots, \sigma_k^q]$$

where $\sigma_1^q,\dots,\sigma_k^q$ is the totality of oriented q-simplexes and F_G is the free abelian G-group using these oriented simplexes as generators, i.e., all finite formal sums

$$c = \sum g_i \sigma_i^q , \qquad g_i \in G .$$

The boundary operator $\partial_q : C_{q,\alpha}(X) \to C_{q-1,\alpha}(X)$ is defined as usual:

for a basic q-chain $\sigma^q = g\langle U_0,\dots,U_q\rangle$

$$\partial_q(g\sigma^q) = \sum_{k=0}^{q} (-1)^k g \langle U_0,\dots,\hat{U}_k,\dots,U_q\rangle$$

where the "hat" denotes omission of that vertex.

The mapping Π_α^β induces a homomorphism (also called Π_α^β) which commutes with the boundary operator

$$\begin{array}{ccc} C_{q,\beta}(X) & \xrightarrow{\Pi_\alpha^\beta} & C_{q,\alpha}(X) \\ \downarrow \partial_q & & \downarrow \partial_q \\ C_{q-1,\beta}(X) & \xrightarrow{\Pi_\alpha^\beta} & C_{q-1,\alpha}(X) \end{array}$$

Let $P = \prod_{\alpha \in \mathcal{B}(X)} C_{q,\alpha}(X)$ be the direct product and let Π_α be projection onto the α-coordinate, i.e., if $x = (x_\alpha) \in P$ then $\Pi_\alpha x = x_\alpha \in C_{q,\alpha}(X)$. The system $\{C_{q,\alpha}(X), \Pi_\alpha^\beta\}$ over the directed set $\mathcal{B}(X)$ is an inverse system of groups and we define

$$C_q(X) = \varprojlim \{C_{q,\alpha}, \Pi_\alpha^\beta\}$$

the inverse limit of this system, i.e., $C_q(x)$ is the set of all $c = (c_\alpha) \in P$ such that $\beta > \alpha \Rightarrow \Pi_\alpha^\beta c_\beta = c_\alpha$. $C_q(X)$ is the projective chain group of X. A boundary operator is induced on the limit and, again calling it ∂, we have the chain complex

$$\rightarrow C_q(X) \xrightarrow{\partial_q} C_{q-1}(X) \rightarrow .$$

Defining the projective cycles $Z_q(X) = \ker \partial_q$ and the projective boundaries $B_q(X) = \operatorname{Im} \partial_{q+1}$ we may form the projective homology groups of X, $H_q(X)$. The following is proved in [1]:

LEMMA 1. If X is compact and G is compact or a field then for each $q \geqslant 0$, $H_q(X)$ is isomorphic to the Čech homology group of X with coefficients G.

We assume, from this point on, that X is compact and G is a compact group or a field.

3. The Invariant Homology Group

Our aim here is to single out from $C_q(X)$ all those chains which represent in a certain sense closed invariant subsets of X. For this purpose for each integer $q \geqslant 0$ and closed subset $Y \subset X$ define $I_q(Y) = \{F \subset Y\colon F$ is closed, invariant and $\dim F \leqslant q\}$. The dimension of a set F, $\dim F$, is taken to be the inductive definition as stated in [4]. Define the direct sum

$$S_q(X) = \bigoplus_{F \in I_q(X)} C_q(F) .$$

Now $F \subset X$ induces the inclusion $\phi_{q,F}\colon C_q(F) \to C_q(X)$ and let $\phi_q\colon S_q(X) \to C_q(X)$ be the map induced on the direct sum by addition: if $s \in S_q(X)$, $s = c_{F_1} + \cdots + c_{F_n}$ then $\phi_q(s) = \sum_{i=1}^{n} \phi_{q,F_i}(c_{F_i})$. Thus we have

$$\begin{array}{ccc} S_q(X) & & S_{q-1}(X) \\ \downarrow \phi_q & & \downarrow \phi_{q-1} \\ C_q(X) & \xrightarrow{\partial} & C_{q-1}(X) \end{array}$$

and we finally define the subgroup of "invariant" chains

$$\hat{C}_q(X) = \text{Im } \phi_q \cap \partial^{-1} \text{ Im } \phi_{q-1} .$$

Intuitively, a chain $c \in \hat{C}_q(X)$ is a q-chain lying on a q-dimensional closed invariant subset of X and whose boundary is a (q-1)-chain lying on a (q-1)-dimensional closed invariant subset. Chains in $C_q(X)$ will be referred to as "ordinary chains" to distinguish them from the "invariant chains." It is easily verified that the boundary operator restricted to $\hat{C}_q$ maps

$$\hat{C}_q(X) \xrightarrow{\partial} \hat{C}_{q-1}(X) ,$$

thus giving rise to a chain complex from which we form the <u>invariant homology groups</u> $\hat{H}_q(X)$. Two invariant q-cycles z_q and z_q' are homologous in X if $z_q - z_q'$ forms the boundary of an invariant (q+1)-chain $c_{q+1} \in \hat{C}_{q+1}(X)$; $z_q - z_q' = \partial c_{q+1}$. Thus we see in Fig. 1, $\hat{H}_0(X) = G \oplus G$ and in Fig. 2, $\hat{H}_0(X) = G$. An immediate consequence of the definitions is

LEMMA 2. $\hat{H}_0(X) = 0$ if, and only if, there are no singularities of the flow Π. Other examples are

(a) the torus T^2 with the irrational flow for which $\hat{H}_0 = \hat{H}_1 = 0$, $\hat{H}_2 = G$;

(b) T^2 with all orbits periodic for which $\hat{H}_0 = 0$, $\hat{H}_1 = G$, $\hat{H}_2 = G$;

(c) T^2 with two periodic orbits p and p' and for all $x \notin p \cup p'$, $\alpha(x) = p$, $\omega(x) = p'$. Then $\hat{H}_0 = 0$, $\hat{H}_1 = G \oplus G$, $\hat{H}_2 = G$.

Under the conditions stated it is clear in example (c) that there are no singularities of the flow and hence $\hat{H}_0 = 0$. Referring to p and p' as 1-cycles we see that $p - p'$ forms the boundary of $\bar{\alpha}$ where α is either of the disjoint open invariant annuli A or A' whose union is $T^2 - p \cup p'$. Thus $p \sim p'$ is one generator of $\hat{H}_1$. The other generator is constructed as follows: Let $a \in A$ and $a' \in A'$ be arbitrary points and let ℓ and ℓ' denote the full orbits through these points. Then $F = \ell \cup \ell' \cup p \cup p' \in I_1(T^2)$ carries a 1-cycle independent of p.

<u>Remark</u>. If for some $Y \subset X$, $Y \in I_n(Y)$ then every n-cycle z on Y is supported by a n-dimensional closed invariant set, namely Y, and

therefore z is an invariant n-cycle. Thus $\hat{H}_n(Y) \cong H_n(Y)$.

We will now justify the use of the phrase "invariant n-cycle" by showing that a *minimal closed carrier* [3] of such a cycle is indeed invariant. A minimal closed carrier of a cycle z is defined as follows: Let z represent a non-zero class $[z] \in H_n(Y)$. A closed subset $K \subset Y$ is a minimal closed carrier of z if $[z]$ is in the image of the inclusion induced map $i_*: H_n(K) \to H_n(Y)$ but if K is replaced by any proper closed subset of K then $[z]$ is no longer in the image.

LEMMA 3. Let z be a non-bounding invariant n-cycle in Y where Y is some closed subset of X, i.e., z represents a non-zero class in $\hat{H}_n(Y)$. Then the minimal closed carrier K of z is invariant; i.e., $K \in I_n(Y)$.

Proof. From the definition of $\hat{H}$ there is an $F \in I_n(Y)$ such that z is a cycle on F. Clearly also z represents a non-zero class $[z] \in H_n(F)$. Let $K \subset F$ be a minimal closed carrier of z, i.e., $[z]$ is in the image of $i_*: H_n(K) \to H_n(F)$ but if K is replaced by any proper closed subset of K then $[z]$ is not in the image. It follows then that if $[z]$ is the class of z in $H_n(K)$ then for any proper closed $K' \subset K$, $[z]$ is not in the image of $i_*: H_n(K') \to H_n(K)$. Now assume K is not invariant. Then for some $t \in R$, $K \cap K \cdot t$ is properly contained in K. Let h_t denote the homeomorphism induced by the flow $h_t: K \to K \cdot t$ and consider the inclusions

$$i_1: K \cap K \cdot t \to K\,, \qquad i_2: K \cap K \cdot t \to K \cdot t\,,$$

$$j_1: K \to K \cup K \cdot t\,, \qquad j_2: K \cdot t \to K \cup K \cdot t\,,$$

$$i = (i_1, -i_2)\,, \quad j = j_1 + j_2 \quad \text{and} \quad i': K \cup K \cdot t \to F\,.$$

Now consider the diagram

$$\begin{array}{ccccc} & & & & H_{n+1}(F, K \cup K \cdot t) \\ & & & & \downarrow \partial \\ H_n(K \cap K \cdot t) & \xrightarrow{i_*} & H_n(K) \oplus H_n(K \cdot t) & \xrightarrow{j_*} & H_n(K \cup K \cdot t) \\ & & & & \downarrow i'_* \\ & & & & H_n(F) \end{array}$$

consisting of a Mayer-Vietoris sequence and the exact sequence of the pair $(F, K \cup K \cdot t)$. From the minimality of K we see that $\lambda = ([z], -h_t*[z])$ is not in the image of $i*$ and therefore $\nu = j_*\lambda \neq 0$. But since h_t is homotopic to the identity via the flow Π, $i'_*\nu = 0$ in $H_n(F)$. But $\dim F \leq n$ implies $H_{n+1}(F, K \cup K \cdot t) = 0$, contradicting the exactness of the vertical sequence and thus proving the lemma.

From Lemma 3 we obtain

THEOREM 1. Let $F \in I_1(X)$ and suppose F contains a Jordan curve γ. Then γ is invariant. In particular if $\hat{H}_0(F) = 0$ then γ is a periodic orbit of the flow Π.

Proof. From the sequence

$$0 \to H_1(\gamma) \to H_1(F)$$

we see that γ carries a non-trivial 1-cycle in F. Clearly γ is a minimal closed carrier of this cycle and hence invariant. $\hat{H}_0 = 0$ implies no singularities in F and the assertion follows.

We may also define a minimal closed carrier for a chain $c \in \hat{C}_n(X)$ as follows: If $\partial c = 0$ then c is a cycle on some n-dimensional subset $Y \subset X$ and is therefore non-bounding on Y. In this case the previous definition applies. So assume $\partial c \neq 0$. By definition of $\hat{C}_n(X)$, ∂c is a cycle on some set $H \in I_{n-1}(X)$ and since $\dim H \leq n-1$, ∂c is non-bounding on H. Let $E \subset H$ be a minimal closed carrier of ∂c and let $F \in I_n(X)$ be such that c is a chain on F. Then c is a relative cycle representing a non-trivial class $[c] \in H_n(F,E)$. A closed subset K such that $E \subset K \subset F$ is defined to be a minimal closed carrier of c if $[c]$ is in the image if $i_*: H_n(K,E) \to H_n(F,E)$ but if K is replaced by any proper closed subset K' such that $E \subset K' \subset K$ then $[c]$ is no longer in the image of i_*. We then have

THEOREM 2. Let c be an invariant n-chain, i.e., $c \in \hat{C}_n(X)$. Then the minimal closed carrier K of c is invariant; $K \in I_n(X)$.

Proof. If $\partial c = 0$ then c is a cycle on some n-dimensional subset $Y \subset X$ and is therefore non-bounding on Y. Lemma 3 then tells us that a minimal closed carrier of c is invariant and we are done. If

$\partial c \neq 0$ then the minimal closed carrier E of ∂c is invariant (by Lemma 3) and the remainder of the proof is analogous to the proof of Lemma 3 where we now employ the diagram

$$\begin{array}{ccccc} & & & & H_{n+1}(F,\ K \cup K \cdot t) \\ & & & & \downarrow \\ H_n(K \cap K \cdot t,\ E) & \longrightarrow & H_n(K, E) \oplus H_n(K \cdot t,\ E) & \longrightarrow & H_n(K \cup K \cdot t,\ E) \\ & & & & \downarrow \\ & & & & H_n(F,\ E) \end{array}$$

composed of a relative Mayer-Vietoris sequence and the exact sequence of the triple $(F,\ K \cup K \cdot t,\ E)$.

4. Exact Sequences

The invariant homology groups are related to the ordinary homology groups of the space by exact sequences. We saw that $\hat{C}_q(X)$ is a subgroup of $C_q(X)$ so that forming the quotient $\tilde{C}_q(X) = C_q(X)/\hat{C}_q(X)$ we have the short exact sequence of chain complexes

$$0 \longrightarrow \hat{C}(X) \longrightarrow C(X) \longrightarrow \tilde{C}(X) \longrightarrow 0$$

where $C(X)$, for example, represents the chain complex

$$\cdots \longrightarrow C_q(X) \xrightarrow{\partial_q} C_{q-1}(X) \longrightarrow \cdots \quad .$$

Passing to the homology of chain complexes one obtains the long exact sequence

$$\cdots \longrightarrow \hat{H}_q(X) \longrightarrow H_q(X) \longrightarrow \tilde{H}_q(X) \longrightarrow \hat{H}_{q-1}(X) \longrightarrow \cdots \ .$$

A cycle z representing a class in $\tilde{H}_q(X)$ is an ordinary q-chain whose boundary is an invariant q-1 cycle. The sequence

$$H_1 \longrightarrow \tilde{H}_1 \longrightarrow \hat{H}_0 \longrightarrow H_0 \longrightarrow \tilde{H}_0$$

for Fig. 1 is

$$0 \longrightarrow G \longrightarrow G \oplus G \longrightarrow G \longrightarrow 0$$

and for Fig. 2 is

$$0 \longrightarrow 0 \longrightarrow G \longrightarrow G \longrightarrow 0 \ .$$

For $A \subset X$ a closed subset we obtain the commutative diagram of chain complexes with exact rows and columns

$$\begin{array}{ccccccccc}
 & & 0 & & 0 & & 0 & & \\
 & & \downarrow & & \downarrow & & \downarrow & & \\
0 & \longrightarrow & \hat{C}(A) & \xrightarrow{i_1} & \hat{C}(X) & \longrightarrow & \hat{C}(X,A) & \longrightarrow & 0 \\
 & & \downarrow{\scriptstyle i_2} & & \downarrow{\scriptstyle i_4} & & \downarrow{\scriptstyle \phi} & & \\
0 & \longrightarrow & C(A) & \xrightarrow{i_3} & C(X) & \longrightarrow & C(X,A) & \longrightarrow & 0 \\
 & & \downarrow & & \downarrow & & \downarrow & & \\
0 & \longrightarrow & \tilde{C}(A) & \xrightarrow{\Psi} & \tilde{C}(X) & \longrightarrow & \tilde{C}(X,A) & \longrightarrow & 0 \\
 & & \downarrow & & \downarrow & & \downarrow & & \\
 & & 0 & & 0 & & 0 & &
\end{array}$$

where $C(X,A) = C(X) / C(A)$, $\hat{C}(X,A) = \hat{C}(X) / \hat{C}(A)$ and $\tilde{C}(X,A)$ will be defined later. The maps ϕ and Ψ are induced by the inclusion maps labeled i. That ϕ and Ψ are 1-1 maps can be proved using the following lemma and diagram chasing.

<u>LEMMA 4</u>. $$i_4 \hat{C}(X) \cap i_3 C(A) = i_1 \circ i_4 \hat{C}(A) .$$

<u>Proof</u>. "$\supset$" is immediate since the maps labeled i are inclusions. To prove "$\subset$" let $n \geqslant 0$ and let $c \in i_4 \hat{C}_n(X) \cap i_3 C_n(A)$. Let K be a minimal closed carrier for c. Then since $c \in i_4 \hat{C}_n(X)$, Theorem 2 gives us $K \in I_n(X)$. But $c \in i_3 C_n(A)$ implies $K \subset A$. Thus $K \in I_n(A)$. Also $c \in i_3 C_n(A)$ implies $\partial c \in i_3 C_{n-1}(A)$ and thus a minimal closed carrier E of ∂c is in A. Finally $c \in i_4 \hat{C}_n(X)$ implies $\partial c \in i_4 \hat{C}_{n-1}(X)$ and by Theorem 2, $E \in I_{n-1}(X)$. Combining these we see $E \in I_{n-1}(A)$. Thus c is a chain on $K \in I_n(A)$ whose boundary is a chain on $E \in I_{n-1}(A)$, i.e., $c \in i_1 \circ i_4 \hat{C}_n(A)$ and the lemma is proved.

The group $\tilde{C}(X,A)$ is defined to be any of the groups in

<u>LEMMA 5</u>. The following groups are isomorphic:

$$\frac{C(X,A)}{\operatorname{Im} \phi} \cong \frac{\tilde{C}(X)}{\operatorname{Im} \Psi} \cong \frac{C(X)}{i_4\hat{C}(X) + i_3C(A)} .$$

Proof. An exercise using the Noether isomorphism theorem.

Passing to the homology of the chain complexes above we obtain the commutative diagram with exact rows and columns

$$\begin{array}{ccccccccc}
 & \downarrow & & \downarrow & & \downarrow & & \downarrow & \\
\to & \hat{H}_q(A) & \to & \hat{H}_q(X) & \to & \hat{H}_q(X,A) & \to & \hat{H}_{q-1}(A) & \to \\
 & \downarrow & & \downarrow & & \downarrow & & \downarrow & \\
\to & H_q(A) & \to & H_q(X) & \to & H_q(X,A) & \to & H_{q-1}(A) & \to \\
 & \downarrow & & \downarrow & & \downarrow & & \downarrow & \\
\to & \tilde{H}_q(A) & \to & \tilde{H}_q(X) & \to & \tilde{H}_q(X,A) & \to & \tilde{H}_{q-1}(A) & \to \\
 & \downarrow & & \downarrow & & \downarrow & & \downarrow & \\
\to & \hat{H}_{q-1}(A) & \to & \hat{H}_{q-1}(X) & \to & \hat{H}_{q-1}(X,A) & \to & \hat{H}_{q-2}(A) & \to \\
 & \downarrow & & \downarrow & & \downarrow & & \downarrow &
\end{array}$$

5. Gradient-Like Flows

The question arises for which flows do the groups $\hat{H}$ and H agree. Consider the case in which $X = M^n$, a compact connected n-manifold and $\Pi: M^n \times R \to M^n$ is gradient-like, i.e., assume (1) the set S of singularities of Π is finite, (2) for all $x \in X$, $\alpha(x) \in S$ and $\omega(x) \in S$, and (3) there are no oriented solution cycles. Here an oriented solution cycle is an oriented Jordan curve γ consisting of a subset (possibly empty) of S and finitely many orbits of Π such that on these orbits the orientation induced by $t \to +\infty$ $(t \in R)$ always agrees with the orientation on γ. The gradient flow $\dot{x} = -\operatorname{grad} f$ where f is a non-degenerate Morse function is an example of such a flow.

It follows (e.g. from [2]) that if s and $s' \in S$ then there exist $\{s_0, s_1, \ldots, s_\ell\} \subset S$ with $s_0 = s$, $s_\ell = s'$ such that for $j = 0, \ldots, \ell-1$, s_j and s_{j+1} are jointed by a transit orbit, i.e., there is an $x_j \in X$ such that $\alpha(x_j) = s_j$ and $\omega(x_j) = s_{j+1}$ (or $\alpha(x_j) = s_{j+1}$ and $\omega(x_j) = s_j$). Thus $\hat{H}_0(X) = H_0(X) = G$. It can also be shown that $\tilde{H}_1(X) = 0$ and therefore we have $\hat{H}_1(X) \to H_1(X) \to 0$. Thus all the homology in dimension one is supported by 1-dimensional closed invariant subsets.

Remarks. Mappings, Excisions and Homotopies will be considered in a later work. Also a useful variation in the definition of $\hat{H}$ can be obtained as follows: In the definition we constrained our n-chains to lie on n-dimensional invariant sets, i.e., sets in $I_n(X)$. We could allow higher dimensional carriers by taking n-chains supported by sets in $I_{j(n)}(X)$ where $j(n) \geq n$ is integer valued. Everything goes through as before except that now an admissibility condition is needed on subspaces A when considering the homology of a pair (X,A) since Theorem 2 no longer holds.

REFERENCES

[1] S. LEFSCHETZ, Algebraic Topology, AMS Colloquium Publ. No. 27, 1942.

[2] R.J. SACKER and G.R. SELL, On the existence of nontrivial recurrent motions (to appear).

[3] R.L. WILDER, Topology of Manifolds, AMS Colloquium Publ. No. 32.

[4] W. HUREWICZ and H. WALLMAN, Dimension Theory. Princeton Univ. Press, 1941.

University of Southern California, Los Angeles, California

LINEAR DIFFERENTIAL EQUATIONS WITH DELAYS: ADMISSIBILITY AND EXPONENTIAL DICHOTOMIES*

J. J. Schäffer

1. Introduction

This contribution is a summary report of research aimed at applying to linear differential equations with delays the methods of functional analysis developed for linear differential equations by Massera and Schäffer (see especially [4]) and for linear difference equations by themselves in [1]. The primary purpose of these investigations is to relate properties of the nonhomogeneous equation such as "admissibility" ("for every second member in some given function space there is a solution in some given function space") and certain forms of conditional stability behaviour ("dichotomies") of the solutions of the homogeneous equation. The irreversibility of the process described by an equation with delays makes it appear advisable to reduce the problem to the simplest kind of irreversible process, that described by a difference equation.

This approach was used in earlier joint work with C. V. Coffman [2], [3] for equations satisfying rather special conditions. G. Pecelli [5] has also obtained results of this nature by constructing a theory paralleling those of differential and difference equations, without reduction to either.

*This work was supported in part by NSF Grants GP-19126 and GP-28999.

A full account, including proofs, of the work summarized here will appear in [7]. Thanks are due to Professor C. V. Coffman for his valuable suggestions in the course of this investigation.

We consider on $[0,\infty)$ an equation of the form

$$\dot{u} + Mu = r \tag{1.1}$$

and the corresponding homogeneous equation

$$\dot{u} + Mu = 0 \tag{1.2}$$

in a finite-dimensional Banach space E; r is a continuous vector-valued function; the "solution" u is defined on $[-1,\infty)$, and M, the "memory functional", takes a continuous function u linearly into a continuous function Mu in such a way that the value of Mu at any given value t of the argument depends on the values of u on $[t-1,t]$ only.

The assumptions of our main result (Theorem 7.3) are that M transforms bounded functions "boundedly" into bounded functions, and that (1.1) has at least one bounded solution for each bounded r --in the tradition of [1], [2], [3], [4], "$(\underset{\sim}{C},\underset{\sim}{C})$ is admissible for (1.1)." The conclusion describes the behaviour of "slices" of length 1 of solutions of (1.2) and its restrictions to $[m,\infty)$ for real $m \geqq 0$: roughly speaking, the slices of bounded solutions tend uniformly exponentially to 0, and there exists a complementary finite-dimensional manifold of solutions of (1.2) whose slices tend uniformly exponentially to infinity and stay away uniformly from those of bounded solutions: this behaviour is a kind of "exponential dichotomy", in the sense of [1].

Reliance on the theory of difference equations allows us to avoid all consideration of possibly unbounded operators and all explicit representations of M --say as a Stieltjes integral--and other, more technical complications of [5]; the use of a compactness argument first presented in [6] allows us to achieve the description of the behaviour of the solutions of (1.2) with no extra assumptions.

We have dealt here only with a concrete example of the "continuous case"; however, the same method is also applicable to the "Carathéodory case", where (1.1), (1.2) only hold locally in $\underset{\sim}{L}^1$, and where boundedness is replaced by membership in translation-invariant spaces of measurable functions.

2. Spaces

Throughout this paper, E will denote a given real or complex finite-dimensional Banach space. The norm in E, as in all normed spaces for which no other symbol is prescribed, is denoted by $\| \ \|$. If X, Y are Banach spaces, $[X \to Y]$ denotes the Banach space of operators (bounded linear mappings) from X to Y, and we set $\tilde{X} = [X \to X]$.

We shall be dealing with sequences and with functions defined on intervals of the real line. We denote by ω the set $\{0,1,...\}$ of all natural numbers, and set $\omega_{[m]} = \{n \in \omega: n \geqq m\}$, $m = 0,1,...$. The notation for intervals of the real line is the usual one.

If m, m' are real numbers (natural numbers) with $m' \geqq m$, and f is a function defined on $[m,\infty)$ (on $\omega_{[m]}$), then $f_{[m']}$ shall denote the restriction of f to $[m',\infty)$ (to $\omega_{[m']}$).

Assume that X is a Banach space. For each natural number m we denote by $\underset{\approx}{s}_{[m]}(X)$ the linear space of all functions $f: \omega_{[m]} \to X$ and by $\underset{\sim}{\ell}^{\infty}_{[m]}(X)$ the Banach space of all bounded ones, with the norm $\mathbf{|}f\mathbf{|} = \sup\{\|f(n)\|: n \in \omega_{[m]}\}$. For each real m we denote by $\underset{\sim}{K}_{[m]}(X)$ the linear space of all continuous functions $f: [m,\infty) \to X$ and by $\underset{\sim}{C}_{[m]}(X)$ the Banach space of all bounded ones among them, with the norm $\mathbf{|}f\mathbf{|} = \sup\{\|f(t)\|: t \in [m,\infty)\}$. In all these notations the subscript is omitted when $m = 0$.

Finally, we denote by $\underset{\sim}{E}$ the Banach space of all continuous functions $f: [-1,0] \to E$, with the norm $[\![f]\!] = \max\{\|f(t)\|: t \in [-1,0]\}$.

The following example illustrates some obvious notational conventions. Suppose that $g \in \underset{\sim}{\ell}^{\infty}(\underset{\sim}{E})$; then $[\![g]\!]$ is the element of $\underset{\sim}{\ell}^{\infty}(R)$ given by $[\![g]\!](n) = [\![g(n)]\!]$, $n = 0,1,...$; and $\mathbf{|}g\mathbf{|} = \mathbf{|}[\![g]\!]\mathbf{|}$ is the norm of g as an element of $\underset{\sim}{\ell}^{\infty}(E)$.

3. Slicing Operations

Let $m \geqq 0$ be a given real number. For each $t \geqq m$ we define the linear mapping $\Pi(t)\colon \underset{\sim}{K}_{[m-1]}(E) \to \underset{\sim}{E}$ by

$$(3.1)\qquad (\Pi(t)f)(s) = f(t+s)\ , \qquad s \in [-1,0], \qquad f \in \underset{\sim}{K}_{[m-1]}(E)\ .$$

Thus $\Pi(t)$ maps f into the "slice" of f between $t-1$ and t, transplanted to $[-1,0]$ for convenience. (Note that indication of m is omitted; this will not cause any confusion.)

When m is an integer and $f \in \underset{\sim}{K}_{[m-1]}(E)$, we define $\varpi f \in \underset{\sim}{S}_{[m]}(\underset{\sim}{E})$ by

$$(3.2)\qquad (\varpi f)(n) = \Pi(n)f\ , \qquad n \in \omega_{[m]}\ , \qquad f \in \underset{\sim}{K}_{[m-1]}(E)\ .$$

Thus ϖ is a linear injective mapping of $\underset{\sim}{K}_{[m-1]}(E)$ into $\underset{\sim}{S}_{[m]}(\underset{\sim}{E})$.

4. The Memory Functional

We now make precise the assumptions on the "memory functional" M that appears in (1.1). It is linear and maps continuous functions into continuous functions, and the value of Mu at t is to depend only on the slice of u between $t-1$ and t. Specifically, we assume the following:

(M_1) $M\colon \underset{\sim}{K}_{[-1]}(E) \to \underset{\sim}{K}(E)$ is a linear mapping such that if $t \in [0,\infty)$ and $u, u' \in \underset{\sim}{K}_{[-1]}(E)$ satisfy $\Pi(t)u = \Pi(t)u'$, then $(Mu)(t) = (Mu')(t)$.

Assumption (M_1) permits, for each real $m \geqq 0$, the "cutting down" of M to a linear mapping $M_{[m]}\colon \underset{\sim}{K}_{[m-1]}(E) \to \underset{\sim}{K}_{[m]}(E)$: Each $u \in \underset{\sim}{K}_{[m-1]}(E)$ satisfies $u = v_{[m-1]}$ for some $v \in \underset{\sim}{K}_{[-1]}(E)$, and we may set $M_{[m]}u = (Mv)_{[m]}$; since $t \geqq m$ implies $\Pi(t)v = \Pi(t)u$, assumption (M_1) shows that $M_{[m]}u$ thus defined does not depend on the choice of v. If $m' \geqq m \geqq 0$, these cut-down memory functionals then satisfy

$$(4.1)\qquad M_{[m']}u_{[m'-1]} = (M_{[m]}u)_{[m']}\ , \qquad u \in \underset{\sim}{K}_{[m]}(E)\ .$$

It is obvious that (M_1) implies the existence, for every $t \in [0,\infty]$, of a linear mapping $\hat{M}(t): \underset{\sim}{E} \to E$ such that

$$(4.2) \qquad (M_{[m]}u)(t) = \hat{M}(t)\Pi(t)u \ , \qquad t \geqq m \geqq 0 \ , \qquad u \in \underset{\sim}{K}_{[m-1]}(E) \ .$$

We shall generally impose the following additional assumption:

(M_2) The restriction of M to $\underset{\sim}{C}_{[-1]}(E)$ is a bounded linear mapping $M_{\underset{\sim}{C}}: \underset{\sim}{C}_{[-1]}(E) \to \underset{\sim}{C}(E)$.

If M satisfies (M_1) and (M_2) it follows at once that $\hat{M}(t)$ is bounded, i.e., in $[\underset{\sim}{E} \to E]$, for each t, with

$$(4.3) \qquad \|M_{\underset{\sim}{C}}\| = \sup \{\|\hat{M}(t)\| : t \in [0,\infty)\} \ .$$

5. Solutions

Henceforth we assume given the space E and the memory functional M satisfying conditions (M_1) and (M_2).

For every $r \in \underset{\sim}{K}(E)$, a solution of (1.1) is a function $u \in \underset{\sim}{K}_{[-1]}(E)$ whose restriction $u_{[0]}$ to $[0,\infty)$ is continuously differentiable (the derivative is $\dot{u}_{[0]} \in \underset{\sim}{K}(E)$) and that satisfies $\dot{u}_{[0]} + Mu = r$ on $[0,\infty)$. More generally, for every real $m \geqq 0$, a solution of $(1.1)_{[m]}$ is a function $u \in \underset{\sim}{K}_{[m-1]}(E)$ whose restriction $u_{[m]}$ is continuously differentiable and that satisfies $\dot{u}_{[m]} + M_{[m]}u = r_{[m]}$ on $[m,\infty)$. In particular, if $m' \geqq m \geqq 0$ and u is a solution of $(1.1)_{[m]}$, then $u_{[m'-1]}$ is a solution of $(1.1)_{[m']}$ on account of (4.1). These definitions and statements of course also apply to the homogeneous equation (1.2).

Existence and uniqueness theorems for the initial value problem follow as usual from Banach's Contractive Mapping Principle, and inequalities for the solutions from Gronwall's Inequality. The inequality $\|(M_{[m]}u)(t)\| \leqq \|M_{\underset{\sim}{C}}\| \|\Pi(t)u\|$, an immediate consequence of (4.2) and (4.3), plays a basic role here. We omit the details. In view of the linearity of the equation, the results are summarized as follows.

LEMMA 5.1. For each real $m \geqq 0$ there exist linear mappings $P(m): \underset{\sim}{E} \to \underset{\sim}{K}_{[m-1]}(E)$ and $Q(m): \underset{\sim}{K}(E) \to \underset{\sim}{K}_{[m-1]}(E)$ such that, for every

$v \in \underset{\sim}{E}$ and every $r \in \underset{\sim}{K}(E)$, the function $u = P(m)v + Q(m)r$ is the unique solution of $(1.1)_{[m]}$ with $\Pi(m)u = v$; and

$$(5.1)\qquad \begin{aligned} &\|(P(m)v)(t)\| \leq \|v\| \exp(\|\underset{\sim}{M_C}\|(t-m)) && t \geq m, \quad v \in \underset{\sim}{E} \\ &\|(Q(m)r)(t)\| \leq \left(\int_m^t \|r(s)\| ds\right)\exp(\|\underset{\sim}{M_C}\|(t-m)), && t \geq m, \quad r \in \underset{\sim}{K}(E)\ . \end{aligned}$$

We note the following corollary of Lemma 5.1 and the preceding discussion on "cutting down" the domain of the equation.

<u>LEMMA 5.2</u>. If u is a solution of $(1.2)_{[m]}$ for some $m \geqq 0$, then

$$\|\Pi(t)u\| \leqq \|\Pi(t_0)u\| \exp(\|\underset{\sim}{M_C}\|(t-t_0))\ , \qquad t \geqq t_0 \geqq m\ .$$

6. <u>The Associated Difference Equation</u>

We construct a certain difference equation in $\underset{\sim}{E}$ in such a way that the values of a solution are the slices of a solution of (1.1). For this purpose, we define the linear mappings

$$(6.1)\qquad \begin{aligned} A(n) &= -\Pi(n)P(n-1): \underset{\sim}{E} \to \underset{\sim}{E} \\ B(n) &= \Pi(n)Q(n-1): \underset{\sim}{K}(E) \to \underset{\sim}{E} \end{aligned} \qquad n = 1,2,\ldots$$

and observe that (5.1) implies

$$(6.2)\qquad \begin{aligned} &A(n) \in \tilde{\underset{\sim}{E}}\ , \quad \|A(n)\| \leqq \exp\|\underset{\sim}{M_C}\|\ , && n = 1,2,\ldots \\ &\|B(n)r\| \leqq \|(\varpi r)(n)\| \exp\|\underset{\sim}{M_C}\|\ , && n = 1,2,\ldots, r \in \underset{\sim}{K}(E)\ . \end{aligned}$$

We set $A = (A(n)) \in \underset{\sim}{\ell}^{\infty}_{[1]}(\tilde{\underset{\sim}{E}})$ and define a linear mapping $B: \underset{\sim}{K}(E) \to \underset{\sim}{s}_{[1]}(\underset{\sim}{E})$ by $(Br)(n) = B(n)r$, $n = 1,2,\ldots$.

With A thus defined, we consider the following difference equations in $\underset{\sim}{E}$:

$$(6.3)\qquad x(n) + A(n)x(n-1) = f(n)\ , \qquad n = 1,2,\ldots$$

$$(6.4)\qquad x(n) + A(n)x(n-1) = 0\ , \qquad n = 1,2,\ldots$$

and their restrictions $(6.3)_{[m]}$, $(6.4)_{[m]}$ to $n = m+1, m+2, \ldots$ for $m \in \omega$. Here $f \in \underset{\sim}{s}_{[1]}(E)$.

The fact that (6.3) and (6.4) are, in some sense, reduced forms of (1.1) and (1.2) is expressed by the following result.

LEMMA 6.1. Let $m \in \omega$ and $r \in \underset{\sim}{K}(E)$ be given. A function $x \in \underset{\sim}{s}_{[m]}(\underset{\sim}{E})$ is a solution of $(6.3)_{[m]}$ with $f = Br$ if and only if $x = \varpi u$ for some solution u of $(1.1)_{[m]}$. In particular, x is a solution of $(6.4)_{[m]}$ if and only if $x = \varpi u$ for some solution u of $(1.2)_{[m]}$.

It is clear that not every $f \in \underset{\sim}{s}_{[1]}(\underset{\sim}{E})$ is of the form $f = Br$. It is still possible, however, to relate equation (6.3) with arbitrary f to equation (1.1). This is the purpose of the principal result of this section.

THEOREM 6.2. For each $f \in \underset{\sim}{s}_{[1]}(\underset{\sim}{E})$ there exists $r \in \underset{\sim}{K}(E)$ such that

$$(6.5) \qquad \|(\varpi r)(n)\| \leqq k^2(\|f(n-2)\| + \|f(n-1)\|) , \qquad n = 1,2,\ldots$$

and such that the solution w of

$$(6.6) \qquad w(n) + A(n)w(n-1) = f(n) - (Br)(n) , \qquad n = 1,2,\ldots$$

with $w(0) = 0$ satisfies

$$(6.7) \qquad \|w(n)\| \leqq k(\|f(n-1)\| + \|f(n)\|) , \qquad n = 1,2,\ldots ,$$

where $f(-1) = f(0) = 0$ and $k = \frac{3}{2} + \|M_{\underset{\sim}{C}}\|$.

The proof of this theorem is obtained by a quite explicit construction of w and r. The main technical difficulty -- specific to the "continuous" case -- lies in ensuring that the slices with integral end-points of r, which are constructed separately, match at the ends so as to build a continuous function.

7. Admissibility and the Solutions of the Homogeneous Equation

The discussion in the preceding section enables us to reduce the consideration of equations (1.1) and (1.2) to analysis of the difference equations (6.3) and (6.4) by means of the theory in [1]. M is still assumed to satisfy (M_1) and (M_2), and A, B are defined by (6.1).

We begin with the non-homogeneous equations. We say that $(\underset{\sim}{C},\underset{\sim}{C})$ is admissible with respect to M --more loosely, with respect to (1.1)-- if for every $r \in \underset{\sim}{C}(E)$ there is a bounded solution u of (1.1). We recall ([1; p. 154]) that, similarly, $(\underset{\sim}{\ell}^{\infty},\underset{\sim}{\ell}^{\infty})$ is admissible with respect to A --or with respect to (6.3) -- if for every $f \in \underset{\sim}{\ell}^{\infty}_{[1]}(E)$ there is a bounded solution x of (6.3).

These properties are connected by the following theorem; in its proof Theorem 6.2 plays a crucial and rather obvious part.

THEOREM 7.1. $(\underset{\sim}{C},\underset{\sim}{C})$ is admissible with respect to M if and only if $(\underset{\sim}{\ell}^{\infty},\underset{\sim}{\ell}^{\infty})$ is admissible with respect to A.

The admissibility of $(\underset{\sim}{\ell}^{\infty},\underset{\sim}{\ell}^{\infty})$ with respect to A implies, under certain additional conditions, an exponential dichotomy of the solutions of the homogeneous equations $(6.4)_{[m]}$ (see [1; Section 7]): roughly speaking, the bounded solutions tend uniformly exponentially to 0, there exists a "complementary" manifold of solutions of (6.4) tending uniformly exponentially to infinity, the two kinds of solutions remain uniformly apart, and together they span all solutions. Since Lemma 6.1 provides a bijective correspondence between solutions of $(1.2)_{[m]}$ and $(6.4)_{[m]}$ (for integral m), Theorem 7.1 will allow us to translate that result into an analogous implication for differential equations with delays. The fact that E is finite-dimensional allows us to make use of the following compactness result.

LEMMA 7.2. If E is finite-dimensional, then $A(n)$ is a compact operator for $n = 1,2,\ldots$.

We now state our main result, to the effect that admissibility of $(\underset{\sim}{C},\underset{\sim}{C})$ with respect to M implies a kind of "exponential dichotomy" of the solutions of $(1.2)_{[m]}$.

THEOREM 7.3. Assume that $(\underset{\sim}{C},\underset{\sim}{C})$ is admissible with respect to M. Then there exist numbers $\nu, N > 0$ such that, for every real $m \geqq 0$, every bounded solution v of $(1.2)_{[m]}$ satisfies

(i) $\|\Pi(t)v\| \leq Ne^{-\nu(t-t_0)}\|\Pi(t_0)v\|$ for all $t \geqq t_0 \geqq m$.

There further exist a finite-dimensional linear manifold $\underset{\sim}{W}$ of solutions of (1.2), and numbers $\nu', N' > 0$, $\lambda_0 > 1$ such that, for every real $m \geq 0$, every solution u of $(1.2)_{[m]}$ is of the form $u = v + w_{[m-1]}$, where v is a bounded solution and $w \in \underset{\sim}{W}$, and such that every solution $w \in \underset{\sim}{W}$ satisfies

(ii) $\|\Pi(t)w\| \geqq N'^{-1} e^{\nu'(t-t_0)}\|\Pi(t_0)w\|$ for all $t \geqq t_0 \geqq 0$,

(iii) $\|\Pi(t)w\| \leqq \lambda_0\|\Pi(t)w - \Pi(t)v\|$ for all $t \geqq m \geqq 0$ and all bounded solutions v of $(1.2)_{[m]}$.

We sketch an outline of the proof. Since $(\underset{\sim}{\ell}^\infty, \underset{\sim}{\ell}^\infty)$ is admissible with respect to A and A is compact-valued (Theorem 7.1 and Lemma 7.2), the work in [1] and [6] allows us to conclude that there is an exponential dichotomy for A and that the subspaces of initial values of bounded solutions of $(6.4)_{[n]}$ are closed and have a fixed finite co-dimension in $\underset{\sim}{E}$ (these subspaces constitute the covariant sequence inducing the exponential dichotomy). This result is then translated by means of Lemma 6.1 into a description of the behaviour of the "integral" slices of solutions of $(1.2)_{[n]}$. Finally, Lemma 5.2 is used to show that this behaviour, described in the statement of the theorem, is common to all slices.

We finally remark that the converse of Theorem 7.3 is valid. This follows from the "converse" theorems in [1] via Lemma 6.1 and Theorem 7.1.

REFERENCES

[1] C.V. COFFMAN and J.J. SCHÄFFER, Dichotomies for linear difference equations, Math. Ann. 172 (1967), 139-166.

[2] C.V. COFFMAN and J.J. SCHÄFFER, Linear Differential Equations with Delays: Admissibility and Conditional Stability, Department of Mathematics, Carnegie-Mellon University, Report 70-2, Pittsburgh, Pennsylvania (1970).

[3] C.V. COFFMAN and J.J. SCHÄFFER, Linear differential equations with delays: admissibility and conditional exponential stability, J. Diff. Eqns. 9 (1971), 521-535.

[4] J.L. MASSERA and J.J. SCHÄFFER, Linear Differential Equations and Function Spaces. Academic Press, New York, 1966.

[5] G. PECELLI, Dichotomies for linear functional-differential equations, J. Diff. Eqns. 9 (1971), 555-579.

[6] J.J. SCHÄFFER, Linear difference equations: closedness of covariant sequences, Math. Ann. 187 (1970), 69-76.

[7] J.J. SCHÄFFER, Linear differential equations with delays: admissibility and conditional exponential stability, II, J. Diff. Eqns. (to appear).

Carnegie-Mellon Institute, Pittsburgh, Pennsylvania

TOPOLOGICAL DYNAMICAL TECHNIQUES FOR DIFFERENTIAL AND INTEGRAL EQUATIONS*

George R. Sell

1. Introduction

During the past six years I have been investigating the connection between topological dynamics and nonautonomous differential equations as well as Volterra integral equations. This viewpoint has proven beneficial (and sometimes essential) for the analysis of certain problems arising in the study of differential and integral equations. In this lecture I would like to examine a few of these problems for the purpose of illustrating the role of the topological dynamical techniques. Since my objective is to paint a broad canvas with soft strokes, the lecture will be semi-expository. The reader may consult the references for some of the technical details which fill in the fine colors.

The first point we notice is that by imbedding the differential or integral equation in a flow, this introduces a new viewpoint or a new language for studying solutions of these equations. One expects that the new language will play a unifying role in the study of a large class of problems. This is indeed what happens here. Certain asymptotic phenomena, for highly diverse equations, can be studied for a common viewpoint.

---

*This research was supported in part by NSF Grant No. GP-27275.

The picture does not end with language, but more importantly, it does include technique. Specifically we are interested in examining how the structure of the theory of topological dynamics can be used to analyze the solutions of differential and integral equations. Many recent results on differential and integral equations were first discovered and proved by topological dynamical techniques. Later, some results were proven by different methods. However, there are a few results where the viewpoint and techniques of topological dynamics seem to be essential for the analysis and proof. One such result, which has applications to space rescue problems, is discussed in Sections 10 and 11.

In the next two sections we will present the appropriate viewpoint for imbedding the solutions of nonautonomous differential equations and Volterra integral equations in a flow.

2. Differential Equations

Consider the differential equation

$$x' = g(x,t)$$

where $x \in W$, $t \in R$ and g belongs to some function space G. We will assume throughout that the function space G has the following properties:

α) For every $g \in G$ and $x \in W$, there is a unique solution $\phi(x,g,t)$ of $x' = g(x,t)$ that satisfies $\phi(x,g,o) = x$.

β) G is closed under translations, that is $g_\tau \in G$ whenever $g \in G$, where $g_\tau(x,t) = g(x,\tau+t)$.

One then constructs a flow on $W \times G$ by

$$\pi(x,g,\tau) = (\phi(x,g,\tau),f_\tau)$$

where $\tau \in R$. (In the classical definition of a flow, cf. [23], one would have to require global existence for the solutions of the differential equation, however with a suitably modified definition, cf. [29] or [36], one can dispense with this technical requirement, provided we assume that $\phi(x,g,t)$ denotes the maximally-defined solution.)

The concept of a flow does, of course, require that the mapping π be continuous. This then imposes an a priori restriction on the topology on the function space G. In particular, π will be continuous precisely when the topology on G is chosen so that

1) the mapping $(g,\tau) \to g_\tau$ is continuous, and
2) the function $\phi(x,g,t)$ depends continuously on its three arguments.

In practice, the first continuity condition above is not very restrictive. The second condition is a standard problem arising in the fundamental theory of differential equations. We refer the reader to the books by Coddington and Levinson [4], Hale [7] and Hartman [9] as well as recent papers by Hale and Cruz [8], Miller and Sell [19], Neustadt [24], Opial [25] and Sell [32]. Typical examples of function spaces which satisfy the two continuity conditions occur when G is a subset of

a) the space of bounded continuous functions with the uniform topology,
b) the space of continuous functions with the topology of uniform convergence on compact sets (local uniform topology),
c) the space of Carathéodory-type functions with the local L_p-topology, $1 \leqslant p < \infty$,
d) the space of Carathéodory-type functions with certain weak topologies.

3. Volterra Integral Equations

Now consider the Volterra integral equation

$$x(t) = f(t) + \int_0^t a(t,s)g(x(s),s)ds , \qquad t \geqslant 0 , \tag{1}$$

where $x(t) \in R^n$, $f \in F$, $g \in G$ and $a \in A$. We assume that the function spaces F, G and A are chosen so that for each triple (f,g,a) there is a unique solution $\phi(t) = \phi(f,g,a;t)$ of (1). With ϕ so determined we define $T_\tau f = T_\tau(f,g,a)$ by

$$T_\tau f(\theta) = f(\tau + \theta) + \int_0^\tau a(\tau+\theta,s)g(\phi(s),s)ds . \tag{2}$$

$T_\tau f(\theta)$ is thus defined for all $\theta \geqslant 0$ and for all τ lying in the interval of definition of ϕ.

One then constructs a semi-flow on $F \times G \times A$ by setting

$$\pi(f,g,a;\tau) = (T_\tau f, g_\tau a_\tau) , \qquad \tau \geqslant 0 ,$$

where $g_\tau(x,t) = g(x,\tau+t)$ and $a_\tau(t,s) = a(\tau+t,\tau+s)$. Here again we assume that the function spaces F, G and A are closed under the appropriate translates and that the topologies on F, G and A are such that the mapping π is continuous. Sufficient conditions on the topologies can be easily formulated when F is the space of continuous functions from R^1 to R^n with local uniform topology. In this case we require that the topologies on G and A be chosen so that

1) the mapping $(g,\tau) \to g_\tau$ is continuous,
2) the mappings $(a,\tau) \to a_\tau$ and $\tau \to a(\tau+\cdot,\cdot)$ are continuous,
3) the mapping $(\xi,g,a) \to y$ where

$$y(t) = \int_0^t a(t,s)g(\xi(s),s)ds , \qquad 0 \leqslant t < \alpha ,$$

is a continuous mapping of $F \times G \times A$ into $C([0,\alpha),R^n)$, and
4) for each triple $(f,g,a) \in F \times G \times A$, the integral equation (1) admits a unique "maximally defined" solution $\phi(f,g,a;t)$ which is "properly continuous."

The reader is referred to Miller and Sell [20] for a precise definition of the last condition, which is merely a technical formulation of the continuity of ϕ.

The reason for introducing the mapping $T_\tau f$ in (2) is so that π will satisfy the semi-group property for a semi-flow. Note that at $\tau = 0$ one has

$$T_0 f(\theta) = f(\theta) ,$$

and at $\theta = 0$ one has

$$T_\tau f(0) = \phi(\tau) .$$

Also note that if Eqn. (1) is really a differential equation in disguise (that is, $a(t,s) =$ constant and $f(t) =$ constant), then $T_\tau f(\theta)$ is independent of θ, that is $T_\tau f(\theta) = \phi(\tau)$ for all τ and θ.

Pairs of spaces (G,A) that satisfy the four continuity conditions above are said to be compatible. The existence of compatible spaces is studied in Miller and Sell [19] and [20]. We will not give examples here, since they are fairly technical. Instead we will merely note that the conditions above allow one to use various weak topologies on G and A. (An illustration of an appropriate weak topology is given in Theorem 1 below.) An interesting phenomenon to observe is that, when the topology on G is weak, it is necessary to have a strong topology on A (and vice versa) in order to preserve compatibility. This is due to certain technical facts which are described in Miller and Sell [19].

4. Limiting Equations and Asymptotic Behavior

One of the questions of great interest in the theory of differential equations and integral equations is to analyze the asymptotic behavior of a solution $\phi(t)$ as $t \to \infty$. Since these equations generate a flow or semi-flow, one can use the theory of orbits and their ω-limit sets to give a qualitative description of the asymptotic behavior of $\phi(t)$. For example, if in Eqn. (1), there is a generalized sequence $\tau_n \to \infty$ and such that

$$T_{\tau_n} f \to F, \quad g_{\tau_n} \to G, \quad a_{\tau_n} \to A,$$

in the respective spaces, then the limiting equation (which may depend on the sequence τ_n) would be

$$X(t) = F(t) + \int_0^t A(t,s)G(X(s),s)ds \,, \quad t \geqslant 0 \,, \tag{3}$$

and, furthermore, one has $x(\tau_n + t) \to X(t)$, that is

$$\phi(f,g,a;\tau_n + t) \to \phi(F,G,A;t) \,, \tag{4}$$

uniformly for t in compact sets.

Presumably, the limiting equation (3) is easier to analyze or study. The limit in (4) then gives important information about the asymptotic behavior of the solution $x(t)$.

This fact can, perhaps, be best illustrated by the following theorem, which analyzes an asymptotically-autonomous, nonlinear renewal equation.

THEOREM 1. Assume that the integral equation

$$x(t) = f(t) + \int_0^t [a(t-s) + r(t,s)][g(x(s)) + h(x(s),s)]ds \tag{5}$$

has a bounded solution $x(t)$ defined for all $t \geq 0$. Assume further that the following conditions are satisfied:

1) $A = \int_0^\infty |a(r)|dr < \infty$.
2) There is a $k \geq 0$ such that $|g(x) - g(y)| \leq k|x-y|$ for all x and y.
3) $Ak < 1$.
4) f is continuous and $\lim f(t) = f_0$ exists as $t \to \infty$.
5) h is bounded and continuous for compact sets $I \subset R$ and $W \subset R^n$, there is a function $\sigma(\tau)$ such that $\sigma(\tau) \to 0$ as $\tau \to \infty$ and

$$\int_I |h(\xi(s),\tau + s)|ds \leq \sigma(\tau)$$

for all continuous functions $\xi: I \to W$.
6) $r(t,s)$ is measurable and for any compact sets I, J in R^+ and any function $\xi \in L_1(I)$ there is a function $\rho(\tau)$ such that $\rho(\tau) \to 0$ as $\tau \to \infty$ and

$$\left|\int_I r(\tau+t,\tau+s)\xi(s)ds\right| \leq \rho(\tau), \quad t \in J .$$

Then Eqn. (5) has a single limit equation, namely,

$$X(t) = f_0 + \int_{-\infty}^t a(t-s)g(X(s))ds . \tag{6}$$

Furthermore, Eqn. (6) has a unique solution $X(t) = x_0$, where x_0 satisfies

$$x_0 = f_0 + Ag(x_0)$$

and $x(t) \to x_0$ as $t \to \infty$.

The proof of the above theorem is based on a straightforward application of the theory presented in Miller and Sell [20]. After one shows that Eqn. (6) does indeed represent the only limiting equation for Eqn. (5) (that is, the limiting equation does not depend on the sequence τ_n), then one uses the Contraction Mapping Theorem to observe that Eqn. (6) has a unique solution $X(t) = x_0$. The limiting behavior then follows from the general theory cited above.

Before going on, let us comment briefly on the assumptions concerning h and r in the last theorem. These are chosen so that, in the appropriate topologies, one has

$$h_\tau \to 0 \quad \text{and} \quad r_\tau \to 0$$

as $\tau \to \infty$. Notice that the condition on r does permit one to use a weak topology for the space A.

5. Modus Operandi

Throughout the remainder of this lecture I would like to focus our attention on the study of differential equations. Many of the concepts we will now examine can, of course, be reformulated for Volterra integral equations. Similarly many of the results we will present can be extended to integral equations. However, some of the extensions are nontrivial.

In addition to the uniqueness condition α) mentioned in Section 2, we will assume that the positive trajectory g_t $(t \geqslant 0)$ lies in a compact set in G. This means that the positive hull

$$H^+(g) = \mathrm{Cl}\{g_t : t \geqslant 0\}$$

is compact and nonempty and that the ω-limit set

$$\Omega_g = \{g^* \in G : g^* = \lim g_{\tau_n} \quad \text{for some sequence} \quad \tau_n \to \infty\}$$

is compact and nonempty. (The compactness condition can easily be translated into conditions on $g(x,t)$, cf. [30] and [32].)

We should observe in passing that the two conditions mentioned above (that is, uniqueness and compactness) are somewhat interrelated, cf. [29, pp. 249-254].

6. Stability Theory

Consider now the differential equation

$$x' = g(x,t) \tag{7}$$

and the associated limiting equations

$$x' = g^*(x,t) \ , \qquad g^* \in \Omega_g \ .$$

It is not difficult to show that stability properties, in particular uniform stability properties, of Eqn. (7) are inherited by every limiting equation. Thus, if Eqn. (7) has a bounded uniformly stable (or uniformly asymptotically stable) solution, then every limiting equation has a bounded uniformly stable (or uniformly asymptotically stable) solution, cf. Sell [30].

This, however, is not the main question one would like to study. Heuristically speaking, the limiting equations are devoid of transients and, presumably, they are somewhat simpler than the given Eqn. (7). Therefore, one expects the analysis of the limiting equation(s) to be easier than that of the given equation, and indeed, this is oftentimes the case. What one would like to study then is the Inverse Limit Problem, that is, one would like a theory of stability of the limiting equation(s), along with certain perturbation properties, which will allow one to conclude something about the solutions of the original equation. A theory of this type is known, as we shall see below, in the asymptotically autonomous case. However, for the general case, the following theorem is the best known result.

<u>THEOREM 2</u>. Assume that g_t lies in a compact set in G for $t \geqslant 0$. Assume further that there is a bounded uniformly stable solution $\phi(x,g,t)$ of Eqn. (7) and that there is an $a > 0$ such that for every limiting equation $g^* \in \Omega_g$ one has

$$(8) \qquad |\phi(y,g^*,t) - \phi(z,g^*,t)| \to 0 \quad \text{as} \quad t \to \infty$$

whenever $|y - z| \leqslant a$, where $(y,g^*) = \lim\, (\phi(x,g,t_n),g_{t_n})$ as $t_n \to +\infty$.

Then the solution $\phi(x,g,t)$ is asymptotically stable.

The proof of this, cf. Sell [30], is based on the theory of attractors in a dynamical system, cf. [1] and [2]. Condition (8) merely asks that the solution $\phi(y,g^*,t)$ of the limiting equation be asymptotically stable. Since $\phi(x,g,t)$ is assumed to be uniformly stable, we already know that $\phi(y,g^*,t)$ is also uniformly stable. However, it is unknown at this time whether one can drop the stability assumption on $\phi(x,g,t)$, and replace that with the weaker condition that the limiting solutions be uniformly stable. This is merely one example of the multitude of stability problems which are included in the Inverse Limit Problem. Very little is known here, and it does appear to be a fruitful area for future research.

If the limiting equation is autonomous, one can say more about the Inverse Limit Problem, as we shall see in the next section. However, except for Theorem 2 above, nothing more is known about this problem even in the asymptotically periodic case. We shall explore the reason for this in Section 8.

7. Perturbation Theory: Asymptotically Autonomous Case

Consider now the differential equations

$$x' = f(x,t) = g(x) + h(x,t) \tag{9}$$

and

$$x' = g(x) , \tag{10}$$

where f, g and h are in G. We consider f and g as given and set $h = f - g$. We then ask whether Eqn. (10) is the limiting equation for Eqn. (9), that is, is $\Omega_f = \{g\}$? It is not hard to see that a necessary and sufficient condition for this to happen is that $h_t \to 0$ in the topology on G.

With the appropriate topologies on G, the following conditions on h are sufficient (and sometimes necessary) conditions for $h_t \to 0$ in G:

a) $|h(x,t)| \to 0$ as $t \to \infty$, uniformly for x in compact sets;

b) $\int_0^\infty |h(x,t)|dt \leq B < \infty$, uniformly for x in compact sets;

c) $\int_t^{t+1} |h(x,s)|ds \to 0$ as $t \to \infty$, uniformly for x in compact sets; and

d) $\sup_{0 \leq \sigma \leq 1} \left| \int_t^{t+\sigma} h(x,s)ds \right| \to 0$ as $t \to \infty$, uniformly for x in compact sets.

These four conditions on h, as well as other similar conditions, are rather standard assumptions appearing in perturbation theory. See for example, [3], [5], [10], [11], [16], [34], [35], and [38].

If $\Omega_f = \{g\}$, that is, if Eqn. (9) is asymptotically autonomous, then the Inverse Limit Problem admits a fairly complete solution. The following theorem of Markus [13] is an illustration:

THEOREM 3. Consider the differential equation

$$x' = f(x,t) = Ax + \tilde{g}(x) + h(x,t) \tag{11}$$

where $h_t \to 0$ in G, $|\tilde{g}(x)| = o(|x|)$ as $|x| \to 0$ and the eigenvalues of A all have negative real part. Then for all $|x|$ sufficiently small one has $\phi(x,f,t) \to 0$ as $t \to \infty$.

In this theorem, the assumptions are chosen so that the linear part of

$$g(x) = Ax + \tilde{g}(x)$$

dominates the solutions. There are many generalizations of the last theorem, cf. for example [38]. The proofs of these theorems are based on Lyapunov's second method. Specifically one uses a converse theorem to construct a Lyapunov function for Eqn. (10), and then one uses standard perturbation arguments to analyze Eqn. (9). Unfortunately, Lyapunov's theory is not helpful for solving the Inverse Limit Problem in anything more general than the asymptotically autonomous case, as we shall now see.

8. Perturbation Theory: Asymptotically Periodic Case

A differential equation

$$x' = f(x,t) \tag{12}$$

is said to be asymptotically periodic if Ω_f is a limit cycle, that is, if Ω_f consists of a single periodic trajectory. In this case, the limiting equations are all of the form

$$x' = g(x,\tau+t) , \qquad 0 \leqslant \tau < T ,$$

where $g(x,t)$ is T-periodic in t.

If it happens that the given Eqn. (12) has the form

$$f(x,t) = g(x,t) + h(x,t) \tag{13}$$

where g is periodic in t and $h_t \to 0$ in G, then Eqn. (12) is asymptotically periodic. Moreover, in this case one can use Lyapunov theory to study the Inverse Limit Problem, cf. [38]. However, the condition given in Eqn. (13) is only a sufficient condition for Eqn. (12) to be asymptotically periodic; it is not a necessary condition, cf. [32]. It is for this reason that Lyapunov theory, as it is currently understood, is not helpful for solving the Inverse Limit Problem for the general asymptotically periodic case.

We repeat, the Inverse Limit Problem does suggest an area for fruitful research.

9. Periodic and Almost Periodic Solutions

Some of the first results, which were discovered by exploiting the topological dynamical viewpoint for nonautonomous differential equations, concern the existence of periodic and almost periodic solutions. The basic idea here is that one can describe the structure of a ω-limit set in terms of stability concepts. While this is an old idea going back to Markov [12] and Nemytskii [22], some of the more recent results of Deysach and Sell [6], Miller [17], Sell [28], and Sibirskii [33] are more suitable for applications to differential equations.

As an illustration, one is able to prove the following statement:

THEOREM 4. [6] Let $x(t)$ be a bounded uniformly stable solution of $x' = f(x,t)$ where f is periodic in t. Then there exists an almost periodic solution of $x' = f(x,t)$.

Many extensions of the above statement are now known and they are reported in Sell [32]. Two such extensions are:

1) One gets the same conclusion if f is almost periodic in t and $x(t)$ is bounded and uniformly asymptotically stable, cf. Miller [17] and Seifert [26].
2) If the solution $x(t)$ in Theorem 4 is uniformly asymptotically stable, then one can prove the existence of a periodic solution, cf. Sell [28].

Recently many of the above results have been extended to differential equations lacking uniqueness, by a careful study of properties of asymptotically autonomous functions, cf. [37].

10. Invariant Measures

Let us now consider the differential equation

$$x' = f(x,t)$$

on $W \times R$, where f is a C^1-function which is uniformly almost periodic in t and

$$\mathrm{div}_x f(x,t) \equiv 0 \ , \quad (x \in W,\ t \in R) \ .$$

We assume that W is a Riemannian manifold so that div_x is defined with respect to the local Riemannian coordinates, and we let μ denote the induced volume measure on W.

The hull of f is given by

$$H(f) = \mathrm{Cl}\{f_\tau : \tau \in R\}$$

where the closure is taken in the local uniform topology. Since f is almost periodic in t, $H(f)$ is the space of a compact, topological Abelian group, and we let ν denote the Haar measure on $H(f)$, cf. [32] and [39].

It is shown in Sell [31] that the product measure $m = \mu \times \nu$ is an invariant measure for the flow (see Section 2) on $W \times H(f)$. Therefore, if W is compact, then by a theorem of Hopf, we see that almost all points in $W \times H(f)$ are Poisson stable in the flow. (Recall that a point p is Poisson stable if for every neighborhood U of p, the trajectory through p enters U infinitely often for future as well as for past time.)

Let us now look at an application of this fact, which will play a crucial role in the control-theoretic problem we study in the next section. It is convenient, for this purpose, to introduce the following terminology.

Let Q be a property which may be attributable to points in $H(f)$. We shall say that the property Q is true with <u>probability one</u> if $\nu(A) = 0$, where A denotes the subset of $H(f)$ for which Q is false. (We can sssume that the Haar measure is normalized so that $\nu(H(f)) = 1$. In other words, we can assume that ν is a probability measure.)

A function $g \in H(f)$ is said to be a <u>world</u>. A world $g \in H(f)$ is said to be <u>good</u> if the set

$$\{x \in W\colon \phi(x,g,t) \text{ is Poisson stable}\}$$

is dense in W.

Under the assumption that W is compact, one then has the following result:

<u>THEOREM 5</u>. A world is good, with probability one.

The proof of the last statement is not difficult, cf. Sell [31]. It uses the Hopf result mentioned above together with

1) the Fubini Theorem, and
2) the fact that every nonempty open set in W has positive measure.

The important thing to note here is not that the proof is simple, but rather, that the viewpoint adopted in Section 2 is crucial. There seems to be no other way to prove Theorem 5.

Before we turn to the control problem, let us note that it is possible to greatly weaken the almost periodicity assumption in Theorem 5. We refer the reader to [31] for the details.

11. A Control Problem

Consider a satellite with zero gravitational mass moving in the gravitational field of n-bodies. We will assume that the satellite has thrust engines which can be used for guidance purposes. The equations of motion for this (n+1)-body problem are of the form

$$(14) \qquad y'' = g(y) , \quad z'' = H(y,z) + u ,$$

where the y-equation denotes the equations of motion for the n-bodies, $H(y,z)$ denotes the gravitational field as it affects the satellite and u denotes control effect of the thrust engines. The only assumption we make about u is that it is piece-wise continuous and satisfies $|u| \leqslant \varepsilon$ where ε is some prescribed positive number.

Let us assume that the solution of the y-equation is known, call it $y(t)$, and further let us assume that $y(t)$ is almost periodic in t. Then Eqn. (14) reduces to

$$(15) \qquad z'' = H(y(t),z) + u .$$

If one makes the standard change of variables $x = (z,z')$ then Eqn. (15) can be written as a first order system

$$(16) \qquad x' = f(x,t) + Bu$$

where B is a suitable matrix, f is a C^1-function which is uniformly almost periodic in t and $\mathrm{div}_x f \equiv 0$ for all x and t.

Assume now that without control (i.e. with $u = 0$) the given satellite which we will call the pursuit satellite will continue to move in a prescribed compact, connected set W in the x space. Suppose now that there is a derelict satellite also afloat on a known course in W. The pursuit satellite, which satisfies Eqn. (16), is supposed to make contact with the derelict. This mission is accomplished if the position and velocity coordinates of the pursuer agree with the position and velocity coordinates of the derelict at one and the same time. We shall call this the Capture Problem.

The Capture Problem is analyzed in the periodic case (that is, when $y(t)$ is periodic in t) in Markus and Sell [14]. The generalization, which includes the almost periodic case, will be treated in a forthcoming paper [15].

Actually the analysis of the problem does not depend heavily on the way the controller u enters in Eqn. (16). One can "solve" this problem in a more general setting where Eqn. (16) is replaced by

$$x' = f(x,t,u)$$

where $f(x,t,0)$ is uniformly almost periodic in t and $\mathrm{div}_x f(x,t,0) \equiv 0$ for all x and t. However, in order to simplify our discussion let us limit our attention to Eqn. (16).

Let $\phi(x_0,t_0,u;t_0+t)$ denote the solution of Eqn. (16) that satisfies $\phi(x_0,t_0,u;t_0) = x_0$. Define the sets of attainability by

$$A(x_0,t_0,t) = \{\phi(x_0,t_0,u;t_0+t): |u| \leqslant \varepsilon\}$$

and

$$A(x_0,t_0) = \cup_{t \geqslant 0} A(x_0,t_0,t) \ .$$

The Capture Problem now reduces to showing that given the initial data (x_0,t_0) for the pursuit satellite, there is a $T \geqslant 0$ such that

$$A(x_0,t_0,T) = W \ . \tag{17}$$

Indeed, if Eqn. (17) holds one merely notes the position and velocity coordinates of the derelict at time (t_0+T) (this is a point in W) and then one chooses the appropriate controller u for steering the pursuer (this gives a point in $A(x_0,t_0,T)$).

The analysis of the Capture Problem is based on two hypotheses:

1) the uniform controllability hypothesis, and

2) the good world hypothesis,

which we now describe.

First, a solution of Eqn. (16) with $u = 0$ is said to be a free solution. Next, we shall say that Eqn. (16) satisfies a uniform controllability hypothesis with $\eta = \eta(t)$ if for $t > 0$ there is an $\eta(t) > 0$ such that $A(x_0,t_0,t)$ contains a ball of radius $\eta(t)$ centered at the free solution $\phi(x_0,t_0,0;t_0+t)$. (The important point here is that η does not depend on x_0 and t_0.)

It can be shown that under the almost periodicity assumption on $y(t)$, Eqn. (16) does satisfy the uniform controllability hypothesis provided x_0 is restricted to a compact region W.

The good world hypothesis refers to the terminology of the last section when applied to Eqn. (16) with u set equal to 0. Actually the precise wording of this hypothesis is somewhat different from that given by Section 10, but we will not go into these details here. Suffice it to say that the analogue of Theorem 5 is still true, that is, the good world hypothesis is satisfied with probability one.

One can then show that the Capture Problem is solvable under the uniform controllability hypothesis and the good world hypothesis. By combining this fact with Theorem 5 we get

THEOREM 6. The Capture Problem for Eqn. (16) is solvable, with probability one.

With this theorem we have presented an interesting result in which the topological dynamical techniques for nonautonomous differential equations plays a fundamental and essential role. We have also completed our picture.

REFERENCES

[1] J. AUSLANDER, N.P. BHATIA, and P. SEIBERT, Attractors in dynamical systems, Bol. Soc. Mat. Mexicana (1964), 55-56.

[2] N.P. BHATIA and G.P. SZEGO, Stability Theory of Dynamical Systems. Springer-Verlag, New York, 1970.

[3] L. CESARI, Asymptotic Behavior and Stability Problems in Ordinary Differential Equations. Springer, Berlin, 1963.

[4] E.A. CODDINGTON and N. LEVINSON, Theory of Ordinary Differential Equations. McGraw-Hill, New York, 1955.

[5] W.A. COPPEL, Stability and Asymptotic Behavior of Differential Equations. Heath, Boston, 1965.

[6] L.G. DEYSACH and G.R. SELL, On the existence of almost periodic motions, Michigan Math. J. 12 (1965), 87-95.

[7] J.K. HALE, Ordinary Differential Equations. Wiley-Interscience, New York, 1969.

[8] J.K. HALE and M.A. CRUZ, Existence uniqueness and continuous dependence for hereditary systems, Ann. Mat. Pura Appl. 85 (1970), 63-81.

[9] P. HARTMAN, Ordinary Differential Equations. Wiley, New York, 1964.

[10] J.P. LASALLE and R.J. RATH, Eventual stability, Proc. 2nd IFAC Congress, Basel, 1963; Butterworth, London, 1964, Vol. II, 556-560.

[11] N. KRASOVSKII, Stability in Motion. Stanford Univ. Press, Stanford, 1963.

[12] A.A. MARKOV, Stabilität in Liapunoffschen Sinne und Fastperiodiztät, Math. Zeit. 36 (1933), 708-738.

[13] L. MARKUS, Asymptotically autonomous differential systems, Contributions to Nonlinear Oscillations, Vol. 3. Princeton Univ. Press, Princeton, New Jersey, 1956.

[14] L. MARKUS and G.R. SELL, Capture and control in conservative dynamical systems, Arch. Rational Mech. Anal. 31 (1968), 271-287.

[15] L. MARKUS and G.R. SELL, Capture and control: the aperiodic case (to appear).

[16] R.K. MILLER, Asymptotic behavior of nonlinear delay equations, J. Diff. Eqns. 1 (1965), 293-305

[17] R.K. MILLER, Almost periodic differential equations as dynamical systems with applications to the existence of a.p. solutions, J. Diff. Eqns. 1 (1965), 337-345.

[18] R.K. MILLER, Lecture Notes on Volterra Integral Equations. Brown Univ., Providence, R.I., 1968.

[19] R.K. MILLER and G.R. SELL, Existence, uniqueness and continuity of solutions of integral equations, Ann. Mat. Pura Appl. 80 (1968), 135-152.

[20] R.K. MILLER and G.R. SELL, Volterra Integral Equations and Topological Dynamics. Amer. Math. Soc. Memoir No. 102, Providence, R.I., 1970.

[21] R.K. MILLER and G.R. SELL, Existence, uniqueness and continuity of solutions of integral equations: an addendum, Ann. Mat. Pura Appl. 87 (1970), 281-286.

[22] V.V. NEMYTSKII, Sur les systèmes de courbes remplissant un space metrique, Mat. Sbornik (N.S.) 6 (1939), 283-292.

[23] V.V. NEMYTSKII and V.V. STEPANOV, Qualitative Theory of Differential Equations. Princeton University Press, Princeton, N. J., 1960.

[24] L.W. NEUSTADT, On the solutions of certain integral-like operator equations: existence, uniqueness and dependence theorems, Arch. Rational Mech. Anal., 38 (1970), 131-160.

[25] Z. OPIAL, Sur la depéndance des solutions d'un système d'équations différentielles de leurs seconds membres: applications aux systèmes presque autonomous, Ann. Polon. Math. 8 (1960), 75-89.

[26] G. SEIFERT, Stability conditions for the existence of almost-periodic solutions of almost-periodic systems, J. Math. Anal. Appl. 10 (1965), 409-418.

[27] G. SEIFERT, Almost periodic solutions for almost periodic systems of ordinary differential equations, J. Diff. Eqns. 2 (1966), 305-319.

[28] G.R. SELL, Periodic solutions and asymptotic stability, J. Diff. Eqns. 2 (1966), 143-157.

[29] G.R. SELL, Nonautonomous differential equations and topological dynamics I: The basic theory, Trans. Amer. Math. Soc. 127 (1967), 241-262.

[30] G.R. SELL, Nonautonomous differential equations and topological dynamics II: Limiting equations, Trans. Amer. Math. Soc. 127 (1967), 263-283.

[31] G.R. SELL, Invariant measures and Poisson stability, Topological Dynamics. Benjamin, New York, 1968, 435-454.

[32] G.R. SELL, Lectures on Topological Dynamics and Differential Equations. Van Nostrand-Rheinhold, Princeton, New Jersey, 1971.

[33] K.S. SIBIRSKII, Uniform approximation of points and properties of motions in dynamical limiting sets, Isv. Akad. Nauk Moldav. SSR, Ser. Estestven. Tekh. Nauk, No. 1 (1963), 38-48.
[34] A. STRAUSS and J.A. YORKE, Perturbation theorems for ordinary differential equations, J. Diff. Eqns. 3 (1967), 15-30.
[35] A. STRAUSS and J.A. YORKE, Perturbing uniform asymptotically stable nonlinear systems, J. Diff. Eqns. 6 (1969), 452-483.
[36] T. URA, Sur le courant extérieur a une région invariante, Prolongements d'une caractéristique at porde de stabilité, Funkcial. Ekvac. 2 (1959), 143-200.
[37] T. YOSHIZAWA, Asymptotically almost periodic solutions of an almost periodic system, Funkcialaj Ekvac. 12 (1969), 23-40.
[38] T. YOSHIZAWA, Stability Theory by Liapunov's Second Method. Math. Soc. Japan, Tokyo, 1966.
[39] L. PONTRJAGIN, Topological Groups. Princeton University Press, Princeton, N. J., 1939.

University of Minnesota, Minneapolis, Minnesota, and
Instituto Matematico "Ulisse Dini", Firenze, Italy

DOUBLE ASYMPTOTIC EXPANSIONS FOR LINEAR ORDINARY DIFFERENTIAL EQUATIONS

Wolfgang Wasow

1. The Problem

The simple differential equation

$$\varepsilon^2 \frac{d^2u}{dx^2} = xu , \tag{1.1}$$

where ε is a small parameter, is a prototype for a surprisingly large number of problems involving asymptotic expansions, including the one I wish to talk about.

The change of variables

$$x = t\varepsilon^{2/3} \tag{1.2}$$

changes (1.1) into Airy's equation

$$\frac{d^2u}{dt^2} = tu , \tag{1.3}$$

whose solution can be represented by convergent series in power of t for all finite t and by well known asymptotic series near $t = \infty$. In terms of the original independent variable x the convergent expansions have the form

$$\sum_{r=0}^{\infty} a_r(x\varepsilon^{-2/3})^r \tag{1.4}$$

with certain coefficients a_r which permit the calculation of the solutions in any disk

$$|x| \leq x_0|\varepsilon|^{2/3}, \quad (x_0 \text{ independent of } \varepsilon) \tag{1.5}$$

in the complex x-plane. Their convergence is, however unsuitably slow when ε is very small, unless x is appropriately small, as well.

The asymptotic expansions have, as always in these theories, different forms in different sectors of the t-plane. To simplify the discussion I assume, from now on, that

$$\varepsilon > 0 .$$

For certain special solutions these asymptotic expansions have the form

$$x^{-1/4} \sum_{r=0}^{\infty} c_r (x^{-3/2}\varepsilon)^r \, e^{\frac{1}{\varepsilon} x^{3/2}}, \quad \text{as} \quad x^{-1}\varepsilon^{2/3} \to 0 \tag{1.6}$$

where the constants c_r and the determinations of the fractional powers of x depend on the sector in the x-plane in a way I will not elaborate here.

For what I want to talk about, the following observations are important.

Observation 1. Formula (1.6) combines two results, which, in the classical theory of asymptotic expansions belong to two related but distinct theories, namely asymptotic solutions for differential equations, as the independent variable goes to infinity, and asymptotic solutions as a parameter that appears in a singular manner in the differential equation tends to zero.

Observation 2. Formula (1.6) is meaningless at $x = 0$, the "turning point" of the differential equation. From the special theory of Airy's equation it is, however, known which expansion of the form (1.4), valid about $x = 0$, represents the same solution of the differential equation (1.1) as a given expansion of the form (1.6). (See, e.g., [7].)

The foregoing two remarks naturally lead to two more general problems.

Problem 1. Let us call expansions for solutions of differential equations with a small parameter ε doubly asymptotic, if the remainder after breaking off at the Nth term is small with increasing order of magnitude as N grows, both with respect to the independent variable, as it tends to infinity, as well as with respect to the parameter ε, as it tends to zero. The problem then is to find such doubly asymptotic expansions for the solutions of some wider classes of differential equations.

Problem 2. Once such doubly asymptotic expansions have been found it turns out that they share with the special expansions (1.6) the feature of degenerating near certain points such as $x = 0$ in example (1.1). I will refer to such points loosely as "turning points", without attempting a general definition. The problem of finding expansions of the same solutions valid at a turning point has been called the "central connection problem".

In the remainder of this talk I will report on some results related to those two problems.

2. Fedoryuk's Expansions

M. V. Fedoryuk has studied many asymptotic problems for classes of second order differential equations similar to (1.1). His emphasis has been on results applicable in large unbounded domains of the x-plane. Doubly asymptotic expansions are one of his main tools in these investigations. Only some special, partial results can be quoted here without elaborate preparation. Consider differential equations of the form

$$\varepsilon^2 \frac{d^2u}{dx^2} = p(x)u \,, \tag{2.1}$$

where $p(x)$ is a polynomial of degree m. Fedoryuk constructs certain unbounded regions in the complex x-plane, called canonical domains by him, in which he can calculate a fundamental system of solutions of the differential equation (2.1) by means of doubly asymptotic expansions. Here is the form of these solutions:

$$u(x,\varepsilon) = p^{-1/4}(x)\hat{u}(x,\varepsilon) \exp\left\{ \frac{1}{\varepsilon} \int_{x_1}^{x} p^{1/2}(\tau)d\tau \right\} , \tag{2.2}$$

where

$$\hat{u}(x,\varepsilon) \sim \sum_{r=0}^{\infty} u_r(x) \left(x^{-\frac{m+2}{2}} \varepsilon \right)^r . \tag{2.3}$$

The last relation means that the error committed by breaking off at the Nth term is less than $K_N |x^{-\frac{m+2}{2}} \varepsilon|^{N+1}$, K_N a constant. The similarity between formula (1.6) and Fedoryuk's more general results above is apparent. (See [1].)

The statement just made is still rather vague. To be more precise, we observe that in (2.2) some decision has to be made as to the branch of $p^{1/2}(\tau)$ to be taken. The canonical domain D in which the representation is valid contains no zeros of $p(x)$ and is simply connected, so that the two determinations of $p^{1/2}(\tau)$ are uniquely defined in D. In fact, formula (2.2) represents two linearly independent solutions, with two different functions $\hat{u}(x,\varepsilon)$, depending on the branch of $p^{1/2}(\tau)$ chosen. The ambiguity in the definition of $p^{-1/4}(\tau)$ in (2.2) is less crucial, but it is a delicate point when the analytic continuation of $u(x,\varepsilon)$ is studied in detail. The asymptotic character of the series (2.3) is uniform in D.

One of the advantages of doubly asymptotic representations of solutions of equation (2.1) is that they determine that solution uniquely, whenever they are valid in regions of the x-plane that contain rays along which the exponent $\frac{1}{\varepsilon}\int_{x_1}^{x} p^{1/2}(\tau)d\tau$ of the representation has negative real part for all large x ("subdominant" behavior). This is trivially true for asymptotic expansions, as $x \to \infty$, in the parameterless theory. It is not true, in general, for expansions with respect to a parameter ε that are valid as $\varepsilon \to 0$. For the special solutions with doubly asymptotic expansions the uniqueness is, however, an easy consequence of the arguments in [8], more particularly of Lemma 3.5 and the proof of Theorem 4.2.

The main disadvantage of the doubly asymptotic expansions is that they are not valid at the turning points, i.e., the zeros of $p(x)$.

Long before Fedoryuk, R. E. Langer had studied equations of the form (2.1) and more general ones from a different view point. He was not interested in doubly asymptotic expansions, but rather in the central connection problem mentioned above. His approach has been generalized by many mathematicians, including myself. The essential idea is to transform the given differential equation by means of asymptotic series, into a decisively simpler form, which permits solutions in terms of well known special functions. In regions containing just one zero of $p(x)$, and that zero of order one, the simplified differential equation is precisely equation (1.1), whose asymptotic theory is completely known. (See [7].)

It is a plausible conjecture that such a reduction to Airy's equation might be possible by means of transformations involving doubly asymptotic series in unbounded regions containing one or more turning points. If this could be done, the unpleasant task of "matching" different types of expansions, which is the bane of many investigations in asymptotic theory, could be obviated. Unfortunately, as I have proved, such a reduction is impossible, at least by series in power of ε, and the matching must be carried out. For equations of the form (1.2) I have done this at first order zeros of $p(x)$. My student, Anthony Leung, has extended the procedure to second order zeros by using results of another student of mine, Roy Lee. (See [2] and [3].)

I am not going to give an account of this work here. The following typical example may show the flavor, not a very attractive one, of the type of results obtained:

The differential equation

$$\varepsilon u'' = (x^3 - 1)u$$

has a first order turning point at $x = 1$. It has a particular solution $u(x,\varepsilon)$ which tends exponentially to zero, as $x \to \infty$ in the sector $|\arg(x-1)| < \frac{\pi}{3}$, and which has a doubly asymptotic expansion of the form (2.2), (2.3) in this sector, except at the turning point $x = 1$. This solution is unique except for an arbitrary constant factor independent of ε. If the solution is normalized by the conditions that it be real on the real axis and that $u_0(+\infty) = 1$ the matching calculation shows that its value at $x = 1$ is

$$u(1,\varepsilon) = 2\sqrt{\pi}(1 + h\varepsilon)\left(\frac{3^{-5/6}}{\Gamma(2/3)}\varepsilon^{-1/6} - \frac{119\cdot 3^{-4/3}}{600\Gamma(1/3)}\varepsilon^{1/6}\right) + O(\varepsilon^{11/6}) ,$$

where h is the constant

$$h = \frac{1}{2}\int_0^\infty \tau^{-1/2} \frac{d}{d\tau} [q^{-2}(\tau) \frac{dq(\tau)}{d\tau}]d\tau ,$$

with

$$q(\tau) = \frac{d\phi(\tau)}{d\tau} ,$$

and $\phi(\tau)$ is implicitly defined by

$$\tau = [\frac{3}{2}\int_1^\phi (t^3 -1)^{1/2} dt]^{2/3}, \quad \phi > 1, \tau > 0 .$$

(See [8].)

3. Differential Equations of Higher Order

The restrictive condition that the coefficient $p(x)$ is a polynomial is not in itself necessary, particularly if the asymptotic study is limited to properly chosen sectors, provided that in this sector $p(x)$ shares with polynomials the properties of having a finite number of zeros, no poles, and polynomial growth, as $x \to \infty$.

A more interesting question is the extension of Fedoryuk's theory to higher order differential equations. Significant results in this direction have been obtained by A. Leung in his dissertation.

Leung considers equations of the form

$$\varepsilon^n u^{(n)} - \varepsilon^{n-1} p_{n-1}(x)u^{(n-1)} - \cdots - \varepsilon p_1(x)u' - p_0(x)u = 0 , \tag{3.1}$$

where the $p_j(x)$ are polynomials. I repeat that the restriction to polynomials could be replaced by milder conditions at the price of more involved arguments, but without necessitating essentially new ideas. Furthermore, no significant difficulties should stand in the way of letting the p_j depend in a holomorphic way on ε. Thus, the interest of the investigation to be reported here goes beyond the special type of the differential equation (3.1).

It is true, however, that Leung introduces a seriously restrictive set of conditions: Let m be the degree of $p_0(x)$, then he must have

$$\text{degree of } p_j(x) < \frac{m}{n}(n-j) , \quad \text{for } j = 1, 2, \ldots, n-1. \tag{3.2}$$

Experts in this field will readily recognize the significance of these inequalities. They simplify the structure of the algebraic functions $\lambda_j(x)$ which are defined as the zeros of the characteristic equation

$$(3.3) \qquad \phi(x,\lambda) \equiv \lambda^n - p_{n-1}(x)\lambda^{n-1} - \cdots - p_1(x)\lambda - p_0(x) = 0$$

associated with the differential equation. Thanks to conditions (3.2) the expansions of the $\lambda_j(x)$ about $x = \infty$ all proceed in powers of the same fractional power $x^{-1/n}$, i.e.,

$$(3.4) \qquad \lambda_j(x) = \sum_{r=0}^{\infty} a_{rj} x^{\frac{m-r}{n}} ,$$

Expressed differently, the conditions (3.2) guarantee that Newton's polygon for the polynomial (3.3) reduces to just one segment.

It is not known, and it might be worth investigating, how the results to be stated will have to be modified if (3.2) is not true.

It turns out that the differential equation (3.1) possesses particular fundamental systems with doubly asymptotic expansions of the following form:

$$(3.5) \qquad u_j(x,\varepsilon) = x^{-\frac{m}{n}\frac{n-1}{2}} \hat{u}(x,\varepsilon) \exp\left\{ \frac{1}{\varepsilon} \int_{x_0}^{x} \lambda_j(\tau)d\tau \right\} , \quad j = 1,2,\ldots,n$$

where

$$(3.6) \qquad \hat{u}_j(x,\varepsilon) \sim \sum_{r=0}^{\infty} u_{jr}(x) \left(x^{-(\frac{m}{n}+1)} \varepsilon \right)^r , \quad j = 1, 2, \ldots, n .$$

The functions $\hat{u}_{jr}(x)$ possess convergent series in ascending powers of $x^{-1/n}$ about $x = \infty$. The constant x_0 is arbitrary.

This result is a true generalization of those of Fedoryuk to equations of degree greater than two. It is, however, significantly weaker in some respects. Fedoryuk's expansions are valid, as $\varepsilon > 0$, in large "canonical" regions of the complex x-plane. Leung's representations are proved by him to be valid only in sectors of angular width less than $\frac{\pi}{n} \frac{1}{(\frac{m}{n}+1)}$ and for large x. These sectors are otherwise arbitrary, but the fundamental system of solutions so represented may vary from sector to sector.

The extension of Leung's result to larger regions has to cope with the difficulty that Fedoryuk's construction of canonical domains is based on an analysis of the family of curves

$$\mathrm{Re}\left[\int_{x_0}^{x} \lambda_j(\tau) - \lambda_k(\tau))d\tau\right] = \text{const } j,k = 1,2,\ldots,n,m \quad \text{in the complex plane.}$$

For $n > 2$ this family is very complicated, so that an extension by analogy would be very intricate. It may be wiser to attempt it only for special equations that arise in some context which makes them particularly interesting.

Incidentally, the appearance of a fractional power of x itself in (3.5), instead of the power of $p(x)$ in (2.2), is only a matter of notation and not otherwise significant.

4. Brief Sketch of Leung's Method

The standard transformation $u = y_1$, $\varepsilon u' = y_2, \ldots, \varepsilon^{n-1}u^{(n-1)} = y_n$ takes the differential equation (3.1) into the equivalent system

$$(4.1) \qquad \varepsilon y' = A(x)y$$

where

$$A(x) = \begin{bmatrix} 0 & 1 & 0 & \cdots & 0 \\ 0 & 0 & 1 & \cdots & 0 \\ 0 & 0 & 0 & \cdots & 1 \\ p_0(x) & p_1(x) & p_2(x) & \cdots & p_{n-1}(x) \end{bmatrix}, \quad y = \begin{bmatrix} y_1 \\ \cdot \\ \cdot \\ y_n \end{bmatrix}.$$

In a formal way the system (4.1) can be asymptotically solved by a modification of a technique of Turrittin [5] which resembles the procedure in Wasow [6]. Its essence is a transformation of (4.1) into a system whose coefficient matrix is formally diagonal. "Formally" means that the new coefficient matrix is a, generally divergent, series in power of ε, and nothing is said about the functions this series represents asymptotically.

There is no time here to explain this algorithm. It results in a matrix series

$$(4.2)\qquad T(x,\varepsilon) \doteq \sum_{r=0}^{\infty} T_r(x)\left(x^{-(\frac{m}{n}+1)}\varepsilon\right)^r ,$$

whose coefficients $T_r(x)$ are convergent series of $x^{-1/n}$, with the property that the transformation

$$(4.3)\qquad y = T(x,\varepsilon)v$$

takes the system (4.1) into

$$(4.4)\qquad \varepsilon v' = x^n D(x,\varepsilon)v$$

where

$$D(x,\varepsilon) \doteq \sum_{r=0}^{\infty} D_r(x)\left(x^{-\frac{m}{n}-1}\varepsilon\right)^r$$

and the $D_r(x)$ are diagonal matrices. The symbol "$\doteq$" means that the series is formal: The left member is only a briefer notation for that series. The transformation is carried out by the usual operations with series without regard to the question of analytic validity.

If the divergence of the series $D(x,\varepsilon)$ is disregarded, its diagonal nature permits the solution of equation (4.4) by an exponential function. Expansion of this exponential function, reordering according to powers of ε and return from v to y by means of (4.3) produces the formulas (3.5), (3.6), except that $\hat{u}_j(x,\varepsilon)$ is not a function but just a name for the divergent series in the right member.

The analytic part of the argument consists in showing that among the infinitely many functions asymptotically represented by the series in the right member of (3.6) there exist some for which (3.5) is truly a solution of the differential equation. It is a priori clear that this is not a trivial matter, because only at this stage do specific sectors of asymptotic validity enter the argument. Leung uses an adaptation of Fedoryuk's procedure for this part of the argument. It is a long and difficult proof, which I cannot begin to sketch here. As is usual in this field it is based on iterations of a Volterra integral equation equivalent to the differential equation. It would be desirable to have a proof which is as simple as the ultimate result.

REFERENCES

[1] M.A. EVGRAFOV and M.V. FEDORYUK, Asymptotic behavior of the equation $w''(z) - p(z,\lambda)w(z) = 0$, as $\lambda \to \infty$ in the complex z-plane, Uspehi Mat. Nauk., 21 (1966), No. 1 (127), 3-50; English transl. Russian Math. Surveys, 21 (1966), 1-48.

[2] ROY LEE, On uniform simplification of linear differential equations in a full neighborhood of a turning point, J. Math. Anal. Appl., 27 (1969), 501-510.

[3] A. LEUNG, Connection formulas for asymptotic solutions of second order turning points in unbounded domains (manuscript). Abstract No. 71T-B46, Notices of the Amer. Math. Soc., 18 (1971), 411-412.

[4] A. LEUNG, Doubly asymptotic expansions for the solutions of nth order linear differential equations with a parameter. Ph.D. Thesis, Univ. of Wisconsin, 1971.

[5] H. TURRITTIN, Asymptotic expansions of solutions of systems of ordinary differential equations, Contributions to the Theory of Nonlinear Oscillation II; Ann. of Math. Studies No. 29 (1952), 81-116, Princeton.

[6] W. WASOW, Turning point problems for systems of linear differential equations, I: The formal theory, Comm. Pure Appl. Math., 14 (1961), 657-673.

[7] W. WASOW, Asymptotic Expansions for Ordinary Differential Equations. Wiley-Interscience Publishers, 1965, 362 pp.

[8] W. WASOW, Simple turning-point problems in unbounded domains, SIAM J. Math. Anal., 1 (1970), 153-170.

University of Wisconsin, Madison, Wisconsin

OSCILLATORY PROPERTY FOR SECOND ORDER DIFFERENTIAL EQUATIONS

Taro Yoshizawa

There are many results on oscillatory property of solutions of differential equations. In this article, we shall discuss oscillatory property of solutions and the existence of a bounded nonoscillatory solution of second order differential equations by applying Liapunov second method. Consider an equation

$$(1)\qquad (r(t)x')' + f(t,x,x') = 0 \qquad (' = \frac{d}{dt}),$$

where $r(t) > 0$ is continuous on $I = [0,\infty)$ and $f(t,x,u)$ is continuous on $I \times R \times R$, $R = (-\infty,\infty)$. To discuss oscillatory property of solutions of (1), we consider an equivalent system

$$(2)\qquad x' = \frac{y}{r(t)}, \qquad y' = -f(t,x,\frac{y}{r(t)}).$$

A solution $x(t)$ of (1) which exists in the future is said to be oscillatory if for every $T > 0$ there is a $t_0 > T$ such that $x(t_0) = 0$. Moreover, the equation (1) is said to be oscillatory if every solution of (1) which exists in the future is oscillatory.

THEOREM 1. Assume that there exist two continuous scalar functions $V(t,x,y)$ and $W(t,x,y)$ defined on $t \geq T$, $0 < x < K$, $|y| < \infty$ and on $t \geq T$, $-K < x < 0$, $|y| < \infty$, respectively, where T can be large and $K > 0$ or $K = \infty$, and assume that $V(t,x,y)$ and $W(t,x,y)$ satisfy the following conditions;

(i) $V(t,x,y) \to \infty$ uniformly for $0 < x < K$ and $-\infty < y < \infty$ as $t \to \infty$, and $W(t,x,y) \to \infty$ uniformly for $-K < x < 0$ and $-\infty < y < \infty$ as $t \to \infty$,

(ii) $\dot{V}_{(2)}(t,x(t),y(t)) \leq 0$ for all sufficiently large t, where $\{x(t),y(t)\}$ is a solution of (2) such that $0 < x(t) < K$ for all large t and

$$\dot{V}_{(2)}(t,x(t),y(t)) = \overline{\lim_{h\to 0^+}} \frac{1}{h}\{V(t+h,x(t+h),y(t+h)) - V(t,x(t),y(t))\},$$

(iii) $\dot{W}_{(2)}(t,x(t),y(t)) \leq 0$ for all sufficiently large t, where $\{x(t),y(t)\}$ is a solution of (2) such that $-K < x(t) < 0$ for all large t and

$$\dot{W}_{(2)}(t,x(t),y(t)) = \overline{\lim_{h\to 0^+}} \frac{1}{h}\{W(t+h,x(t+h),y(t+h)) - W(t,x(t),y(t))\}.$$

Then the solution $x(t)$ of (1) such that $|x(t)| < K$ for all large t is oscillatory. Moreover, if $K = \infty$, the equation (1) is oscillatory.

<u>Proof</u>. Let $x(t)$ be a solution of (1) which is defined on $[t_0,\infty)$ and bounded by K for all large t, and suppose that $x(t)$ is not oscillatory. Then $x(t)$ is either positive or negative for all large t. Now assume that $0 < x(t) < K$ for all $t \geq \sigma$ where $\sigma \geq T$. By the condition (i), if t is sufficiently large, say $t \geq t_1$, we have

$$V(\sigma,x(\sigma),y(\sigma)) < V(t,x,y)$$

for all $0 < x < K$, $|y| < \infty$. However, by the condition (ii), we have

$$V(t,x(t),y(t)) \leq V(\sigma,x(\sigma),y(\sigma))$$

for all $t \geq \sigma$, if necessary, choosing a large σ. This contradicts $V(t_1,x(t_1),y(t_1)) > V(\sigma,x(\sigma),y(\sigma))$. When we assume that $-K < x(t) < 0$ for all large t, we have also a contradiction by using $W(t,x(t),y(t))$. Thus we see that $x(t)$ is oscillatory.

To apply this theorem, the following lemmas play an important role. In the following, a scalar function $v(t,x,y)$ will be called a Liapunov function for (2), if $v(t,x,y)$ is continuous in (t,x,y) in the domain of definition and is locally Lipschitzian in (x,y). Moreover, we define $\dot{v}_{(2)}(t,x,y)$ by

$$\dot{v}_{(2)}(t,x,y) = \overline{\lim_{h\to 0^+}} \frac{1}{h} \{v(t+h, x+h\frac{y}{r(t)}, y-hf(t,x,\frac{y}{r(t)})) - v(t,x,y)\}.$$

<u>LEMMA 1</u>. For $t \geq T^*$, $x > 0$, $|y| < \infty$, where T^* can be large, we assume that there exists a Liapunov function $v(t,x,y)$ which satisfies the following conditions;

(i) $yv(t,x,y) > 0$ for $t \geq T^*$, $x > 0$, $y \neq 0$,

(ii) $\dot{v}_{(2)}(t,x,y) \leq -\lambda(t)$, where $\lambda(t)$ is a continuous function defined on $t \geq T^*$ and $\underline{\lim}_{t\to\infty} \int_T^t \lambda(s)ds \geq 0$ for all large T.

Moreover, we assume that there is a τ and a $w(t,x,y)$ for all large T such that $\tau \geq T$ and $w(t,x,y)$ is a Liapunov function defined on $t \geq \tau$, $x > 0$, $y < 0$, which satisfies the following conditions;

(iii) $y \leq w(t,x,y)$ and $w(\tau,x,y) \leq b(y)$, where $b(y)$ is continuous, $b(0) = 0$ and $b(y) < 0$ $(y \neq 0)$,

(iv) $\dot{w}_{(2)}(t,x,y) \leq -\rho(t)w(t,x,y)$, where $\rho(t) \geq 0$ is continuous and

$$\int_\tau^\infty \frac{1}{r(t)} \exp\{-\int_\tau^t \rho(s)ds\}dt = \infty .$$

Then, if $\{x(t),y(t)\}$ is a solution of (2) such that $x(t) > 0$ for all large t, we have $y(t) \geq 0$ for all large t.

We can obtain a similar lemma for a solution $\{x(t),y(t)\}$ of (2) such that $x(t) < 0$ for all large t. For the proof of Lemma 1 and the details, see [5].

<u>PROPOSITION 1</u>. For the equation (1) we assume that

(i) $$\int_0^\infty \frac{dt}{r(t)} = \infty ,$$

(ii) for $t \geq 0$ and $x \geq 0$, there exist continuous functions $a(t)$ and $\alpha(x)$ such that

(3) $$\varliminf_{t\to\infty} \int_T^t a(s)ds \geqslant 0 \quad \text{for all large } T$$

and that $x\alpha(x) > 0$ $(x \neq 0)$, $\alpha'(x) \geqslant 0$ and for all large t, $x \geqslant 0$, $|u| < \infty$

$$a(t)\alpha(x) \leqslant f(t,x,u) ,$$

(iii) for $t \geqslant 0$ and $x \leqslant 0$, there exist continuous functions $b(t)$ and $\beta(x)$ such that

$$\varliminf_{t\to\infty} \int_T^t b(s)ds \geqslant 0 \quad \text{for all large } T$$

and that $x\beta(x) > 0$ $(x \neq 0)$, $\beta'(x) \geqslant 0$ and for all large t, $x \leqslant 0$, $|u| < \infty$

$$f(t,x,u) \leqslant b(t)\beta(x) .$$

Then, if $\int_0^\infty a(t)dt = \infty$ and $\int_0^\infty b(t)dt = \infty$, the equation (1) is oscillatory. Moreover, if we have

(4) $$\int_0^\infty a(t)dt < \infty, \qquad \int_0^\infty \left(\frac{1}{r(s)} \int_s^\infty a(u)du\right) ds = \infty$$

and

(5) $$\int_0^\infty b(t)dt < \infty, \qquad \int_0^\infty \left(\frac{1}{r(s)} \int_s^\infty b(u)du\right) ds = \infty ,$$

then all bounded solutions of (1) are oscillatory. In addition to the conditions above, if

(6) $$\int_\varepsilon^\infty \frac{du}{\alpha(u)} < \infty , \qquad \int_{-\varepsilon}^{-\infty} \frac{du}{\beta(u)} < \infty \quad \text{for some } \varepsilon > 0,$$

the equation (1) is oscillatory.

Proof. Under our assumptions, if we consider a function $v(t,x,y) = \frac{y}{\alpha(x)}$ for large t, this function satisfies the conditions in Lemma 1 with $\lambda(t) = a(t)$. Since the condition (3) implies that for all large T, there is a τ such that $\tau \geqslant T$ and $\int_\tau^t a(s)ds \geqslant 0$ for all $t \geqslant \tau$, a function $w(t,x,y) = y + \alpha(x)\int_\tau^t a(s)ds$ defined on $t \geqslant \tau$, $x > 0$, $y < 0$ satisfies the conditions in Lemma 1 with $\rho(t) \equiv 0$. Thus we can see that if $\{x(t), y(t)\}$ is a solution of (2) such that $x(t) > 0$ for all large t, then $y(t) \geqslant 0$ for all large t. We can also see that for a solution such that $x(t) < 0$ for all large t, $y(t) \leqslant 0$ for all large t.

In the case where we assume that $\int_0^\infty a(t)dt = \infty$ and $\int_0^\infty b(t)dt = \infty$, for large t, if we define $V(t,x,y)$ and $W(t,x,y)$ by

$$V(t,x,y) = \begin{cases} \dfrac{y}{\alpha(x)} + \displaystyle\int_0^t a(s)ds & (y \geqslant 0) \\ \displaystyle\int_0^t a(s)ds & (y < 0) \end{cases}$$

and

$$W(t,x,y) = \begin{cases} \displaystyle\int_0^t b(s)ds & (y > 0) \\ \dfrac{y}{\beta(x)} + \displaystyle\int_0^t b(s)ds & (y \leqslant 0) \end{cases},$$

we can see that these functions satisfy the conditions in Theorem 1 for $K = \infty$, and hence the equation (1) is oscillatory.

In the case where we assume (4) and (5), letting $K > 0$ be a constant, set

$$V(t,x,y) = \int_x^K \frac{du}{\alpha(u)} + \int_0^t \left(\frac{1}{r(s)} \int_s^\infty a(u)du\right) ds$$

for $t \geqslant 0$, $0 < x < K$ and $|y| < \infty$. For a solution $x(t)$ of (2) which satisfies $0 < x(t) < K$ for all large t, there is a $\sigma > 0$ such that $0 < x(t) < K$ and $y(t) \geqslant 0$ for $t \geqslant \sigma$, and hence

$$\dot{V}_{(2)}(t,x(t),y(t)) = \frac{1}{r(t)} \left\{- \frac{y(t)}{\alpha(x(t))} + \int_t^\infty a(u)du\right\} .$$

If we set $V^*(t,x,y) = - \dfrac{y}{\alpha(x)} + \int_t^\infty a(u)du$, we have $\overline{\lim}_{t\to\infty} V^*(t,x(t),y(t)) \leqslant 0$. On the other hand, we have $\dot{V}^*_{(2)}(t,x,y) \geqslant 0$, and hence $V^*(t,x(t),y(t)) \leqslant 0$, which implies that $\dot{V}_{(2)}(t,x(t),y(t)) \leqslant 0$ for $t \geqslant \sigma$. For $t \geqslant 0$, $-K < x < 0$ and $|y| < \infty$, define $W(t,x,y)$ by

$$W(t,x,y) = \int_x^{-K} \frac{du}{\beta(u)} + \int_0^t \left(\frac{1}{r(s)} \int_s^\infty b(u)du\right) ds .$$

Then the conclusion follows from Theorem 1, because K is arbitrary.

In addition, when we assume (6), we can set $K = \infty$ in $V(t,x,y)$ and $W(t,x,y)$ above, and hence the equation (1) is oscillatory.

The result above contains Coles' result [2] and Macki and Wong's result [3].

Remark. It is clear that we can combine the conditions on $a(t)$ and $b(t)$. The Liapunov's method is also applicable to obtain Bobisud's [1] and Opial's [4] results, see [5].

Now we shall discuss the existence of a bounded nonoscillatory solution of (1). The following theorems will be applied. Consider an equation of the second order

$$x'' = F(t,x,x'), \tag{7}$$

where $F(t,x,y)$ is continuous on $I \times R \times R$. Let $\underline{\omega}(t)$ and $\overline{\omega}(t)$ be two functions defined on I, twice differentiable and bounded on I with their derivatives. We assume that $\underline{\omega}(t) \leq \overline{\omega}(t)$,

$$\overline{\omega}''(t) \leq F(t,\overline{\omega}(t),\overline{\omega}'(t)) \tag{8}$$

and

$$\underline{\omega}''(t) \geq F(t,\underline{\omega}(t),\underline{\omega}'(t)) \tag{9}$$

for all $t \geq 0$.

THEOREM 2. Suppose that there exist two Liapunov functions $V(t,x,y)$ and $W(t,x,y)$ defined on $0 \leq t < \infty$, $\underline{\omega}(t) \leq x \leq \overline{\omega}(t)$, $y \geq K$ and on $0 \leq t < \infty$, $\underline{\omega}(t) \leq x \leq \overline{\omega}(t)$, $y \leq -K$, respectively, where $K > 0$ can be large, and assume that $V(t,x,y)$ and $W(t,x,y)$ satisfy the following conditions;

(i) $V(t,x,y) \leq b(y)$ and $W(t,x,y) \leq b(|y|)$, where $b(r) > 0$ is continuous,

(ii) $V(t,x,y) \to \infty$ as $y \to \infty$, $W(t,x,y) \to \infty$ as $y \to -\infty$, uniformly for t, x,

(iii) in the interior of their domains of definition

$$\dot{V}(t,x,y) = \overline{\lim_{h\to 0^+}} \frac{1}{h} \{V(t+h, x+hy, y+hF(t,x,y)) - V(t,x,y)\} \geq 0$$

and

$$\dot{W}(t,x,y) = \overline{\lim_{h\to 0^+}} \frac{1}{h} \{W(t+h, x+hy, y+hF(t,x,y)) - W(t,x,y)\} \leq 0$$

or

(iii)' in the interior of their domains of definition

$$\dot{V}(t,x,y) \leq 0 \qquad \text{and} \qquad \dot{W}(t,x,y) \geq 0 .$$

Then the equation (7) has a solution $x(t)$ such that $\underline{\omega}(t) \leqslant x(t) \leqslant \overline{\omega}(t)$ and $x'(t)$ is bounded for all $t \geqslant 0$.

THEOREM 3. Under the assumptions in Theorem 2, if $\underline{\omega}(0) = \overline{\omega}(0)$ and

$$\dot{V}(t,x,y) \leqslant 0 \quad \text{and} \quad \dot{W}(t,x,y) \leqslant 0$$

in the interiors of their domains of definition, then the equation (7) has a solution $x(t)$ such that $\underline{\omega}(t) \leqslant x \leqslant \overline{\omega}(t)$ and $x'(t)$ is bounded for all $t \geqslant 0$.

For the proofs, see [6].

In discussing the existence of a bounded nonoscillatory solution of (1), we assume that the derivative of $r(t)$ is continuous, and consequently the equation (1) can be written as

$$x'' + \frac{r'(t)}{r(t)} x' + \frac{1}{r(t)} f(t,x,x') = 0. \tag{10}$$

PROPOSITION 2. Suppose that there exist functions $b(t)$ and $\beta(x)$ which satisfy the following conditions;

(i) $b(t)$ is continuous on I and $b(t) \geqslant 0$ for $t \geqslant T$, where T can be large,

(ii) $\beta(x)$ is continuous on $x \geqslant 0$,

(iii) for $t \geqslant T$, $x > 0$ and all y,

$$f(t,x,y) \leqslant b(t)\beta(x) . \tag{11}$$

Moreover, we assume that there is a $c > 0$ such that $f(t,c,0) \geqslant 0$ for $t \geqslant T$. Then, if

$$0 < \varepsilon \leqslant r(t) \leqslant \rho \qquad \text{for some } \varepsilon,\rho \text{ and all } t \geqslant 0 \tag{12}$$

and

$$\int_0^\infty tb(t)dt < \infty \tag{13}$$

or if, there is an $A > 0$ such that

$$\left|\frac{r'(t)}{r(t)}\right| < A \qquad \text{for } t \geqslant 0 \tag{14}$$

and we have

$$\int_0^\infty \left(\frac{1}{r(s)}\int_s^\infty b(u)du\right)ds < \infty \,, \tag{15}$$

the equation (10) has a bounded nonoscillatory solution.

Proof. Under the conditions (12) and (13), the condition (13) implies that $\int_0^\infty b(t)dt < \infty$, and consequently $\int_s^\infty b(u)du$ exists and is small if s is sufficiently large, because $b(t) \geqslant 0$ eventually. Since $\varepsilon \leqslant r(t) \leqslant \rho$, we have

$$\int_0^\infty \left(\frac{1}{r(s)}\int_s^\infty b(u)du\right)ds < \infty \quad \text{and} \quad \frac{1}{r(t)}\int_t^\infty b(u)du < \infty \,.$$

There is an $L > 0$ such that $\beta(c) \leqslant \frac{L}{2}$ and there is a $\delta > 0$ such that $\beta(x) \leqslant L$ if $|x - c| \leqslant \delta$. Choose $t_0 \geqslant T$ so large that

$$0 \leqslant L\int_{t_0}^t \left(\frac{1}{r(s)}\int_s^\infty b(u)du\right)ds \leqslant \delta \qquad \text{for all} \quad t \geqslant t_0 \,.$$

For $t_0 \leqslant t < \infty$, define $\underline{\omega}(t)$ and $\overline{\omega}(t)$ by

$$\underline{\omega}(t) = c \qquad \text{and} \qquad \overline{\omega}(t) = c + L\int_{t_0}^t \left(\frac{1}{r(s)}\int_s^\infty b(u)du\right)ds \,.$$

Then $0 < \underline{\omega}(t) \leqslant \overline{\omega}(t) \leqslant c + \delta$ for all $t \geqslant t_0$, and $\underline{\omega}(t), \overline{\omega}(t)$ are bounded with their derivatives.

Clearly we have $\underline{\omega}''(t) \geqslant -\frac{r'(t)}{r(t)}\underline{\omega}'(t) - \frac{1}{r(t)} f(t,\underline{\omega}(t),\underline{\omega}'(t))$. On the other hand, $\overline{\omega}'(t) = \frac{L}{r(t)}\int_t^\infty b(u)du$ and $\overline{\omega}''(t) = -\frac{r'(t)}{r^2(t)} L\int_t^\infty b(u)du - \frac{L}{r(t)} b(t)$. Thus, using (11), we have

$$\begin{aligned} -\frac{r'(t)}{r(t)}\overline{\omega}'(t) - \frac{1}{r(t)} f(t,\overline{\omega}(t),\overline{\omega}'(t)) &\geqslant -\frac{r'(t)}{r^2(t)} L\int_t^\infty b(u)du - \frac{1}{r(t)} b(t)\beta(\overline{\omega}(t)) \\ &\geqslant -\frac{r'(t)}{r^2(t)} L\int_t^\infty b(u)du - \frac{L}{r(t)} b(t) \\ &\geqslant \overline{\omega}''(t) \,, \end{aligned}$$

since $c \leqslant \overline{\omega}(t) \leqslant c + \delta$ for all $t \geqslant t_0$ and hence $\beta(\overline{\omega}(t)) \leqslant L$ for $t \geqslant t_0$.

For $t \geq t_0$, $\underline{\omega}(t) \leq x \leq \overline{\omega}(t)$ and $y \geq K$, define $V(t,x,y)$ by

$$V(t,x,y) = L\int_{t_0}^{t} b(s)ds + r(t)y$$

and for $t \geq t_0$, $\underline{\omega}(t) \leq x \leq \overline{\omega}(t)$ and $y \leq -K$, define $W(t,x,y)$ by

$$W(t,x,y) = -L\int_{t_0}^{t} b(s)ds - r(t)y \ .$$

Then it is clear that $V(t,x,y)$ and $W(t,x,y)$ satisfy the conditions (i) and (ii) in Theorem 2. Since $\beta(x) \leq L$ for $\underline{\omega}(t) \leq x \leq \overline{\omega}(t)$, we have

$$\begin{aligned}\dot{V}(t,x,y) &= Lb(t) + r'(t)y + r(t)\{-\frac{r'(t)}{r(t)}y - \frac{1}{r(t)}f(t,x,y)\} \\ &= Lb(t) - f(t,x,y) \\ &\geq Lb(t) - b(t)\beta(x) \geq Lb(t) - Lb(t) = 0\end{aligned}$$

and we have also $\dot{W}(t,x,y) \leq 0$. Therefore, it follows from Theorem 2 that the equation (10) has a solution $x(t)$ such that

$$0 < c \leq x(t) \leq c + \delta \qquad \text{for all } t \geq t_0$$

and that $x'(t)$ is bounded for all $t \geq t_0$.

Under the conditions (14) and (15), we can use the same $\underline{\omega}(t)$ and $\overline{\omega}(t)$, since (14) and (15) imply that $\frac{1}{r(t)}\int_t^\infty b(u)du < \infty$. Moreover, (14) and (15) imply that $\int_0^\infty \frac{b(s)}{r(s)} ds < \infty$, and hence it is sufficient to consider

$$V(t,x,y) = y + Ax + L\int_{t_0}^{t} \frac{b(s)}{r(s)} ds$$

and

$$W(t,x,y) = -y + Ax - L\int_{t_0}^{t} \frac{b(s)}{r(s)} ds \ .$$

<u>Remark</u>. Assuming the existence of functions $a(t)$ and $\alpha(x)$ such that $a(t)\alpha(x) \leq f(t,x,y)$ for $t \geq T$, $x < 0$ and all y, we can obtain a result similar to Proposition 2.

By applying Theorem 3, we shall now prove the following proposition.

PROPOSITION 3. Suppose that there exist two functions $b(t)$ and $\beta(x,y)$ which satisfy the following conditions;

(i) $b(t)$ is continuous on I and $b(t) \geq 0$ for $t \geq T$, where T can be large,

(ii) $\beta(x,y)$ is continuous on $x \geq 0$ and $y \geq 0$,

(iii) for $t \geq T$, $x > 0$ and $y \geq 0$,

$$f(t,x,y) \leq b(t)\beta(x,y) . \tag{16}$$

Moreover, we assume that there is a $c > 0$ such that

$$f(t,c,0) \geq 0 \quad \text{for } t \geq T , \tag{17}$$

and that for some $M > 0$ such that $c < M$

$$f(t,x,y) \geq 0 \quad \text{for } t \geq T, \quad M \geq x \geq c \quad \text{and} \quad y > K, \tag{18}$$

$$f(t,x,y) \leq 0 \quad \text{for } t \geq T, \quad M \geq x \geq c \quad \text{and} \quad y < -K, \tag{19}$$

where K can be large. If $0 < \varepsilon \leq r(t) \leq \rho$ for some ε,ρ and all $t \geq 0$ and

$$\int_0^\infty tb(t)dt < \infty \tag{20}$$

or if there is an $A > 0$ such that

$$-A < \frac{r'(t)}{r(t)} \quad \text{for all } t \geq 0 \tag{21}$$

and we have

$$\int_0^\infty \left(\frac{1}{r(s)} \int_s^\infty b(u)du\right) ds < \infty, \tag{22}$$

then the equation (10) has a bounded nonoscillatory solution.

Proof. In both cases, we have (22) and the fact that $\frac{1}{r(t)} \int_t^\infty b(u)du \to 0$ as $t \to \infty$. For the c, $\beta(c,0) \leq \frac{L}{2}$ for some L. Since $\beta(x,y)$ is continuous, there is a $\delta > 0$ such that if $|x - c| \leq \delta$ and $0 \leq y \leq \delta$, we have $\beta(x,y) \leq L$. Choose $t_0 \geq T$ so large that

$$L \int_{t_0}^t \left(\frac{1}{r(s)} \int_s^\infty b(u)du\right) ds \leq \min(M-c,\delta) \quad \text{for } t \geq t_0$$

and

$$L \frac{1}{r(t)} \int_t^\infty b(u)du \leqslant \delta \qquad \text{for} \quad t \geqslant t_0 .$$

Then, in the same way as in the proof of Proposition 2, we can see that

$$\underline{\omega}(t) \equiv c \qquad \text{and} \qquad \overline{\omega}(t) = c + L \int_{t_0}^t \left(\frac{1}{r(s)} \int_s^\infty b(u)du \right) ds$$

satisfy the conditions (8) and (9) for $t \geqslant t_0$.

In the case where we have $\varepsilon \leqslant r(t) \leqslant \rho$, for $t \geqslant t_0$, $\underline{\omega}(t) \leqslant x \leqslant \overline{\omega}(t)$ and $y > K$, define $V(t,x,y)$ by $V(t,x,y) = r(t)y$. Then we have

$$\begin{aligned} \dot{V}(t,x,y) &= r'(t)y + r(t)\{- \frac{r'(t)}{r(t)} y - \frac{1}{r(t)} f(t,x,y)\} \\ &= - f(t,x,y) \\ &\leqslant 0 . \end{aligned}$$

For $t \geqslant t_0$, $\underline{\omega}(t) \leqslant x \leqslant \overline{\omega}(t)$ and $y < -K$, $W(t,x,y) = -r(t)y$ satisfies $\dot{W}(t,x,y) \leqslant 0$. In the case where we have the condition (21), $V(t,x,y) = y - Ax$ and $W(t,x,y) = -y + Ax$ are Liapunov functions that we desire. Therefore, it follows from Theorem 3 that the equation (10) has a bounded nonoscillatory solution.

<u>Remark</u>. Assuming the existence of functions $a(t)$ and $\alpha(x,y)$ such that $a(t)\alpha(x,y) \leqslant f(t,x,y)$ for $t \geqslant T$, $x < 0$ and $y \leqslant 0$, we can obtain a result similar to Proposition 3.

Now consider the equation (1). We assume that there exist continuous functions $a(t)$, $b(t)$, $\alpha(x)$ and $\beta(x)$ which satisfy the following conditions;

(i) $a(t)$ and $b(t)$ are nonnegative for $t \geqslant T$, where T can be large,

(ii) $x\alpha(x) > 0$ and $x\beta(x) > 0$ for $x \neq 0$, and $\alpha'(x) \geqslant 0$, $\beta'(x) \geqslant 0$,

(iii) $a(t)\alpha(x) \leqslant f(t,x,u) \leqslant b(t)\beta(x)$ for $t \geqslant T$, $|x| < \infty$ and $|u| < \infty$.

Moreover, we assume that the derivative of $r(t)$ is continuous and that $0 < \varepsilon \leqslant r(t) \leqslant \rho$ for some ε,ρ and all $t \geqslant 0$.

Under the assumptions above, the following results follow immediately from Propositions 1 and 2 with the remark.

PROPOSITION 4. A necessary and sufficient condition in order that all bounded solutions of (1) are oscillatory is that

$$\int_0^\infty ta(t)dt = \infty \quad \text{and} \quad \int_0^\infty tb(t)dt = \infty .$$

PROPOSITION 5. A necessary and sufficient condition in order that the equation (1) has a bounded nonoscillatory solution is that

$$\int_0^\infty ta(t)dt < \infty \quad \text{or} \quad \int_0^\infty tb(t)dt < \infty .$$

PROPOSITION 6. In addition to the assumptions above, if

$$\int_\varepsilon^\infty \frac{du}{\alpha(u)} < \infty \quad \text{and} \quad \int_{-\varepsilon}^{-\infty} \frac{du}{\beta(u)} < \infty \qquad \text{for some } \varepsilon > 0,$$

a necessary and sufficient condition in order that the equation (1) is oscillatory is that

$$\int_0^\infty ta(t)dt = \infty \quad \text{and} \quad \int_0^\infty tb(t)dt = \infty .$$

For the equation (1), we assume that there exist continuous functions $a(t)$, $b(t)$, $\alpha(x)$ and $\beta(x)$ mentioned above. Moreover, we assume that the derivative of $r(t)$ is continuous, $\int_0^\infty \frac{dt}{r(t)} = \infty$ and there is an $A > 0$ such that $\left|\frac{r'(t)}{r(t)}\right| < A$ for $t \geq 0$. Then, by Proposition 1, if we have the condition A that

$$(23) \qquad \begin{cases} \int_0^\infty a(s)ds = \infty \\ \text{or} \\ \int_0^\infty a(s)ds < \infty, \quad \int_0^\infty \left(\frac{1}{r(s)} \int_s^\infty a(u)du\right) ds = \infty \end{cases}$$

and

$$(24)\qquad \begin{cases} \int_0^\infty b(s)ds = \infty \\ \text{or} \\ \int_0^\infty b(s)ds < \infty, \quad \int_0^\infty \left(\frac{1}{r(s)} \int_s^\infty b(u)du\right) ds = \infty , \end{cases}$$

all bounded solutions of (1) are oscillatory. On the other hand, if we have the condition B that

$$(25)\qquad \int_0^\infty a(s)ds < \infty \quad \text{and} \quad \int_0^\infty \left(\frac{1}{r(s)} \int_s^\infty a(u)du\right) ds < \infty$$

or

$$(26)\qquad \int_0^\infty b(s)ds < \infty \quad \text{and} \quad \int_0^\infty \left(\frac{1}{r(s)} \int_s^\infty b(u)du\right) ds < \infty ,$$

the equation (1) has a bounded nonoscillatory solution. Therefore the condition A is a necessary and sufficient condition in order that all bounded solutions of (1) are oscillatory, and the condition B is a necessary and sufficient condition in order that the equation (1) has a bounded nonoscillatory solution.

REFERENCES

[1] L.E. BOBISUD, Oscillation of nonlinear differential equations with small nonlinear damping, SIAM J. Appl. Math., 18 (1970), 74-76.

[2] W.J. COLES, Oscillation criteria for nonlinear second order equations, Ann. Mat. Pura Appl., 82 (1969), 123-133.

[3] J.W. MACKI and J.S.W. WONG, Oscillation of solutions to second-order nonlinear differential equations, Pacific J. Math., 24 (1968), 111-117.

[4] Z. OPIAL, Sur une critère d'oscillation des integrales de l'equation différentielle $(Q(t)x')' + f(t)x = 0$, Ann. Polon. Math., 6 (1959), 99-104.

[5] T. YOSHIZAWA, Oscillatory property of solutions of second order differential equations, Tohoku Math. J., 22 (1970), 619-634.

[6] T. YOSHIZAWA, Stability Theory by Liapunov's Second Method. The Mathematical Society of Japan, Tokyo, 1966.

Tohoku University, Sendai, Japan

II

SEMINAR PAPERS

A PICTORIAL STUDY OF AN INVARIANT TORUS IN PHASE SPACE OF FOUR DIMENSIONS

R. Baxter

H. Eiserike

A. Stokes

This is a report on the behavior of the invariant manifolds of a particular fourth-order equation, as a parameter in the equation is varied over large intervals. The equation in question is an example in Hale's book on nonlinear oscillations [1], and consists of two coupled Van der Pol equations. The investigation was conducted with the aid of a computer graphics device at Goddard Space Flight Center.

Briefly, the behavior is as follows: For the parameter ε small, $0 < \varepsilon < .595$, the manifold in question is an asymptotically stable torus, with the flow on the torus being simply a rotation. For $\varepsilon > .595$ (and at least for ε up to 1) the torus persists, and remains asymptotically stable, but now on the torus the flow changes. There appear 4 periodic solutions, 2 asymptotically stable (three multipliers less than one), and 2 saddle-stable (one multiplier greater than one, two less than one). Thus, there are now four sub-manifolds, and the torus itself is now described by the union of these four limit cycles with the four families of solutions connecting each saddle-stable cycle to each stable cycle.

In the neighborhood of the torus, the flow changes, as solutions end up being pushed away from the unstable cycles, and attracted to the

2 stable cycles. Effectively what happens is that above $\varepsilon = .595$, the torus "disappears", that is, becomes very difficult to observe, and what one "sees" is two stable limit cycles. And so the behavior, in engineering terms, of the system in this region of phase space has changed, that is, if one were to view the solution as the output of some electronic circuit, below $\varepsilon = .595$, the output would be one of many almost periodic solutions on the torus, while above this value, the output would definitely oscillate, and be restricted to one of two oscillations. This qualitative change in the phase portrait has in this sense some points in common with that described by Sacker [2] in which a stable torus and an unstable limit cycle merge and form a single stable cycle.

In the transition described here, the only thing that changes is the nature of the flow on the torus, which itself persists throughout. And yet the torus no longer plays a significant role in the phase portrait of the system. Of far greater importance are the 4 cycles, in particular the 2 stable cycles. The torus now appears as almost an accident, introduced only because the four families of solutions emanating from the saddle-stable cycles at $t = -\infty$ are attracted to the stable cycles as $t \to +\infty$. And in fact, if one altered the domains of attraction of the stable cycles, so that these four families could be attracted elsewhere, the demise of the torus would be indeed difficult to detect, and of very little note.

It is important to observe that our investigation of this equation, in four dimensions, and over the parameter interval $[0,1]$, was only made possible by the use of the graphics output device on the IBM 360-91 at the Goddard Space Flight Center. It is difficult to conceive of alternate methods of investigating the above phenomena. Further details on the technical aspects of using this output device are to appear in a paper by R. Baxter, who wrote the graphics subroutine used here, and H. Eiserike, who did the programming over the course of our work.

The equation studied was the following

$$\text{(1)} \qquad \ddot{z}_1 + \mu_1^2 z_1 = \varepsilon(1 - z_1^2 - az_2^2 - bz_1^2 z_2^2)\dot{z}_1 ,$$

$$\ddot{z}_2 + \mu_2^2 z_2 = \varepsilon(1 - z_2^2 - \alpha z_1^2)\dot{z}_2 ,$$

where we set $\mu_1 = 1$, $\mu_2 = \frac{50}{23}$, $a = .25$, $\alpha = .8$, $b = 1.5$. For ε near 0, it is known (see Hale [1]), that there exists an asymptotically stable torus T, given by

$$z_i = f_i(\theta_1,\theta_2,\varepsilon) \sin \mu_i\theta_i, \; \dot{z}_i = \mu_i f_i(\theta_1,\theta_2,\varepsilon) \cos \mu_i\theta_i , \quad i = 1,2,$$

with f_1, f_2 periodic in θ_1, θ_2 separately, and f_1, f_2 approach positive constants as $\varepsilon \to 0$.

For the computer display, it was found convenient to introduce polar coordinates $(r_1,\theta_1,r_2,\theta_2)$, defined by

$$z_i = r_i \sin \mu_i\theta_i, \; \dot{z}_i = \mu_i r_i \cos \mu_i\theta_i, \quad i = 1,2 .$$

The first type of pictures obtained were of the three-dimensional "half-space" of R^4 defined by $\theta_1 \equiv 0(2\pi)$. The coordinates displayed on the screen were $(r_1,z_2,\dot{z}_2)$ in the usual (x,y,z) orientation.

By beginning with a set of initial conditions near the cycle $(f_1(0,\theta_2,0),0,f_2(0,\theta_2,0),\theta_2)$ for various values of θ_2, and integrating forward in time, a number of solutions were obtained which were distributed around the torus T. Then points on these solutions satisfying the condition $\theta_1 \equiv 0(2\pi)$ were selected (interpolating if necessary) and displayed on the screen in $(r_1,z_2,\dot{z}_2)$ space. In this manner, a cross-section of the torus T was seen. As ε increased, new initial conditions were chosen from the cross-section last displayed. This is how the pictures 1 to 4 were made.

Around $\varepsilon = .6$, however, the cross-section no longer appeared to be homeomorphic to a circle, and soon reduced to two points. It appeared that upon integrating forward in time, the solutions were now describing two limit cycles, or perhaps one limit cycle intersected the space $\theta_1 \equiv 0(2\pi)$ twice.

To study this further, a second type of display was used. Here, a projection from R^4 to R^3 was developed, which had the property that a torus in R^4 described by $r_1 = c_1 > 0$, $r_2 = c_2 > 0$ would map into a torus in R^3. For this, a point with coordinates $(r_1,\theta_1,r_2,\theta_2)$ was mapped into a point (x,y,z) determined as follows: First the coordinates (r_1,θ_1) were used to determine a point P_0 in the (x,y) plane, in the usual fashion. Then, using the coordinates (r_2,θ_2) the point (x,y,z) was chosen in the plane determined by P_0 and the z-axis, at the end of a vector from P_0, of length r_2, making an angle θ_2 with the (x,y)-plane. (See Figure 1.) In practice, some scaling of r_2 was needed, so that our "doughnuts" would have a hole in the center. Using this projection, many solutions were mapped from R^4 to R^3, with the results as given in pictures 5 to 12. The initial conditions for these solutions were chosen as in the earlier sequence of pictures.

After many attempts to explain the abrupt transition seen above, it was decided to return to the earlier cross-sections of the torus, and program the stroboscopic map S of the cross-section into itself. That is, display the cross-section $\theta_1 \equiv 0(2\pi)$, select a point on the cross-section with a light pen, integrate the solution with this initial condition until $\theta_1 \equiv 0(2\pi)$ again. With this program, it was discovered that for $\varepsilon > .6$, the map S had four fixed points on the cross-section, two stable, and two unstable. The coordinates of the two stable points were easily obtained by iterating S. By observing the location on the cross-section at which S reversed direction, an approximation of the coordinates of the unstable points was obtained.

With this information, it was determined that four limit cycles of (1) existed, for each $\varepsilon > .595$. Of these, two were asymptotically orbitally stable, with three characteristic exponents having negative real parts, and two saddle stable, with two characteristic exponents having negative real parts, and one with a positive real part. (A table of exact values of a point on each limit cycle for a range of values of ε is given in the Appendix, with information on the exponents.*)

---

*This was obtained through the kind assistance of Carmelo Velez, of Goddard, who performed a differential correction on the initial coordinates to force the orbit to close. This method allows one to locate even a saddle-stable orbit, given a good initial guess, without being affected by its instability.

Returning to the (x,y,z)-space by means of the projection, these four limit cycles were displayed (see pictures 13-18). The two stable ones, of course, had been seen earlier in pictures 10 to 12.

At this point, it was no longer clear whether or not the torus T still existed. After some effort, a program was written to select initial conditions belonging to the families of solutions which approach the saddle-stable limit cycles as $t \to -\infty$. The solutions with these initial conditions were then integrated forward in time through several multiples of 2π. After some experimentation, it was concluded that the ω-limit sets of these families were the two stable limit cycles. Experimentation was needed, because these families are difficult to work with numerically. The following phenomena were used in arriving at the above conclusion: i) Integrating backward in time, the solutions ended up on the saddle-stable cycles. Thus, the two-parameter family of solutions had been correctly determined. ii) If one introduced a parameter which measured the distance of the chosen initial condition from the unstable limit cycle, it was discovered that a continuous behavior was exhibited as follows: After integrating through several multiples of 2π, for small values of the parameter the solution was still close to the unstable limit cycle. As the parameter was increased, the solution would move away from the unstable cycle, and approach a stable cycle. By proper choice of this parameter, a set of solutions was obtained which illustrates how these families connect the four cycles, and so make a torus. See pictures 19 to 23. Observe the similarity between picture 8 and picture 19, although they were obtained by completely different means.

In (ii) above, "small" was interpreted in relation to the size of the exponent with positive real part. As ε increased, so did this exponent, making the cycle more unstable, and requiring a closer approach to the limit cycle before beginning.

This experimentation served to convince us that the families of solutions defined by the positive exponents of the unstable limit cycles did indeed approach the unstable cycles as $t \to -\infty$, and the stable cycles as $t \to +\infty$. Thus, the stable torus T still existed for $\varepsilon > .6$, even though the earlier pictures made it seem to have disappeared. This apparent disappearance is now explained in terms of the appearance of the stable and unstable cycles (sub-manifolds of T) for $\varepsilon > .595$.

REFERENCES

[1] J.K. HALE, Oscillations in Non-Linear Systems. McGraw-Hill, New York, 1963.

[2] R.J. SACKER, On Invariant Surfaces and Bifurcation of Periodic Solutions of Ordinary Differential Equations, NYU Report IMM-NYU 333, Oct., 1964, New York University.

RB: Goddard Space Flight Center, Greenbelt, Maryland

HE: Goddard Space Flight Center, Greenbelt, Maryland

AS: Georgetown University, Washington, D. C.

APPENDIX

Initial Conditions for One Stable Cycle, and One Unstable Cycle*

ε	Stable Cycle ($\dot{z}_2 = 0$)			Largest Multiplier	Unstable Cycle ($\dot{z}_2 = 0$)			Largest Multiplier
	z_1	$\dot{z}_1$	z_2	(in Absolute Value, $\neq 1$)	z_1	$\dot{z}_1$	z_2	(in Absolute Value, $\neq 1$)
.595	.87496	.37045	-1.71214	.794	.79832	.41030	-1.759018	1.205
.64	.98585	.31233	-1.63482	.461	.68706	.47448	-1.81724	1.920
.78	1.10433	.23974	-1.54194	.486	.56182	.57362	-1.86763	3.220
.90	1.15306	.19998	-1.50326	.537	.50118	.64299	-1.88465	4.223

*The other cycles are obtained via the transformation $L: (z_1, \dot{z}_1, z_2, \dot{z}_2) \to (z_1, \dot{z}_1, -z_2, \dot{z}_2)$. The equation is invariant under L, and the stable (unstable) cycle is mapped into the other stable (unstable) cycle.

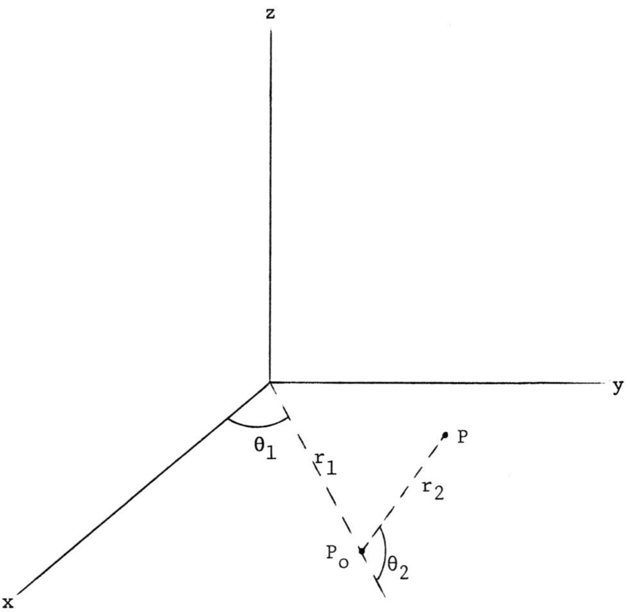

Fig. 1

Pictures 1, 2, 3, 4

These display a section of the stable torus at $\theta_1 = 0$, for $\varepsilon = .01$, .30, .55, and .65 respectively. The axes are labeled x, z, $\dot{z}$, and in fact are r_1, z_2, $\dot{z}_2$.

1

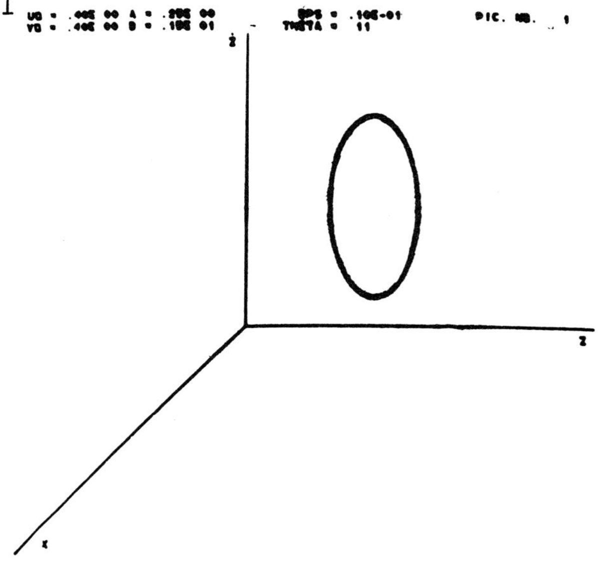

2

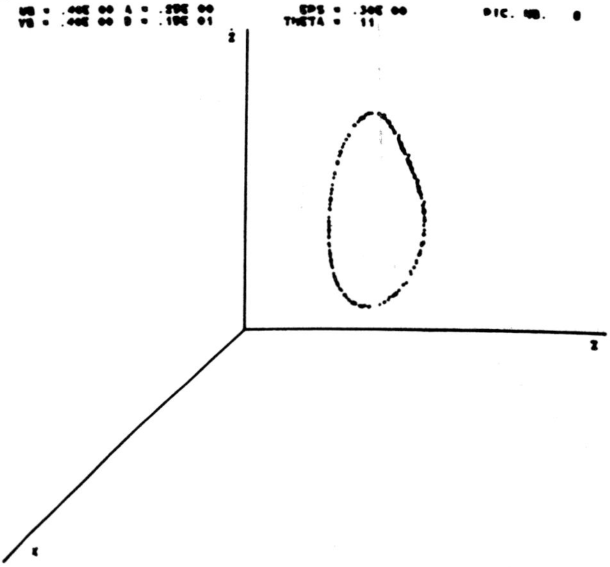
THETA = 11
z
z
x

3

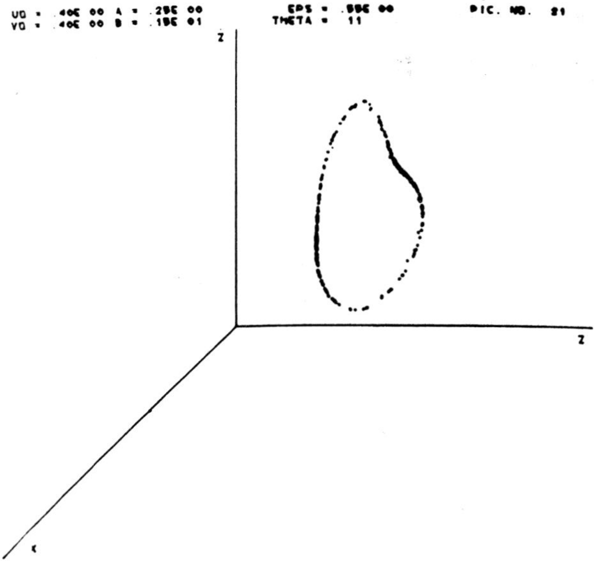
U0 = .40E 00 A = .25E 00
V0 = .40E 00 B = .15E 01
EPS = .55E 00
THETA = 11
PIC. NO. 21
z
z
x

4

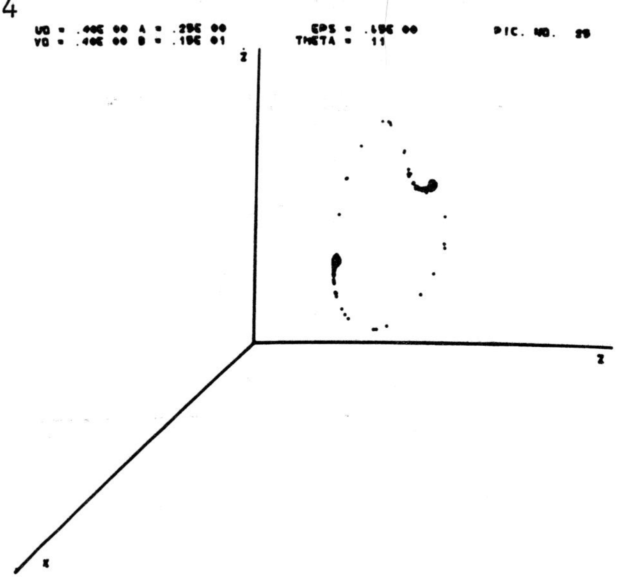

<u>Pictures 5, 6, 7, 8, 9</u>

These display a projection of solutions in R^4 into R^3. The axes are labeled x, z, ż, and in fact are x, y, z, with no direct interpretation in terms of the R^4-coordinates. Pictures 5 and 6 display two views of the torus for $\varepsilon = .01$. Pictures 7, 8 and 9 give a view of the torus for $\varepsilon = .30$, $.60$ and $.63$ respectively. In picture 9 the torus has already become difficult to observe.

5

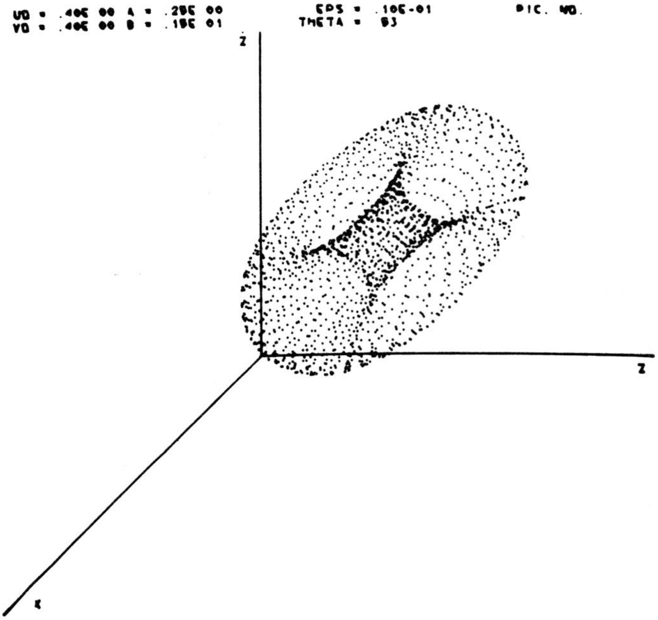

6

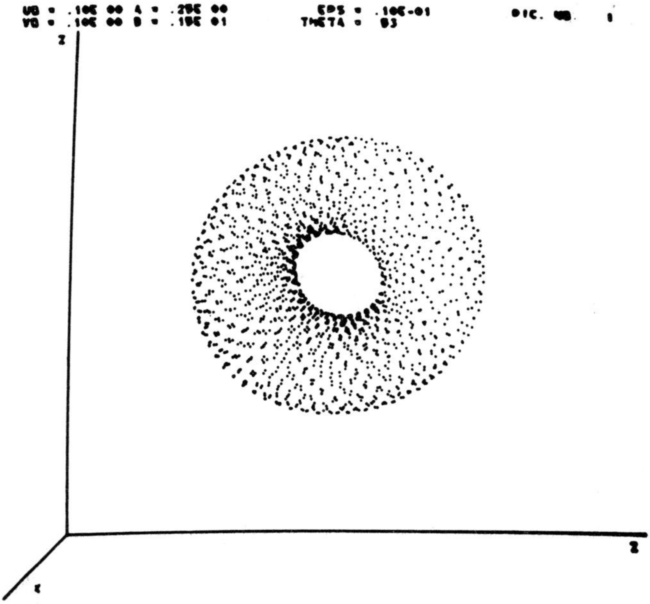
U0 = .10E 00 A = .25E 00
V0 = .10E 00 B = .15E 01
EPS = .10E-01
THETA = 83
PIC. NO. 1
Z
X

7

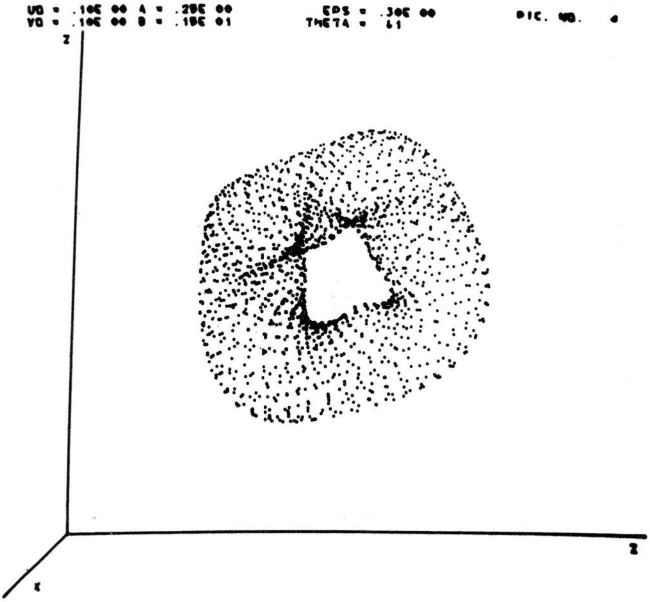
U0 = .10E 00 A = .25E 00
V0 = .10E 00 B = .15E 01
EPS = .30E 00
THETA = 61
PIC. NO.
Z
X

8

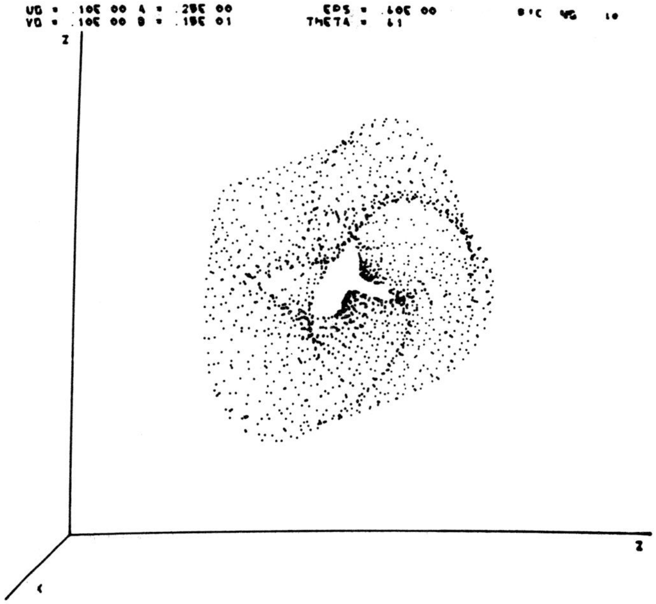
UO = .10E 00 A = .25E 00
VO = .10E 00 B = .15E 01
EPS = .60E 00
THETA = 61
Z
Z

9

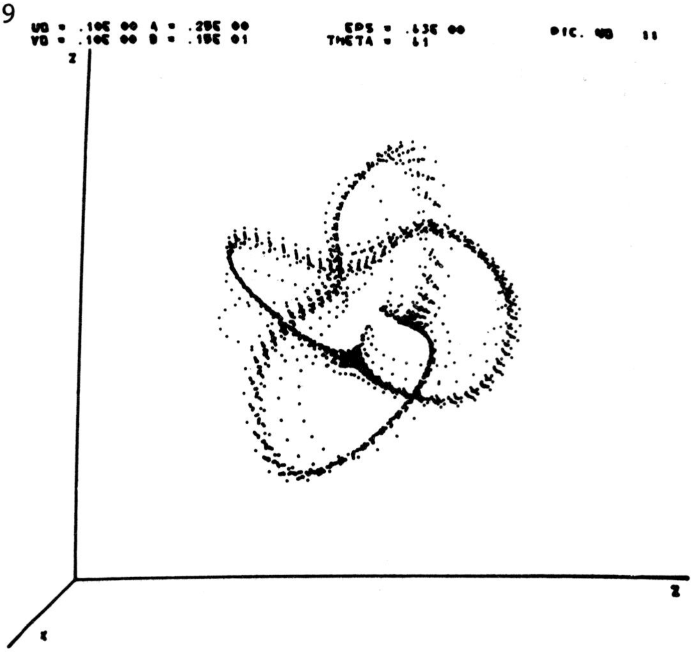
UO = .10E 00 A = .25E 00
VO = .10E 00 B = .15E 01
EPS = .63E 00
THETA = 61
Z
Z

Pictures 10, 11, 12

These display the two stable limit cycles, projected from R^4 to R^3. In 10, $\varepsilon = .65$. In 11 and 12, $\varepsilon = .80$, with two views given.

10

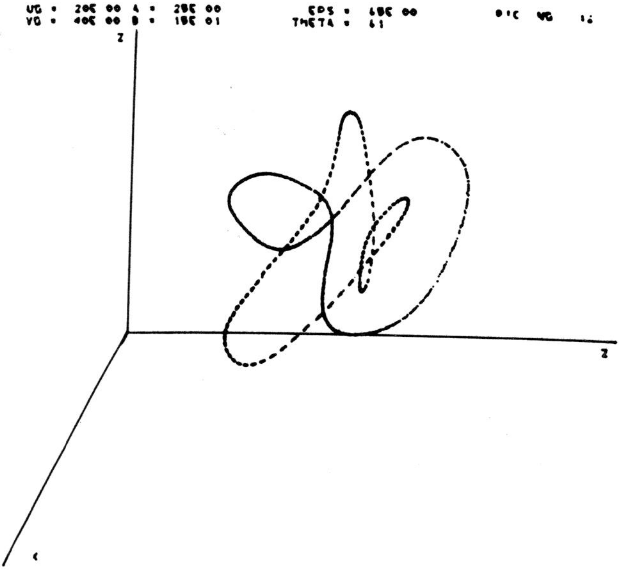

11

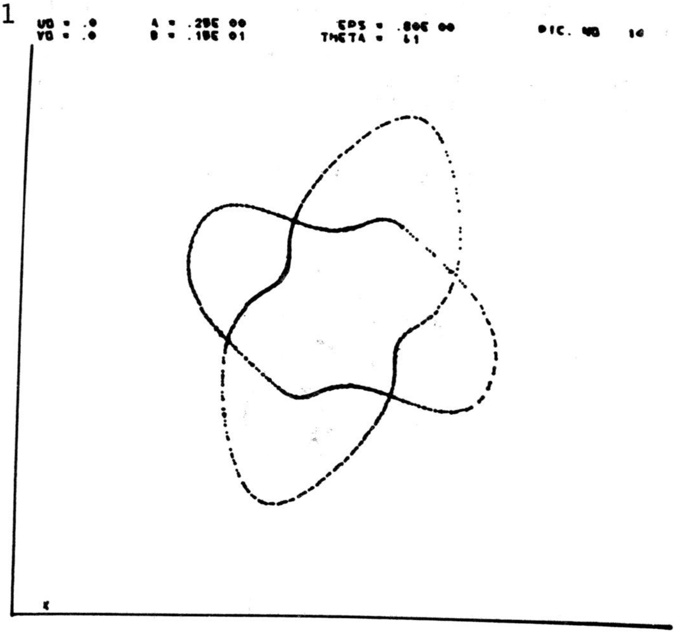

12

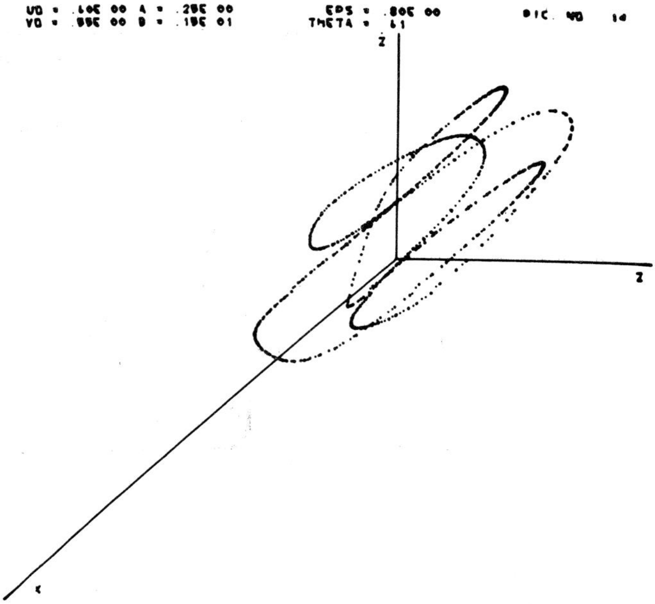

Pictures 13, 14, 15, 16, 17, 18

These display the four limit cycles projected into R^3. In 13, 14, $\varepsilon = .60$, for 15, 16, $\varepsilon = .64$, and $\varepsilon = .78$ in 17, 18. Two views are given in each case. By comparing 12 with 18, one can see that the unstable cycles have been added in, and are distinct from the stable cycles. But in 13, 14, the unstable and stable cycles are very close.

13

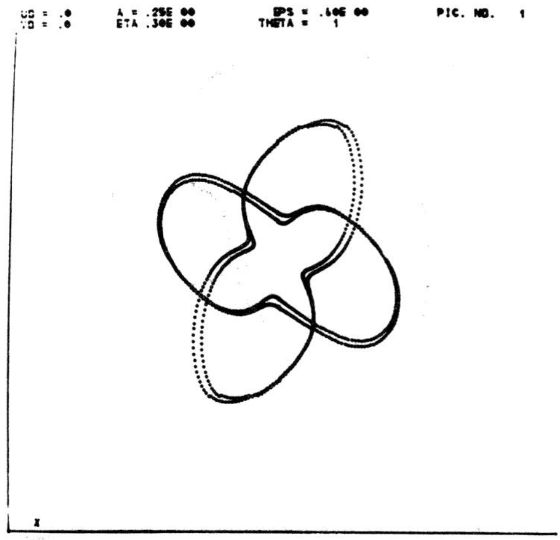

14

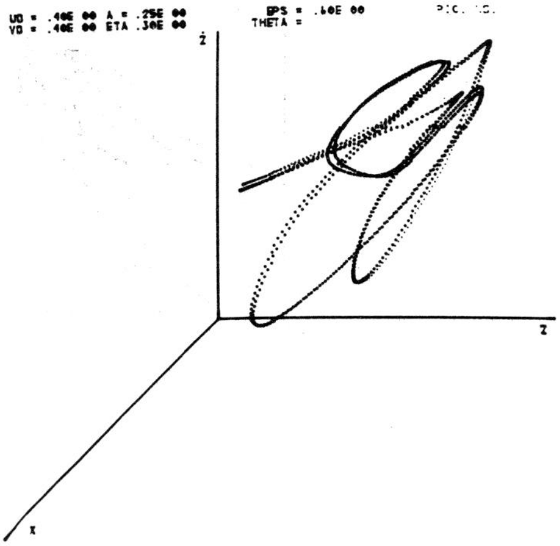

15

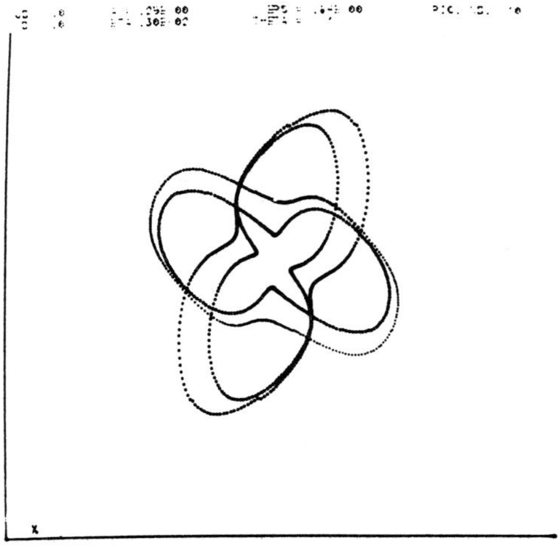

16

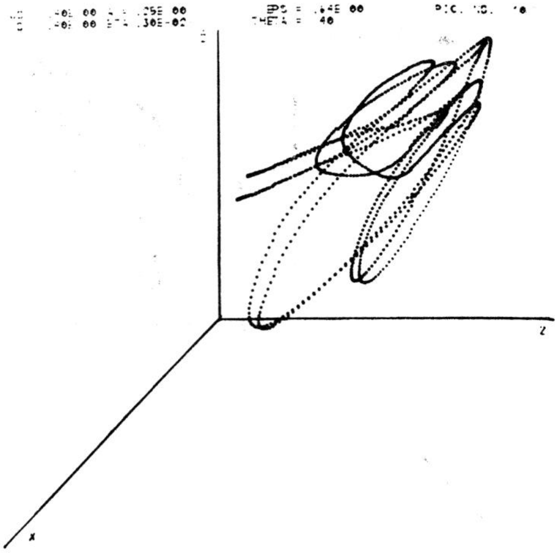

17

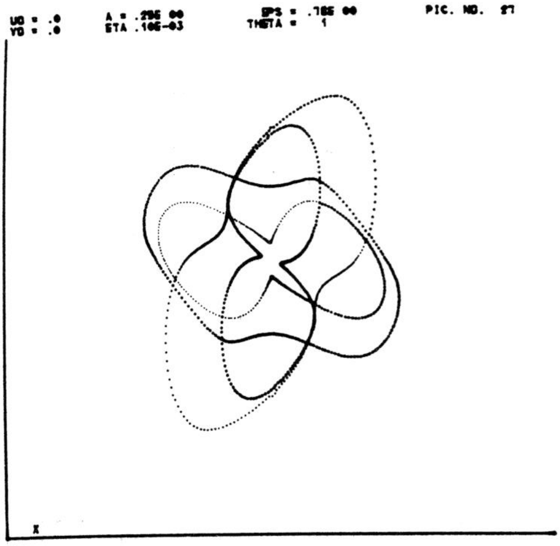

18

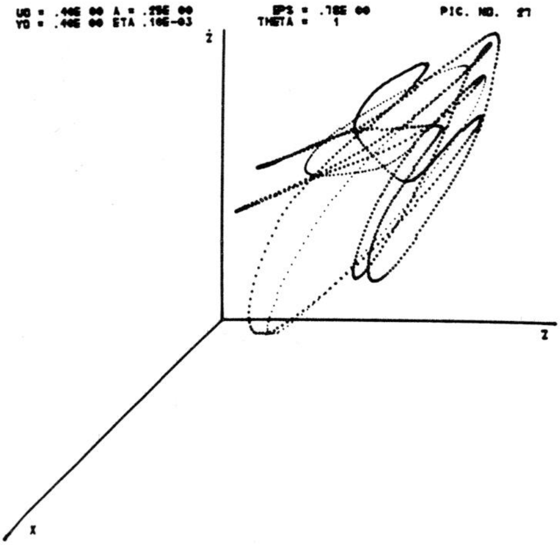

Pictures 19, 20, 21, 22, 23

The stable torus in projected into R^3, for $\varepsilon = .60$ in 19 and 20 (two views), for $\varepsilon = .64$ in 21, 22 (also two views), and for $\varepsilon = .78$ in 23.

19

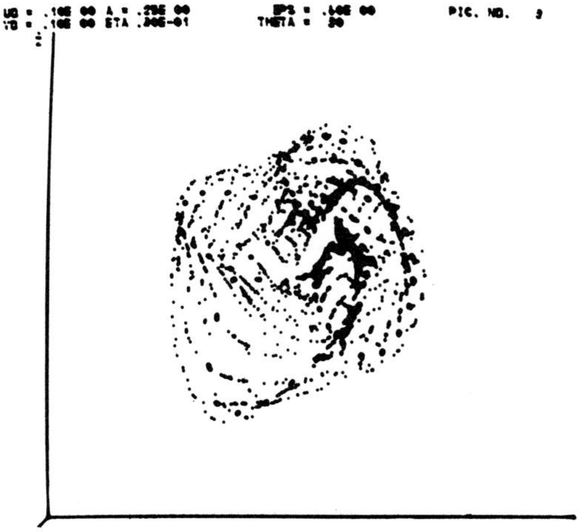

20

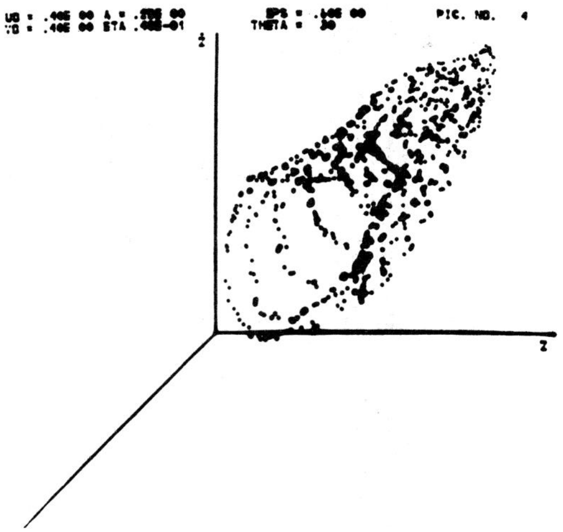

21

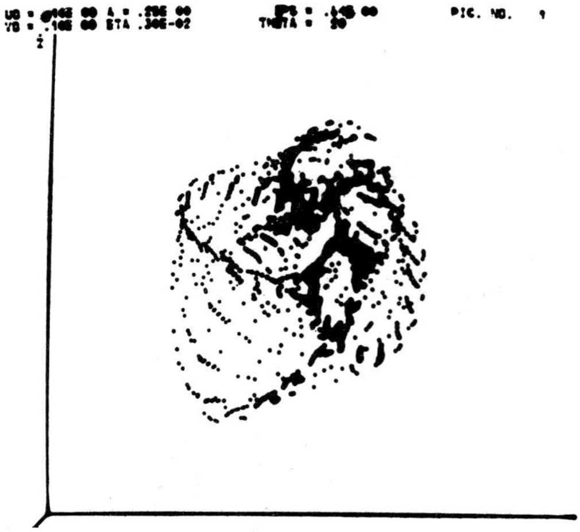

22

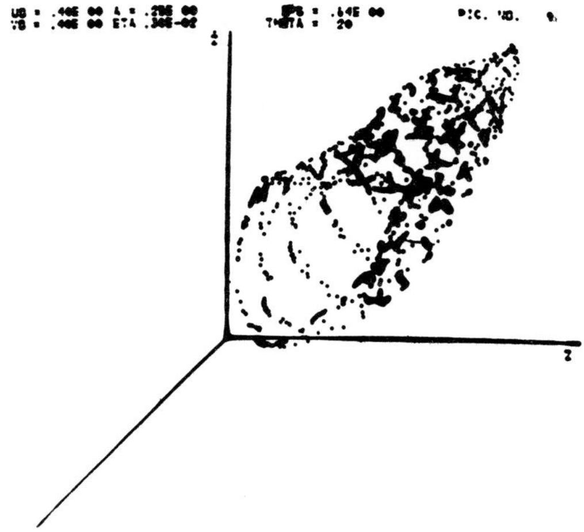

23

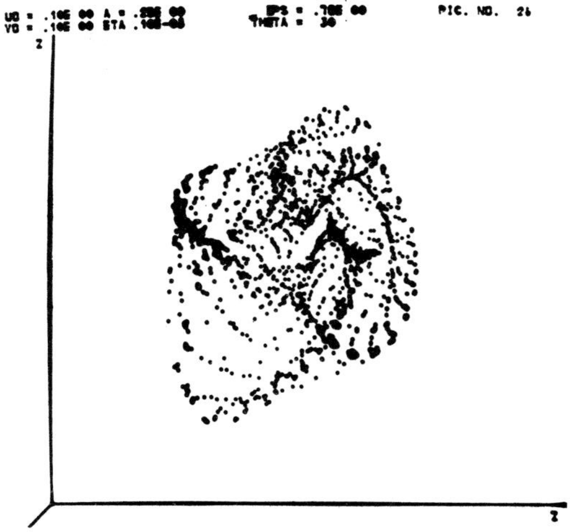

AUTONOMOUS PERTURBATIONS OF SOME HAMILTONIAN SYSTEMS, ALL OF WHOSE SOLUTIONS ARE PERIODIC

Melvyn S. Berger

Many ingenious methods have been devised for the determination of periodic solutions of autonomous Hamiltonian systems. Among such methods are: fixed point theory, various implicit function theorems, and the so-called majorant method. Yet there is one classical method which has been rather overlooked in recent years, namely the calculus of variations, or "infinite dimensional critical point theory" in modern terminology. Here we hope to show that this last method is well adapted for the study of Hamiltonian systems, and we do this by treating a relatively simple problem that cannot be successfully studied by any of the other methods mentioned above.

(A) Statement of Problems to be Considered

We shall investigate the periodic solutions of the following systems of N autonomous ordinary differential equations for $\beta > 0$:

(1) $$\ddot{x} + \beta^2 x + \varepsilon \nabla F(x) = 0 \quad (\text{where } |\nabla F(x)| = o(|x|))$$

(2) $$\ddot{x} + \varepsilon B(x)\dot{x} + \beta^2 x = 0$$

(3) $$\ddot{x} + \frac{x}{|x|^3} + \varepsilon \nabla V(x) = 0 \quad (\text{for } N = 2, 3) .$$

Here $x(t)$ denotes an N-vector function of x, $F(x)$ and $V(x)$ are C^2 real-valued functions of x, and $B(x) = (b_{ij}(x))$ is an $N \times N$

antisymmetrix matrix of x such that the system (2) is Hamiltonian and $|b_{ij}(x)| = O(|x|)$. We consider those periodic solutions of (1)-(3) that are near the unperturbed solutions of (1)-(3) with $\varepsilon = 0$. Indeed when $\varepsilon = 0$, all the solutions of (1)-(3) are periodic, and so we attempt to show that some of these periodic solutions are preserved under (autonomous Hamiltonian) perturbations. Problems of the type considered here were taken up by Moser [1]; however his results are based on stringent nondegeneracy hypotheses. Here we extend some of Moser's results by removing this nondegeneracy. The problems given above have considerable significance in mechanics. The system (1) represents a nonlinearly coupled system of harmonic oscillators whose periodic solutions cannot be found by Liapunov's classic theorem [2, pp. 216-232], since the required irrationality conditions are not satisfied. The system (2) is a simple version of a regularized dynamical system which arises in celestial mechanics. The system (3) represents an autonomous perturbation of the system associated with Kepler's problem of two-body motion, under Newtonian attraction.

(B) Preliminary Information

In order to apply the calculus of variations to the problems raised in (A), the following steps are necessary:

Step I. Determination of an extremal principle P for the desired periodic solutions.

Step II. Selection of an appropriate admissible class C of functions for P.

Step III. Solution of the problem P within the class C.

The difficulties connected with Steps I - III are rather formidable. For most principles P, the critical points associated with the desired periodic solutions are of "minimax type", so that one must add appropriate "isoperimetric conditions" to P to obtain a classic minimum problem. On the other hand, if these additional isoperimetric conditions are not selected correctly, the resulting critical points will not satisfy the desired differential equation. The selection of the admissible class C also takes some care. In general our first requirement for the class C is that it consist of T-periodic, N-vector functions $x(t) = (x_1(t), \ldots, x_N(t))$ such that $x_i(t)$ is absolutely continuous and $x_i'(t)$ is square integrable over the period T for each i, $(i = 1,\ldots,N)$. The set of such functions $\{x(t)\}$ forms a Hilbert space H_T with respect to the inner product

$$(4) \qquad (x(t),y(t)) = \int_0^T [\dot{x}(t)\cdot\dot{y}(t) + x(t)\cdot y(t)]dt \ .$$

Furthermore the following inequality holds if $\int_0^T x(t)dt = 0$,

$$(5) \qquad \max\left\{\sup_{[0,T]} |x(t)|,\ \|x\|_{L_2[0,T]}\right\} \leqslant K_T\|\dot{x}\|_{L_2[0,T]}$$

where K_T is some constant independent of $x \in H_T$. Also

$$(6) \qquad x_n \to x \text{ weakly in } H_T, \text{ implies } x_n \to x \text{ uniformly on } [0,T].$$

(C) <u>Results for the System</u> (1): $\ddot{x} + \beta^2 x + \varepsilon\nabla F(x) = 0$
<u>Where</u> $|\nabla F(x)| = o(|x|)$

Here we shall outline proofs of the following results:

<u>THEOREM 1</u>. If $U(x) = \frac{1}{2}\beta^2 x\cdot x + \varepsilon F(x)$ with $F(x) \in C^2(\mathbb{R}^N)$, then (1) possesses a one-parameter family of nontrivial periodic solutions $x_R(t)$ for R sufficiently small with the following properties:

(i) As $R \to 0$, $|x_R(t)| \to 0$ uniformly on $[0,T]$, and the period of $x_R(t)$, $\tau(R) \to 2\pi/\beta$.

(ii) R denotes the mean-value of $U(x)$ over a period $\tau(R)$.

(iii) If $U(x) \to \infty$ as $|x| \to \infty$ and is convex, then $x_R(t)$ exists for all $R > 0$.

<u>THEOREM 2</u>. With the notations and assumptions of Theorem 1, and if $U(x)$ is even, then (1) possesses at least N distinct one-parameter families $x_{R,j}(t)$ $(j = 1,\ldots,N)$ of periodic solutions with the property (i). Furthermore, the period of $x_{R,j}(t)$, $\tau_j(R)$, and the mean-value $\tilde{T}_{R,j}$ of $T(x_j) = \frac{1}{2}\dot{x}_j\cdot\dot{x}_j$ over $\tau_j(R)$, satisfy the relation $(\tilde{T}_{R,j})^2 T_j(R) = 4\pi^2 R$.

<u>Sketch of the Proof of Theorem 1</u>. This result is obtained by using the exact method of our paper [3]. After the change of variables $t = \lambda s$,

one finds $2\pi\lambda$-periodic solutions of (1) by the following variational parinciple:

(π_1) Minimize $\frac{1}{2}\int_0^{2\pi} x^2(s)ds$ subject to the constraints $\int_0^{2\pi} U(x(s))ds = R$, $\int_0^{2\pi} \nabla U(x(s))ds = 0$ where R is a parameter running over $(0,\infty)$ (provided the function $U(x)$ is convex for $|x| \leq g(R)$ where g is some function depending only on U and such that $g(R) \to \infty$ as $R \to \infty$).

Here we have altered the procedure of [3] slightly by defining (π_1) over the entire interval $[0,2\pi]$, but this causes no change in the proof.

Consequently only the behavior as $R \to 0$ remains to be studied. If $x_R(T)$ denotes the solution of (π_1), one notes that the convexity of $U(x)$ implies that $|x_R(t)|$ tends to zero uniformly over a period. This is proven from (5) by noting that the mean value of $U(x_R(t))$ and $|\dot{x}_R(t)|^2 \to 0$ with R, while the mean value of $x_R(t)$ is uniquely determined by the relation $\int_0^{2\pi} \nabla U(x_R(s))ds = 0$ in terms of that part of $x_R(t)$ with mean value zero. On the other hand, if one compares (π_1) with the approximate quadratic variational problem

$(\tilde{\pi}_1)$ Minimize $\int_0^{2\pi} |\dot{x}^2(s)|ds$ subject to the constraints $\int_0^{2\pi} x(s)ds = 0$ and $\int_0^{2\pi} \beta(x \cdot x) = R$, a constant,

one easily finds after reparametrization that $\tau(R)$ (the period of $x_R(t)$) $\to 2\pi/\beta$ as $R \to 0$.

<u>Sketch of the Proof of Theorem 2</u>. This result is quite a bit harder to prove and depends on our previous results in [4] and [5]. The result is simplest to discuss in terms of operator theory in Hilbert space. Suppose L is a linear self-adjoint compact operator with eigenvalues $0 < \lambda_1^2 \leq \lambda_2^2 \leq \lambda_3^2 \leq \cdots$ and $\lambda_n^2 \to \infty$. Then we compare the linear eigenvalue problem $x = \lambda Lx$ with a nonlinear perturbation

$$x = \lambda\{Lx + Nx\} \tag{7}$$

where Nx is a completely continuous higher order gradient map such that $N(-x) = -N(x)$. Then our result (Corollary 5 of [5]) shows that the dimension of the eigenspace (N, say) of any λ_i^2 $(i = 1,2,\ldots)$ is preserved under perturbation by $N(x)$, in the sense that there are still N distinct one-parameter families $(x_{R,j}(t),\lambda_j(R))$ $(j = 1,\ldots,N)$ of nontrivial solutions of (7) bifurcating from $(0,\lambda_j^2(R))$. These families are characterized by a minimax principle similar to that of Fischer for the eigenvalues of a self-adjoint compact operator. Furthermore each solution is a critical point of the functional $G(u)$ on the sphere $\|x\|^2 = R$, where $\nabla G(x) = Lx + Nx$. Now retranslating this general result to the problem at hand, using the procedure of [4], one obtains N distinct one-parameter families of periodic solutions of (1), $(x_{R,j}(t),\tau_j(R))$ bifurcating from $(0, 2\pi/\beta)$. Here the appropriate Hilbert space H is the one consisting of odd 2π-periodic N-vector functions, $x(s)$, that are absolutely continuous and possess a square integrable derivative $x'(s)$ over a period. The linear equation $\ddot{x} + \lambda^2\beta^2 x = 0$ has eigenvalues $2\pi k/\beta$ $(k = 1,2,\ldots)$ in H and the multiplicity of $2\pi/\beta$ is N. The relation between $\tau_j(R)$ and $\tilde{T}_{R,j}$ of Theorem 2 is found by making the change of variables $t = \lambda s$ in the integral $\|x\|_H^2 = \int_0^{2\pi} \dot{x}^2(s)ds = R$. Of course one must show that after reparametrization these families remain distinct. This follows since all the families considered have periods near $2\pi/\beta$, so that none of the families can be a multiple covering (or a reparametrization) of another family.

(D) Results for the System (2): $\ddot{x} + \varepsilon B(x)\dot{x} + \beta^2 x = 0$

Here we prove the following analogue of Theorem 2.

THEOREM 3. If $B(x)$ is an antisymmetric matrix (odd in x) and such that the Euler-Lagrange derivative of $\int_0^T B(x)\,\dot{x}\,dt$ is $B(x)\dot{x}$ with $|B(x)| = O(|x|)$, then the conclusions of Theorem 2 hold.

Outline of Proof. By using the argument of Theorem 2, it suffices to represent (2) as an operator equation of the form (7) in a Hilbert space H. Let H be the Hilbert space defined in Theorem 2. Then after the change of variables $t = \lambda s$ in (2), (2) becomes

$$x_{ss} + \lambda B(x)x_s + \lambda^2\beta^2 x = 0 . \tag{2'}$$

By the method of [4], (2') can be rewritten as an operator equation of the form

$$(8) \qquad x = \lambda Nx + \lambda^2 Lx$$

where L is a self-adjoint compact operator $H \to H$ defined implicitly by the formula $(Lx,y) = \beta^2 \int_0^\pi x\cdot y \, ds$ and Nx is a completely continuous gradient map defined by the formula $(Nx,y) = \int_0^\pi \varepsilon B(x(s))x_s(s)\cdot y(s)\, ds$. Now setting $z = \lambda L^{\frac{1}{2}}x$, (8) can be rewritten as the system

$$(9) \qquad \begin{cases} z = \lambda L^{\frac{1}{2}} x \\ x = \lambda(L^{\frac{1}{2}} z + Nx) \end{cases} .$$

We consider this system defined on $H \times H$, and it clearly satisfies the abstract conditions discussed in the proof of Theorem 2, for systems of the form (7). The remaining steps in the proof are now analogous to the proof of Theorem 2.

(E) <u>Results for the System</u> (3): $\ddot{x} + \frac{x}{|x|^3} + \varepsilon\nabla V(x) = 0 \quad (N = 2,3)$

Discussions with Professor R. Arensdorf have been essential in proving

<u>THEOREM 4</u>. If $\nabla V(x) = O(|x|)$ is a C^1 function, then (3) has a one-parameter family of periodic solutions $x_i(h)$, where the parameter h varies over $(0,-\infty)$ and denotes the total energy of $x_i(h)$.

<u>Outline of Proof for $N = 2$</u>. Let the system (3) have fixed energy h. Then by regularization theory, one finds that periodic solutions of (3) with total energy h are in (1-1) correspondence with the periodic solutions of

$$(10) \qquad 4\ddot{u} = hu + \varepsilon\nabla W(u) \qquad \text{where } W(u) = |u|^2 V(u^2)$$

$$(11) \qquad \text{provided that} \quad 2|u'|^2 - u\bar{u}\{\varepsilon V(u^2) + h\} = 1 .$$

Here we have used the obvious notation $u = u_1 + iu_2$, $|u|^2 = u_1^2 + u_2^2$. Now we apply Theorem 1 to (10) to define a one-parameter family of solutions $(u_R(t),\tau_R)$ for (10), and we find that value of R such that (11) holds by taking the mean value of (11) over the period $\tau_i(R)$ and noting that the mean value of $u\bar{u}\{\varepsilon V(u^2) + h\}$ over the period $\tau_i(R)$ is R.

Remarks. 1. If $N = 3$, the results of [6, pp. 441-451] can be used to transform the perturbed Kepler problem into a system of the form (1). These results are essential to prove Theorem 4 for $N = 3$.

2. Clearly Theorem 2 can be used to prove much stronger results for (3) than that given above. This work will be carried out in a subsequent paper.

3. A stronger result than Theorem 4 will be found in Moser's paper [1], however as already mentioned this result requires that a strong nondegeneracy hypothesis on the periodic solutions of (3) be satisfied.

REFERENCES

[1] J. MOSER, Regularization of Kepler's Problem and the Averaging method on a manifold, Comm. Pure Appl. Math., 18 (1970), 609-636.
[2] V.V. NEMYTSKII and V.V. STEPANOV, Qualitative Theory of O.D.E. Princeton Univ. Press, Princeton, N.J., 1960.
[3] M.S. BERGER, Periodic solutions of second order dynamical systems and isoperimetric variational problems, Amer. Jour. of Math, 1971, 1-10.
[4] M.S. BERGER, On periodic solutions of second order Hamiltonian systems, Jour. Math. Anal. and Applications, 29 (1970), 512-522.
[5] M.S. BERGER, Multiple solutions of nonlinear operator equations arising from the calculus of variations, Proc. Sym. Pure Math. A.M.S., 18 (1970), 10-28.
[6] R.F. ARENSDORF, Regularization Theory for the elliptic restricted three body problem, Jour. Diff. Eqn., 6 (1969), 420-451.

Belfer Graduate School, Yeshiva University, New York, New York

SOME BOUNDARY VALUE PROBLEMS FOR NONLINEAR ORDINARY DIFFERENTIAL EQUATIONS ON INFINITE INTERVALS

Melvyn S. Berger

The classical Sturm Liouville theory for (linear) two point boundary value problems defined on a bounded interval has recently been extended to a nonlinear context by various authors. Here we obtain some results on the nonlinear analogue of the discrete spectrum of (nonlinear) ordinary differential operators (possibly containing singular coefficients) defined on unbounded intervals. Such problems arise naturally in many contexts in modern mathematical physics.

(A) Formulations

Consider the simple differential operator $Ax = x_{ss} + q(s)x^{2\beta+1}$ defined on some infinite interval $\Omega = (-\infty,\infty)$ or $(0,\infty)$, say. Now a real number λ is in the discrete spectrum $\sigma_d(A)$ of A if there is a function $y(s) \not\equiv 0$ (i.e. an eigenfunction) belonging to $L_2(\Omega)$ such that $Ay + \lambda y = 0$. If $\beta = 0$, the discrete spectrum of the linear operator A has been well studied for many years in connection with problems arising from Schrödinger's equation. The results discussed here also have relevance to mathematical phycics, since it one sets $u = e^{i\lambda t}w$ with λ and w real in the nonlinear Schrödinger equation in $\mathbb{R}^N$,

(1) $$i\frac{\partial u}{\partial t} = \Delta u + q|u|^{2\beta}u ,$$

or in the nonlinear wave equation in $\mathbb{R}^N$,

(2) $$u_{tt} = \Delta u + q|u|^{2\beta} u - m^2 u ,$$

then one obtains an equation of the form

(3) $$\Delta w + qw^{2\beta+1} + \gamma w = 0 .$$

Now if one considers radially symmetric solutions of (3) in the case $N = 3$, or L_2 solutions of (3) in the case $N = 1$, provided q is radially symmetric, an equation of the form $Ax + \lambda x = 0$ results. In particular, for $N = 3$, setting $x(s) = sw(|y|)$ in (3) where $s = |y|$ denotes the length of the vector $y \in \mathbb{R}^N$, a simple computation shows that the solutions of (3) are in (1-1) correspondence with the solutions of

(4) $$x_{ss} + \frac{q(s)}{s^{2\beta}} x^{2\beta+1} + \gamma x = 0.$$

The problem considered here is the determination of the discrete spectrum $\sigma_d(A)$ for various types of functions $q(s)$ defined on Ω, with $\beta > 0$.

More generally, we shall also consider the discrete spectrum for higher order ordinary differential operators

(5) $$Bx = (-1)^{M+1} x^{2M} + q(s)x^{2\beta+1}$$

subject to appropriate boundary conditions of infinite intervals Ω with $\beta > 0$.

(B) Statement of Results

THEOREM 1. If $q(s)$ is a positive function with $0 < a \leq q(s) \leq b$ where a, b are finite constants and

(6) $$Ax = x_{ss} + q(s)\, x^{2\beta+1} ,$$

then $\sigma_d(A) = (-\infty,0)$ for any $\beta > 0$. A similar result holds if $q(s)$ is positive, smooth, and $q(s) \to 0$ as $|s| \to \infty$.

For equations of the form (4), we obtain

THEOREM 2. If $0 < q(s) \leq \frac{C}{s^{2\beta}}$ where C is a positive constant, then for A defined on $[0,\infty)$ with $x(0) = 0$, $\sigma_d(A) = (-\infty,0)$ for $0 < 2\beta < 4$.

The result of Theorem 2 is sharp in the following sense:

THEOREM 3. If $q(s) = \frac{C}{s^{2\beta}}$, where C is a positive constant, then $\sigma_d(A) = \emptyset$, the empty set, for $\beta \geq 2$.

Each of these results can be extended to higher order differential operators B as defined in [5]. For example, we have the following analogue of Theorem 2.

THEOREM 4. If $q(s)$ is a positive, smooth function such that $q(s) \to 0$ as $|s| \to \infty$, and B is the differential operator defined by (5), then $\sigma_d(B) = (-\infty,0)$.

(C) Effects of Nonlinearity on the Discrete Spectrum

For $\beta > 1$, the operators A and B are nonlinear and this fact has crucial implications in the qualitative nature of the discrete spectrum. Here we wish to point out the following 3 "nonlinear peculiarities":

(1) Exponential decay of eigenfunctions. In general we required only that an eigenfunction $y(s)$ of A be in $L_2(\Omega)$. However, even if $q(s)$ itself has no decay, we have $y(s) = O(e^{-\alpha|s|})$ as $|s| \to \infty$, in general. This can be seen by rewriting the equation $Ay + \lambda y = 0$ in the form

$$\ddot{y} + \lambda y + q(s)\eta(s)y = 0 \quad \text{where} \quad \eta(s) = y^{2\beta} . \tag{7}$$

So, when (7) is regarded as a linear equation (since $y \in L_2(\Omega)$, and the potential $q(s)\eta(s)$ has some decay at infinity), y also must have a somewhat stronger decay at ∞ than merely square integrability. Clearly this argument can be iterated to yield exponential decay.

(2) The discrete spectrum is generally continuous, if it exists.

(3) Growth conditions on β control the existence of the discrete spectrum.

(D) Comments on Proofs

The above results are somewhat difficult to prove because of the lack of compactness criteria analogous to the Arzela-Ascoli and Rellich theorems for bounded intervals. Thus two distinct approaches to the desired proofs come to mind: (i) Proof of analogues of the Theorems for a sequence of bounded intervals $\Omega_n \to \Omega$ (as $n \to \infty$) and the determination of appropriate a priori bounds on eigenfunctions $\{x_n\}$ so that as $n \to \infty$ the limit of eigenfunctions in again an eigenfunction. (ii) Proof of extensions of Rellich's compactness theorems analogous to those of Molcanov and Birman [3] to handle the case of unbounded intervals directly. We used approach (i) to prove Theorem 1, and (ii) for Theorem 2. Since this latter approach is perhaps more interesting, we sketch it here.

Outline of Proof of Theorem 2. Let $\dot{W}_{1,2}(0,\infty)$ denote the closure of C^∞ real-valued functions defined on $(0,\infty)$ with compact support in the norm $\|u\|_{1,2}^2 \equiv \int_0^\infty [u^2(s) + \dot{u}^2(s)]ds$. Then we have the following:

LEMMA 1 [1]. The linear map $Lu(s) = s^{-\frac{\sigma}{\sigma+2}} u$ is a bounded compact mapping from $\dot{W}_{1,2}(0,\infty)$ to $L_{\sigma+2}(0,\infty)$ for $0 < \sigma < 4$. Thus for each fixed $\lambda > 0$, we maximize $\int_0^\infty \frac{|u|^{\sigma+2}}{s^\sigma}$ among the functions in $\dot{W}_{1,2}(0,\infty)$ subject to the constraint $\int_0^\infty (\dot{u}^2 + \lambda u^2)ds = \text{const.} > 0$. By [2, pp. 113], the maximum is attained and, after scaling, gives rise to nontrivial eigenfunctions $v \in L_2(0,\infty)$ of A as defined in (6) for $0 < 2\beta < 4$.

Outline of Proof of Theorem 3. The negative result is obtained by invoking the following result from partial differential equations.

LEMMA 2 [1]. Any solution $v(x)$ (vanishing exponentially at infinity) of $\Delta v + f(v) = 0$ in $\mathbb{R}^3$ satisfies the identity $6\int_{\mathbb{R}^3} F(v) = \int_{\mathbb{R}^3} vf(v)$ where $F'(v) = f(v)$ and $F(0) = 0$.

Now apply Lemma 2 to equation (3) with q = positive constant $= \alpha^2$ (say) and $f(v) = \alpha^2 v^{2\beta+1} + \gamma v$ so that $F(v) = \alpha^2(2\beta+2)^{-1} v^{2\beta+2} + \frac{\gamma}{2} v^2$, and we find that (3) has nontrivial exponentially decaying solutions only if $0 < \beta < 4$ and $\gamma \leq 0$, or $\beta = 4$ and $\gamma = 0$. Consequently, a simple argument shows that (4) has eigenfunctions only if $0 < 2\beta < 4$ and

$\gamma < 0$, or $\gamma = 0$ and $2\beta = 4$. However this latter case is impossible since the equation $x_{ss} + \frac{\alpha^2}{s^4} x^5 = 0$ can be explicitly solved.

(E) Some Open Problems

We hope the circle of ideas discussed here will open a fruitful area of research for the study of "nonlinear spectral theory." To this end, we list a few open problems.

(i) Study of the "continuous spectrum" for nonlinear differential operators A and B. The continuous spectrum for A and B in the linear case with $\beta = 0$ have been well studied. Do analogous results hold for $\beta > 0$?

(ii) Oscillation properties of nonlinear eigenfunctions. Do resilts analogous to Titchmarch [4, pp. 107-115] hold in the nonlinear case when $\beta > 0$?

(iii) Completeness of the eigenfunctions of a nonlinear differential operator. Is some notion of completeness of the eigenfunction expansion for the linear case carried over for $\beta > 0$?

(iv) Extensions for elliptic partial differential operators. Results analogous to Theorems 2, 3, and 4 have been obtained for non-linear elliptic partial differential operators by us and M. Schechter in [5]. However analogues for the first part of Theorem 1 are still unproved and apparently quite difficult.

REFERENCES

[1] M.S. BERGER, On the existence and structure of stationary states for a nonlinear Klein-Gordan equation, Journal of Functional Analysis (to appear).

[2] M.S. BERGER and M.S. BERGER, Perspectives in Nonlinearity. W.A. Benjamin Publishers, New York, N. Y., 1968.

[3] M.S. BIRMAN, On the spectrum of singular boundary value problems, Amer. Math. Soc. Trans. (2), 53 (1966), 23-80.

[4] E.C. TITCHMARCH, Eigenfunction Expansions, (Pt. I). Second Edition, Oxford Univ. Press, Oxford, 1962.

[5] M.S. BERGER and M. SCHECHTER, Embedding theorems and quasilinear elliptic boundary value problems for unbounded domains (to appear in Transactions of the Amer. Math. Soc.).

Belfer Graduate School, Yeshiva University, New York, New York

POSITIVELY STABLE DYNAMICAL SYSTEMS*

Nam P. Bhatia

Our purpose here is to announce the following definition and theorem. (For related definitions and notation, see Section 2.)

DEFINITION. A dynamical system (X,π) is said to be positively stable whenever every positive trajectory closure is positively stable.

MAIN THEOREM. Let the dynamical system (X,π) have a metric phase space X which is not the union of two disjoint open invariant sets and let X be not minimal. Then (X,π) is positively stable if and only if the following conditions hold.

(A) The positive limit set $L^+(x)$ is non-empty compact minimal and positively stable for each $x \in X$,

(B) The union M of compact minimal subsets of X is a closed globally asymptotically stable set.

The proof follows the techniques introduced in [1].

Remarks. 1. Notions similar to the one introduced in the above definition are discussed in [2], [3], [4], [5], [6] and [7]. All of these however assume local compactness of the phase space. The results in [6] and [7] are for dynamical systems defined in the plane. Our most

*This research was partially supported by the National Science Foundation under Grant No. NSF-GP-27284.

striking difference is the absence of the assumption of local compactness on the phase space. As in [5] the general results in [2], [3], [4] apply to discrete dynamical systems. The theorem above applies to both continuous and discrete dynamical systems.

2. It is possible to introduce a notion intermediate between that of positive stability and Taro Ura's notion of a dynamical system of characteristic 0^+ [3-7] and obtain results completely analogous to those obtained in locally compact spaces, but without such an assumption.

3. It is possible to obtain similar results for arbitrary topological transformation groups [8]. The basic preparation for this is contained in [9].

4. For the basic theory of dynamical systems see [10].

5. One may introduce new definitions, e.g. requiring every positive trajectory $C^+(x)$ to be positively stable, or requiring every positive prolongation $D^+(x)$ to be positively stable. Examples of the first case have been studied deeply for differentiable dynamical systems in the two and three dimensional Euclidean space [11], [12], where every trajectory is periodic and stable.

1. Basic Definitions and Notation

Throughout this paper X denotes a metric space with metric ρ. R and Z denote the set of reals and the set of integers, respectively, endowed with their usual topological and algebraic structure. G denotes either R or Z. G^+ is the set of non-negative numbers in G. The boundary, closure, and complement of a set $M \subset X$ is denoted by ∂M, $\overline{M}$, and $X - M$, respectively.

A global dynamical system is defined as follows:

1.1 DEFINITION. The pair (X,π) is a (global) dynamical system if and only if π maps $X \times G$ into X and satisfies the three following conditions (the image of (x,t) under π is denoted by $x\pi t$ or simply xt when no confusion occurs):

1.1.1 $xo = x$ for all $x \in X$ (identity axiom)

1.1.2 $(xt)s = x(t+s)$ for all $x \in X$ and $t,s \in G$ (group axiom)

1.1.3 π is continuous (continuity axiom).

A global dynamical system is always assumed given. For $M \subset X$ and $E \subset G$ the set $xt : x \in M,\ t \in E$ is denoted by ME. If $M = \{x\}$ or $E = \{t\},\}$ then we use xE and Mt to denote $\{x\}E$ and $M\{t\}$, respectively.

1.2 DEFINITION. For $x \in X$, the positive trajectory $C^+(x)$, the positive limit set $L^+(x)$, and the positive prolongation $D^+(x)$ is defined by

1.2.1 $C^+(x) = xG^+$,

1.2.2 $L^+(x) = \cap\{\overline{C^+(y)} : y \in C^+(x)\}$,

1.2.3 $D^+(x) = \cap\{\overline{UG^+} : U$ is a neighborhood of $x\}$.

1.3 DEFINITION. A set $M \subset X$ is invariant (positively invariant) if $M' = MG$ or $(M = MG^+)$.

1.4 DEFINITION. A set $M \subset X$ is minimal (positively minimal) if it is closed and invariant (positively invariant) and contians no non-empty proper subset with these properties.

1.5 DEFINITION. A set $M \subset X$ is said to be (positively) stable if every neighborhood of M contains a (positively) invariant neighborhood. It is asymptotically stable if it is positively stable and has a neighborhood U such that for each $x \in U$ and each neighborhood V of M there is a $T \in G$ such that $xt \in V$ for $t \geqq T$.

1.6 NOTATION ON SEQUENCES. For a sequence $x_n \in X$ we write $x_n \to x$ when the sequence converges to x and $x_n \to \infty$ if the sequence has no convergent subsequence. For a given sequence x_n, $P(x_n)$ denotes the set $\{x_n : n = 1,2,\ldots\}$. Note that if $x_n \to x$, then $P(x_n) \cup \{x\}$ is compact, and if $x_n \to \infty$, then $P(x_n)$ is closed.

2. Characterization of Positively Stable Systems

We start by characterization of a positively stable minimal set and then prove the main theorem. Theorem 2.1 below follows from results proved in [1] but we give a direct proof.

2.1 THEOREM. Let $M \subset X$ be minimal. Then M is positively stable if and only if either M is open or $\overline{xG^+}$ is compact and positively stable for all $x \in X$ such that $\overline{xG^+} \cap M \neq \phi$ (moreover $\overline{xG^+} = xG^+ \cup M$). In the positive case M is open or compact.

Proof. The "if" part is immediate, so we prove only the "only if" part. Let M be positively stable and let M be not open. Then $M = \partial M$. let $x \in X$ and $\overline{xG^+} \cap M \neq \phi$. First consider the case that $x \in M$. Then M being closed and invariant, we have $\overline{xG^+} \subset M$. Since $M \subset L^+(x)$, we conclude $M = \overline{xG^+}$. Thus we need only prove that M is compact. If M is not compact, then there is a sequence x_n in M with $x_n \to \infty$. By continuity of the map π and the fact that $M = \partial M$, we can choose sequences y_n in $X - M$, t_n in G^+ such that $\rho(y_n, x_1) < \frac{1}{n}$, $\rho(y_n, t_n, x_n) < \frac{1}{n}$ for each n. Consequencly $y_n t_n \to \infty$ and $y_n t_n$ in $X - M$. Clearly, the open set $X - P(y_n t_n)$ is a neighborhood of M which contains no positively invariant neighborhood of M. Thus M cannot be positively stable. This contradiction shows that M must be compact. Next consider the case that $x \notin M$. We first show that $\overline{xG^+}$ is compact. Otherwise, there is a sequence t_n in G^+, such that $xt_n \to \infty$ (and necessarily $t_n \to +\infty$). Clearly $xt_n \notin M$ as $x \notin M$. Thus $X - P(xt_n)$ is a neighborhood of M. We claim that this contains no positively invariant neighborhood of M. To see this note that for a $y \in M$ there is a sequence $\tau_k \in G^+$ with $x\tau_k \to y \in M$. We may choose a subsequence t_{n_k} of t_n such that $t_{n_k} > \tau_k$ for each k. Then $x\tau_k$ is ultimately in every neighborhood of M, but $xt_{n_k} = x\tau_k(t_{n_k} - \tau_k) \notin X - P(xt_n)$. Thus M is not positively stable. This contradiction shows that $\overline{xG^+}$ is compact (consequently M is compact and stability of M implies $\overline{xG^+} = \overline{xG^+} \cup M$). To see that $\overline{xG^+}$ is also positively stable, let if possible, U be a neighborhood of $\overline{xG^+}$ and let there be a sequence $x_n \to y \in \overline{xG^+}$ and a sequence t_n in G^+ with $x_n t_n \notin U$ (note that this is a necessary and sufficient condition for a set to be positively unstable). Then $t_n \to +\infty$ necessarily. For any $z \in M$, there is a sequence $\tau_k \to +\infty$ with $x\tau_k \to z$, and we may choose a subsequence x_{n_k} of x_n such that $x_{n_k}\tau_k \to z$ and $t_{n_k} > \tau_k$. But this shows that M is positively unstable. This contradiction completes the proof that $\overline{xG^+}$ is compact and positively stable. The theorem is proved.

2.2 Proof of Main Theorem. Using Theorem 2.1, the "if" part follows from condition (A). We prove the "only if" part. Let (X,π) be positively stable and let X be not minimal. Since the union of positively stable sets is positively stable we conclude that every closed positively invariant subset of X is positively stable. In particular, for any $x \in X$, the sets $L^+(x)$, $L^-(x)$ and $\overline{xG}$ are positively stable. We first claim that $L^-(x)$ and $L^+(x)$ are also minimal. Consider, for example, the set $L^+(x)$. If $L^+(x)$ is not minimal, then there is a closed invariant proper subset M. Such an M is positively stable. However, we may choose points $y \in M$ and $z \in L^+(x) - M$, and sequences $\tau_n \to +\infty$, $t_n \to +\infty$, $t_n > \tau_n$, such that $x\tau_n \to y$ and $xt_n \to z$. Indeed, $xG^+ \cap M = \phi$ (otherwise $\overline{xG^+} \subset M$ and so $L^+(x) \subset M$) and consequently $X - P(xt_n) \cup \{z\}$ is a neighborhood of M which contains no positively invariant neighborhood of M. This contradicts stability of M. Thus $L^+(x)$ is minimal and positively stable. Similarly, $L^-(x)$ is minimal. To complete the proof of (A) we need show that $L^+(x) \neq \phi$. So let $L^+(x) = \phi$ and first consider the case that $L^-(x) = \phi$ also. Then xG is minimal and open and hence $xG = X$. This contradiction shows that $L^-(x) \neq \phi$. Now consider the case that $L^-(x) \neq \phi$. Since $L^-(x)$ is minimal and $L^-(x) \neq X$, we conclude $L^-(x)$ is compact since it is positively stable. But then $x \notin L^-(x)$ (otherwise $\overline{xG^+}$ is a compact subset of $L^-(x)$ and hence $L^+(x) \neq \phi$). However, then $X - \{x\}$ is a neighborhood of $L^-(x)$ which contains no positively invariant neighborhood, and we have a contradiction to positive stability of $L^-(x)$. We have thus shown that $L^+(x) \neq \phi$ for each $x \in X$ and condition (A) holds by Theorem 2.1. To see finally that (B) also holds, we only prove the non-trivial assertion that the union M of compact minimal sets is closed. Indeed, if M is not closed then there is a sequence $x_n \in M$ such that $x_n \to x \notin M$. Since $y \in L^+(y)$ if and only if $y \in M$, we have $x \notin L^+(x)$. From the fact that $x_n \in L^+(x_n)$ for each n, and $x_n \to x$, given any $y \in L^+(x)$ we can find sequences $\tau_n \to +\infty$, $t_n \to +\infty$, such that $t_n > \tau_n$, $x_n\tau_n \to y$, and $\rho(x_n, x_nt_n) < \frac{1}{n}$ for each n. Then $x_nt_n \to x$ and clearly the neighborhood $X - P(x_nt_n) \cup \{x\}$ of $L^+(x)$ does not contain any positively invariant neighborhood. This contradicts positive stability of $L^+(x)$. Thus $x \in L^+(x)$ and hence $x \in M$ and M is closed. The theorem is proved.

We close the paper by characterizing a positively stable system (X,π) for which X is minimal.

2.3 THEOREM. If X is minimal, then (X,π) is positively stable if and only if X is positively minimal or $L^+(x) = \phi = L^-(x)$ and $X = xG$ for any $x \in X$. (The easy proof is deleted.)

REFERENCES

[1] N.P. BHATIA, Characteristic properties of stable sets and attractors, Symposite Matematica VI, Academic Press, 1971, pp. 155-166.

[2] V.V. NEMYTSKII and V.V. STEPANOV, Qualitative Theory of Differential Equations. Princeton University Press, 1960.

[3] N.P. BHATIA and O. HAJEK, Theory of dynamical systems, part II, Technical Note BN606, May 1969, IFDAM, University of Maryland.

[4] N.P. BHATIA and O. HAJEK, Local semi-dynamical systems, Lecture Notes in Mathematics #90, Springer-Verlag, 1969.

[5] N.P. BHATIA, Semi-systems of characteristic 0^+, Technical Note BN-699, May 1971, IFDAM, University of Maryland.

[6] S. AHMAD, Dynamical systems of characteristic 0^+, Pacific J. Math. 32 (1970), 561-574.

[7] RONALD A. KNIGHT, Characterization of certain classes of planar dynamical systems, Ph.D. Thesis, May 1971, Oklahoma State University, Stillwater, Oklahoma.

[8] W.H. GOTTSCHALK and G.A. HEDLUND, Topological dynamics, AMS Colloquium Publication, Vol. 36, Providence, 1955.

[9] O. HAJEK, Prolongations in topological dynamics, Seminar on Differential Equations and Dynamical Systems II, Lecture Notes in Mathematics 144, Springer-Verlag, 1970, pp. 79-89.

[10] N.P. BHATIA and G.P. SZEGÖ, Stability theory of dynamical systems, Gundlehren der Mathematischen Wissenschaften #161, Springer-Verlag, 1970.

[11] L. MARKUS, Parallel dynamical systems, Topology 8 (1969), pp. 47-57.

[12] D. MONTGOMERY and L. ZIPPIN, Topological Transformation Groups. Interscience, New York, 1955.

University of Maryland Baltimore Campus, Baltimore, Maryland, and University of Maryland, College Park, Maryland

A NONLINEAR PREDATOR-PREY PROBLEM*

Fred Brauer

1. A problem which has been of some interest in biology is to describe the population of two species, one of which preys on the other. This occurs either in a predator-prey form, or a host-parasite form. It is assumed that an increase in the prey population produces an increase in the predator population while an increase in the predator population produces a decrease in the prey population. If the growth rates are also assumed to be linear in the populations, this leads to a system of differential equations

$$\begin{aligned} x' &= x(\lambda - ax - by) \\ y' &= y(\mu + cx - dy) \ , \end{aligned} \tag{1}$$

where x is the size of the prey population and y is the size of the predator population. The first study of this type was the Lotka-Volterra model [2], which is the special case $a = d = 0$ of (1). Later, the more complicated model (1) with $a \geq 0$, $b > 0$, $c > 0$, $d \geq 0$ was introduced (see, for example [3], [4]). In this paper, we discuss the behavior of solutions of a system of differential equations in which the linear expressions $\lambda - ax - by$ and $\mu + cx - dy$ in (1) are replaced

*This work was supported in part by the National Science Foundation, Contract No. GP-11495, and by the U.S. Army Research Office, Contract No. DA-31-124-ARO-D-462.

by nonlinear functions with conditions on the partial derivatives to correspond to the assumptions made as to the signs of the coefficients in the linear case. Two problems are considered - the behavior of solutions near a critical point (Section 2), and the approximate location of critical points of the nonlinear problem by examining a linearized problem (Section 3).

2. We consider the system of two first order differential equations

$$x' = x\,f(x,y)\ , \qquad y' = y\,g(x,y)\ , \tag{2}$$

where f and g are defined and continuously differentiable for $x \geqslant 0$, $y \geqslant 0$. The predator-prey problem (1) mentioned in the preceding section is of this form, with the functions $f(x,y)$ and $g(x,y)$ which describe the growth rates of the prey and predator species respectively linear in x and y. We assume that there is an equilibrium point (x_0,y_0) with $x_0 > 0$ and $y_0 > 0$ such that $f(x_0,y_0) = g(x_0,y_0) = 0$. In the linear case, x_0 and y_0 can be found explicitly. We also make assumptions on the growth rates at this equilibrium which correspond to hypotheses made in the linear case, namely that

$$\begin{aligned} -a &= f_x(x_0,y_0) \leqslant 0\ , & -b &= f_y(x_0,y_0) < 0 \\ c &= g_x(x_0,y_0) > 0\ , & -d &= g_y(x_0,y_0) \leqslant 0 \end{aligned} \tag{3}$$

with a and d not both zero. The assumption that a and d are not both zero excludes the Lotka-Volterra model, but it will be possible to draw some conclusions in this case also.

Using $f(x_0,y_0) = g(x_0,y_0) = 0$ and (3), we may write

$$f(x,y) = -a(x - x_0) - b(y - y_0) + h_1(x - x_0, y - y_0)$$

$$g(x,y) = c(x - x_0) - d(y - y_0) + h_2(x - x_0, y - y_0)\ ,$$

where $h_i(x - x_0, y - y_0) = O[(x - x_0)^2 + (y - y_0)^2]$ for $i = 1, 2$ as $x \to x_0$, $y \to y_0$. The change of variable $u = x - x_0$, $v = y - y_0$ transforms (2) to the system

$$u' = (u + x_0)[-au - bv + h_1(u,v)]$$
$$v' = (v + y_0)[\ cu - dv + h_2(u,v)]\ ,$$

or

$$(4) \qquad \begin{aligned} u' &= -ax_0u - bx_0v + p_1(u,v) \\ v' &= \ \ cy_0u - dy_0v + p_2(u,v)\ , \end{aligned}$$

where $p_i(u,v) = O(u^2 + v^2)$ as $u,v \to 0$. The linear part of (4) is

$$(5) \qquad \begin{aligned} \hat{u}' &= -ax_0\hat{u} - bx_0\hat{v} \\ \hat{v}' &= \ \ cy_0\hat{u} - dy_0\hat{v}\ . \end{aligned}$$

The eigenvalues of the coefficient matrix of (5) are given by

$$\lambda = \frac{-(ax_0 + dy_0) \pm \sqrt{(ax_0 - dy_0)^2 - 4bc\ x_0y_0}}{2}\ .$$

Since $x_0 > 0$, $y_0 > 0$, $a \geq 0$, $b > 0$, $c > 0$, $d \geq 0$ and a and d are not both zero, it is easy to see that these eigenvalues both have negative real part. They are real and negative if

$$(ax_0 - dy_0)^2 - 4bc\ x_0y_0 \geq 0$$

and they are non-real if

$$(ax_0 - dy_0)^2 - 4bc\ x_0y_0 < 0\ .$$

The elementary theory of critical points of two-dimensional linear systems [1, Chapter 15] shows that the origin is an asymptotically stable critical point of (5). It is an improper node if $(ax_0 - dy_0)^2 - 4bc\ x_0y_0 > 0$, a proper node if $(ax_0 - dy_0)^2 - 4bc\ x_0y_0 = 0$, and a spiral point if $(ax_0 - dy_0)^2 - 4bc\ x_0y_0 < 0$. It follows [1, Chapter 15] that the origin is an asymptotically stable critical point for the nonlinear system (4), an improper node if $(ax_0 - dy_0)^2 - 4bc\ x_0y_0 > 0$, a proper node if $(ax_0 - dy_0)^2 - 4bc\ x_0y_0 = 0$, and a spiral point if $(ax_0 - dy_0)^2 - 4bc\ x_0y_0 < 0$.

In the spiral point case, by making the further change of variable

$$u = \xi \exp[-\tfrac{1}{2}(ax_0 + dy_0)t]\ , \quad v = \eta \exp[-\tfrac{1}{2}(ax_0 + dy_0)t]$$

and studying the resulting system for ξ, η, it is not difficult to show that the solutions of (4) are described asymptotically by the solutions of (5); in particular they have the same amplitude and period. Thus

$$u(t) = A\,e^{-\frac{1}{2}(ax_0+dy_0)t}\left[\cos\left(\frac{\sqrt{4bc\;x_0y_0 - (ax_0-dy_0)^2}}{2}\,t+\delta\right) + o(1)\right]$$

$$v(t) = A\,e^{-\frac{1}{2}(ax_0+dy_0)t}\left[\sin\left(\frac{\sqrt{4bc\;x_0y_0 - (ax_0-dy_0)^2}}{2}\,t+\delta\right) + o(1)\right]$$

as $t \to \infty$.

We may now summarize our results as follows:

THEOREM 1. Consider the system

$$x' = x\,f(x,y)\,, \qquad y' = y\,g(x,y)\,,$$

where f and g are continuously differentiable functions for $x \geq 0$, $y \geq 0$. We assume that

(i) there exist $x_0 > 0$, $y_0 > 0$ such that

$$f(x_0,y_0) = g(x_0,y_0) = 0$$

(ii) $$-a = f_x(x_0,y_0) \leq 0\,, \quad -b = f_y(x_0,y_0) < 0$$
$$c = g_x(x_0,y_0) > 0\,, \quad -d = g_y(x_0,y_0) \leq 0\,,$$

with a and d not both zero.

Then there is a region in the quadrant $x \geq 0$, $y \geq 0$ such that solutions of (2) beginning in this region tend to the equilibrium $x = x_0$, $y = y_0$. If $(ax_0 - dy_0)^2 < 4bc\,x_0y_0$, they spiral about the equilibrium with asymptotic period $4\pi/\sqrt{4bc\,x_0y_0 - (ax_0 - dy_0)^2}$. In this case, the functions $x(t)$ and $y(t)$, representing prey and predator populations respectively, are one quarter cycle out of phase.

In the case $a = d = 0$ studied by Lotka and Volterra, the analysis already carried out shows that the origin is a center for the linear system (5), and the period of oscillations is $2\pi/\sqrt{bc\,x_0y_0}$. Here the nonlinear system (4) may have a center or a spiral point.

3. In the preceding section, we have assumed the existence of an equilibrium point in the first quadrant. When the functions f and g are linear, this equilibrium can be calculated explicitly as the intersection of the lines $f = 0$ and $y = 0$. In the nonlinear case, it is natural to expect that there should be an equilibrium point which is approximated by the equilibrium point of the linearized problem.

We write

$$
\begin{aligned}
f(x,y) &= a_1 - b_1 x - c_1 y + p(x,y) \\
g(x,y) &= a_2 + b_2 x - c_2 y + q(x,y) ,
\end{aligned}
\tag{6}
$$

where $p(x,y)$ and $q(x,y)$ will be assumed small. Consider the linear expressions

$$
\begin{aligned}
f_0(x,y) &= a_1 - b_1 x - c_1 y \\
g_0(x,y) &= a_2 + b_2 x - c_2 y .
\end{aligned}
\tag{7}
$$

It is easy to verify that if $a_1 c_2 - a_2 c_1 > 0$, then there is a unique point $(\hat{x},\hat{y})$ with $\hat{x} > 0$, $\hat{y} > 0$ such that $f_0(\hat{x},\hat{y}) = g_0(\hat{x},\hat{y}) = 0$, namely

$$
\hat{x} = \frac{a_1 c_2 - a_2 c_1}{b_1 c_2 + b_2 c_1} , \qquad \hat{y} = \frac{a_1 b_2 + a_2 b_1}{b_1 c_2 + b_2 c_1}
\tag{8}
$$

THEOREM 2. Let f and g be given by (6) with $|p(x,y)| < \varepsilon$, $|q(x,y)| < \varepsilon$. Suppose $a_1 c_2 - a_2 c_1 > 0$, so that there is a point $(\hat{x},\hat{y})$ given by (8) in the first quadrant with $f_0(\hat{x},\hat{y}) = g_0(\hat{x},\hat{y}) = 0$. Then there is a point (x_0,y_0) in the parallelogram whose vertices are

$$
\left(\hat{x} + \frac{\varepsilon(c_1 - c_2)}{b_1 c_2 + b_2 c_1} , \hat{y} - \frac{\varepsilon(b_1 + b_2)}{b_1 c_2 + b_2 c_1}\right), \quad \left(\hat{x} + \frac{\varepsilon(c_1 + c_2)}{b_1 c_2 + b_2 c_1} , \hat{y} + \frac{\varepsilon(b_2 - b_1)}{b_1 c_2 + b_2 c_1}\right),
$$

$$
\left(\hat{x} + \frac{\varepsilon(c_2 - c_1)}{b_1 c_2 + b_2 c_1} , \hat{y} + \frac{\varepsilon(b_1 + b_2)}{b_1 c_2 + b_2 c_1}\right), \quad \left(\hat{x} - \frac{\varepsilon(c_1 + c_2)}{b_1 c_2 + b_2 c_1} , \hat{y} + \frac{\varepsilon(b_1 - b_2)}{b_1 c_2 + b_2 c_1}\right)
$$

such that $f(x_0,y_0) = g(x_0,y_0) = 0$.

Proof. On the line $f_0(x,y) = \varepsilon$, $f(x,y) = \varepsilon + p(x,y) > 0$ and on the line $f_0(x,y) = -\varepsilon$, $f(x,y) = -\varepsilon + p(x,y) < 0$. Thus on each line $g_0(x,y) = \kappa$, $-\varepsilon \leqslant \kappa \leqslant \varepsilon$, there is a point at which $f = 0$. Thus there is a branch of the curve $f(x,y) = 0$ between the lines $f_0(x,y) = \varepsilon$ and $f_0(x,y) = -\varepsilon$, necessarily continuous, running from the line

$g_0(x,y) = \varepsilon$ to the line $g_0(x,y) = -\varepsilon$. By a similar argument, there is a branch of the curve $g(x,y) = 0$ between the lines $g_0(x,y) = \varepsilon$ and $g_0(x,y) = -\varepsilon$ running from the line $f_0(x,y) = \varepsilon$ to the line $f_0(x,y) = -\varepsilon$. These two curves intersect in a point (x_0,y_0), which lies in the parallelogram bounded by the lines $f_0(x,y) = \pm\varepsilon$ and $g_0(x,y) = \pm\varepsilon$, whose vertices are the points given in the statement of the theorem.

We remark that if we assume not only that $p(x,y)$ and $q(x,y)$ are small, but also that $p_x(x,y)$, $p_y(x,y)$, $q_x(x,y)$, and $q_y(x,y)$ are small, then the values of the partial derivatives of f and g at (x_0,y_0) are close to their values at $(\hat{x},\hat{y})$, and the type of critical point at (x_0,y_0) can be determined by neglecting higher order terms and computing as if f and g were linear and as if the critical point were at $(\hat{x},\hat{y})$.

4. It should be possible to extend Theorems 1 and 2 to more complicated predator-prey problems involving more than two species. It may not be possible to describe the qualitative nature of the critical points in more than two dimensions, but it should be possible to determine the approximate location of the critical points and to determine whether they are asymptotically stable.

Another problem of biological interest is the problem of two species competing for the same food supply. This may be described by a system of the form (2) with the only change in (3) that c is replaced by $-c$. The eigenvalues of the coefficient matrix of (5) are now

$$\lambda = \frac{-(ax_0 + dy_0) \pm \sqrt{(ax_0 - dy_0)^2 + 4bc\, x_0y_0}}{2},$$

which are both real. One eigenvalue is always negative, and both are negative if

$$\sqrt{(ax_0 - dy_0)^2 + 4bc\, x_0y_0} < ax_0 + dy_0 ,$$

or

$$bc < ad .$$

In this case, solutions tend to the critical point. If $bc > ad$, the critical point is a saddle point, and one species will eventually die out. In the competing species problem, the question of whether there is a critical point (x_0,y_0) with $x_0 > 0$, $y_0 > 0$ is more difficult than in the predator-prey problem. It is still true that critical points can be found approximately by looking at the linear terms, but the lines $f_0(x,y) = 0$ and $g_0(x,y) = 0$ now both have negative slope and may not intersect in the first quadrant. Conditions for such an intersection are given, for example, in [3, pp. 53-58]. What we have shown is that the treatment of the linear problem given there describes the nonlinear problem qualitatively. There should be no serious problem in extending this to higher dimensions.

REFERENCES

[1] E.A. CODDINGTON and N. LEVINSON, Theory of Ordinary Differential Equations. McGraw-Hill, 1955.
[2] A.J.LOTKA, Elements of Mathematical Biology. Dover, 1956 (reprint of A.J. Lotka, Elements of Physical Biology, Williams & Wilkins, 1924).
[3] E.C. PIELOU, An Introduction to Mathematical Ecology. Wiley, 1969.
[4] J.M. SMITH, Mathematical Ideas in Biology. Cambridge University Press, 1968.

University of Wisconsin, Madison, Wisconsin

LIE ALGEBRAS AND LINEAR DIFFERENTIAL EQUATIONS

Roger W. Brockett*

Abdolhossein Rahimi**

1. Differential Equations

In this paper we study certain symmetry properties possessed by the solutions of linear differential equations. This is accomplished by use of some basic ideas from the theory of finite dimensional linear systems together with the work of Wei and Norman [1] on the use of Lie algebraic methods in differential equation theory. Our study is also strongly motivated by the results of reference [2] which provided a link between the present paper and a number of questions about the controllability of systems for which the control enters multiplicatively.

Let $\mathbb{R}^{n\times m}$ denote the set of real n by m matrices. By a Lie algebra $\mathscr{L}$ in $\mathbb{R}^{n\times n}$ we understand a subset of $\mathbb{R}^{n\times n}$ which is a real vector space having the property that if A and B belong to $\mathscr{L}$ then so does $[A,B] = AB - BA$. Given an arbitrary subset $\mathscr{N}$ of $\mathbb{R}^{n\times n}$ we denote by $\{\mathscr{N}\}_A$ the smallest Lie algebra which contains $\mathscr{N}$. We denote the identity matrix by I and introduce the square matrix

---

*This work was supported in part by the U.S. Office of Naval Research under the Joint Services Electronics Program by Contract N00014-67-A-0298-0006 and by the National Aeronautics and Space Administration under Grant NGR 22-007-172.

**This work was supported by the National Science Foundation Grant Number NSF GS-14152 DSR Project 72103.

$$J = \begin{bmatrix} 0 & I \\ -I & 0 \end{bmatrix}.$$

Letting prime denote transpose, we say that a matrix A is Hamiltonian if

$$JA = (JA)' .$$

We call a matrix P symplectic if $P'JP = J$.

Wei and Norman [1] have observed that for $|t|$ small the solution of differential equations of the type

$$\dot{x}(t) = \left(\sum_{i=1}^{\nu} a_i(t)A_i \right) x(t)$$

can be expressed as

$$x(t) = e^{H_1 g_1(t)} e^{H_2 g_2(t)} \cdots e^{H_\mu g_\mu(t)} x(0)$$

where $\{H_i\}$ is a basis for the Lie algebra of n by n matrices generated by $\{A_i\}$ and where the g_i satisfy a set of nonlinear differential equations. In this paper we investigate some aspects of this theory in the special case where $\nu = 2$.

2. Lie Algebras and Rational Functions

We begin by establishing two results on the Lie algebra generated by a pair of n by n matrices.

In order to avoid undue repetition let us agree to call a matrix of rational functions $G(s)$ regular if it is square and approaches zero as $|s|$ approached infinity. Our first point is that it is possible to associate a Lie algebra with each regular matrix of rational functions in a natural way. This correspondence goes as follows. It is well known [3, 4] that every regular matrix of rational functions can be expressed as

$$G(s) = C(Is - A)^{-1}B$$

with $C \in \mathbb{R}^{m\times n}$, $A \in \mathbb{R}^{n\times n}$, $B \in \mathbb{R}^{n\times m}$. Moreover it is always possible to pick A, B and C such that

$$\mathrm{rank}(B, AB, \ldots, A^{n-1}B) = \mathrm{rank}(C; CA; \ldots; CA^{n-1}) = n$$

where (,) denotes a column partition and (;) a row partition. In this case we say that the triple [A,B,C] is a minimal realization of G(s). Now minimal realizations are not uniquely determined by G(s), but if [A,B,C] and [F,G,H] are two minimal realizations then there exists a unique nonsingular P such that $PAP^{-1} = F$, $PB = G$ and $CP^{-1} = H$. This result first stated by Kalman [3], is known in system theory as the state space isomorphism theorem (see e.g. [5] for an introductory account in the present notation). We now come to the Lie algebra. Given a regular matrix G(s) we find a minimal realization [A,B,C], and construct $\{A,BC\}_A$, the Lie algebra of n by n matrices generated by A and BC. This collection depends on the particular realization but if [F,G,H] is a second minimal realization of G(s) then $F = PAP^{-1}$ and $GH = PBCP^{-1}$ so that the Lie algebras are isomorphic. That is $\{A,BC\}_A$ and $\{F,GH\}_A$ are matrix representations of the same abstract Lie algebra. We call this abstract algebra the Lie algebra associated with G(s). This Lie albegra reflects the symmetry properties of G(s) as the following theorems make clear.

THEOREM 1. Let A, B and C belong to $\mathbb{R}^{n\times n}$, $\mathbb{R}^{n\times m}$ and $\mathbb{R}^{m\times n}$ respectively. Suppose [A,B,C] is a minimal realization of G(s) and suppose B and C are of rank m. Then:

i) There exists a nonsingular matrix P such that PAP^{-1} and $PBCP^{-1}$ are both Hamiltonian if and only if there exists a nonsingular symmetric matrix T such that

$$TG(s) = G'(-s)T \ .$$

ii) There exists a nonsingular P such that PAP^{-1} and $PBCP^{-1}$ are both skew-symmetric if and only if there exists a nonsingular skew-symmetric T such that $TG(s) = G'(-s)T$ and

$$TG(s) = \sum_{i=1}^{\nu} R_i \frac{s}{s^2+\lambda_i^2} \quad ; \qquad R_i = R_i' \geq 0 \ .$$

Proof. (Hamiltonian Case) Suppose that A and BC are Hamiltonian. Then we have $JBC = C'B'J'$ and in view of the rank conditions $JB = C'T$ for $T = B'J'C'(CC')^{-1}$. Note T is nonsingular. Clearly $C'TC = C'T'C$ so T is symmetric. Thus (recall $J^2 = -I$)

$$
\begin{aligned}
TG(s) &= TC(Is - A)^{-1}B \\
&= TCJ'(Is - JAJ')^{-1}JB \\
&= B'(-Is - A')^{-1}C'T \\
&= G'(-s)T .
\end{aligned}
$$

On the other hand, suppose that for some symmetric nonsingular T we have $TG(s) = G'(-s)T$. Thus

$$
\begin{aligned}
TC(Is - A)^{-1}B &= B'(-Is - A')^{-1}C'T \\
&= -B'(Is + A')^{-1}C'T .
\end{aligned}
$$

Since both sides are minimal realizations it follows from the state space isomorphism theorem referred to above, that there exists a nonsingular matrix P such that

$$PAP^{-1} = -A' \quad ; \quad PB = C'T \quad ; \quad TCP^{-1} = -B' \; ;$$

thus upon transposition and rearrangement we get

$$P'AP'^{-1} = -A' \; ; \quad P'B = -C'T \; ; \quad CP'^{-1} = -B' \; .$$

Now by uniqueness of P (compare with [4,6]) we see that $P = -P'$. Thus there exists a nonsingular Q such that $Q'JQ = P$. Finally we see that $[QAQ^{-1}, QB, CQ^{-1}]$ is a realization such that QAQ^{-1} and $QBCQ^{-1}$ are Hamiltonian. See references [4] and [6] for additional insight into arguments of this type.

(Skew-Symmetric Case) Suppose that A and BC are skew-symmetric. Then we have $BC = -C'B'$ and in view of the rank conditions $B = C'T$ for $T = (B'C')(CC')^{-1}$. Note T is nonsingular. Clearly $C'TC = -C'T'C$ and so T is skew-symmetric. Thus

$$
\begin{aligned}
TG(s) &= TC(Is - A)^{-1}B \\
&= B'(Is - A)^{-1}B \\
&= B'(Is + A')^{-1}B \\
&= B'(-Is - A')^{-1}C'T \\
&= G'(-s)T .
\end{aligned}
$$

All zeros of $\det(Is - A)^{-1}$ are on the imaginary axis since $A = -A'$. The partial fraction expansion of $(Is - A)^{-1}$ has only terms of multiplicity one since A is normal. Clearly the residues of $TG(s) = TC(Is - A)^{-1}C'T'$ at these poles are symmetric and nonnegative definite.

On the other hand, suppose that for some nonsingular skew-symmetric T we have $TG(s) = -G'(-s)T$ with $TG(s)$ given by the partial fraction expansion displayed in the theorem statement. Expand each R_i as the sum of dyads and renumber (if necessary) the λ's so that

$$TG(s) = \sum_{i=1}^{\mu} b_i b_i' \frac{s}{s^2 + \lambda_i^2}$$

with each b_i being an m by m vector. Now let A, B, and C be given by

$$A = \begin{bmatrix} 0 & \lambda_1 & 0 & 0 & \cdot \\ -\lambda_1 & 0 & 0 & 0 & \cdot \\ 0 & 0 & 0 & \lambda_2 & \cdot \\ 0 & 0 & -\lambda_2 & 0 & \cdot \\ \cdot & \cdot & \cdot & \cdot & \cdot \end{bmatrix} ; \quad B = \begin{bmatrix} b_1' \\ 0 \\ b_2' \\ 0 \\ \vdots \end{bmatrix} ; \quad TC = [\, b_1 \;\; 0 \;\; b_2 \;\; 0 \;\; \cdots] .$$

Then $TC(Is - A)^{-1}B = TG(s)$, $A = -A'$ and $BC = -C'B'$. (Compare with Theorem 2 of [7] from which one can see a relationship between this result and the structure of lossless electrical networks.) ∎

We now characterize the conditions under which the representation of the Lie algebras obtained this way are irreducible. We call a set of matrices $\{A_1, A_2, \ldots, A_n\}$ <u>irreducible</u> if there exists no nonsingular P such that all the PA_iP^{-1} are in block triangular form:

$$PA_iP = \begin{bmatrix} F_i & G_i \\ 0 & H_i \end{bmatrix} ; \quad F_i = \nu \text{ by } \nu ; \quad H_i = \mu \text{ by } \mu .$$

We recall the matrix form of Schur's lemma which says that a set of n by n matrices are irreducible if and only if there exists no nonsingular matrix which is not a scalar multiple of the identity and which commutes with all the matrices in the set.

THEOREM 2. Let $G(s)$ be a given regular matrix of rational functions and let $[A,B,C]$ be a minimal realization of $G(s)$. Let A belong to $\mathbb{R}^{n\times n}$ and let B and C belong to $\mathbb{R}^{n\times m}$ and $\mathbb{R}^{m\times n}$ respectively. Suppose B and C are of rank m. Then the Lie algebra $\{A,BC\}_A$ is irreducible if and only if the set of m by m matrices $G(\mathbb{C})$ is irreducible ($\mathbb{C}$ is the field of complex numbers and $G(\mathbb{C})$ is its image under $G(\cdot)$).

Proof. Suppose that $TG(s) = G(s)T$ for some constant matrix T which is invertible and not a multiple of the identity. Let $[A,B,C]$ be a minimal realization of $G(s)$. Then since

$$C(Is - A)^{-1}B = T^{-1}C(Is - A)^{-1}BT$$

we see that $[A,BT,T^{-1}C]$ is also a minimal realization of $G(s)$. By the state space isomorphism theorem we know there exists P such that

$$PB = BT$$
$$CP^{-1} = T^{-1}C$$
$$PAP^{-1} = A \ .$$

Since B is of full rank P cannot be a multiple of the identity if T is not.

On the other hand, if $\{A,BC\}$ is reducible then there exists a nonsingular P, unequal to a multiple of the identity, such that $PAP^{-1} = A$ and $PBCP^{-1} = BC$. However, since B is one to one and C is onto, it follows that $PB = BT$ for some nonsingular T and $CP^{-1} = RC$ for some nonsingular R. Thus $BTRC = BC$ and since B and C have left and right inverses respectively, we see that $T = R^{-1}$ and thus $TG(s) = G(s)T$. Now P is not a multiple of the identity, and so $PB \neq \alpha B$ (this would violate uniqueness of P in the state space isomorphism theorem). Thus T is not a multiple of the identity and $G(\mathbb{C})$ is reducible. ∎

We note that in particular, if BC is a dyad and $\mathrm{rank}(B, AB, \ldots, A^{n-1}B) = \mathrm{rank}(C; CA; \ldots; CA^{n-1}) = n$, then the representation $\{A,BC\}_A$ is irreducible and it is equivalent to a Hamiltonian algebra if and only if $g(s) = g(-s)$. In particular, the algebra associated with $1/s^n$ depends on whether n is even or odd. It has been shown by direct construction in [8] that it is the full $n(n+1)/2$ dimensional Hamiltonian algebra if n is even. Probably the Lie algebra associated with $s^{n-1}/(s^n+1)$ is the full n^2 dimensional algebra of zero trace matrices for every n (see [8]). We observe that to generate skew-symmetric algebras we can use a $G(s)$ of the form

$$G_n(s) = \frac{1}{s}\begin{bmatrix} 1 & 1 \\ -1 & -1 \end{bmatrix} + \frac{1}{s^2+1}\begin{bmatrix} 1 & s \\ -s & 1 \end{bmatrix} + \cdots + \frac{1}{s^2+n^2}\begin{bmatrix} n & s \\ -s & n \end{bmatrix}$$

for the odd dimensional case and

$$G_n(s) = \frac{1}{s^2+1}\begin{bmatrix} 1 & s \\ -s & 1 \end{bmatrix} + \frac{1}{s^2+4}\begin{bmatrix} 2 & s \\ -s & 2 \end{bmatrix} + \cdots + \frac{1}{s^2+n^2}\begin{bmatrix} n & s \\ -s & n \end{bmatrix}$$

for n even.

3. An Application to Stability

As is well known, the symplectic matrices form a group and the eigenvalues of symplectic matrices occur in reciprocal pairs. That is to say, if λ is an eigenvalue of a symplectic matrix then so is $1/\lambda$. This observation together with the basic ideas of Floquet theory enables one to show that for $0 \leq t < \infty$ all solutions of

$$\dot{x}(t) = (A(t) + \varepsilon B(t))x(t) \ ; \quad A(t+T) = A(t) \ ; \quad B(t+T) = B(t)$$

are bounded for ε sufficiently small provided $A(t)$ and $B(t)$ are Hamiltonian and the solution of the equation

$$\dot{x}(t) = A(t)x(t)$$

has distinct characteristic multipliers all lying on the unit circle (see reference [9]). This together with Theorem 1 yields the following theorem.

THEOREM 3. Let p and q be polynomials with the degree of p larger than that of q. Suppose $k(t)$ is periodic with period T. Then there exists $\varepsilon > 0$ such that for $|k(t)| < \varepsilon$ all solutions of

$$p(D)x(t) + k(t)q(D)x(t) = 0 \; ; \quad D = \frac{d}{dt}$$

are bounded provided i) $q(s)/p(s) = q(-s)/p(-s)$ where ii) $p(s) = (s^2+\lambda_1^2)(s^2+\lambda_2^2)\ldots(s^2+\lambda_n^2)$ with λ_j all real and nonzero mod $2\pi/T$ with $(\lambda_i-\lambda_j)$ nonzero mod $2\pi/T$.

Proof. Under the given hypothesis there exists a realization of $[A,B,C]$ of $q(s)/p(s)$ such that A and BC are Hamiltonian. Thus we can express the evolution equations in first order form as

$$\dot{x}(t) = (A + k(t)D)x(t)$$

with A and D Hamiltonian. By hypothesis e^{AT} has all its eigenvalues on the unit circle, and none are repeated. Thus by the perturbation result quoted, there exists $\varepsilon > 0$ such that if $|k(t)| < \varepsilon$ for all t and $k(t+T) = k(t)$ then we have stability.

REFERENCES

[1] J. WEI and E. NORMAN, On global representations of the solutions of linear differential equations as a product of exponentials, Proc. Am. Math. Soc., April 1964.

[2] R.W. BROCKETT, System theory on group manifolds and coset spaces (submitted for publication).

[3] R.E. KALMAN, Mathematical description of linear dynamical systems, SIAM J. Cont. 1, No. 2 (1963), 152-192.

[4] D.C. YOULA and P. TISSI, n-port synthesis via reactance extraction - part I, IEEE International Convention Record, April 1965.

[5] R.W. BROCKETT, Finite Dimensional Linear Systems. J. Wiley, New York, 1970.

[6] R.W. BROCKETT and R.A. SKOOG, A new perturbation theory for the synthesis of nonlinear networks, Symposium on Appl. Math., Vol. 22, American Math. Soc., Providence, Rhode Island (to appear).

[7] B.D.O. ANDERSON, A system theory criterion for positive real matrices, SIAM J. Cont. 5, No. 2 (1967), 171-182.

[8] ABDOLHOSSEIN RAHIMI, Lie Algebraic Methods in Linear System Theory, Ph.D. Thesis, Dept. of Electrical Engineering, M.I.T., 1970.

[9] I.M. GEL'FAND and V.B. LIDSKII, On the structure of the regions of stability of linear canonical systems of differential equations with periodic coefficients, Amer. Math. Soc. Translations - Series 2, Vol. 8 (1958), 143-181.

RWB: Harvard University, Cambridge, Massachusetts

AR: Massachusetts Institute of Technology, Cambridge, Massachusetts

AN ALGORITHM FOR COMPUTING LIAPUNOV FUNCTIONALS FOR SOME DIFFERENTIAL-DIFFERENCE EQUATIONS

Richard Datko

In this note an algorithm is developed for determining the stability behavior of linear autonomous differential-difference equations with a single lag. The technique, which is originally due to Repin [6], is to treat the solutions to the differential-difference equation as the range of a strongly continuous semi-group of operators of class C_0 and to attempt to find Hermitian forms H and R such that a particular algebraic relationship holds between the forms H and R and the infinitesimal generator, A, of the semi-group. Loosely speaking, given an Hermitian form H, which is positive in a certain sense, we attempt to find an Hermitian form R such that the equation $RA + A^*R = -H$ is satisfied. To do this we assume that R has a particular representation in terms of unknown matrices and using our knowledge of H and A we hope to determine these matrices. This process leads to the solution of a two point boundary value problem for a system of linear ordinary differential equations. If the problem has a solution and the resulting Hermitian form R is positive we prove the differential-difference equation is exponentially stable. If R is indefinite we prove the system is unstable but has no periodic solutions.

This note is divided into two sections. Section one establishes the necessary theoretical justification for the algorithm given in Section two. Theorem 1 in Section 1 is known and is implicit in the work of Hale [3]. However it is proved since the statement given here is, to the best of our knowledge, not found in the literature. Theorem 2 is a Liapunov type theorem on stability and instability of differential-difference equations in terms of generalized energies and we believe it is new.

Section two is devoted to the presentation of the algorithm and an example is given demonstrating the computation of the Liapunov functional for scalar differential-difference equations.

1. Notation and Preliminary Results

X will denote the complex Banach space of continuous linear mappings from the closed interval $[-\tau,0]$ into the complex Euclidean n-space. The norm in X will be given by the expression $|\phi| = \sup_{-\tau\leq t\leq 0} \|\phi(t)\|$, where $\|\phi(t)\|$ is the usual Euclidean norm in C^n.

An Hermitian form on X is a continuous sesqui-linear mapping H from the Cartesian product of $X \times X$ into the complex plane such that if ϕ, ψ and μ are any three points in X and a is any complex number, then:

(i) $$H(\phi+\psi,\mu) = H(\phi,\mu) + H(\psi,\mu) ,$$

(ii) $$H(\phi,\psi) = \overline{H(\psi,\phi)} ,$$

(iii) $$H(a\phi,\psi) = aH(\phi,\psi) \quad \text{and}$$

(iv) $$H(\phi,a\psi) = \bar{a}H(\phi,\psi) .$$

If in addition H also satisfies the condition:

(v) $$H(\phi,\phi) > 0 \quad \text{for all } \phi \neq 0 \text{ in } X ,$$

then H is said to be positive and this fact is denoted by the expression $H > 0$.

In this note we consider some properties of the differential-difference equation

$$\dot{x}(t) = A_1x(t) + A_2x(t-\tau) , \tag{1}$$

where A_i, $i = 1,2$ are real $n \times n$ matrices, τ is a fixed constant greater than zero and $x(\cdot)$ is an absolutely continuous mapping from $[0,\infty)$ into C^n. By a solution of (1) with initial value $\phi \in X$ we mean an absolutely continuous $x(\cdot)$ such that for $t > 0$ $x(\cdot)$ satisfies (1) and

$$x(t) = \phi(t) \quad \text{if} \quad -\tau \leqslant t \leqslant 0 . \tag{2}$$

It is well known (see e.g. [3] or [5]) that the solutions of (1) satisfying (2) generate a strongly continuous semi-group of operators $T(t)$ of class C_0 defined on X and that for $t \geqslant \tau$ this semi-group is compact. We shall denote the infinitesimal generator of $T(t)$ by the symbol A. Moreover, A is a closed unbounded operator defined on X with a dense domain, $\mathcal{D}(A)$. In fact if $\phi \in \mathcal{D}(A)$ then $A\phi$ is given by

$$\begin{cases} (A\phi)(s) = \dfrac{d\phi}{ds}(s) & \text{if} \quad -\tau \leqslant s < 0 \\ (A\phi)(0) = \dfrac{d\phi}{ds}(0) = A_1\phi(0) + A_2\phi(-\tau) \end{cases} . \tag{3}$$

<u>Some Properties of $T(t)$ and A</u>

<u>Prop. 1</u>. If $t \geqslant \tau$, then $T(t)$ is a compact operator.

<u>Prop. 2</u>. (See [4], p. 467). The point spectrum of $T(t)$ and A satisfy the relation $P\sigma(T(t)) = \exp t\, P\sigma(A)$, plus, possibly the point $\lambda = 0$. If μ is in $P\sigma(T(t))$ for some fixed $t > 0$ where $\mu \neq 0$ and if $\{\alpha_n\}$ is the set of roots of $e^{\alpha t} = \mu$, then at least one of the points α_n lies in $P\sigma(A)$. The null manifold $N[\mu I - T(t)]$ is the closed extension of the linearly independent manifolds $N[\alpha_n I - A]$, where n ranges over all α_n in $P\sigma(A)$.

<u>Prop. 3</u>. From Prop. 1 and Prop. 2 we deduce that the finite part of the spectrum of A consists of isolated points in the point spectrum and that at most a finite number of points in $P\sigma(A)$ can lie in any given right half plane. Moreover from the general theory of compact operators and Prop. 1 and Prop. 2 the generalized eigenspace associated with any $\lambda \in P\sigma(A)$ is finite dimensional.

Prop. 4. Assume that no point of $\sigma(A)$ lies on the imaginary axis. Let $\Lambda = \{\lambda : \lambda \in \sigma(A)$ and $\operatorname{Re} \lambda > 0\}$. Then there exists a direct sum decomposition of X into two closed subspaces E and F, where E is finite dimensional or empty, such that $T(t)|_E$ can be defined for all $t \in R$, $T(t)E \subseteq E$ for all t in R and $T(t)F \subseteq F$ for all $t \geqslant 0$. Moreover $\sigma(A|_E) = \Lambda$ and $\sigma(A|_F) = \sigma(A) - \Lambda$.

From Property 4 and the fact that the spectral radii of $T(t)|_E$ and $T(t)|_F$ are determined solely by their point spectra we can deduce the following result.

THEOREM 1. Assume no point of $\sigma(A)$ lies on the imaginary axis and let Λ be the set described in Prop. 4 and E and F the corresponding direct sum decomposition of X. If $\Lambda \neq \emptyset$, then there exists $\alpha_1 > 0$ and $K_1 \geqslant 1$ such that for all points $\phi \in E$, $|T(t)\phi| \leqslant K_1 e^{\alpha_1 t}|\phi|$ if $t \leqslant 0$. There also exists constants $K_2 \geqslant 1$ and $\alpha_2 > 0$ such that $|T(t)\psi| \leqslant K_2 e^{-\alpha_2 t}|\psi|$ for all $\psi \in F$ if $t \geqslant 0$.

Proof. Since $T(t)E \subseteq E$ for all $t \in R$ and E is finite dimensional it follows (see e.g. [2]) that $T(t)$ restricted to E is a uniformly continuous group of operators on E and hence $T(t)|_E = \exp(A|_E)t$. Thus on E, $T(t)$ behaves like the fundamental solution of a system of linear ordinary differential equations on a finite dimensional complex Euclidean space. Since $\sigma(A|_E)$ has only a finite number of points, all of which have real parts greater than zero, the first conclusion of the theorem is a consequence of the usual theory of linear autonomous ordinary differential equations in a finite Euclidean space.

To prove the second conclusion, observe that $T(t)|_F$ is a semigroup acting on a Banach space which satisfies Prop. 1, 2, and 3. Moreover, since $\sigma(A|_F) = \sigma(A) - \Lambda$ it follows that $\sigma(A|_F)$ lies strictly in some left half plane $P = \{z: \operatorname{Re} z \leqslant -\alpha,\ \alpha > 0\}$. But this implies (see e.g. [2]) that if $\alpha_2 = \frac{\alpha}{2}$ there is a $K_2 \geqslant 1$ such that $|T(t)\psi| \leqslant K_2 e^{-\alpha_2 t}|\psi|$ for all $\psi \in F$ and $t \geqslant 0$. This completes the proof of the theorem.

The conclusions of Theorem 1 lead to the following definitions.

Definition 1. The system (1) is said to have saddle point behavior if the spectrum of the infinitesimal generator A has no points on the imaginary axis.

Definition 2. The system (1) is said to be exponentially stable if there exist constants $K \geq 1$ and $\alpha > 0$ such that $|T(t)\phi| \leq Ke^{-\alpha t}|\phi|$ for all ϕ in X and $t \geq 0$.

The following is an immediate corollary to Theorem 1.

COROLLARY. System (1) has saddle point behavior if and only if it has no periodic solutions. It is exponentially stable if and only if the spectrum of A lies in the left half plane $P = \{z\colon \operatorname{Re} z < 0\}$.

The following theorem is a "Liapunov" type stability result for system (1), the Hermitian form R described in the theorem being the Liapunov functional.

THEOREM 2. If the system (1) has saddle point behavior, then there exist Hermitian forms R and H, with $H > 0$, such that for all and ψ in $\mathcal{D}(A)$ the following relation is satisfied

$$R(A\phi,\psi) + R(\phi,A\psi) = -H(\phi,\psi) . \tag{4}$$

R is greater than zero if and only if $T(t)$ is exponentially stable.

Proof. The second conclusion of the theorem is a special case of Theorem 3 in [1].

To prove the first conclusion, assume that the system (1) has saddle point behavior and that $\Lambda = \{\lambda\colon \lambda \in \sigma(A)$ and $\operatorname{Re}\lambda > 0\}$ is a nonempty set. Let E and F be the direct sum decomposition of X described in Prop. 4. Since E is a finite dimensional complex Banach space we can induce an inner product on E which shall be denoted by $(\cdot,\cdot)_E$. On F we define the inner product $(\cdot,\cdot)_F$ by

$$(\phi,\psi)_F = \int_{-\tau}^{0} (\phi(s),\psi(s))ds \tag{5}$$

where $(\cdot,\cdot)$ is the usual inner product on C^n. For each $\phi \in X$ we denote by ϕ_E and ϕ_F the respective projections of ϕ on E and F.

Define $H > 0$ by

$$(6) \qquad H(\phi,\psi) = (\phi_E,\psi_E)_E + (\phi_F,\psi_F)_F .$$

Clearly H is Hermitian and $H > 0$. Next let R be given by

$$(7) \qquad R(\phi,\psi) = \int_0^{-\infty} (T(t)\phi_E, T(t)\psi_E)_E \, dt + \int_0^{\infty} (T(t)\phi_F, T(t)\psi_F)_F \, dt .$$

Because of Theorem 1 both of the integrals in (7) are well defined for all ϕ and ψ in X. If ϕ and ψ are in $\mathcal{D}(A)$, then it is easy to verify that

$$(8) \qquad R(A\phi,\psi) + R(\phi,A\psi) = \int_0^{-\infty} \frac{d}{dt}(T(t)\phi_E, T(t)\psi_E)_E \, dt + \int_0^{\infty} \frac{d}{dt}(T(t)\phi_F, T(t)\psi_F)_F \, dt = -H(\phi,\psi) .$$

Finally, assume that (4) is satisfied for all ϕ and ψ in $\mathcal{D}(A)$. Let λ be an eigenvalue of A with a corresponding eigenvector $\phi \neq 0$. Then since $A\phi = \lambda\phi$

$$(9) \qquad R(A\phi,\phi) + R(\phi,A\phi) = 2\mathrm{Re}\,\lambda\, R(\phi,\phi) = -H(\phi,\phi)$$

from which it follows that $\mathrm{Re}\,\lambda \neq 0$. Thus the system (1) has, by Definition 1, saddle point behavior.

<u>COROLLARY</u>. If Hermitian forms R and H, with $H > 0$, satisfy (4) and $R(\phi,\phi) < 0$ for some ϕ in X, then $\sigma(A)$ has at least one eigenvalue λ_0 with $\mathrm{Re}(\lambda_0) > 0$.

<u>Proof</u>. The proof is an immediate consequence of Theorem 2.

2. The Computation of Liapunov Functionals

In this section we present an algorithm for finding Liapunov functionals for the system (1).

Assume that the Hermitian form R can be represented as

$$(10)\quad \begin{aligned} R(\phi,\psi) &= (Q\phi(0),\psi(0)) + \int_{-\tau}^{0} (B(s)\phi(s),\psi(0))ds \\ &+ \int_{-\tau}^{0} (B^*(s)\phi(0),\psi(s))ds + \int_{-\tau}^{0} (G(s)\phi(s),\psi(s))ds \\ &+ \int_{-\tau}^{0}\int_{s}^{0} [(D(s,r)\phi(r),\psi(s)) + (D^*(s,r)\phi(s),\psi(r))]drds \ . \end{aligned}$$

Here we assume that Q, $B(s)$, $G(s)$ and $D(s,r)$ are unknown real $n \times n$ matrices which are continuously differentiable in all their arguments and that Q and G are symmetric. Using (3) we compute, for ϕ and ψ in $\mathcal{D}(A)$,

$$(11)\quad \begin{aligned} &R(A\phi,\psi) + R(\phi,A\psi) \\ &= ((QA_1 + A_1^*Q + B(0) + B^*(0) + G(0))\phi(0),\psi(0)) \\ &+ ((QA_2 - B(-\tau))\phi(-\tau),\psi(0)) + ((A_2^*Q - B^*(-\tau))\phi(0),\psi(-\tau)) \\ &- (G(-\tau)\phi(-\tau),\psi(-\tau)) \\ &+ \int_{-\tau}^{0} ((-\frac{dB}{ds}(s) + A_1^*B(s) + D^*(s,0))\phi(s),\psi(0))ds \\ &+ \int_{-\tau}^{0} ((-\frac{dB^*(s)}{ds} + B^*(s)A_1 + D(s,0))\phi(0),\psi(s))ds \\ &+ \int_{-\tau}^{0} ((A_2^*B(s) - D(-\tau,s))\phi(s),\psi(-\tau))ds \\ &- \int_{-\tau}^{0} (\frac{dG}{ds}(s)\phi(s),\psi(s))ds \\ &+ \int_{-\tau}^{0}\int_{s}^{0} ((\frac{\partial D}{\partial r}(s,r) - \frac{\partial D}{\partial s}(s,r)\phi(r),\psi(s))drds \\ &+ \int_{-\tau}^{0}\int_{s}^{0} ((\frac{\partial D^*}{\partial r}(s,r) - \frac{\partial D^*}{\partial s}(s,r))\phi(s),\psi(r))drds \ . \end{aligned}$$

We assume $H > 0$ has the form

$$(12) \qquad H(\phi,\psi) = (\phi(0),\psi(0)) + \int_{-\tau}^{0}(\phi(s),\psi(s))ds .$$

Thus equation (4) will be satisfied for all ϕ and ψ in $\mathcal{D}(A)$ if and only if the following equations can be solved.

$$(13) \qquad QA_1 + A_1^*Q + B(0) + B^*(0) + G(0) = -I$$

$$(14a) \qquad QA_2 - B(-\tau) = 0$$

$$(14b) \qquad A_2^*Q - B^*(-\tau) = 0$$

$$(15) \qquad -G(-\tau) = 0$$

$$(16) \qquad \frac{dG}{ds}(s) = I$$

$$(17a) \qquad -\frac{dB}{ds}(s) + A_1^*B(s) + D^*(s,0) = 0$$

$$(17b) \qquad -\frac{dB^*}{ds}(s) + B^*(s)A_1 + D(s,0) = 0$$

$$(18a) \qquad B^*(s)A_2 - D^*(-\tau,s) = 0$$

$$(18b) \qquad A_2^*B(s) - D(-\tau,s) = 0$$

$$(19a) \qquad \frac{\partial D}{\partial r}(s,r) - \frac{\partial D}{\partial s}(s,r) = 0$$

$$(19b) \qquad \frac{\partial D^*}{\partial r}(s,r) - \frac{\partial D^*}{\partial s}(s,r) = 0$$

The solution of equations (13)-(19) can be reduced to the following set of relations:

$$(20) \qquad G(s) = (s+\tau)I$$

$$(21) \qquad QA_1 + A_1^*Q + B(0) + B^*(0) = -(1+\tau)I$$

$$(22) \qquad QA_2 - B(-\tau) = 0$$

$$(23) \qquad D(s,r) = A_2^*B(-\tau-s+r)$$

(24) $$\begin{cases} \dfrac{dB(s)}{ds} = A_1^* B(s) + B_1(s) A_2 \\ \dfrac{dB_1}{ds}(s) = -A_2^* B(s) - B_1(s) A_1 \end{cases}$$

(25) $$B_1(s) = B^*(-\tau - s)$$

(26) $$B_1\left(\frac{-\tau}{2}\right) = B^*\left(\frac{-\tau}{2}\right) .$$

Notice that (24) is a linear system of $2n$ ordinary differential equations and that (26) reduces this to looking at an n-parameter family of solutions. Equations (21) and (22) can be looked upon as two point boundary conditions for this n-parameter family. Thus if (20)-(26) has a solution the system (1) has saddle point behavior and if the resulting Hermitian form R is positive the system is exponentially stable.

Conjecture. The system (20)-(26) will have a solution if (1) is exponentially stable.

Example. Consider the scalar differential-difference equation

(27) $$\dot{x}(t) = -x(t) - bx(t-\tau) , \quad b > 1 \quad \text{and} \quad \tau \geq 0 .$$

Equations (20)-(24) in this case reduce to

(28) $$G(s) = s + \tau$$

(29) $$-2Q + 2B(0) = -(1+\tau)$$

(30) $$-bQ - B(-\tau) = 0$$

(31) $$D(s,r) = -bB(-\tau - s + r)$$

(32) $$\frac{dB}{ds}(s) = -B(s) - bB_1(s)$$

(33) $$\frac{dB_1}{ds}(s) = bB(s) + B_1(s) .$$

The general solution of $B(s)$ can be written in the form

$$c_1 \cos \lambda s + c_2 \sin \lambda s \ , \qquad \lambda = \sqrt{b^2-1} \ . \tag{34}$$

From (25) we obtain the relation $B_1(s) = B(-\tau-s)$. This leads to the one parameter family of solutions

$$B(s) = \frac{c}{\lambda + b \sin \lambda\tau} (-b \cos \lambda(\tau+s) + \cos \lambda s + \lambda \sin \lambda s) \ , \tag{35}$$

if $\lambda + b \sin \lambda\tau \neq 0$. If $\lambda + b \sin \lambda\tau = 0$ an analysis similar to the one carried out below shows that (27) has saddle point behavior. Thus we shall assume below that $\lambda + b \sin \lambda\tau \neq 0$. With this assumption we find that the solution of (29) and (30) leads to the matrix equation

$$\begin{pmatrix} -2 & 2(-b \cos \lambda\tau + 1) \\ -b & b - \cos \lambda\tau + \lambda \sin \lambda\tau \end{pmatrix} \begin{pmatrix} Q \\ c \end{pmatrix} = \begin{pmatrix} -(1+\tau) \\ 0 \end{pmatrix} . \tag{36}$$

Equation (36) has a solution if and only if the determinate of the matrix on the left is not equal to zero. That is, if

$$-2\lambda(\lambda \cos \lambda\tau + \sin \lambda\tau) \neq 0 \ . \tag{37}$$

If this is the case then

$$Q = \frac{(1+\tau)\ (b - \cos \lambda\tau + \lambda \sin \lambda\tau)}{2\lambda(\lambda \cos \lambda\tau + \sin \lambda\tau)}$$

and

$$B(s) = \frac{-\ (1+\tau)b}{2\lambda(\lambda \cos \lambda\tau + \sin \lambda\tau)} (-b \cos \lambda(\tau+s) + \cos \lambda s + \lambda \sin \lambda s).$$

Thus

$$\begin{aligned} R(\phi,\psi) = \\ & \frac{(1+\tau)}{2\lambda(\lambda \cos \lambda\tau + \sin \lambda\tau)}\Big[(b - \cos \lambda\tau + \lambda \sin \lambda\tau)\phi(0)\overline{\psi}(0) \\ & \quad - b \int_{-\tau}^{0} \big(-b \cos \lambda(\tau+s) + \cos \lambda s \\ & \qquad + \lambda \sin \lambda s\big)\big(\phi(s)\overline{\psi}(0) + \phi(0)\overline{\psi}(s)\big)ds \\ & \quad + \int_{-\tau}^{0} \int_{s}^{0} b^2\big(-b \cos \lambda(r-s) + \cos \lambda(r-s-\tau) \\ & \qquad + \lambda \sin \lambda(r-s-\tau)\big)\big(\phi(r)\overline{\psi}(s) \\ & \qquad + \phi(s)\overline{\psi})r)\big)drds\Big] \\ & + \int_{-\tau}^{0} (s+\tau)\phi(s)\overline{\psi}(s)ds \ . \end{aligned} \tag{38}$$

Since equation (36) has no solution whenever $\lambda \cos \lambda\tau + \sin \lambda\tau = 0$, we might expect to find periodic solutions of (27) for those values of τ satisfying this condition. This is indeed the case, and the periodic solutions are linear combinations of $\sin \lambda t$ and $\cos \lambda t$.

Another question to ask is when is the form R, given by equation (38), positive. This can be answered as follows. Observe that, in a certain sense, the form R is continuous as a function of τ. When $\tau = 0$ the form is positive and hence (27) is exponentially stable. As τ varies from zero R must remain positive until a point τ_0 is reached at which point equations (29) and (30) are not solvable. At this point periodic solutions of (27) exist. This point is the smallest positive value of τ_0 which satisfies the equation

$$\tau_0 = \frac{1}{\lambda} \tan^{-1}(-\lambda) , \quad \lambda = \sqrt{b^2-1} . \tag{39}$$

(Here negative values of the tangent are assumed to lie in second quadrant.) Using Liapunov functionals, it is not difficult to verify that for all τ above the graph of (39) the equation (27) is not exponentially stable.

Thus if b is fixed in (27) and τ is allowed to vary, then (27) will generate an exponentially stable semi-group if τ lies strictly below the graph of (39). For the point τ_0 on the graph periodic solutions of (27) will exist, and for τ above the graph (27) will have eigenvalues in the right half plane $\operatorname{Re} z > 0$ except at the set of points $\{\tau_0 + \frac{2\pi n}{\lambda}\}$, $n = 1,2,\ldots,$ at which periodic solutions of (27) will exist.

REFERENCES

[1] R. DATKO, An extension of a theorem of A. M. Lyapunov to semi-groups of operators, J. Math. Anal. Appl. 24 (1968), 290-295.

[2] N. DUNFORD and J.T. SCHWARTZ, Linear Operators, Vol. 1. Wiley (Interscience), New York, 1958.

[3] J.K. HALE, Linear functional equations with constant coefficients, Cont. Diff. Eqs. 1 (1963), 291-317.

[4] E. HILLE and R.S. PHILLIPS, Functional Analysis and Semi-groups, A.M.S. Colloquium Publications, Vol. XXXI, 1957.

[5] N.N. KRASOVSKII, Stability in Motion. Stanford University Press, Stanford, 1963.

[6] IU. M. REPIN, Quadratic Liapunov functionals for systems with delay, P.M.M. 29 (1965), 564-566 (in English translation 669-672).

Georgetown, University, Washington, D.C.

PERIODIC SOLUTIONS TO HAMILTONIAN SYSTEMS WITH INFINITELY DEEP POTENTIAL WELLS

William B. Gordon

1. Variational techniques have recently been used to obtain information concerning the existence of periodic solutions to conservative Hamiltonian systems with convex potential wells. (See Berger [1], [2], [3] and Gordon [4].) One considers systems of the type

$$\ddot{x} + \nabla V(x) = 0 \tag{1}$$

and attempts to obtain periodic solutions in a neighborhood of a local min. of V by exhibiting such solutions as solutions to an isoperimetric problem in the calculus of variations. I.e., periodic solutions to (1) are exhibited as critical points of a certain functional E restricted to a manifold of the type $F = \text{constant}$, where F is a certain other functional defined on an appropriate function space. The critical points are then "located" by standard methods, e.g., the method of steepest descent. In this article, we shall give an account of similar results which have been obtained for Hamiltonian systems with infinitely deep potential wells.

Let $x = (x_1, x_2)$ denote a general point in R^2, and let $V = V(x)$ be a function on R^2 which is C^2 everywhere except on a non-empty, discrete set S of points at which V has infinitely deep wells. I.e. we suppose that $V(x) \to -\infty$ as $x \to p$ for each p in S. We further assume:

(A) For each p in S, $(x-p)\cdot\nabla V(x) > 0$ for every x sufficiently close to p.

(B) Every p in S has a deleted neighborhood N_p on which there is defined a C^2 function U_p which has an infinitely deep well at p and satisfies

$$(*) \qquad -V(x) \geqslant |\nabla U_p(x)|^2 , \quad \text{for all } x \text{ in } N_p .$$

With these assumptions we have the following

THEOREM. Every non-trivial element of the fundamental group of $R^2 - S$ contains an infinite number of periodic solutions to (1). More precisely, let γ be a smooth closed path which winds around certain of the singularities in a given sense and in a non-trivial way; i.e., we suppose that γ cannot be continuously deformed into a point without crossing a singularity. Then for any $\varepsilon > 0$, γ can be continuously deformed into a periodic solution of (1) which winds around the same singularities in the same sense, and intersects the ε-neighborhoods of each of the given singularities. Moreover, γ does not cross any of the singularities during the deformation process.

Remark. An unfortunate feature of this theorem is that the gravitational case is excluded by assumption (B), which is a condition on the steepness of the walls of the potential wells. To see why this is true, suppose that V behaves like $-1/r$ in a neighborhood of p, $(r = |x-p|)$, and that U_p satisfies (*). Then U_p can behave no worse than $\sqrt{r}$, which contradicts the requirement that U_p have an infinitely deep well at p.

2. In order to extend the theorem to higher dimensional configuration spaces R^n, it is necessary that the set S of singularities satisfy a certain geometric condition, viz.,

(D) That $R^n - S$ be multiply connected and that any closed path which winds around S cannot be continuously moved off to infinity without either crossing S or having its arc length become infinite.

For example, in R^3, S could consist of two non-parallel lines, but it would not be sufficient for S to consist of a single line, or a number of parallel lines.

If this condition (D) is fulfilled, and if, moreover, the decrease of $V(x)$ is sufficiently rapid as $x \to S$, then the theorem will generalize to R^n.

3. The proof of the theorem is similar to the proof in [4], and a detailed exposition will be given elsewhere. Here we shall merely attempt to describe the roles that are played by assumptions (A) and (B).

(i) In the proof we obtain a 2π-periodic solution to

$$\ddot{x} + \lambda \nabla V(x) = 0 \tag{2}$$

where λ is a non-zero number which appears as a Lagrange multiplier in the isoperimetric variational problem. Then, depending on the sign of λ, we set $y(t) = x(t\sqrt{\pm\lambda})$, and thus obtain a periodic solution to (1) or (1*) where

$$x - \nabla V(x) = 0 \ . \tag{1*}$$

We want to exclude this latter possibility, and this as the function of assumption (A). Now, we actually obtain periodic solutions (to (1) or (1*)) which have arbitrarily small arc length consistent with the condition that the solutions wind around a given set of singularities. By making the arc length sufficiently small we can insure that the solution be concave around at least one of the singularities p; i.e., that the acceleration vector $\ddot{x}$ of the solution point in the general direction of p when $x(t)$ is close to p. But according to assumption (A), $\nabla V(x)$ points away from p when x is close to p. Hence we must have $\ddot{x} = -\nabla V(x)$, i.e. (1).

To put the matter another way, the points p in S are "attractors" for system (1) and "repulsors" for system (1*). Hence assumption (A) can be replaced with any condition which excludes the possibility of having periodic solutions which wind around and come close to the system of repulsors described by (1*).

(ii) For a smooth 2π-periodic map f from R to R^2 given in Euclidean coordinates by $x = x(t)$, let

$$E(f) = \int_0^{2\pi} V(x(t))dt \ , \qquad J(f) = \frac{1}{2}\int_0^{2\pi} |\dot{x}(t)|^2 dt \ .$$

Roughly speaking, we attempt to obtain periodic solutions to (2) by maximizing $E(f)$ subject to the constraint $J(f) = c$, where c is a fixed positive number. (Note that the cycles in $J^{-1}(c)$ are bounded above in arc length by $\sqrt{4\pi c}$.) More precisely, let $f_* = f_*(t)$ be an arbitrary cycle which winds around a given set of singularities, $J(f_*) = c$, and let $f_\tau = f_\tau(t)$ be the family of cycles given by

$$(3) \qquad \frac{\partial f_\tau}{\partial \tau} = \nabla E(f_\tau) \ , \qquad f_0 = f_* \ , \qquad \tau \geqslant 0 \ ,$$

where ∇E is the gradient of E restricted to an appropriate manifold of 2π-periodic maps satisfying $J(f) = c$. The purpose of assumption (B) is to prevent the family f_τ from crossing any singularity as $\tau \to \infty$. Once this is established, the proof proceeds very much as in [4]: One shows that a solution to (3) is defined for all $\tau \geqslant 0$, and contains a zero of ∇E in its closure.

In order to show that the family f_τ is bounded away from the set of singularities, we proceed as follows. Suppose we wish to obtain a periodic solution which winds around, say, two of the singularities p_1, p_2, Let f_0 be a cycle which winds around p_1 and p_2. Using assumption (B) we construct a function U which is defined in a sufficiently large neighborhood of p_1 and p_2, has infinitely deep wells at p_1 and p_2, and satisfies the relation $-V \geqslant |\nabla U|^2$. Let Σ be the surface $z = U(x_1,x_2)$, and let $\hat{f}_\tau$ be the projection of f_τ onto Σ, where f_τ is defined by (3). Along the trajectory (3) we have

$$J(f_\tau) = J(f_0) = \text{constant}; \qquad |E(f_\tau)| \leqslant \text{constant}$$

and it is easy to show that these relations imply that

$$\text{arc length } (\hat{f}_\tau) \leqslant \text{constant}.$$

Hence $\hat{f}_\tau$ does not fall infinitely far down either well of U, and this proves that f_τ does not come arbitrarily close to p_1 or p_2.

REFERENCES

[1] MELVYN S. BERGER, On periodic solutions to second order Hamiltonian systems, J. Math. Anal. Appl. 29 (1970), 512-522.

[2] __________, Multiple solutions of nonlinear operator equations arising from the calculus of variations, Proc. Symposia in Pure Math., Vol. XVIII, Part 1, Am. Math. Soc. (1970), 10-27.

[3] __________, Periodic solutions to second order dynamical systems and isoperimetric variational problems, Am. J. Math., to appear.

[4] WILLIAM B. GORDON, A theorem on the existence of periodic solutions to Hamiltonian systems with convex potential, J. Diff. Eqs. 10 (1971), to appear.

Mathematics Research Center, Naval Research Laboratory, Washington, D.C.

VARIATIONAL PROBLEMS WITH DELAYED ARGUMENT

A. Halanay

1. The Problem: Necessary Conditions

Let $I \subset R$, $I \supset [x_0, x_1]$, $G \subset R^n$, $D \subset R^n$, open regions. The variational problem is defined by a function

$$\mathcal{L} : I \times G \times G \times D \times D \to R .$$

We shall assume that $\mathcal{L}$ is continuous and C^2 with respect to the last four arguments.

Consider the class of functions $y: [x_0 - \tau, x_1] \to G$, continuous and with piece-wise continuous derivatives with values in D; let a continuous function with piece-wise continuous derivative $\phi: [-\tau, 0] \to G$, $\phi': [-\tau, 0] \to D$ and $y_1 \in G$ be given. A function y is admissible if $y(x_0 + s) = \phi(s)$ $y'(x + s) = \phi'(s)$ for $s \in [-\tau, 0]$ and $y(x_1) = y_1$. For an admissible function the number

$$I(y) = \int_{x_0}^{x_1} \mathcal{L}(x, y(x), y(x-\tau), y'(x), y'(x-\tau))dx$$

is well defined.

The function $\tilde{y}$ is optimal if it is admissible and if $I(\tilde{y}) \leqslant I(y)$ for all admissible functions y. This problem is a special case of an optimal control system with time lag. It corresponds to the control system $y' = u$, $y(x_0 + s) = \phi(s)$, $u(x_0 + s) = \phi'(s)$,

$$I(u) = \int_{x_0}^{x_1} \mathcal{L}(x,\ y(x),\ y(x-\tau),\ u(x),\ u(x-\tau))dx\ , \quad y(x_1) = y_1\ .$$

We may use the general theory of necessary conditions for such systems [1]. The adjoint system is

$$\begin{aligned}\psi'(x) &= \lambda_0 \frac{\partial\mathcal{L}}{\partial y}(x,\ \tilde{y}(x),\ \tilde{y}(x-\tau),\ \tilde{u}(x),\ \tilde{u}(x-\tau)) \\ &\quad + \lambda_0 \frac{\partial\mathcal{L}}{\partial z}(x+\tau,\ \tilde{y}(x+\tau),\ \tilde{y}(x),\ \tilde{u}(x+\tau),\ \tilde{u}(x)) \\ \psi(x) &\equiv 0 \quad \text{for} \quad x > x_1\ , \quad \psi_i(x_1) = -\lambda_i\ .\end{aligned}$$

The Hamiltonian is

$$H = \psi u - \lambda_0 \mathcal{L}(x,y,z,u,v)$$

and the necessary conditions given by the maximum principle are

$$\begin{aligned}&\psi(x)u - \lambda_0\mathcal{L}(x,\ \tilde{y}(x),\ \tilde{y}(x-\tau),\ u,\ \tilde{u}(x-\tau)) \\ &\qquad + \psi(x+\tau)\tilde{u}(x+\tau) - \lambda_0\mathcal{L}(x+\tau,\ \tilde{y}(x+\tau),\ \tilde{y}(x),\ \tilde{u}(x+\tau),\ u) \\ &\leqslant \psi(x)\tilde{u}(x) - \lambda_0\mathcal{L}(x,\ \tilde{y}(x),\ \tilde{y}(x-\tau),\ \tilde{u}(x),\ \tilde{u}(x-\tau)) \\ &\qquad + \psi(x+\tau)\tilde{u}(x+\tau) - \lambda_0\mathcal{L}(x+\tau,\ \tilde{y}(x+\tau),\ \tilde{y}(x),\ \tilde{u}(x+\tau),\tilde{u}(x)).\end{aligned}$$

Since we have no restrictions for u we get

$$\begin{aligned}&\psi(x) - \lambda_0[\frac{\partial\mathcal{L}}{\partial u}(x,\ \tilde{y}(x),\ \tilde{y}(x-\tau),\ \tilde{y}'(x),\ \tilde{y}'(x-\tau)) \\ &\qquad + \frac{\partial\mathcal{L}}{\partial v}(x+\tau,\ \tilde{y}(x+\tau),\ \tilde{y}(x),\ \tilde{y}'(x+\tau),\ \tilde{y}'(x))] = 0\ .\end{aligned}$$

If $\lambda_0 = 0$ then $\psi(x) = 0$ hence $\psi_i(x_1) = \lambda_i = 0$, a contradiction. It follows that $\lambda_0 \neq 0$ hence we may consider $\lambda_0 = 1$. We get the Euler-Lagrange equation

$$\begin{aligned}&\frac{d}{dx}[\frac{\partial\mathcal{L}}{\partial u}(x,\ \tilde{y}(x),\ \tilde{y}(x-\tau),\ \tilde{y}'(x),\ \tilde{y}'(x-\tau)) \\ &\quad + \frac{\partial\mathcal{L}}{\partial v}(x+\tau,\ \tilde{y}(x+\tau),\ \tilde{y}(x),\ \tilde{y}'(x+\tau),\ \tilde{y}'(x))] \\ &= \frac{\partial\mathcal{L}}{\partial y}(x,\ \tilde{y}(x),\ \tilde{y}(x-\tau),\ \tilde{y}'(x),\ \tilde{y}'(x-\tau)) \\ &\quad + \frac{\partial\mathcal{L}}{\partial z}(x+\tau,\ \tilde{y}(x+\tau),\ \tilde{y}(x),\ \tilde{y}'(x+\tau),\ \tilde{y}'(x))\ .\end{aligned}$$

The maximum principle gives also

$$\begin{aligned} 0 \leq\ & \mathcal{L}(x,\tilde{y}(x),\tilde{y}(x-\tau),u,\tilde{u}(x-\tau)) + \mathcal{L}(x+\tau,\tilde{y}(x+\tau),\tilde{y}(x),\tilde{u}(x+\tau),u) \\ & - \mathcal{L}(x,\tilde{y}(x),\tilde{y}(x-\tau),\tilde{u}(x),\tilde{u}(x-\tau)) - \mathcal{L}(x+\tau,\tilde{y}(x+\tau),\tilde{y}(x),\tilde{u}(x+\tau),\tilde{u}(x)) \\ & - \frac{\partial\mathcal{L}}{\partial u}(x,\tilde{y}(x),\tilde{y}(x-\tau),\tilde{u}(x),\tilde{u}(x-\tau))(u-\tilde{u}(x)) \\ & - \frac{\partial\mathcal{L}}{\partial v}(x+\tau,\tilde{y}(x+\tau),\tilde{y}(x),\tilde{u}(x+\tau),\tilde{u}(x))(u-\tilde{u}(x)) . \end{aligned}$$

For

$$M(u) = \mathcal{L}(x,\tilde{y}(x),\tilde{y}(x-\tau),u,\tilde{u}(x-\tau)) + \mathcal{L}(x+\tau,\tilde{y}(x+\tau),\tilde{y}(x)\tilde{u}(x+\tau),u)$$

we deduce $M(u) - M(\tilde{u}(x)) - \frac{\partial M}{\partial u}(\tilde{u}(x))(u-\tilde{u}(x)) \geq 0$ hence $\frac{\partial^2 M}{\partial u \partial u}(u(x)) \geq 0$ and we get the Legendre condition

$$\frac{\partial^2\mathcal{L}}{\partial u\partial u}(x,\tilde{y}(x),\tilde{y}(x-\tau),\tilde{u}(x),\tilde{u}(x-\tau)) + \frac{\partial^2\mathcal{L}}{\partial v\partial v}(x+\tau,\tilde{y}(x+\tau),\tilde{y}(x),\tilde{u}(x+\tau),\tilde{u}(x)) \geq 0.$$

These necessary conditions were obtained by D. K. Hughes [2].

2. The Legendre Transformation and the Hamiltonian System

We shall assume in the following the regularity condition

$$\det\left[\frac{\partial^2\mathcal{L}}{\partial u\partial u}(x,y,z_1,u,v) + \frac{\partial^2\mathcal{L}}{\partial v\partial v}(x+\tau,z_2,y,w,u)\right] \neq 0$$

for all $x \in [x_0,x_1]$, $y,z_1,z_2 \in G$, $u,v,w \in D$.

We define by the implicit function theorem $u(x,y,z_1,z_2,v,w,\psi)$ from the equation

$$\psi - \frac{\partial\mathcal{L}}{\partial u}(x,y,z_1,u,v) - \frac{\partial\mathcal{L}}{\partial v}(x+\tau,z_2,y,w,u) = 0 .$$

In the domain of definition of u we define the Hamiltonian

$$\begin{aligned} \mathcal{H}(x,y,z_1,z_2,v,w,\psi) = {} & \psi u(x,y,z_1,z_2,v,w,\psi) \\ & - \mathcal{L}(x,y,z_1,u(x,y,z_1,z_2,v,w,\psi),v) \\ & - \mathcal{L}(x+\tau,z_2,y,w,u(x,y,z_1,z_2,v,w,\psi)) . \end{aligned}$$

According to the definition of the function u

$$\frac{\partial \mathcal{H}}{\partial \psi}(x,y,z_1,z_2,v,w,\psi)$$

$$= u(x,y,z_1,z_2,v,w,\psi) + \psi\frac{\partial u}{\partial \psi}(x,y,z_1,z_2,v,w,\psi)$$

$$- \frac{\partial \mathcal{L}}{\partial u}(x,y,z_1,u(x,y,z_1,z_2,v,w,\psi),v)\frac{\partial u}{\partial \psi}(x,y,z_1,z_2,v,w,\psi)$$

$$- \frac{\partial \mathcal{L}}{\partial v}(x+\tau,z_2,y,w,u(x,y,z_1,z_2,v,w,\psi))\frac{\partial u}{\partial \psi}(x,y,z_1,z_2,v,w,\psi)$$

$$= u(x,y,z_1,z_2,v,w,\psi) \; ;$$

$$\frac{\partial \mathcal{H}}{\partial y}(x,y,z_1,z_2,v,w,\psi)$$

$$= \psi\frac{\partial u}{\partial y}(x,y,z_1,z_2,v,w,\psi) - \frac{\partial \mathcal{L}}{\partial y}(x,y,z_1,u(x,y,z_1,z_2,v,w,\psi),v)$$

$$- \frac{\partial \mathcal{L}}{\partial u}(x,y,z_1,u(x,y,z_1,z_2,v,w,\psi),v)\frac{\partial u}{\partial y}(x,y,z_1,z_2,v,w,\psi)$$

$$- \frac{\partial \mathcal{L}}{\partial z}(x+\tau,z_2,y,w,u(x,y,z_1,z_2,v,w,\psi))$$

$$- \frac{\partial \mathcal{L}}{\partial v}(x+\tau,z_2,y,w,u(x,y,z_1,z_2,v,w,\psi))\frac{\partial u}{\partial y}(x,y,z_1,z_2,v,w,\psi)$$

$$= -\frac{\partial \mathcal{L}}{\partial y}(x,y,z_1,u(x,y,z_1,z_2,v,w,\psi),v)$$

$$- \frac{\partial \mathcal{L}}{\partial z}(x+\tau,z_2,y,w,u(x,y,z_1,z_2,v,w,\psi)) \; .$$

We shall consider the Hamiltonian system

$$y'(x) = \frac{\partial \mathcal{H}}{\partial \psi}(x,y(x),y(x-\tau),y(x+\tau),y'(x-\tau),y'(x+\tau),\psi(x))$$

$$\psi'(x) = -\frac{\partial \mathcal{H}}{\partial y}(x,y(x),y(x-\tau),y(x+\tau),y'(x-\tau),y'(x+\tau),\psi(x)) \; ,$$

and we shall establish the relationship between the solutions of this system and the solutions of the Euler-Lagrange system - the extremals of the variational problem.

Let $\tilde{y}$ be a solution of the Euler-Lagrange system, $\tilde{u} = \tilde{y}'$; define

$$\tilde{\psi}(x) = \frac{\partial \mathcal{L}}{\partial u}(x,\tilde{y}(x),\tilde{y}(x-\tau),\tilde{u}(x),\tilde{u}(x-\tau)) + \frac{\partial \mathcal{L}}{\partial v}(x+\tau,\tilde{y}(x+\tau),\tilde{y}(x),\tilde{u}(x+\tau),\tilde{u}(x)).$$

Taking into account the definition of $u(x, y, z_1, z_2, v, w, \psi)$ we have

$$\tilde{\psi}(x)-\frac{\partial\mathcal{L}}{\partial u}(x,\tilde{y}(x),\tilde{y}(x-\tau),u(x,\tilde{y}(x),\tilde{y}(x-\tau),\tilde{y}(x+\tau),\tilde{u}(x-\tau),\tilde{u}(x+\tau),\tilde{\psi}(x)),\tilde{u}(x-\tau))$$
$$-\frac{\partial\mathcal{L}}{\partial v}(x+\tau,\tilde{y}(x+\tau),\tilde{y}(x),\tilde{u}(x+\tau),u(x,\tilde{y}(x),\tilde{y}(x-\tau),\tilde{y}(x+\tau),\tilde{u}(x-\tau),\tilde{u}(x+\tau),\tilde{\psi}(x))$$
$$= 0 ,$$

and the uniqueness property in the theorem of implicit functions gives

$$u(x,\tilde{y}(x),\tilde{y}(x-\tau),\tilde{y}(x+\tau),\tilde{u}(x-\tau),\tilde{u}(x+\tau),\tilde{\psi}(x)) \equiv \tilde{u}(x) = \tilde{y}'(x) ,$$

hence

$$\tilde{y}'(x) = \frac{\partial\mathcal{H}}{\partial\psi}(x,\tilde{y}(x),\tilde{y}(x-\tau),\tilde{y}(x+\tau),\tilde{y}'(x-\tau),\tilde{y}'(x+\tau),\tilde{\psi}(x)) .$$

Moreover, from the Euler-Lagrange equation

$$\tilde{\psi}'(x) = \frac{\partial\mathcal{L}}{\partial y}(x,\tilde{y}(x),\tilde{y}(x-\tau),\tilde{u}(x),\tilde{u}(x-\tau))$$
$$+ \frac{\partial\mathcal{L}}{\partial z}(x+\tau,\tilde{y}(x+\tau),\tilde{y}(x),\tilde{u}(x+\tau),\tilde{u}(x))$$
$$= -\frac{\partial\mathcal{H}}{\partial y}(x,\tilde{y}(x),\tilde{y}(x-\tau),\tilde{y}(x+\tau),\tilde{y}'(x-\tau),\tilde{y}'(x+\tau),\tilde{\psi}(x)) ,$$

hence $(\tilde{y}, \tilde{\psi})$ is a solution of the Hamiltonian system.

Consider now a solution $(\tilde{y}, \tilde{\psi})$ of the Hamiltonian system and define $\tilde{u}(x) = u(x,\tilde{y}(x),\tilde{y}(x-\tau),\tilde{y}(x+\tau),\tilde{y}'(x-\tau),\tilde{y}'(x+\tau),\tilde{\psi}(x))$. From

$$\tilde{y}'(x) = \frac{\partial\mathcal{H}}{\partial\psi}(x,\tilde{y}(x),\tilde{y}(x-\tau),\tilde{y}(x+\tau),\tilde{y}'(x-\tau),\tilde{y}'(x+\tau),\tilde{\psi}(x))$$
$$= u(x,\tilde{y}(x),\tilde{y}(x-\tau),\tilde{y}(x+\tau),\tilde{y}'(x-\tau),\tilde{y}'(x+\tau),\tilde{\psi}(x))$$
$$= \tilde{u}(x)$$

and from the definition of $u(x,y,z_1,z_2,v,w,\psi)$,

$$\tilde{\psi}(x) = \frac{\partial\mathcal{L}}{\partial u}(x,\tilde{y}(x),\tilde{y}(x-\tau),\tilde{y}'(x),\tilde{y}'(x-\tau))$$
$$+ \frac{\partial\mathcal{L}}{\partial v}(x+\tau,\tilde{y}(x+\tau),\tilde{y}(x),\tilde{y}'(x+\tau),\tilde{y}'(x)) .$$

Since $\tilde{\psi}$ is a solution of the Hamiltonian system we get

$$\tilde{\psi}'(x) = \frac{\partial\mathcal{L}}{\partial y}(x,\tilde{y}(x),\tilde{y}(x-\tau),\tilde{y}'(x),\tilde{y}'(x-\tau))$$
$$+ \frac{\partial\mathcal{L}}{\partial z}(x+\tau,\tilde{y}(x+\tau),\tilde{y}(x),\tilde{y}'(x+\tau),\tilde{y}'(x)) ,$$

hence y is a solution of the Euler-Lagrange system. We deduce that the necessary conditions for extremum may be stated by using the Hamiltonian system. In order to get the optimal solution we have to solve a boundary value problem for the Hamiltonian system:
$\tilde{y}(x_0+s) = \phi(s)$, $\tilde{y}'(x_0+s) = \phi'(s)$, $\tilde{y}(x_1) = y_1$.

3. An Example

We shall now consider the example of the problem defined by

$$\mathcal{L}(x,y,z,u,v) = (u-v)^2, \quad x_0 = 0, \quad x_1 = 3, \quad \tau = 1,$$
$$\phi(s) = -s, \quad y_1 = 2.$$

This example was also considered by Hughes.

We have $\frac{\partial\mathcal{L}}{\partial u}(x,y,z_1,u,v) = 2(u-v)$,

$$\frac{\partial\mathcal{L}}{\partial v}(x+1,z_2,y,w,u) = -2(w-u) \quad \text{for } x \in [0,2]$$

$$\frac{\partial\mathcal{L}}{\partial v}(x+1,z_2,y,w,u) = 0 \quad \text{for } x \in [2,3]$$

$$\frac{\partial^2\mathcal{L}}{\partial u\partial u}(x,y,z_1,u,v) = 2$$

$$\frac{\partial^2\mathcal{L}}{\partial v\partial v}(x+1,z_2,y,w,u) = 2 \quad \text{for } x \in [0,2]$$

$$\frac{\partial^2\mathcal{L}}{\partial v\partial v}(x+1,z_2,y,w,u) = 0 \quad \text{for } x \in (2,3],$$

hence the regularity condition is satisfied.

The function $u(x,y,z_1,z_2,v,w,\psi)$ is defined by

$$\psi - 2(u-v) + 2(w-u) = 0 \quad \text{for } x \in [0,2]$$

$$\psi - 2(u-v) = 0 \quad \text{for } x \in [2,3],$$

hence

$$u(x,y,z_1,z_2,v,w,\psi) = \frac{1}{4}(\psi+2v+2w) \quad \text{for } x \in [0,2]$$

$$u(x,y,z_1,z_2,v,w,\psi) = \frac{1}{2}(\psi+2v) \quad \text{for } x \in (2,3]$$

$$\mathcal{H}(x,y,z_1,z_2,v,w,\psi) = \frac{1}{4}\psi(\psi+2v+2w) - [\frac{1}{4}(\psi+2v+2w)-v]^2 - [w-\frac{1}{4}(\psi+2v+2w)]^2 \quad \text{for } x \in [0,2]$$

$$\mathcal{H}(x,y,z_1,z_2,v,w,\psi) = \frac{1}{2}\psi(\psi+2v) - [\frac{1}{2}(\psi+2v)-v]^2 \quad \text{for } x \in (2,3],$$

hence the Hamiltonian system is

$$y'(x) = \frac{1}{4}[\psi(x)+2y'(x-1)+2y'(x+1)] + \frac{1}{4}\psi(x) - \frac{1}{8}(\psi(x)-2y'(x-1)+2y'(x+1)) + \frac{1}{8}(2y'(x+1)-\psi(x)-2y'(x-1)) \quad \text{for } x \in [0,2]$$

$$\psi'(x) = 0$$

$$y'(x) = \frac{1}{2}[\psi(x)+2y'(x-1)] + \frac{1}{2}\psi(x) - \frac{1}{2}\psi(x) \quad \text{for } x \in (2,3]$$

$$\psi'(x) = 0,$$

hence

$$y'(x) = \frac{1}{4}\psi(x) + \frac{1}{2}y'(x-1) + \frac{1}{2}y'(x+1) \quad \text{for } x \in [0,2]$$

$$\psi'(x) = 0$$

$$y'(x) = \frac{1}{2}\psi(x) + y'(x-1) \quad \text{for } x \in (2,3]$$

$$\psi'(x) = 0 .$$

The initial conditions are $\phi(s) = -s$, $\phi'(s) = -1$; if we denote $y' = u$ we have for u the finite-difference equation

$$u(x) = \frac{1}{4}c + \frac{1}{2}(u(x-1) + u(x+1)) \quad \text{for } x \in [0,2]$$

$$u(x) = -1 \quad \text{for } x \in [-1,0]$$

$$u(x) = \frac{1}{2}c + u(x-1) \quad \text{for } x \in (2,3]$$

By using the conditions $y(0) = 0$, $y(3) = 2$ we get $c = \frac{5}{7}$ and

$y'(x) = \frac{1}{14}$ for $x \in [0,1]$, $y'(x) = \frac{11}{14}$ for $x \in [1,2]$, $y'(x) = \frac{8}{7}$ for $x \in [2,3]$. The corresponding value of the functional is $2 - \frac{3}{14}$. Remark that for the same example Hughes gives another solution; for the solution of Hughes the value of the functional is $\frac{55}{2}$. The example considered is a particular case of the problem of minimizing a quadratic functional; we may consider the situation (not the most general)

$$\mathcal{L}(x,y,z,u,v) = \frac{1}{2}(A_{11}y^2 + 2A_{12}yz + A_{22}z^2 + B_{11}u^2 + 2B_{12}uv + B_{22}v^2).$$

For such problem the regularity condition is $B_{11}(x) + B_{22}(x+\tau) \neq 0$ and the Hamiltonian system is

$$y'(x) = \frac{\psi(x) - B_{12}(x)y'(x-\tau) - B_{12}(x+\tau)y'(x+\tau)}{B_{11}(x) + B_{22}(x+\tau)}$$

$$\psi'(x) = -[A_{11}(x)+A_{22}(x+\tau)]y(x) + A_{12}(x)y(x-\tau) + A_{12}(x+\tau)y(x+\tau)$$

(with the usual convention that B_{12}, B_{22}, A_{22}, A_{12} are zero for $x > x_1$). We may expect that this system may be considered as an integrodifferential equation; indeed, the first equation is a finite-difference equation for y' and from it we may obtain y' as a function of ψ, then an integral representation of y as function of ψ and that gives for ψ an integrodifferential equation.

As a final remark we would like to insist on the interest that may present a general approach to the study of the class of systems represented by the Hamiltonian systems associated to variational problems with time lag.

REFERENCES

[1] A. HALANAY, Optimal controls for systems with time lag, SIAM J. Control, 6, No. 2 (1968), 215-234.
[2] D.K. HUGHES, Variational and optimal control problems with delayed argument, J. Opt. Theory and Appl., 2, No. 1 (1968), 1-14.

Academie, Republique Socialiste de Roumanie, Bucharest, Romania

FREQUENCY-DOMAIN CRITERIA FOR DISSIPATIVITY

A. Halanay

1. In a joint paper with Prof. I. Barbalat we tried to obtain by using frequency-domain methods simple and effective criteria of dissipativity for third and fourth order nonlinear differential equations.

Considering the equation

$$x''' + \alpha x'' + f(x'') + \beta x' + \gamma x = p(t,x,x',x'')$$

and assuming $\alpha > 0$, $\gamma > 0$, $\alpha\beta - \gamma > 0$, $\frac{\gamma-\alpha\beta}{\beta} \leqslant \frac{f(z)}{z} \leqslant \frac{\gamma-\alpha\beta}{\beta} + \frac{\beta^2}{\gamma} + \varepsilon$ one may prove that it is dissipative. The result is different from what one could expect since in the linear case $f(z) = hz$ one has dissipativity for all $\frac{\gamma-\alpha\beta}{\beta} < h$. It is the same situation as in the case $\xi' = \eta - \phi(x) + p$, $\eta' = -\xi + \zeta$, $\zeta' = -\gamma\xi$ studied by Müller. We tried to obtain better results by assuming f to be monotone and using the Brockett-Willems-Popov type condition, but we did not succeed. It was only by using the Iakubovič result for f differentiable and $0 \leqslant \frac{f(z)}{z} \leqslant \mu_0$, $0 \leqslant f'(z) \leqslant \mu_0$, for large $|z|$,

$$(*) \qquad \underline{\lim}_{|z|\to\infty} \frac{1}{z^2}\left[\frac{1}{2} z f(z) - \int_0^z f(\zeta)d\zeta\right] \geqslant 0$$

that we deduced that one has dissipativity for all μ_0. Indeed, the Iakubovič result gives as a condition of dissipativity

$$\frac{\tau_1}{\mu_0} + \mathrm{Re}[(\tau_1 + \frac{\theta}{\mu_0} i\omega)\gamma(i\omega)] + \frac{\tau_2}{\mu_0}\omega^2[1 + \mu_0 \mathrm{Re}\,\gamma(i\omega)] > 0$$

where $\gamma(i\omega) = -\dfrac{\omega^2}{\gamma - \alpha\omega^2 + i\omega(\beta-\omega^2)}$; this condition is written

$$\frac{\tau}{\mu_0} - \frac{\tau\gamma\omega^2}{(\gamma-\alpha\omega^2)^2 + \omega^2(\beta-\omega^2)^2} + \frac{\tau\alpha\omega^4}{(\gamma-\alpha\omega^2)^2 + \omega^2(\beta-\omega^2)^2}$$

$$- \frac{\tilde{\theta}\beta\omega^4}{(\gamma-\alpha\omega^2)^2 + \omega^2(\beta-\omega^2)^2} + \frac{\tilde{\theta}\omega^6}{(\gamma-\alpha\omega^2)^2 + \omega^2(\beta-\omega^2)^2} + \omega^2$$

$$- \frac{\mu_0\gamma\omega^2}{(\gamma-\alpha\omega^2)^2 + \omega^2(\beta-\omega^2)^2} + \frac{\mu_0\alpha\omega^6}{(\gamma-\alpha\omega^2)^2 + \omega^2(\beta-\omega^2)^2} > 0 .$$

Choose $\tilde{\theta}$ such that $\tilde{\theta} + \mu_0 > 0$, $\tilde{\theta}\beta + \mu_0\gamma < 0$ (hence $\tilde{\theta}$ will be negative) and then $0 < \tau < \frac{1}{\gamma}\min_{\omega}\{(\gamma-\omega^2)^2 + \omega^2(\beta-\omega^2)^2\}$; the condition for dissipativity will be satisfied without further conditions on μ_0. Similar situations arise in the study of some equations of fourth order.

For the equation

$$x^{iv} + ax''' + bx'' + cx' + dx + f(x) = p(t,x,x',x'',x''')$$

assume

$$a > 0,\quad ab-c > 0,\quad (ab-c)c - a^2d > 0,\quad d > 0\ ,$$

$$0 \leqslant \frac{f(x)}{x} \leqslant \mu_0\ ,\quad 0 \leqslant f'(x) \leqslant \mu_0\ ,\quad \text{for large } |x|$$

and again (*).

The Iakubovič condition is written

$$\tau_1 + \tau_2\omega^2 + \frac{\mu_0(\tau_1 + \tau_2\omega^2)(\omega^4 - b\omega^2 + d) + \omega^2(c-a\omega^2)}{(\omega^4 - b\omega^2 + d)^2 + \omega^2(c - a\omega^2)^2} > 0 .$$

Choose $\theta = -\dfrac{\mu_0 a}{c}[\tau_1 + \dfrac{\tau_2 c}{a}]$ and then $\dfrac{\tau_2}{\tau_1}$ large enough. Then we see that for such choice of parameters

$$\max_{\omega}\left[\frac{(a\omega^2-c)\dfrac{\theta\omega^2}{\mu_0(\tau_1+\tau_2\omega^2)} - \omega^4 + b\omega^2 - d}{(\omega^4-b\omega^2+d)^2 + \omega^2(c-a\omega^2)^2}\right] = \frac{a^2}{(ab-c)c - a^2d}$$

and the maximum value is taken for $\omega^2 = \frac{c}{a}$.

We deduce that we have dissipativity for

$$\mu_0 < \frac{(ab-c)c - a^2 d}{a^2}$$

and we obtained the same sector as in the linear case.

For the equation

$$x^{iv} + ax''' + f(x''') + bx'' + cx' + dx = v(t,x,x',x'',x''')$$

$$0 \leqslant \frac{f(z)}{z} \leqslant \mu_0 , \quad 0 \leqslant f'(z) \leqslant \mu_0 , \quad \text{for large } |z|$$

and again with (*) we have to choose

$$\frac{\theta}{\mu_0} = \frac{c\omega_1^2(c - a\omega_0^2)(\tau_1 + \tau_2\omega^2)}{\omega_0^2(c\omega_1^2 - ad)(\omega_0^2 - \omega_1^2)}$$

where ω_0^2, ω_1^2 are the roots of $\lambda^2 - b\lambda + d$, $\tau_2 > \frac{c-a\omega_0^2}{a\omega_0^2(\omega_1^2-\omega_0^2)}$, $\tau_1 > 0$, and we deduce again the dissipativity for $0 \leqslant \mu_0 < \frac{c}{d}\omega_1^2 - a$, the same result as in the linear case.

2. These examples show the usefulness of the result of Iakubovič, and motivates its generalisation. In a joint work with I. Barbalat we obtained the following general result.

Consider the system

$$\frac{dx}{dt} = Ax - B\phi(\sigma) + f(t,x), \quad \sigma = R^*x, \quad \phi(\sigma) = \mathrm{col}(\phi_i(\sigma_i)),$$

$$-\alpha_1^i \leqslant \phi_i'(\alpha) \leqslant \alpha_2^i , \quad 0 \leqslant \frac{\phi_i(\alpha)}{\alpha} \leqslant \mu_0^i$$

for $|\alpha|$ large enough, $\mu_0^i \leqslant \alpha_2^i$, $\alpha_1^i \geqslant 0$.

Suppose A is stable and assume there exist diagonal matrices $D_1 > 0$, $D_2, D_3 \geqslant 0$ such that

$$D_1 + \mathrm{Re}[D_1 \operatorname{diag}(\mu_0^j) + i\omega D_2)G(i\omega)]$$
$$+ \omega^2[D_3(E + \mathrm{Re}(\operatorname{diag}(\alpha_2^j - \alpha_1^j)G(i\omega)) - G^*(-i\omega)D_3 \operatorname{diag}(\alpha_1^j\alpha_2^j)G(i\omega)] > 0$$

for all real ω ; $G(i\omega) = R^*(i\omega E - A)^{-1}B$.

Then the system is dissipative if ϕ satisfies the complementary condition

$$\underline{\lim}_{|\lambda|\to\infty} D_2 \frac{1}{|\lambda|^2} \operatorname{diag}\Big[\int_0^\lambda \phi_j(\alpha)d\alpha - \frac{1}{2}\phi_j(\lambda)\lambda\Big] \geq 0.$$

To prove this result let $D_2 = \operatorname{diag}(\theta^i)$, $\theta^i = \theta_1^i - \theta_2^i$, $\theta_1^i \geq 0$, $\theta_2^i \geq 0$, $D_1 = \operatorname{diag}(\tau_1^j)$, $D_3 = \operatorname{diag}(\tau_2^j)$, $R = \operatorname{col}(r_j)$, $B = \operatorname{col}(b_j)$ and consider the Popov system

$$\frac{dx}{dt} = Ax + \Sigma b_i\xi_i\,, \qquad \frac{d\xi_i}{dt} = \mu_i\,,$$

$$\begin{aligned}\eta(0,t_1) = {} & \Sigma\tau_1^i \int_0^{t_1} \xi_i(\xi_i + \mu_0^i r_i^* x)dt + \Sigma\theta_1^i \int_0^{t_1} \xi_i \frac{dr_i^* x}{dt}\, dt \\ & - \Sigma\theta_2^i \int_0^{t_1} (\xi_i + \mu_0^i r_i^* x) \frac{dr_i^* x}{dt}\, dt \\ & + \Sigma\tau_2^i \int_0^{t_1} \Big(\mu_i - \alpha_1^i \frac{dr_i^* x}{dt}\Big)\Big(\mu_i + \alpha_2^i \frac{dr_i^* x}{dt}\Big)dt - \varepsilon\int_0^{t_1} |x(t)|^2 dt,\end{aligned}$$

where $\varepsilon > 0$.

The frequency domain condition and the stability of A allow us to use the Popov general form of the Iakubovič-Kalman-Popov lemma, and obtain

$$\eta(0,t_1) = y^*(t_1)Ny(t_1) - y^*(0)Ny(0) + \int_0^{t_1} |V\mu + W^*y|^2\, dt\,,$$

where $y = \begin{pmatrix} x \\ \xi \end{pmatrix}$. If $N = \begin{pmatrix} N_{11} & N_{12} \\ N_{12}^* & N_{22} \end{pmatrix}$ we may consider the Liapunov function

$$\begin{aligned}V(x) = {} & x^* N_{11} x - x^* N_{12}\phi(R^* x) - \phi(R^* x) N_{12}^* x + \phi^*(R^* x) N_{22}\phi(R^* x) \\ & + \Sigma\theta^j \int_0^{r_j^* x} \phi_j(\alpha)d\alpha + \Sigma \frac{\mu_0^j}{2}\theta_2^j\, x^* R R^* x\,.\end{aligned}$$

For this function we have

$$\frac{dV}{dt} \leq -\varepsilon|x|^2 + \Sigma\tau_1^i\, K_1^i(\lambda_0) + \Sigma\tau_2^i\, K_2^i(\lambda_0)$$

where λ_0 is such that the inequalities on ϕ are satisfied for $|\alpha| \geq \lambda_0$ and the derivative of V is computed along solutions of $\frac{dx}{dt} = Ax - B\phi(\sigma)$.

By a device of Iakubovič one may prove that $\lim_{|x|\to\infty} V(x) = \infty$, and dissipativity follows by standard arguments.

Making use of all parameters involved in this condition seems not to be easy and we have yet no examples of applications of the general statement.

We would like to mention that the condition of Iakubovič does not give the linear sector for the equation

$$x^{iv} + ax''' + bx'' + f(x'') + cx' + dx = v(t,x,x',x'',x''') ,$$

and we feel it would be of interest to obtain new criteria by assuming still more on f.

REFERENCE

[1] V.A. IAKUBOVIČ, Frequency domain conditions for absolute stability and dissipativity of control systems with one differentiable non-linearity (Russian), Dokl. AN SSSR T. 160, No. 1 (1965).

Academie, Republique Socialiste de Roumanie, Bucharest, Romania

NOTE ON ANALYTIC SOLUTIONS OF NONLINEAR ORDINARY DIFFERENTIAL EQUATIONS AT AN IRREGULAR TYPE SINGULARITY*

P. F. Hsieh

1. Introduction

For the analytic study of a system of differential equations, one usually tries to reduce the given equations to a system as simple as possible, and then investigates the solutions of the simplified system of equations. The simplification in turn is performed in two steps. The first step is to find a transformation given in a sort of formal expression, which is in a formal power series of the independent variable for a simple case or that of a certain function in a more complicated case. The second step is to inquire into the analytic meaning of the formal transformation; namely to seek out a function whose convergent or asymptotic expansion agrees with the formal transformation. Naturally, it is desirable to find the domain of validity of the transformation as large as possible.

Since the transformation is for differential equations, it must itself satisfy certain differential equations. However, it is satisfactory to obtain a particular solution satisfying some properties for the purpose of transformation. Thus the problem of analytic simplification of a system of differential equations is reduced to find a particular analytic solution of another system of differential equations, which, more often than not, resembles the original system.

*This work is partially supported by NSF Grant GP-14595.

In the study of nonlinear differential equations at an irregular type singularity, one often encounters a problem of constructing an analytic solution of the following equation

$$(1.1) \qquad \begin{cases} x^{\sigma+1}y' = f(x, y, z) \\ xz' = g(x, z) \end{cases}$$

where σ is a positive integer, x is a complex variable, y and f are m-column vectors while z and g are n-column vectors. Further, f and g are holomorphic in

$$0 < |x| < a, \quad \|y\| \leq b, \quad \|z\| \leq c$$

and

$$0 < |x| < a, \quad \|z\| \leq c$$

respectively,

$$f(0, 0, 0) = 0, \qquad g(0, 0) = 0 ,$$

and both $f_y(0, 0, 0)$ and $g_z(0, 0)$ are nonsingular. Here $\|y\| = \max_{j=1}^{m} \{|y_j|\}$, and f_y denote the Jacobian matrix. Let $\{\nu_1, \nu_2, \ldots, \nu_m\}$ and $\{\mu_1, \mu_2, \ldots, \mu_n\}$ be eigenvalues of $f_y(0, 0, 0)$ and $g_z(0, 0)$ respectively and assume:

$$(1.2) \qquad 0 < \operatorname{Re} \mu_1 \leq \operatorname{Re} \mu_2 \leq \cdots \leq \operatorname{Re} \mu_n .$$

Suppose that

$$(1.3) \qquad \mu_k \neq \text{positive integers.}$$

Then, thanks to (1.2) and (1.3), the system (1.1) can be reduced to

$$(E) \qquad \begin{cases} x^{\sigma+1}y' = f(x, y, z) \\ xz' = (I_n(\mu) + D)z + \Sigma' x^{\ell} z^H b_{\ell H} \end{cases}$$

where $\mu = \operatorname{col}(\mu_1, \ldots, \mu_n)$ and $I_n(\mu) = \operatorname{diag}(\mu_1, \ldots, \mu_n)$, $H = (h_1, h_2, \ldots, h_n)$, $z^H = z_1^{h_1}, z_2^{h_2}, \ldots, z_n^{h_n}$, with z_k the kth component of z, h_k non-negative integers,

$$(1.4) \qquad D = \begin{pmatrix} 0 & & & & \mathbf{0} \\ \delta_2 & \ddots & & & \\ & \ddots & \ddots & & \\ \mathbf{0} & & \delta_n & & 0 \end{pmatrix}, \qquad b_{\ell H} = \begin{pmatrix} b_{1\ell H} \\ \vdots \\ b_{n\ell H} \end{pmatrix}$$

satisfying

$$(1.5) \qquad \delta_k \neq 0 \quad \text{only if} \quad \mu_{k-1} = \mu_k ,$$

and

$$(1.6) \qquad b_{k\ell H} \neq 0 \quad \text{only if} \quad \begin{cases} H = (h_1, \ldots, h_{k-1}, 0, \ldots, 0) \\ \mu_k = \ell + h_1\mu_1 + \cdots + h_{k-1}\mu_{k-1} . \end{cases}$$

Thus, Σ' denotes a finite sum.

As a matter of fact, (1.1) can be reduced formally to (E) by the result of M. Hukuhara [3]. The analyticity of this reduction follows from the main theorem of this paper. Thus, instead of constructing analytic solutions of (1.1), we shall find those for (E).

If we assume, instead of (1.3), more restricted conditions

$$(1.7) \qquad \mu_k \neq \ell + q_1\mu_1 + \cdots + q_n\mu_n , \qquad k = 1, 2, \ldots, n,$$

for all non-negative integers ℓ and q_i, except the trivial case, $q_k = 1$, $\ell = q_i = 0$ $(i \neq k)$, then (1.1) can be further simplified to

$$(1.8) \qquad \begin{cases} x^{\sigma+1}y' = f(x, y, z), \\ xz' = I_n(\mu)z . \end{cases}$$

An analytic expression of solutions of (1.8) is studied by M. Iwano [4], so our result can be regarded as an extension of Iwano's.

2. Main Theorem

Since the second equation of (E) is a polynomial in the right-hand side and satisifes (1.4), (1.5) and (1.6), we can integrate it componentwise. Let $z = W(x) = \mathrm{col}(W_1(x), \ldots, W_n(x))$ be a general solution. Then each component has the following form:

$$(2.1) \qquad W_k(x) = x^{\mu_k}\{C_k + \mathrm{poly}(C_1, C_2, \ldots, C_{k-1}, \log x)\}, \quad (k = 1, 2, \ldots, n),$$

where $C_1, C_2, \ldots, C_n$ are arbitrary constants.

Corresponding to each eigenvalue ν_j of $f_y(0, 0, 0)$, we consider a monomial

$$\Lambda_j(x) = \frac{-\nu_j}{\sigma x^\sigma} , \quad (j = 1, 2, \ldots, m) .$$

A sector $\underline{\theta} < \arg x < \overline{\theta}$ in x-plane is said to be <u>proper</u> with respect to $\{\Lambda_j(x) | \ j = 1, 2, \ldots, m\}$ if

$$|\arg \Lambda_j(x)| < \frac{3\pi}{2} \quad (j = 1, 2, \ldots, m)$$

for all x in $\underline{\theta} < \arg x < \overline{\theta}$ and, also, for each j, there is a direction in $\underline{\theta} < \arg x < \overline{\theta}$ such that $\exp\{\mathrm{Re}\ \Lambda_j(x)\}$ tends to infinity exponentially as x tends to 0 along this direction.

It is noteworthy that a proper sector has an angle greater than $\frac{\pi}{\sigma}$. Furthermore we can fix the eigenvalues $\{\nu_j\}$ in a suitable Reimann surface so that the corresponding proper sector will contain any preassigned direction in its interior.

Let q denote $(q_1, \ldots, q_n)$ with non-negative integers q_k and $|q| = q_1 + q_2 + \cdots + q_n$. We shall prove the following

<u>THEOREM M</u>. Suppose that $f_y(0, 0, 0)$ is non-singular and (1.2) and (1.3) are satisfied. Then, for any sector $\underline{\theta} < \arg x < \overline{\theta}$ proper with respect to $\{\Lambda_j(x)| \ j = 1, 2, \ldots, m\}$, the equation (E) has a solution

$$y = F(x, W(x)) , \qquad z = W(x) \tag{2.1}$$

where $W(x) = \mathrm{col}(W_1(x), \ldots, W_n(x))$ given in (2.1), $F(x,w)$ is holomorphic in

$$0 < |x| < a_0, \quad \underline{\theta} < \arg x < \overline{\theta} , \quad \|w\| < c_0 \tag{2.2}$$

and admits a uniformly convergent power series expansion

$$F(x,w) = F_0(x) + \sum_{|q|=1}^{\infty} w^q F_q(x) \tag{2.3}$$

with $F_0(x)$ and $F_q(x)$ holomorphic in

$$0 < |x| < a_0 , \quad \underline{\theta} < \arg x < \overline{\theta} \tag{2.4}$$

admitting asymptotic expansions in powers of x as x tends to 0 in $\underline{\theta} < \arg x < \overline{\theta}$.

This theorem will be proved in subsequent sections.

3. Leading Term and the First Existence Theorem

We shall find, in the next two sections, a formal series

$$y \sim F_0(x) + \sum_{|q|=1}^{\infty} W(x)^q F_q(x) , \tag{3.1}$$

satisfying properties prescribed above.

In doing this, we shall repeatedly apply an existence theorem, proved in [1, 2, 5].

Consider an m-system of nonlinear equations

$$x^{\sigma+1} y' = f(x, y) \tag{3.2}$$

where f is holomorphic and bounded in

$$0 < |x| \leqslant a, \quad \underline{\theta} < \arg x < \overline{\theta}, \quad \|y\| \leqslant b , \tag{3.3}$$

and admits a uniformly convergent power series of y with coefficients holomorphic in

$$0 < |x| \leqslant a, \quad \underline{\theta} \leqslant \arg x \leqslant \overline{\theta} \tag{3.4}$$

ans possessing asymptotic expansions in power series of x as x tends to 0 in $\underline{\theta} < \arg x < \overline{\theta}$. Let $\nu_1, \nu_2, \ldots, \nu_n$ be eigenvalues of $f_y(0, 0)$ and assume that they are all different from zero. Suppose that (3.2) has a formal solution

$$y \sim \sum_{h=0}^{\infty} x^h g_h \tag{3.5}$$

where g_h are m-vectors and in particular $\|g_0\| < b$. The existence theorem is as following.

THEOREM A. If $\underline{\theta} < \arg x < \overline{\theta}$ is a proper sector with respect to $\{\Lambda_j(x) \mid j = 1, 2, \ldots, m\}$, then (3.3) has a unique solution $\Phi(x)$ which is holomorphic and bounded in

$$(3.6) \qquad 0 < |x| < a_0' , \quad \underline{\theta} < \arg x < \overline{\theta}$$

where $0 < a_0' \leq a$, and admits the asymptotic expansion (3.5) as x tends to 0 in (3.6).

Now, substituting (3.1) into (E), we have

$$(3.7) \qquad x^{\sigma+1}F_0'(x) = f(x, F_0(x), 0) .$$

Since $f(0, 0, 0) = 0$ and $f_y(0, 0, 0)$ is nonsingular, there is a unique formal solution of (3.7)

$$(3.8) \qquad F_0 \sim \sum_{j=1}^{\infty} x^h F_{oh} .$$

Applying Theorem A, we have a solution $F_0(x)$ of (3.7) such that it is holomorphic and bounded in

$$(3.9) \qquad 0 < |x| < a' , \quad \underline{\theta} < \arg x < \overline{\theta} , \quad (0 < a' \leq a)$$

and admits the formal series (3.8) as its asymptotic expansion as x tends to 0 in (3.9).

4. Formal Solution

Put

$$(4.1) \qquad y = \zeta + F_0(x) ,$$

and substitute (4.1) and $z = W(x)$ into the first equation of (E). Then ζ satisfies

$$(4.2) \qquad x^{\sigma+1}\zeta' = \hat{f}(x, \zeta, W(x))$$

where $\hat{f}(x, \zeta, w)$ is holomorphic in

$$(4.3) \qquad 0 < |x| < a_1, \quad \|\zeta\| \leq b_1, \quad \|w\| \leq c_1$$

for suitable positive constants a_1, b_1, and c_1, and expressible as a uniformly convergent power series of ζ

$$\hat{f}(x, \zeta, w) = \hat{f}_0(x, w) + A(x, w)\zeta + \sum_{|p|\geq 2} \hat{f}_p(x, w)\zeta^p$$

with $\hat{f}_0(x, 0) \equiv 0$, $A(0, 0) = f_y(0, 0, 0)$ and $p = (p_1, \ldots, p_m)$.

Put

$$\zeta \sim \sum_{|q|=1}^{\infty} W(x)^q F_q(x) \tag{4.4}$$

and substitute into (4.2). Then, we have

$$x^{\sigma+1}\zeta' = A(x, 0)W(x) + \sum_{|q|=2}^{\infty} [A(x, 0) + K_q(x, F_{q'}(x))]W(x)^q \tag{4.5}$$

where K_q is a polynomial of $F_{q'}$ for $|q'| < |q|$.

On the other hand, differentiate (4.4) and by the use of the second equation of (E), we have

$$x^{\sigma+1}\zeta' = x^{\sigma+1} \sum_{|q|=1}^{\infty} W(x)^q F_q'(x) + x^{\sigma} \sum_{|q|=1}^{\infty} \left[\sum_{k=1}^{n} q_k \left(\mu_k + \frac{\delta_k W_{k-1}(x)}{W_k(x)} + \frac{1}{W_k} \sum{}' x^{\ell} W(x)^{H} b_{k\ell H}\right)\right] W(x)^q F_q(x) . \tag{4.6}$$

Let e_j denote the unit m-row vector with 1 for the jth component and 0 for the rest. Then (4.6) can be rearranged as

$$x^{\sigma+1}\zeta' = x^{\sigma} \sum_{|q|=1} \left[xF_q'(x) + \left(\sum_{k=1}^{n} q_k\mu_k\right) F_q(x) + \sum_{k=2}^{n} q_k \delta_k F_{q-e_{k-1}+e_k}(x) + \tilde{K}_q(x, F_{q'}(x))\right] W(x)^q \tag{4.7}$$

where $\tilde{K}_q$ is a polynomial of $F_{q'}$ with $|q'| < |q|$ or

$$q' = q - e_i + e_k \qquad (i < k) \qquad \text{when } |q'| = |q|. \tag{4.8}$$

Compare the coefficients of $W(x)^q$ in (4.5) and (4.7), we have

$$x^{\sigma+1} F_q'(x) = [A(x, 0) - x^{\sigma} \sum_{k=1}^{n} q_k \mu_k] F_q(x) + G_q(x, F_{q'}(x)) \tag{4.9}$$

where G_q is a polynomial of $F_{q'}$ with $|q'| < |q|$ or q' satisfying (4.8).

Now, establish a linear order of determining $F_q(x)$ by stipulating that $F_{q'}(x)$ is determined before $F_q(x)$ if either $|q'| < |q|$ or the first nonzero element of $q' - q$ is negative when $|q'| = |q|$. Then, by the fact that $A(0, 0)$ is nonsingular, we can determine, successively, a unique formal solution of the linear equation (4.9)

$$\hat{F}_q \sim \sum_{h=0}^{\infty} x^h F_{qh} . \tag{4.10}$$

By the use of Theorem A, (4.9) has a unique solution holomorphic and bounded in

$$0 < |x| < a_2 , \quad \underline{\theta} < \arg x < \overline{\theta} , \quad (0 < a_2 \leq a_1) \tag{4.11}$$

and admits the asymptotic expansion (4.10) as x tends to 0 in $\underline{\theta} < \arg x < \overline{\theta}$. Moreover, the same sector (4.11) is valid for all $F_q(x)$, since (4.9) gives the similar linear equations with the same leading term $A(0, 0)$ for all q.

Thus, we have a unique formal solution (3.1) for (E) with coefficients holomorphic and bounded in (4.11).

5. Analytic Solution and the Second Existence Theorem

The problem of finding the analytic solution (2.1) for (E) can be summarized in the following

THEOREM C. If $\underline{\theta} < \arg x < \overline{\theta}$ is proper with respect to $\{\Lambda_j(x) | j = 1, 2, \ldots, m\}$, then there exists a function $F(x, W(x))$ which is holomorphic and bounded in

$$0 < |x| < a_0, \quad \underline{\theta} < \arg x < \overline{\theta} , \quad \|W(x)\| < c_0 \tag{5.1}$$

and admits a uniformly convergent power series (2.3) whenever (x, w) is in (5.1) with coefficients holomorphic in (2.4) possessing asymptotic expansion (3.8) and (4.10), respectively, as x tends to 0 in (2.4).

A theorem similar to Theorem C is proved recently by Iwano [6] by the use of Tychonoff-type fixed point theorem. If we have the condition (1.7) instead of (1.3), then Theorem C is reduced to Theorem B which is proved recently by Iwano [5] and Hsieh [1, 2]. In Theorem B, $W(x)$ is replaced by $V(x) = I_n(x^{\mu})v$, where v is an arbitrary column vector. Iwano uses the fixed point theorem in [5] while Hsieh uses a successive approximations method in [1, 2]. We shall sketch the proof of Theorem C by means of successive approximations here.

For the sake of simplicity, we introduce the following notations:

$\lambda_k = \operatorname{Re} \mu_k > 0, \quad k = 1, 2, \ldots, n ,$

$\nu = \operatorname{col}(\nu_1, \ldots, \nu_m)$ where $\{\nu_j\}$ are eigenvalues of $f_y(0, 0, 0)$,

$e^{\Lambda(x)} = \operatorname{col}(e^{\Lambda_1(x)}, \ldots, e^{\Lambda_m(x)})$,

$\|\nu\|' = \min_{j=1}^{m} \{|\nu_j|\}$,

$|||w||| = \max_{k=1}^{n} \{|w_k|^{1/\lambda_k}\}$,

$[y] = \operatorname{col}(|y_1|, \ldots, |y_m|)$,

$[y| \leqslant [\tilde{y}]$ means $|y_j| \leqslant |\tilde{y}_j|$ for each j,

$q = (q_1, \ldots, q_n)$, $\lambda = (\lambda_1, \ldots, \lambda_n)$ then $|\lambda q| = \lambda_1 q_1 + \cdots + \lambda_n q_n$.

We shall make changes of variable in the first equation of (E). Let N be a positive integer. Put

(5.2) $$y = \sum_{|\lambda q|<N} W(x)^q F_q(x) + \eta$$

and

(5.3) $$\eta = I_m(e^{\Lambda(x)})Y .$$

Then, Y satisfies

(5.4) $$x^{\sigma+1}Y' = I_m(e^{-\Lambda(x)})\hat{g}(x, I_m(e^{\Lambda(x)})Y, W(x))$$

where $\hat{g}(x, \eta, w)$ is an m-vector holomorphic and bounded for

(5.5) $$0 < |x| < a_N, \quad \underline{\theta} < \arg x < \overline{\theta}, \quad \|\eta\| < b_N, \quad \|w\| < c_N$$

and satisfies

(5.6) $$\|\hat{g}(x, \eta, w)\| \leqslant H\|\eta\| + B\cdot\max\{|x|^N, |||w|||^N\}$$

with H a constant independent of N. Moreover, $g(x, \eta, w)$ satisfies a Lipschitz condition

$$(5.7) \qquad \|\hat{g}(x, \eta^1, w) - \hat{g}(x, \eta^2, w)\| \leq H\|\eta^1 - \eta^2\|$$

for (5.5). Without loss of generality, $f_y(0, 0, 0)$ can be assumed to be in a Jordan canonical form. Consequently, $\hat{g}_\eta(0, 0, 0)$ is nilpotent and H can be taken as small as we wish. Then, the proof of Theorem C is reduced to solving the following

PROBLEM C. If H is small, then there exists a solution $\phi_N(x, W(x))$ of (5.4) such that for suitably chosen a_N', c_N', and K_N

(i) $\phi_N(x, w)$ is a holomorphic and bounded m-vector function for

$$(5.8)_N \qquad 0 < |x| < a_N', \qquad \underline{\theta} < \arg x < \overline{\theta}, \qquad \|w\| < c_N',$$

(ii) $\phi_N(x, w)$ satisfies the inequality

$$(5.9) \qquad [\phi_N(x, w)] \leq K_N \max\{|x|^N, \|\|w\|\|^N\}[e^{-\Lambda(x)}]$$

for (x, w) in $(5.8)_N$.

Moreover, a solution of (5.4) satisfying

$$(5.10) \qquad [Y(x, W(x))] = O(\max\{|x|^N, \|\|W(x)\|\|^N\})[e^{-\Lambda(x)}]$$

is unique.

In fact, from (5.2) and (5.3), the function

$$(5.11) \qquad y_N(x, W(x)) = \sum_{|\lambda q|<N} W(x)^q F_q(x) + I_m(e^{\Lambda(x)})\phi_N(x, W(x))$$

is a solution of the first equation (E), provided that $(x, W(x))$ is in the domain $(4.9)_N$. Let N' be an integer greater than N. Then

$$(5.12) \qquad I_m(e^{-\Lambda(x)})\{y_{N'}(x, W(x)) - \sum_{|\lambda q|<N} W(x)^q F_q(x)\}$$

is a solution of (5.4) if $(x, W(x))$ is in the common part of $(5.8)_N$ and $(5.8)_{N'}$. Moreover, this solution satisfies (5.10). Hence, by the uniqueness of solution, the expression (5.12) must coincide with $\phi_N(x, W(x)) = y_{N'}(x, W(x))$, and consequently,

$$y_N(x\ W(x)) = y_{N'}(x, W(x))$$

provided that $(x, W(x))$ is in the common part of $(5.8)_N$ and $(5.8)_{N'}$. This proves that the solution $y_N(x, W(x))$ is independent of N. We denote it by $F(x, W(x))$.

By analytic continuation, $F(x, w)$ is defined in (2.2) with $a_0 = \sup_N a_N$ and $c_0 = \sup_N c_N$. Obviously, the solution $F(x, W(x))$ has the asymptotic expansion (3.1) as $W(x)$ tends to zero. On the other hand, $w = 0$ is an interior point of (2.2). By Cauchy's theorem, $F(x, W(x))$ can be expanded into a uniformly convergent power series of $W(x)$. By virtue of the uniqueness of asymptotic expansions, the asymptotic expansion (3.1) must coincide with the uniformly convergent expansion. This proves that the formal solution (3.1) is uniformly convergent to $F(x, W(x))$ whenever $(x, W(x))$ is in (2.2).

Thus, the proof of Theorem C subjects to the solution of Problem C.

6. Sketch of Solving Problem C

Due to the complication and length of solving Problem C, only a sketch will be given here.

First, in order to control the behavior of $W(x)$, the domain $(5.8)_N$ is replaced by a stable domain of the form

$$(6.1) \qquad 0 < |x| < a''_N \omega(\arg x), \qquad \underline{\theta} < \arg x < \overline{\theta}, \qquad [w] < c''_N[\chi(\arg x)]$$

where $\omega(\tau)$ is a scalar function, $\chi(\tau)$ is a n-vectorial function defined by Iwano [6]. The domains $(5.8)_N$ and (6.1) are equivalent in the sense that one contains the other by the suitable choice of a'_N, c'_N, a''_N, and c''_N.

Let (x_1, w^1) be an arbitrary point in (6.1) and choose $C_1, \ldots, C_n$ in (2.1) such that

$$(6.2) \qquad W_k(x_1) = w_k^1 \qquad (k = 1, 2, \ldots, n)$$

where w_k^1 are components of w^1. Then Problem C can be solved if we can find m-vector path Γ_{x_1} with components Γ_{jx_1} such that:

(i) Each curve Γ_{jx_1} joins the point x_1 with the origin and is contained in the domain

$$(6.2) \qquad 0 < |x| < a''_N\, \omega(\arg x)], \qquad \underline{\theta} < \arg x < \overline{\theta}$$

except for the origin.

(ii) As x moves on Γ_{jx_1}, we have

$$[W(x)] < c_N''[\chi(\arg x)] , \quad \underline{\theta} < \arg x < \overline{\theta} . \tag{6.4}$$

(iii) If a_N'' is sufficiently small, then

$$\int_{\Gamma_{jx_1}} |x|^{-\sigma-1}(\max\{|x|^N, |||W(x)|||^N\})e^{-\mathrm{Re}\Lambda_j(x)}ds_j \leq \frac{2}{\|v\|' \sin 2\sigma\varepsilon}(\max\{|x|^N, |||W(x)|||^N\}) , \tag{6.5}$$

$(j = 1, 2, \ldots, m)$, where ε is a positive constant such that $4H \leq \sin 2\sigma\varepsilon$ and s_j is the arc length of Γ_{jx_1} measured from the origin.

If the paths Γ_{jx_1} are found, then the task of solving Problem C is reduced to that of solving the integral equations:

$$Y(x_1, w^1) = \int_{\Gamma_{jx_1}} x^{-\sigma-1} I_m(e^{-\Lambda(x)})\hat{g}(x, I_m(e^{\Lambda(x)})Y(x,W(x)),W(x))dx \tag{6.6}$$

such that $Y(x, w)$ satisfies the properties described in Problem C.

The paths Γ_{jx_1} are given in detail by Iwano [6], and the existence of desired solutions are also proved there by the use of Tychonoff fixed point theorem. However the desired solutions are also constructed by successive approximations defined as follows:

$$Y^{(0)}(x_1, w^1) = 0 \tag{6.6}$$

$$Y^{(\alpha+1)}(x_1,w^1) = \int_{\Gamma_{jx_1}} x^{-\sigma-1} I_m(e^{-\Lambda(x)})\hat{g}(x,I_m(e^{\Lambda(x)})Y^{(\alpha)}(x,W(x)),W(x))dx .$$

Here the paths Γ_{jx_1} are taken exactly the same as those given by Iwano [6]. Thus these integrals involve improper contour integration. Also, the difficulty arises in proving the analyticity of the functions defined in (6.8) with respect to (x_1, w^1) when w^1 and x_1 satisfy the relation (6.2). Nevertheless, a similar argument as that used by the author in [1, 2] can be applied to construct the functions $Y^{(\alpha)}(x, w)$ and prove that they converge to $\phi_N(x, w)$ described in Problem C.

REFERENCES

[1] P.F. HSIEH, Successive approximations methods for solutions of nonlinear differential equations at an irregular type singular point, to appear as an NRL-MRC Report and also in Comm. Math. Univ. Sancti Pauli.

[2] P.F. HSIEH, Analytic simplification of a system of ordinary differential equations at an irregular type singularity, to appear as an NRL-MRC Report and also in Comm. Math. Univ. Sancti Pauli.

[3] M. HUKUHARA, Intégration formelle d'un système d'équations differentielles non linéaires dans le voisinage d'un point singulier, Ann. Math. Pura Appl. (4) 19 (1940), 35-44.

[4] M. IWANO, A method to construct analytic expressions for bounded solutions of nonlinear ordinary differential equations with an irregular singular point, Funk. Ekv., 10 (1967), 75-105.

[5] M. IWANO, Analytic expressions for bounded solutions of nonlinear ordinary differential equations with an irregular type singular point, Ann. Mat. Pura Appl. (4), 82 (1969), 189-256.

[6] M. IWANO, Bounded solutions and stable domains of nonlinear ordinary differential equations, Analytic Theory of Differential Equations. Lecture Notes in Mathematics, No. 183, Springer-Verlag (1971), 59-127.

Western Michigan University, Kalamazoo, Michigan, and
Mathematics Research Center, Naval Research Laboratory, Washington, D.C.

STABILITY OF COMPACTNESS FOR FUNCTIONAL DIFFERENTIAL EQUATIONS

G. Stephen Jones

Let X be a metric space, let R^+ denote the nonnegative real numbers, and let $F : R^+ \times X \to X$ be a semi-dynamical system such as might be generated by an autonomous functional differential equation or some other equation of evolution. We shall say that F has <u>compactness stability</u> on $S \subset X$ if there exists a compact K in X such that $F(t,K) \subset K$ for all $t \in R^+$, and for every $\varepsilon > 0$ there exists $t_1 \in R^+$ such that $F(t,S) \subset N_\varepsilon(K)$ for all $t \geq t_1$. An important property of semi-dynamical systems with compactness stability on a closed set S is that all critical points in S and all periodic motions which pass through S are contained in a compact set.

In this paper we shall use some techniques developed in [1] and [2] to show that a general class of neutral functional differential equations with infinite hereditary dependence are compactness stable. For this same class of equation we shall also present new results concerning the existence of periodic solutions and critical points.

Henceforth X denotes a closed convex subset of the space of all continuous functions $z : R^- \to R^n$ where R^- denotes the nonpositive reals, n is a positive integer, and R^n denotes the n-dimensional real vectors. Let $|\cdot|$ denote a standard R^n norm and assume the topology on X determined by the sequence of semi-norms

$$\|z\|_1^k = \sup\{|z(\theta)| : -k \leq \theta \leq (k-1)\}$$

for $k = 1,2,\ldots$. We observe that for any $\tau > 0$ the sequence of semi-norms

$$\|z\|_\tau^k = \sup\{|z(\theta)| : -k\tau \leq \theta \leq -(k-1)\tau\} ,$$

$k = 1,2,\ldots$, defines the same topology on X as $\{\|\cdot\|_1^k\}$. Furthermore, X is a Fréchet space using the quasi-norm

$$(1) \qquad \|z\| = \sum_{i=1}^{\infty} \left(\sum_{j=1}^{\infty} a_j \|z\|_1^j (1 + \|z\|_1^j)^{-1} \right)^i ,$$

where for all j, $a_j > 0$ and $\sum_{j=1}^{\infty} a_j = 1$. For each $\varepsilon > 0$ and $U \subset X$, an ε-neighborhood of U is defined by the formula

$$N(U,\varepsilon) = \{x : \|x-y\| < \varepsilon \text{ for some } y \in U\} .$$

For each $\tau > 0$ and every positive integer k, we also define semi-ε-neighborhoods of U by the formula

$$N_\tau^k(U,\varepsilon) = \{x : \|x-y\|_\tau^k < \varepsilon \text{ for some } y \in U\} .$$

Let $R^\#$ denote the set of nonvoid subsets of the real line R. A function $\alpha : R \times R^n \to R^\#$ such that $\alpha(t,\xi)$ is closed and $t \in \alpha(t,\xi) \subset (-\infty,t]$ for every $t,\xi \in R \times R^n$ we call a lag function. α is said to be autonomous if for each $\xi \in R^n$

$$\alpha(t,\xi) = \{t + \tau : \tau \in \alpha(0,\xi)\} .$$

The maximal lag function is the lag function defined by the formula $\alpha(t,\xi) = (-\infty,t]$ for all t,ξ in $R \times R^n$. The minimal lag function is defined by the formula $\alpha(t,\xi) = \{t\}$ for all t,ξ in $R \times R^n$. Both the maximal and the minimal lag functions are trivially autonomous.

For an arbitrary continuous function $x : R \to R^n$ we denote the restriction of x to $\alpha(t,x(t))$ by $x_t(\alpha)$. If the deviative function $\dot{x}$ is defined we denote its restriction to $\alpha(t,x(t))$ by $\dot{x}_t(\alpha)$. If α represents the maximal lag function, then the notation $x_t(\alpha)$ is further simplified to x_t. If α is the minimal lag function then $x_t(\alpha)$ is expressed by $x(t)$.

Let α and β be two autonomous lag functions and let f be a n-dimensional vector functional defined for each pair $x_t(\alpha)$, $\dot{x}_t(\beta)$. The general class of functional differential equations under consideration here may be expressed in the form

$$\dot{x}(t) = f(x_t(\alpha), \dot{x}_t(\beta)) . \tag{2}$$

To formulate the initial data problem for this equation in the most convenient way, we consider an arbitrary function $\xi : R \to R^n$ and some point $t_0 \geq 0$. A function x such that $x(t) = \xi(t)$ for $t \leq t_0$ and satisfying (2) on some interval $[t_0, t_0+\tau)$, $\tau > 0$, is called a <u>solution</u> corresponding to the <u>initial function</u> ξ at t_0.

Equations generally represented by equations (2) are considered in [3], [4], [5], [6] and elsewhere in regard to existence, uniqueness, continuity with respect to initial data, and continuation of solutions. Without any specific detail, therefore, we assume conditions sufficient to assure existence and uniqueness of solutions for all $t \in R^+$ and their continuous dependence on initial functions considered. In order to avoid over-burdening technical details, we also restrict our attention to the case where α and β are the maximal lag function. That is, we consider only equations of the form

$$\dot{x}(t) = f(x_t, \dot{x}_t) . \tag{3}$$

Our assumptions imply that our functional differential equation generates a semi-dynamical system $F : R^+ \times X \to X$ where the topology for X is as we have already specified. We shall impose conditions on this generated system which imply compactness stability and additional conditions which imply the existence of critical points. Afterwards we will demonstrate that quite general classes of autonomous functional differential equations satisfy these imposed conditions.

Let Ω be the set of all nonvoid compact subsets of X and let $\tau > 0$ be arbitrary but fixed. For each $z \in X$, $K \in \Omega$ and the integers $k = 1,2,\ldots$, we define

$$v_\tau^k(z,K) = \inf\{\|z-w\|_\tau^k : w \in K\} .$$

For arbitrary $U \subset X$ we define

$$\mu_\tau^k(U,K) = \sup\{v_\tau^k(z,K) : z \in U\}$$

and

$$(4) \qquad \lambda_\tau^k(U) = \inf\{\mu_\tau^k(U,K) : K \in \Omega\} .$$

A function $G : R^+ \times X \times X \to X$ is called a splitting of the flow $F : R^+ \times X \to X$ if for all $t \in R^+$ and $x \in X$, $G(t,x,x) = F(t,x)$. F is said to be a Type 1 flow if $F(t,\cdot)$ is $\|\cdot\|$-bounded on $\|\cdot\|$-bounded sets and F has a splitting $G : R^+ \times X \times X \to X$ such that for every closed $\|\cdot\|$-bounded set $S \subset X$ and all $\tau > 0$ sufficiently small the following conditions are satisfied.

(1) For arbitrary $x \in S$, $U \subset S$, and all positive integers k

$$\lambda_\tau^k(G(k\tau,x,U)) = 0 .$$

(2) There exists a sequence of continuous functions $L_k : R^+ \to R^+$ such that $L_k(t) \to 0$ as $t \to 0$ such that

$$\|G(t,x,y) - G(t,z,y)\|_\tau^k \leq L_k(t) \|x - z\|_\tau^k$$

for all x, y and z in S and all positive integers k.

We denote by $X^\#$ the set of all nonvoid subsets of X. A semi-dynamical system $F : R^+ \times X \to X$ is said to constrain a set $S \subset X$ over an interval $I \subset R^+$ if $F(t,S) \subset S$ for all $t \in I$. We can now state our first theorem.

THEOREM 1. Let $F : R^+ \times X \to X$ be a Type 1 semi-dynamical system which constrains a closed $\|\cdot\|$-bounded set $S \subset X$ over some interval $I \subset R^+$. Then F is compactness stable on S. Furthermore, all critical points of F in S and all periodic motions passing through S are contained in a compact set K.

As an application for Theorem 1 let us consider functional differential equations of the form

$$(5) \qquad \dot{x}(t) = f(x_t)$$

where $f : X \to R^n$ maps $\|\cdot\|$-bounded sets into bounded sets in R^n and where conditions are sufficient to assure existence and uniqueness of solutions on R^+ and their continuous dependence on continuous initial data. Choosing an arbitrary function $z \in X$ we have that the solution x of equation (5) corresponding to z as an initial function at 0 is such that

$$x(t) = z(t) \quad \text{for} \quad t \leqslant 0$$

and

$$x(t) = z(0) + \int_0^t f(x_\tau)d\tau \quad \text{for} \quad t > 0 .$$

If $F : R^+ \times X \to X$ is the flow generated by (5), then F is given by the formula

$$F(t,z)(s-t) = z(s) \quad \text{for} \quad s \leqslant 0$$

and

$$F(t,z)(s-t) = z(0) + \int_0^t f(x_t)ds \tag{6}$$

for $s \in (0,t]$. For arbitrary $\tau > 0$, S an arbitrary closed $\|\cdot\|$-bounded set, and each positive integer k let

$$K_\tau^k = \{y : y = x|[-k\tau, -(k-1)\tau], x \in F(k\tau,S)\} .$$

Since f is bounded on $\|\cdot\|$-bounded sets it is clear from formula (6) that there exists $b_1 > 0$ and $a_1 > 0$ such that for all $y \in K_\tau^1$, $\|y\|_\tau^1 \leqslant b_1$ and $|y(s_1) - y(s_2)| \leqslant a_1|s_1 - s_2|$ for all $s \in [-\tau,0]$. By Arzela's Theorem it follows that K_τ^1 is $\|\cdot\|_\tau^1$-compact and therefore $\lambda_\tau^1(F(\tau,S)) = 0$. Since $F(\tau,S)$ is $\|\cdot\|$-bounded we can replace S by $F(\tau,S)$ and K_τ^1 by K_τ^2 and apply the same argument to conclude that $\lambda_\tau^2(F(2\tau,S)) = 0$. In general $F((k-1)\tau,S)$ is $\|\cdot\|$-bounded and by repeated application of the same argument we have that $\lambda_\tau^k(F(k\tau,S)) = 0$ for all positive integers k. $\lambda_\tau^k(F(k\tau,S)) = 0$ clearly implies $\lambda_\tau^k(F(m\tau,S)) = 0$ for all $m \geqslant k$, so it follows that F is Type 1. Using Theorem 1, therefore, we have the following theorem.

<u>THEOREM 2</u>. Let $F : R^+ \times X \to X$ be the semi-dynamical system generated by equation (5) and suppose F constrains a closed $\|\cdot\|$-bounded set

$S \subset X$ over some interval $I \subset R^+$. Then F is compactness stable on S. Furthermore, all critical points of F in S and all periodic solutions of equation (5) passing through S are contained in a compact set K.

Let $h : X^\# \to X^\#$ be the operator which associates with each set $U \subset X$ its closed convex hull $h(U)$. Composing h with a semi-dynamical system $F : R^+ \times X \to X$ we define the convex closure of F to be the flow $\mathcal{F}: R^+ \times X^\# \to X^\#$ defined by the formula $\mathcal{F}(t,U) = \bigcup_{n=1}^{\infty} (hF)^n(\frac{t}{n},U)$ for each $t \in R^+$ and $U \subset X$. F is said to be convexly semi-autonomous if for every $\|\cdot\|$-bounded set $U \subset X$ there exists a $\|\cdot\|$-bounded set $V \subset X$ such that $\mathcal{F}(t,U) \subset F(t,V)$ for all $t \in R^+$. We observe that every linear system F is trivially convexly semi-autonomous. Using $\mathcal{F}$ we can say more about Type 1 systems.

THEOREM 3. Let $F : R^+ \times X \to X$ be a Type 1 convexly semi-autonomous semi-dynamical system which constrains a closed convex $\|\cdot\|$-bounded set $S \subset X$ over some interval $I \subset R^+$. For each $t \in R^+$ let $\Psi_t : X^\# \to X^\#$ be defined by the formula $\Psi_t(U) = h(F(t,U))$ for all $U \subset X$. Then for all t sufficiently large $\Psi_t(S) \subset S$. If

$$(7) \qquad \Gamma = \{t \in R^+ : \Psi_t^k(S) \cup \Psi_t^{k+1}(S) \subset S \text{ for some power } k\},$$

then for each $t \in \Gamma$ the system F has a periodic motion of period t.

Interpreting Theorem 3 in terms of equation (5) we have the following result.

THEOREM 4. Let $F : R^+ \times X \to X$ be the semi-dynamical system generated by equation (5) and suppose F is convexly semi-autonomous and constrains a closed $\|\cdot\|$-bounded set $S \subset X$ over some interval $I \subset R^+$. If Γ is defined by formula (7), then for each $t \in \Gamma$ equation (5) has a periodic solution of period t.

Let $\mathcal{F}: R^+ \times X^\# \to X^\#$ be the convex closure of a semi-dynamical system $F : R^+ \times X \to X$. If for some set $S \subset X$, $\mathcal{F}(t,S) \subset S$ for all t in an interval $I \subset R^+$, then F is said to convexly constrain S

on I. We observe that if S is convex and if F is linear and constrains S, then F convexly constrains S.

THEOREM 5. Let $F : R^+ \times X \to X$ be a Type 1 convexly semi-autonomous semi-dynamical system which convexly constrains a closed convex $\|\cdot\|$-bounded set $S \subset X$ over some interval $I \subset R^+$. Then F has a critical point.

Interpreting Theorem 5 in terms of equation (5) yields our next result.

THEOREM 6. Let $F : R^+ \times X \to X$ be the semi-dynamical system generated by equation (5) and suppose F is convexly semi-autonomous and convexly constrains a closed $\|\cdot\|$-bounded set $S \subset X$. Then equation (5) has a critical point.

A semi-dynamical system $L : R^+ \times X \to X$ is said to be convex on a set $S \subset X$ if $L(t,S)$ is convex for all $t \in R^+$. We observe that if L is linear then L is convex on every convex set $S \subset X$. Let $F : R^+ \times X \to X$ and $H : R^+ \times X \to X$ be two semi-dynamical systems and let $\mathcal{F} : R^+ \times X^\# \to X^\#$ and $\mathcal{H} : R^+ \times X^\# \to X^\#$ be the convex closures of F and H respectively. If $S \subset X$ and $F(t,S) \subset H(t,S)$ for all $t \in R^+$, then H is said to dominate F on S. If $\mathcal{F}(t,S) \subset \mathcal{H}(t,S)$ for all $t \in R^+$, then H is said to convexly dominate F on S.

THEOREM 7. Let $F : R^+ \times X \to X$ be a semi-dynamical system and let $L : R^+ \times X \to X$ be a convex Type 1 semi-dynamical system on $S \subset X$ which convexly dominates F on S where S is closed, convex, and $\|\cdot\|$-bounded. If L constrains S over some interval $I \subset R^+$, then F is compactness stable on S and has a critical point in S.

Consider now a functional differential equation

$$\dot{x}(t) = \ell(x_t) \tag{8}$$

where $\ell : X \to R^n$ maps $\|\cdot\|$-bounded sets into bounded sets in R^n and where conditions are such that equation (8) generates a convex semi-dynamical system on closed convex sets $U \subset X$. Using Theorem 7 we have the following result.

THEOREM 8. Let $F : R^+ \times X \to X$ and $L : R^+ \times X \to X$ be the semi-dynamical systems generated by equation (5) and equation (8) respectively and let L convexly dominate F on a closed convex $\|\cdot\|$-bounded set S in X. If L constrains S over some interval $I \subset R^+$, then equation (5) has a critical point.

Now let $r > 0$ and X be the set of all functions $z : R^- \to R^n$ such that $z|[-r,0]$ is continuous and $z(s) = 0$ for $s < -r$. Let $D : X \to R^n$ be continuous and such that

$$D(z) = \int_{-r}^{0} [d_\theta \eta(\theta)] z(\theta) \ ,$$

$$\det[\eta(0) - \eta(0-)] \neq 0 \ ,$$

and

$$\left| \int_{-s}^{0} [d_\theta \eta(\theta)] z(\theta) \right| \leq \gamma(s) \ \|z\|$$

for all $z \in X$ and $s \in R^+$ where $\eta(\theta)$ is an $n \times n$ matrix function of bounded variation in θ and $\gamma(s)$ is a continuous scalar function for $s \in R^+$ with $\gamma(0) = 0$. Let $f : R^+ \times X \to R^n$ be continuous and map $\|\cdot\|$-bounded sets into bounded sets of R^n. We assume further that the operator D is such that there exists $a > 0$ and $b \geq 1$ such that solutions of the equation

$$D(y_t) = D(y_o) \ , \qquad y_0 = z \in X \ ,$$

satisfy the inequality

$$\|y_t\| \leq be^{-at} \|z\| + b|D(z)|$$

for all $t \in R^+$ and $z \in X$. Considering functional differential equations of the form

$$\text{(9)} \qquad \frac{d}{dt} (D(x_t)) = f(x_t)$$

it is shown in [7] that the semi-dynamical system $F : X \to X$ generated by equation (9) has a linear splitting $G(t,x,y) = J(t,x) + H(t,y)$ where $\|J(t,u) - J(t,v)\| \leq be^{-at}\|u - v\|$ for all $t \in R^+$ and $u,v \in X$. Furthermore for every finite interval $I \subset R^+$ and closed and bounded set $S \subset X$, there exists $M > 0$ such that

$$|H(t,u)(s + \varepsilon) - H(t,u)(s)| \leq M\varepsilon$$

for all $\varepsilon > 0$, $t \in I$, and $u \in X$. It follows easily that F is a Type 1 system. Accordingly Theorem 2, Theorem 4, and Theorem 6 hold with equation (5) replaced by equation (9).

If the semi-dynamical system generated by equation (9) can be shown to be convexly dominated on a closed convex set $S \subset X$ by the semi-dynamical system generated by the equation

(10) $$\frac{d}{dt}(D(x_t)) = \ell(x_t)$$

where $\ell: R^+ \times X \to R^n$ has the properties specified in equation (8), then Theorem 8 holds with equation (9) replacing equation (5) and equation (10) replacing equation (8).

It is also true that the theorems we have presented are applicable to state dependent hereditary functional differential equations of the form

$$\dot{x}(t) = f(x(\alpha_t))$$

where $\alpha : R \times R^n \to R^\#$ is a general continuous lag function such as discussed in more detail in [3]. To develop this extended application here, however, would take us too far afield.

To prove these theorems we shall draw on the Liapunov type theory developed in [1] and [2] and based on the notion of a functional called a compactness gauge defined on $X^\#$. A compactness gauge is positive definite in the sense that it takes on only nonnegative values and vanishes only on a specified family of compact subsets of X. More specifically, if $\overline{R}^+$ denotes the extended positive real numbers, a functional $\lambda : X^\# \to \overline{R}^+$ is defined to be a <u>compactness gauge</u> if the following properties are satisfied:

(1) $\lambda(U) = \lambda(\overline{U})$ and $U \subset V$ implies $\lambda(U) \leqslant \lambda(V)$.

(2) $\overline{U} \in \Omega$ if and only if U is nonvoid and $\lambda(U) = 0$.

(3) If $\{U_n\} \subset X^{\#}$ is a sequence of closed nonvoid sets such that $U_{n+1} \subset U_n$, $n = 1,2,\ldots$, and $\lim_{n\to\infty} \lambda(U_n) = 0$, then $\bigcap_{n=1}^{\infty} U_n$ is nonvoid.

Ω is called the null set of λ. A compactness gauge is said to be <u>convex</u> if for every $U \subset X$, $\lambda(U) = \lambda(h(U))$. Various constructions leading to more generally defined compactness gauges with various null sets and with desired properties of various kinds are presented in [1] and [2]. They are used in the referenced papers to define functionally compact operators and to develop an associated fixed point theory.

A function $\Psi : X^{\#} \to X^{\#}$ is said to be <u>regular</u> if $U \subset V$ implies $\Psi(U) \subset \Psi(V)$. We denote the power mapping of Ψ by Ψ^n, $n = 1,2,\ldots,$ and for each $U \subset X$ define

$$\omega(\Psi,U) = \bigcap_{n=1}^{\infty} \Psi^n(U) .$$

Let $\{X\} \subset X^{\#}$ denote the set of all singleton sets $\{x\} \in X^{\#}$ and let $\Psi : X^{\#} \to X^{\#}$ be regular, have closed images, and be upper semi-continuous when restricted to $\{X\}$. Then Ψ is defined to be <u>strongly functionally compact</u> at $S \in X^{\#}$ if there exists a compactness gauge $\lambda : X^{\#} \to \overline{R^+}$ such that $\lambda(\Psi(S)) < +\infty$ and such that nonvoid closed $U \subset S$ with $\Psi(U) \subset U$ imply $\omega(\Psi,U)$ is nonvoid and $\lambda(\omega(\Psi,U)) = 0$. If Ψ is strongly functionally compact at $S \in X^{\#}$ and $\lambda : X^{\#} \to \overline{R^+}$ is a compactness gauge such that $\lambda(\Psi(S)) < +\infty$ and such that $U \subset S$ and $\Psi(U) \subset U$ imply $\omega(\Psi,U)$ is nonvoid and $\lambda(\omega(\Psi,U)) = 0$, then λ is called a <u>strong Liapunov compactness gauge</u> for Ψ at S.

A function $f : X \to X$ is said to be strongly functionally compact on $S \subset X$ if the function $\Psi : X^{\#} \to X^{\#}$ defined by the formula $\Psi(U) = \overline{f(U)}$ for all $U \subset X$ is strongly functionally compact at S.

It is convenient at this time to state and prove a series of lemmas which we shall use to construct proofs for the theorems we have presented.

LEMMA 1. Let $F : R^+ \times X \to X$ be a semi-dynamical system, $S = \overline{S} \subset X$, $t_1 \in R^+$, and $U = \omega(\overline{F(t_1,\cdot)},S)$. If $F(t_1,S) \subset S$, then $F(t,U) \subset U$ for all points $t \in R^+$ such that $F(t,S) \subset S$. If $F(t,S) \subset S$ for all t is an interval $I \subset R^+$, then there exists $t_2 \in R^+$ such that $F(t,S) \subset S$ and $F(t,U) \subset U$ for all $t \geq t_2$.

Proof. Let t_0 be an arbitrary point in R^+ such that $F(t_0,S) \subset S$. Then

$$\begin{aligned} F(t_0,U) &= F(t_0, \bigcap_{n=1}^{\infty} \overline{F(nt_1,S)}) \\ &\subset \bigcap_{n=1}^{\infty} F(t_0, \overline{F(nt_1,S)}) \\ &\subset \bigcap_{n=1}^{\infty} \overline{F(t_0, F(nt_1,S))} \\ &= \bigcap_{n=1}^{\infty} \overline{F(nt_1, F(t_0,S))} \\ &\subset \bigcap_{n=1}^{\infty} \overline{F(nt_1,S)} = U . \end{aligned}$$

Hence $F(t,U) \subset U$ for all $t \in R^+$ such that $F(t,S) \subset S$.

Let $F(t,S) \subset S$ for all t in an interval $[a,b]$, $b > a$. Then for every positive integer k, $F(t,S) \subset S$ for all $t \in [ka,kb]$. Hence $F(t,S) \subset S$ for all $t \in \bigcup_{k=1}^{\infty} [ka,kb]$. But $kb \geq (k+1)a$ for $k \geq \frac{a}{b-a}$, so setting $t_2 = \frac{a^2}{b-a}$ we have that $[t_2,\infty) \subset \bigcap_{k=1}^{\infty} [ka,kb]$. Thus $F(t,S) \subset S$ for all $t \geq t_2$ and this in turn implies $F(t,U) \subset U$ for all $t \geq t_2$ and our proof is complete. ∎

LEMMA 2. Referring to formulas (1) and (4) for each $\tau > 0$ let $\lambda_\tau : X^\# \to \overline{R}^+$ be defined by the formula

$$\lambda_\tau(U) = \sum_{i=1}^{\infty} \Big(\sum_{j=1}^{\infty} a_j \lambda_\tau^j(U)(1 + \lambda_\tau^j(U))^{-1} \Big)^i \tag{11}$$

for all $U \subset X$. For each $\tau > 0$, λ_τ is a convex compactness gauge. Furthermore, if $\{U_n\} \subset X^\#$ is a sequence of closed sets such that

$U_{n+1} \subset U_n$ for all n and $\lim_{n\to\infty} \lambda_\tau(U_n) = 0$, then for every $\varepsilon > 0$ there exists an integer m and $K \in \Omega$ such that for all $n \geq m$, $U_n \subset N(K,\varepsilon)$.

Proof. Consider arbitrary sets U and V such that $U \subset V$. For arbitrary $\tau > 0$ and every positive integer k we have that

$$\mu_\tau^k(U,K) = \sup\{v_\tau^k(z,K) : z \in U\} \leq \sup\{v_\tau^k(z,K) : z \in V\} = \mu_\tau^k(V,K)$$

for every $K \in \Omega$. Hence clearly $\lambda_\tau^k(U) \leq \lambda_\tau^k(V)$ for $k = 1,2,\ldots$, so it follows that $\lambda_\tau(U) \leq \lambda_\tau(V)$ and in particular $\lambda_\tau(U) \leq \lambda_\tau(\overline{U})$.

Suppose for some set $U \subset X$, $\lambda_\tau(\overline{U}) > \lambda_\tau(U)$. Then for some positive integer k, $\lambda_\tau^k(\overline{U}) > \lambda_\tau^k(U)$, and there must exist $K \in \Omega$ such that

$$\mu^k(U,K) < \lambda_\tau^k(\overline{U}) \leq \mu_\tau^k(\overline{U},K) .$$

This in turn implies there exists $x \in \overline{U}$ such that $v_\tau^k(x,K) > \mu_\tau^k(U,K)$, and consequently there exists $\varepsilon > 0$ such that $v_\tau^k(x,K) - v_\tau^k(z,K) \geq \varepsilon$ for all $z \in U$. But there exists a sequence $\{z_i\} \subset U$ such that $z_i \to x$, so by the continuity of $v_\tau^k(\cdot,K)$ at x no such ε can exist. By contradiction, therefore, $\lambda(\overline{U})$ cannot be greater than $\lambda(U)$ and must therefore equal $\lambda(U)$. Thus λ_τ satisfies property (1) required of a compactness gauge.

Now consider a sequence $\{U_n\}$ of closed sets such that $U_{n+1} \subset U_n$ for $n = 1,2,\ldots$ and $\lim_{n\to\infty} \lambda_\tau(U_n) = 0$. It follows that $\lim_{n\to\infty} \lambda_\tau^j(U_n) = 0$ for $j = 1,2,\ldots$, and for an arbitrarily chosen $\delta > 0$ we choose r such that $\sum_{j=r}^{\infty} a_j < \delta$ and choose U_m such that $\lambda_\tau^j(U_m) < \delta$ for $j = 1,2,\ldots,p$ where $p > \max\{r, \frac{r}{\tau}\}$. There exists $K_\delta^j \in \Omega$ for $j = 1,2,\ldots,p$ such that $\mu_\tau^j(U_m,K_\delta^j) < 2\delta$ or equivalently $U_m \subset N_\tau^j(K_\delta^j,2\delta)$. Defining $K_\delta = \bigcup_{j=1}^{p} K_\delta^j$ it follows by simple calculation that $U_m \subset N(K_\delta,6\delta)$. For arbitrary ε we are free to choose $\delta = \frac{\varepsilon}{6}$, so we have demonstrated that for every $\varepsilon > 0$ there exists m and $K \in \Omega$ such that for all $n \geq m$, $U_n \subset N(K,\varepsilon)$.

To verify that property (2) for a compactness gauge is satisfied by each λ_τ, we observe that $\overline{U} \in \Omega$ by definition implies $U \neq \emptyset$ and $\overline{U}$ is compact. $\overline{U}$ compact by formula (4) clearly implies for $j = 1,2,\ldots$ that $\mu_\tau^j(\overline{U},\overline{U}) = \lambda_\tau^j(\overline{U}) = 0$, so

$$\lambda_\tau(U) = \lambda_\tau(\overline{U}) = \sum_{i=1}^{\infty} \left(\sum_{j=1}^{\infty} a_j \lambda_\tau^j(\overline{U})(1 + \lambda_\tau^j(U))^{-1} \right)^i = 0 .$$

Hence $\overline{U} \in \Omega$ implies $U \neq \emptyset$ and $\lambda_\tau(U) = 0$.

Now suppose $U \neq \emptyset$ and $\lambda_\tau(U) = 0$. If we trivially define a sequence $\{U_n\}$ of sets by set $U_n = U$ for $n = 1,2,\ldots$, then by our previous argument we have that for each $\varepsilon > 0$ there exist $K_\varepsilon \in \Omega$ such that $U \subset N(K_\varepsilon,\varepsilon)$. We can show that $\overline{U} \in \Omega$ by demonstrating that $\overline{U}$ is totally bounded. Let $\rho > 0$ be arbitrary and setting $\varepsilon = \frac{\rho}{5}$, choose $K_\varepsilon \in \Omega$ such that $U \subset N(K_\varepsilon,\varepsilon)$. K_ε is compact so we can choose a finite set of points $\{y_1,\ldots,y_r\} \subset K_\varepsilon$ such that $K_\varepsilon \subset \bigcup_{j=1}^{r} N(y_j, \frac{\rho}{5})$ and hence $\overline{U} \subset \bigcup_{j=1}^{r} N(y_j, \frac{2\rho}{5})$. If $\overline{U} \cap N(y_j, \frac{2\rho}{5}) \neq \emptyset$ choose a point $x_j \in \overline{U} \cap N(y_j, \frac{2\rho}{5})$. If $x \in \overline{U}$, then clearly $x \in N(x_j, \frac{4\rho}{5})$ for some chosen point x_j and it follows that $\overline{U}$ is totally bounded. Thus we have demonstrated that λ_τ has property (2) required of a compactness gauge.

Consider again an arbitrary sequence $\{U_n\}$ of closed sets such that $U_{n+1} \subset U_n$ for $n = 1,2,\ldots$ and $\lim_{n\to\infty} \lambda_\tau(U) = 0$. $U_n = \bigcap_{i=1}^{n} U_i$, so $\lim_{n\to\infty} \lambda_\tau(U) = 0$ implies $\lim_{n\to\infty} \lambda_\tau(\bigcap_{i=1}^{n} U_i) = 0$. Nonvoid $\bigcap_{i=1}^{\infty} U_i \notin \Omega$ would imply by property (2) that $\lambda_\tau(\bigcap_{i=1}^{\infty} U_i) = r > 0$. But $\lim_{n\to\infty} \lambda_\tau(\bigcap_{i=1}^{n} U_i) = 0$ implies there exists k such that $\lambda_\tau(\bigcap_{i=1}^{k} U_i) < \frac{r}{2}$ and since $\bigcap_{i=1}^{\infty} U_i \subset \bigcap_{i=1}^{k} U_i$, $\lambda_\tau(\bigcap_{i=1}^{\infty} U_i) \leqslant \lambda_\tau(\bigcap_{i=1}^{k} U_i) < \frac{r}{2}$. Hence clearly $\bigcap_{i=1}^{\infty} U_i \in \Omega$ if $\bigcap_{i=1}^{\infty} U_i$ is nonvoid.

Suppose $\bigcap_{i=1}^{\infty} U_i$ is void. Then no sequence $\{y_i\}$ with $y_i \in U_i$, $i = 1,2,\ldots$ can be Cauchy. That is, there exist $\varepsilon^* > 0$ such that for every sequence $\{y_i\}$, $y_i \in U_i$, $\operatorname{Diam}\{y_i : i \geqslant n\} \geqslant \varepsilon^*$ for all n. Choose any sequence $\{x_i\}$, $x_i \in U_i$. $\lim_{i\to\infty} \lambda_\tau(U_i) = 0$ and $U_{i+1} \subset U_i$

imply there exist a compact set K and an integer m such that $U_i \subset N(K, \frac{\varepsilon^*}{6})$ for all $i \geq m$. Choosing an $\frac{\varepsilon^*}{6}$-net $\{z_1,\ldots,z_p\}$ for K, it follows that for every point x_i, $i \geq m$, $\|x_i - z_j\| < \frac{\varepsilon^*}{3}$ for some $j \leq p$. It follows that there exists an infinite sequence $\{x_{\nu_i}\} \subset \{x_i\}$ such that for some fixed index j_0, $\|x_{\nu_i} - z_{j_0}\| < \frac{\varepsilon^*}{3}$ for $i = 1,2,\ldots$. Hence $\|x_{\nu_i} - x_{\nu_j}\| < \frac{2}{3}\varepsilon^*$ for all indices ν_i,ν_j. Now define a new series $\{y_m\}$, $m = 1,2,\ldots$ as follows:

$$\begin{aligned} y_m &= x_{\nu_1} && \text{for } 1 \leq m < \nu_1 \\ y_m &= x_m && \text{for } m \in \{\nu_i : i = 1,2,\ldots\} \\ y_m &= x_{\nu_i} && \text{for } \nu_{i-1} < m < \nu_i,\ i = 1,2,\ldots . \end{aligned}$$

Clearly $y_m \in U_m$ for $m = 1,2,\ldots$ and $\|y_{m_1} - y_{m_2}\| < \frac{2\varepsilon^*}{3}$ for all m_1 and m_2. Hence $\text{Diam}\{y_m : m \geq 1\} < \frac{2\varepsilon^*}{3}$ and by contradiction, therefore, we can only conclude that $\bigcap_{i=1}^{\infty} U_i$ is nonvoid and therefore in Ω. This established property (3) for compactness gauge and we have demonstrated that λ_τ is a compactness gauge for each $\tau > 0$.

To complete our proof we have only to prove that for every $\tau > 0$ and $U \subset X$ that $\lambda_\tau(U) = \lambda_\tau(h(U))$. Let x be an arbitrary point in $h(U)$. Then for some n we have that $x = \sum_{i=1}^{n} \alpha_i x_i$ where for each i, $x_i \in U$, $\alpha_i \geq 0$, and $\sum_{i=1}^{n} \alpha_i = 1$. For each $y \in X$ and each integer j clearly

$$\|x - y\|_\tau^j = \|\sum_{i=1}^{n} \alpha_i x_i - y\|_\tau^j \leq \sum_{i=1}^{n} \alpha_i \|x_i - y\|_\tau^s .$$

Consequently for arbitrary $K \in \Omega$

$$\begin{aligned} \nu_\tau^j(x,K) &= \inf\{\|x-y\|_\tau^j : y \in K\} \\ &\leq \sum_{i=1}^{n} \alpha_i \inf\{\|x_i-y\|_\tau^j : y \in K\} \\ &\leq \sum_{i=1}^{n} \alpha_i \nu_\tau^j(x_i,K) . \end{aligned}$$

Thus for all $x \in h(U)$, $\nu_\tau^j(x,K) \leq \mu_\tau^j(U,K)$ which clearly implies $\mu_\tau^j(h(U),K) \leq \mu_\tau^j(U,K)$. Since $\mu_\tau^j(h(U),K) \leq \mu_\tau^j(U,K)$ for arbitrary $K \in \Omega$ it follows that

$$\lambda_\tau^j(h(U)) \leq \lambda_\tau^j(U) .$$

$\lambda_\tau^j(h(U)) \leq \lambda_\tau^j(U)$ for $j = 1,2,\ldots$ obviously implies

$$\lambda_\tau(h(U)) \leq \lambda_\tau(U) .$$

On the other hand $U \subset h(U)$ so property (1) implies $\lambda_\tau(U) \leq \lambda_\tau(h(U))$ and we can only conclude that $\lambda_\tau(h(U)) = \lambda_\tau(U)$. Thus we have demonstrated that λ_τ for every $\tau > 0$ is a convex compactness gauge and our proof is complete. ∎

LEMMA 3. Let $F : R^+ \times X \to X$ be a Type 1 semi-dynamical system. If S is a closed $\|\cdot\|$-bounded set in X and λ_τ is as defined by formula (11), then for all τ sufficiently small $\lim_{t\to\infty} \lambda_\tau(F(t,S)) = 0$. Furthermore, for all τ sufficiently small λ_τ is a strong Liapunov compactness gauge for $F(t,\cdot)$ for any $t \geq \tau$ on S.

Proof. Since $F : R^+ \times X \to X$ is a Type 1 semi-dynamical system it follows that $F(t,\cdot)$ is $\|\cdot\|$-bounded on $\|\cdot\|$-bounded sets. Hence given any $K \in \Omega$, clearly $\mu_\tau^k(F(t,S),K) < \infty$ and therefore $\lambda_\tau^k(F(t,S)) < \infty$. Let $t \in R^+$ be arbitrary but fixed; we define $\lambda_\tau^k(F(t,S)) = M_k$ for $k = 1,2,\ldots$. For ε such that $0 < \varepsilon < \frac{1}{2}$ choose $m \ni \sum_{j=m}^{\infty} a_j \leq \varepsilon$ and define

$$M = \max\{M_k : k = 1,\ldots,m\} .$$

$$\begin{aligned} \frac{\lambda_\tau(F(t,S))}{1+\lambda_\tau(F(t,S))} &= \sum_{j=1}^{m} a_j \frac{\lambda_\tau^j(F(t,S))}{1+\lambda_\tau^j(F(t,S))} \\ &\leq \sum_{j=1}^{m-1} a_j \left(\frac{M}{1+M}\right) + \sum_{j=m}^{\infty} a_j \\ &\leq \frac{M}{1+M} + \frac{1}{1+M}\varepsilon = \frac{M+\varepsilon}{M+1} . \end{aligned}$$

Hence

$$\lambda_\tau(F(t,S)) \leqslant \frac{M+\varepsilon}{1-\varepsilon} < \infty \quad .$$

$F : R^+ \times X \to X$ has a splitting $G : R^+ \times X \times X \to X$ such that for all $\tau > 0$ sufficiently small, $x \in S$, $U \subset S$,

$$\lambda_\tau^k(G(k\tau,x,U) = 0$$

for $k = 1,2,\ldots$. Also there exists a sequence of continuous functions $L_k : R^+ \to R^+$, $k = 1,2,\ldots$, such that $L_k(t) \to 0$ as $t \to \infty$ and

$$\|G(t,x,y) - G(t,z,y)\|_\tau^k \leqslant L_k(t) \|x - z\|_\tau^k$$

for all x, y and z in S and all k. Choose a fixed $x \in S$, let y range over S, and w range over an arbitrary set $K \in \Omega$.

$$\begin{aligned}\|G(t,y,y)-G(t,x,x)-w\|_\tau^k &\leqslant \|G(t,y,y)-G(t,x,y)\|_\tau^k + \|G(t,x,y)-G(t,x,x)-w\|_\tau^k \\ &\leqslant L_k(t)\|y-x\|_\tau^k + \|G(t,x,y)-G(t,x,x)-w\|_\tau^k ,\end{aligned}$$

so clearly

$$\nu_\tau^k(G(t,y,y),G(t,x,x)+K) \leqslant L_k(t)\|y-x\|_\tau^k + \nu_\tau^k(G(t,x,y),G(t,x,x)+K) ,$$

and

$$\mu_\tau^k(G(t,S,S),G(t,x,x)+K) \leqslant L_k(t)\mathrm{diam}(S) + \mu_\tau^k(G(t,x,S),G(t,x,x)+K) .$$

Since K was chosen arbitrarily in Ω, it follows that

$$\lambda_\tau^k(G(t,S,S)) \leqslant L_k(t)\mathrm{diam}(S) + \lambda_\tau^k(G(t,x,S)) ,$$

and consequently

$$\lambda_\tau^k(F(t,S)) \leqslant L_k(t)\mathrm{diam}(S) + \lambda_\tau^k(G(t,x,S)) .$$

By our Type 1 condition (1) it follows that

$$\lambda_\tau^k(F(t,S)) \leqslant L_k(t)\mathrm{diam}(S)$$

for all $t \geqslant k\tau$.

Now for arbitrary $\varepsilon \in (0, \frac{1}{2})$ choose m to be such that $\sum_{j=m}^{\infty} a_j < \frac{\varepsilon}{3}$ and let $L(t) = \max\{L_1(t), L_2(t), \ldots, L_{m-1}(t)\}$.

Obviously $L(t) \to 0$ as $t \to \infty$ and

$$\frac{\lambda_\tau(F(t,S))}{1+\lambda_\tau(F(t,S))} \leq \sum_{j=1}^{m-1} a_j \lambda_\tau^j(F(t,S)) + \frac{\varepsilon}{3}$$

$$\leq L(t)\mathrm{diam}(S) + \frac{\varepsilon}{3}$$

for all $t \geq m\tau$. Accordingly we can choose $t_1 > m\tau$ such that $L(t)\mathrm{diam}(S) < \frac{\varepsilon}{3}$ for all $t \geq t_1$ and

$$\lambda_\tau(F(t,S)) < \varepsilon .$$

Hence we have demonstrated that $\lim_{t\to\infty} \lambda_\tau(F(t,S)) = 0$.

Now suppose $t_0 > \tau$ and nonvoid closed $U \subset S$ are such that $F(t_0,U) \subset U$ and for every $V \subset S$ define $\Psi(V) = \overline{F(t_0,V)}$. Using continuity and the semi-group property of F it follows that

$$\Psi^n(U) \subset \overline{F(nt_0,U)}$$

for $n = 1,2,\ldots$. Hence $\lambda_\tau(\Psi^n(U)) \leq \lambda_\tau(F(nt_0,U))$ for $n = 1,2,\ldots$ which in turn implies

$$\lim_{n\to\infty} \lambda_\tau(\Psi^n(U)) = 0 .$$

Since $\Psi^{n+1}(U) \subset \Psi^n(U)$ for $n = 1,2,\ldots$ it follows from the fact that λ_τ is a compactness gauge that

$$\bigcap_{n=1}^{\infty} \Psi^n(U) = \omega(\Psi,U) \neq \emptyset .$$

Hence $F(t_0,\cdot)$ is strongly functionally compact on S and accordingly λ_τ is a strong Liapunov compactness gauge for $F(t_0,\cdot)$ on S and our proof is complete. ∎

<u>LEMMA 4</u>. Let $F : R^+ \times X \to X$ be a semi-dynamical system which convexly constrains a closed convex set $S \subset X$ on an interval $[t_1,t_2]$, $t_2 > t_1 > 0$. If n is any positive integer such that $\frac{t_1}{n} < t_2 - t_1$ and $V_n = \omega((hF)^n(\frac{t_1}{n},\cdot),S)$, then

$$(hF)^n(\tfrac{t_1}{n},V_n) \cup (hF)^n(\tfrac{t_1}{n},hF(\tfrac{t_1}{n},V_n)) \subset V_n .$$

Proof. We observe first that

$$\begin{aligned}(hF)^n(\tfrac{t_1}{n},V_n) &= (hF)^n(\tfrac{t_1}{n},\bigcap_{i=1}^{\infty}(hF)^{in}(\tfrac{t_1}{n},S))\\ &\subset \bigcap_{i=1}^{\infty}(hF)^n(\tfrac{t_1}{n},(hF)^{in}(\tfrac{t_1}{n},S))\\ &= \bigcap_{i=1}^{\infty}(hF)^{in}(\tfrac{t_1}{n},(hF)^n(\tfrac{t_1}{n},S))\\ &\subset \bigcap_{i=1}^{\infty}(hF)^{in}(\tfrac{t_1}{n},S) = V_n .\end{aligned}$$

Also

$$\begin{aligned}(hF)^n(\tfrac{t_1}{n},hF(\tfrac{t_1}{n},V_n)) &= (hF)^{n+1}(\tfrac{t_1}{n},V_n)\\ &= (hF)^{n+1}(\tfrac{t_1}{n},\bigcap_{i=1}^{\infty}(hF)^{in}(\tfrac{t_1}{n},S))\\ &\subset \bigcap_{i=1}^{\infty}(hF)^{n+1}(\tfrac{t_1}{n},(hF)^{in}(\tfrac{t_1}{n},S))\\ &= \bigcap_{i=1}^{\infty}(hF)^{in}(\tfrac{t_1}{n},(hF)^{n+1}(\tfrac{t_1}{n},S))\\ &= \bigcap_{i=1}^{\infty}(hF)^{in}(\tfrac{t_1}{n},(hF)^n(\tfrac{t_1}{n},hF(\tfrac{t_1}{n},S))) .\end{aligned}$$

But since F convexly constrains S on $[t_1,t_2]$ and $t_1 < \frac{t_1}{n}(n+1) < t_2$, it follows that

$$(hF)^n(\tfrac{t_1}{n},hF(\tfrac{t_1}{n},S)) \subset S ,$$

and

$$\begin{aligned}(hF)^n(\tfrac{t_1}{n},hF(\tfrac{t_1}{n},V_n)) &\subset \bigcap_{i=1}^{\infty}(hF)^{in}(\tfrac{t_1}{n},(hF)^n(\tfrac{t_1}{n},hF(\tfrac{t_1}{n},S)))\\ &\subset \bigcap_{i=1}^{\infty}(hF)^{in}(\tfrac{t_1}{n},S) = V_n\end{aligned}$$

and our proof is complete. ∎

LEMMA 5. Let $F : R^+ \times X \to X$ be a semi-dynamical system, $t_1 \in R^+$, $U \subset X$, and k be a positive integer. If $V = \omega((hF)^k(t_1,U))$ and $(hF)^k(t_1,U) \subset U$, then $(hF)^k(t_1,V) \subset V$.

Proof. We observe that

$$\begin{aligned}(hF)^k(t_1,V) &= (hF)^k(t_1, \bigcap_{n=1}^{\infty}(hF)^n(t_1,U)) \\ &\subset \bigcap_{n=1}^{\infty}(hF)^k((hF)^n(t_1,U)) \\ &= \bigcap_{n=1}^{\infty}(hF)^n((hF)^k(t_1,U)) .\end{aligned}$$

Hence if $(hF)^k(t_1,U) \subset U$ clearly

$$(hF)^k(t_1,V) \subset \bigcap_{n=1}^{\infty}(hF)^n(t_1,U) = V . \qquad \blacksquare$$

LEMMA 6. Let $F : R^+ \times X \to X$ be a semi-dynamical system, $t_1 \in R^+$, $U \subset X$, and k be a positive integer. If $F(kt_1,U) \cup F((k+1)t_1,U) \subset U$, then $\bigcup_{n=m}^{\infty} F(nt_1,U) \subset U$ and $\bigcap_{n=0}^{m-1} F(nt_1,U)$ is nonvoid for all $m \geq k^2$. If $(hF)^k(t_1,U) \cup (hF)^{k+1}(t_1,U) \subset U$, then $\bigcup_{n=m}^{\infty}(hF)^n(t_1,U) \subset U$ and $\bigcap_{n=0}^{m-1}(hF)^n(t_1,U)$ is nonvoid for all $m \geq k^2$.

Proof. $F(kt_1,U) \subset U$ and $F((k+1)t_1,U) \subset U$ imply $F(rt,U) \subset U$ for all r which can be expressed in the form $r = a_1k + a_2(k+1)$ where a_1 and a_2 are arbitrary nonegative integers. Similarly $(hF)^k(t_1,U) \subset U$ and $(hF)^{k+1}(t_1,U) \subset U$ imply $(hF)^r(t_1,U) \subset U$ for all r of the form $a_1k + a_2(k+1)$. But every integer $m \geq k^2$ can be expressed in this form. Hence for all $m \geq k^2$, $F(kt,U) \cup F(k(t+1),U) \subset U$ implies

$$\bigcup_{n=m}^{\infty} F(nt_1,U) \subset U \tag{12}$$

and $(hF)^k(t_1,U) \cup (hF)^{k+1}(t_1,U) \subset U$ implies

$$\bigcup_{n=m}^{\infty} (hF)^n(t_1,U) \subset U .$$

Clearly (12) implies

$$\bigcup_{n=m+1}^{\infty} F(nt_1,U) \subset F(t_1,U)$$

$$\bigcup_{n=m+2}^{\infty} F(nt_1,U) \subset F(2t_1,U)$$

$$\vdots$$

$$\bigcup_{n=2m-1}^{\infty} F(nt_1,U) \quad F((m-1)t_1,U) .$$

Hence $\bigcup_{n=2m-1}^{\infty} F(nt_1,U) \subset \bigcap_{n=0}^{m-1} F(nt_1,U)$, so $\bigcap_{n=0}^{m-1} F(nt_1,U)$ is nonvoid if $F(kt,U) \cup F((k+1)t,U) \subset U$. Moreover $(hF)^k(t_1,U) \cup (hF)^{k+1}(t_1,U) \subset U$ implies $F(kt,U) \cup F((k+1)t,U) \subset U$ implies $\bigcap_{n=0}^{m-1} F(nt_1,U)$ is nonvoid, which finally implies $\bigcap_{n=0}^{m-1} (hF)^n(t_1,U)$ is nonvoid, so our proof is complete. ⌘

<u>LEMMA 7</u>. Let $F : R^+ \times X \to X$ be a semi-dynamical system and let $L : R^+ \quad X \to X$ be a convex Type 1 semi-dynamical system on $S \subset X$ which convexly dominates F on S where S is closed, convex, and $\|\cdot\|$-bounded. For every $t \in R^+$ such that $L(t,S) \subset S$, $K_n = \bigcap_{k=1}^{\infty} (hF)^{kn}(\frac{t}{n},S)$ for $n = 1,2,...$ is a sequence of nonvoid compact convex sets.

<u>Proof</u>. Each set K_n is obviously convex, so we need only to show that each is nonvoid and compact. Consider an arbitrary fixed $t > 0$ such that $L(t,S) \subset S$ and let $\mathcal{L}: R^+ \times X^{\#} \to X^{\#}$ be defined by the formula $\mathcal{L}(U) = \overline{L(t,U)}$ for all $U \subset X$. Since L is Type 1 it follows from Lemma 3 that $\bigcap_{k=1}^{\infty} \mathcal{L}^k(S)$ is nonvoid and compact, and for $\tau < t$, $\lim_{k\to\infty} \lambda_\tau(\mathcal{L}^k(t,S)) = 0$. Since L convexly dominates F on S, we have that $(hF)^{kn}(\frac{t}{n},S) \subset \mathcal{L}^k(t,S)$ for $k = 1,2,...$ so $\lim_{k\to\infty} \lambda_\tau((hF)^{kn}(\frac{t}{n},S)) \to 0$. Furthermore, $(hF)^n(\frac{t}{n},S) \subset \overline{L(t,S)} \subset S$ implies $(hF)^{(k+1)n}(\frac{t}{n},S) \subset (hF)^{kn}(\frac{t}{n},S)$ for $k = 1,2,...$. It follows from the fact that λ_τ is a compactness gauge that $\bigcap_{k=1}^{\infty} (hF)^{kn}(\frac{t}{n},S)$ is a nonvoid compact set and our proof is complete. ⌘

LEMMA 8. Let $V_1, V_2, \ldots, V_k$ be a finite set of subsets of X such that $\bigcap_{i=1}^{k} V_i$ is nonvoid. Then $\bigcup_{i=1}^{k} h(V_i)$ is a retract of $h(\bigcup_{i=1}^{k} h(V_i))$.

Lemma 8 is a special case of a theorem of Nussbaum presented in [8] and will not be proved here.

LEMMA 9. Let $F : R^+ \times X \to X$ be a semi-dynamical system and let $K \subset X$ be compact. Suppose there exists a sequence $\{t_n\} \subset R^+$ such that $\{t_n\} \to 0$ as $n \to \infty$ and a sequence $\{x_n\} \subset K$ such that $F(t_n, x_n) = x_n$ for all n. Then F has a critical point in K.

Proof. Since K is compact clearly the sequence $\{x_n\}$ must contain a subsequence $\{y_n\}$ which converges to a point $x^* \in K$ and such that for each n, $F(s_n, y_n) = y_n$ for some $s_n > 0$ such that $s_n \leq t_n$. Suppose x^* is not a critical point of F. Then there must exist $s \in R^+$ such that $F(s, x^*) \neq x^*$ and we can choose a closed neighborhood $N(F(s, x^*))$ of $F(s, x^*)$ not containing x^*. By continuity we can choose a neighborhood $N_0(x^*)$ of x^* with $N_0(x^*) \cap N(F(s, x^*))$ void and $\varepsilon > 0$ such that $x \in N_0(x^*)$ and $|t - s| < \varepsilon$ imply $F(t, x) \in N(F(s, x^*))$. Finally we can choose n such that $F(s_n, y_n) = y_n$, $y_n \in N_0(x^*)$, and for some positive integer k, $|ks_n - s| < \varepsilon$. But $F(ks_n, y_n) = y_n$ so y_n must be also contained in $N(F(s, x^*))$. But this is a contradiction since $N_0(x^*) \cap N(F(s, x^*))$ is void. Therefore we can only conclude that our supposition that $F(s, x^*) \neq x^*$ cannot hold and we have that $F(t, x^*) = x^*$ for all $t \in R^+$. We thus have that $x^* \in K$ is a critical point of F and our proof is complete. ∎

THEOREM 1. Let $F : R^+ \times X \to X$ be a Type 1 semi-dynamical system which constrains a closed $\|\cdot\|$-bounded set $S \subset X$ over some interval $I \subset R^+$. Then F is compactness stable on S. Furthermore, all critical points of F in S and all periodic motions passing through S are contained in a compact set K.

Proof. Since F constrains S over an interval we have by Lemma 1 that there exists $t_1 > 0$ such that $F(t, S) \subset S$ for all $t \geq t_1$ and if $U = \omega(\overline{F(t_1, \cdot)}, S)$, then $F(t, U) \subset U$ for all $t \geq t_1$. Since F is a Type 1 semi-dynamical system, we have by Lemma 3 that for $\tau > 0$ and sufficiently small that λ_τ is a strong Liapunov compactness gauge for F on S. Since $F(t_1,S) \subset S$ it follows that $U \in \Omega$. By Lemma 2 it follows that for every $\varepsilon > 0$ and t sufficiently large that $F(t,S) \subset N_\varepsilon(U)$. Now let $K = \{F(t,x) : t \in [0,t_1]$ and $x \in U\}$. Clearly K is compact and $F(t,K) \subset K$ for all $t \in R^+$. Furthermore, since $N_\varepsilon(U) \subset N_\varepsilon(K)$ it follows that for all $\varepsilon > 0$ and t sufficiently large that $F(t,S) \subset N_\varepsilon(K)$. Hence F is compactness stable on S.

If for some $t_0 > 0^+$ and some $x \in S$ we have that $F(t_0,x) = x$, then we observe that $F^k(t_0,x) = F(kt_0,x) = x$ for all positive integers k. Clearly, therefore, since for each $\varepsilon > 0$ we have that $F(t,S) \subset N_\varepsilon(K)$ for all t sufficiently large, it follows that all critical points in S and all periodic motions passing through S must be contained in K and our proof is complete. ∎

THEOREM 3. Let $F : R^+ \times X \to X$ be a Type 1 convexly semi-autonomous semi-dynamical system which constrains a closed convex $\|\cdot\|$-bounded set $S \subset X$ over some interval $I \subset R^+$. For each $t \in R^+$ let $\Psi_t : X^\# \to X^\#$ be defined by the formula $\Psi_t(U) = h(F(t,U))$ for all $U \subset X$. Then for all t sufficiently large $\Psi_t(S) \subset S$. If

$$\Gamma = \{t \in R^+ : \Psi_t^k(S) \cup \Psi_t^{k+1}(S) \subset S \text{ for some power } k\} ,$$

then for each $t \in \Gamma$ the system F has a periodic motion of period t.

Proof. If $\mathscr{F} : R^+ \times X^\# \to X^\#$ denotes the closed convex hull of F, then clearly $(hF)^n(t,S) \subset \mathscr{F}(nt,S)$ for all positive inters n and all $t \in R^+$. Furthermore since F is convexly semi-autonomous there exists a closed $\|\cdot\|$-bounded set $Y \subset X$ such that $\mathscr{F}(nt,S) \subset F(nt,Y)$ for all n and t. Hence

$$(hF)^n(t,S) \subset \mathscr{F}(nt,S) \subset F(nt,Y) ,$$

so by Lemma 3 we have for $t > 0$ and τ sufficiently small that $\lim_{n\to\infty} \lambda_\tau(F(nt,Y)) = 0$ which in turn implies $\lim_{n\to\infty} (hF)^n(t,S) = 0$.

Let $t \in \Gamma$ and for all integers $n \geq 0$ define

$$U_n = \bigcap_{i=0}^{n-1} (hF)^i(t,S) .$$

Since by Lemma 3 we have that λ_τ for τ sufficiently small is a compactness gauge it follows that

$$V = \bigcap_{n=1}^{\infty} U_n = \bigcap_{i=0}^{\infty} (hF)^i(t,S)$$

is a nonvoid compact set. V is also clearly convex.

For $t \in \Gamma$ we have for some integer $k > 0$ that $(hF)^k(t,S) \cup (hF)^{k+1}(t,S) \subset S$ so Lemma 5 implies $(hF)^k(t,V) \cup (hF)^{k+1}(t,V) \subset V$. By Lemma 6 it follows that for $m = k^2$ that

$$\bigcup_{n=m}^{\infty} (hF)^n(t,V) \subset V$$

and $\bigcap_{n=0}^{m-1} (hF)^n(t,V)$ is nonvoid. Thus by Lemma 8 we have that

$$K = \bigcup_{n=0}^{m-1} (hF)^n(t,V)$$

is a nonvoid compact retract of a convex set.

$$\begin{aligned} F(t,K) &= F\Big(\bigcup_{n=0}^{m-1}((hF)^n(t,V))\Big) \\ &\subset \bigcup_{n=0}^{m-1} (hF)^{n+1}(t,V) \\ &\subset \bigcup_{n=0}^{m-1} (hF)^n(t,V) = K, \end{aligned}$$

so it follows from the fact that the fixed point property is preserved under retraction and the Tychonov fixed point theorem that there exists $x^* \in K$ such that $F(t,x^*) = x^*$. From the semi-group property of F it follows that $F(t,x^*)$ is a periodic motion and our proof is complete. ∎

THEOREM 5. Let $F : R^+ \times X \to X$ be a Type 1 convexly semi-autonomous semi-dynamical system which convexly constrains a closed convex $\|\cdot\|$-bounded set $S \subset X$ over some interval $I \subset R^+$. Then F has a critical point.

<u>Proof</u>. By hypothesis, F convexly constrains and therefore constrains S on an interval $[t_1,t_2]$, $0 < t_1 < t_2$. It follows from Theorem 1

that F is compactness stable on S. Also since F is Type 1, we have by Lemma 1 and Lemma 3 that $V = \omega(\overline{F(t_1,\cdot)},S)$ is a nonvoid compact set.

Choose an arbitrary positive integer n such that $\frac{t_1}{n} < t_1 - t_2$ and let $V_n = \omega((hF)^n(\frac{t_1}{n},\cdot),S)$. Since F convexly constrains S on $[t_1,t_2]$ we have by Lemma 4 that

$$(hF)^n(\tfrac{t_1}{n},V_n) \cup (hF)^n(\tfrac{t_1}{n},hF(\tfrac{t_1}{n},V_n)) \subset V_n \ . \tag{13}$$

By Lemma 6 it follows that for $m \geqslant n^2$ that

$$\bigcup_{k=m}^{\infty} (hF)^k(\tfrac{t_1}{n},V_n) \subset V_n$$

and

$$\bigcap_{k=0}^{m-1} (hF)^k(\tfrac{t_1}{n},V_n) \neq \emptyset \ . \tag{14}$$

If $\mathscr{F}: R^+ \times X^\# \to X^\#$ denotes the closed convex hull of F, then $(hF)^k(\frac{t_1}{n},S) \subset \mathscr{F}(\frac{kt_1}{n},S)$ for all positive integers k. Furthermore since F is convexly semi-autonomous there exists a closed $\|\cdot\|$-bounded set $Y \subset X$ such that $\mathscr{F}(\frac{kt_1}{n},S) \subset F(\frac{kt_1}{n},Y)$ for all k. Hence $(hF)^k(\frac{t_1}{n},S) \subset \mathscr{F}(\frac{kt_1}{n},S) \subset F(\frac{kt_1}{n},Y)$, so by Lemma 3 we have for τ sufficiently small that $\lim_{n\to\infty} \lambda_\tau(F(\frac{kt_1}{n},Y)) = 0$ which in turn implies $\lim_{n\to\infty} \lambda_\tau((hF)^k(\frac{t_1}{n},S)) = 0$.

For all integers $k \geqslant 1$ define

$$U_k = \bigcap_{i=0}^{k-1} (hF)^i(\tfrac{t_1}{n},S) \ .$$

By (14) each U_k is a nonvoid closed set and $U_{k+1} \subset U_k$, so Lemma 3 implies

$$V_n = \bigcap_{k=1}^{\infty} U_k = \bigcap_{i=0}^{\infty} (hF)^i(\tfrac{t_1}{n},S)$$

is a nonvoid compact convex set. Now (14) and Lemma 8 imply $\bigcup_{i=0}^{n-1} (hF)^i(\frac{t_1}{n},V_n)$ is a retract of a compact convex set. Since

$$F(\tfrac{t_1}{n},\bigcup_{i=0}^{n-1}(hF)^i(\tfrac{t_1}{n},V_n)) \subset \bigcup_{i=0}^{n-1}(hF)^i(\tfrac{t_1}{n},V_n) \ ,$$

it follows from the conservation of the fixed point property under retraction and the Tychonov fixed point theorem that there exists $x_n \in V_n$ such that $F(\frac{t_1}{n}, x_n) = x_n$. Since F is convexly constrained on S it follows that x_n for each integer $n \geq 1$ is contained in the compact set V. Accordingly it follows from Lemma 9 that F has a critical point and our proof is complete. ∎

THEOREM 7. Let $F : R^+ \times X \to X$ be a semi-dynamical system and let $L : R^+ \times X \to X$ be a convex Type 1 semi-dynamical system on $S \subset X$ which convexly dominates F on S where S is closed, convex, and $\|\cdot\|$-bounded. If L constrains S over some interval $I \subset R^+$, then F is compactness stable on S and has a critical point in S.

Proof. Since L constrains S on an interval I it follows from Lemma 1 that there exists $t_1 > 0$ such that $L(t,S) \subset S$ for all $t \geq t_1$ and if $U = \omega(L(t_1,\cdot),S)$, then $L(t,U) \subset U$ for all $t \geq t_1$. Since L is a Type 1 semi-dynamical system, we have by Lemma 3 that for $\tau > 0$ and sufficiently small that λ_τ is a strong Liapunov compactness gauge for L on S. $L(t_1,S) \subset S$ implies $U \in \Omega$ and the convexity of L implies that U is convex. If $\mathcal{F}: R^+ \times X^\# \to X^\#$ denotes the closed convex hull of F, then since L convexly dominates F on S we have that $\mathcal{F}(t,U) \subset L(t,U) \subset U$ for all $t \geq t_1$. Thus Lemma 2 implies that for arbitrary $\varepsilon > 0$ that $\mathcal{F}(t,S) \subset L(t,S) \subset N_\varepsilon(U)$ for all t sufficiently large, so F is compactness stable.

Choosing an arbitrary positive integer n we have from the fact that L convexly dominates F on S that $(hF)^m(\frac{t_1}{n}, U) \subset U$ for all $m \geq n$. Lemma 6 implies $\bigcap_{i=0}^{n-1} (hF)^i(\frac{t_1}{n}, U)$ is nonvoid and by Lemma 8 we have that $\bigcup_{i=0}^{n-1} (hF)^i(\frac{t_1}{n}, U)$ is a retract of a compact convex set. Since $(hF)^n(\frac{t_1}{n}, U) \subset U$ it follows that

$$F(\frac{t_1}{n}, \bigcup_{i=0}^{n-1} (hF)^i(\frac{t_1}{n}, U)) \subset \bigcup_{i=0}^{n-1} (hF)^i(\frac{t_1}{n}, U) .$$

Since the fixed point property is preserved under retraction we have by the Tychonov fixed point theorem that there exists $x_n \in U$ such that

$F(\frac{t_1}{n}, x_n) = x_n$ for all integers $n \geqslant 1$. Hence by Lemma 9 F has a critical point in U and our proof is complete. ∎

REFERENCES

[1] G.S. JONES, A functional approach to fixed point analysis of non-compact operators, Tech. Note BN-677, Univ. of Maryland, 1970.

[2] G.S. JONES, Functionally compact operators, Tech. Note BN- , Univ. of Maryland, 1971.

[3] G.S. JONES, Hereditary structure in differential equations, Math Systems Theory, Vol. 1, No. 3. Springer-Verlag, New York, 1967, pp. 263-278.

[4] R.D. DRIVER, Existence theory for delay-difference systems, Contributions to Differential Equations, Vol. 1, 1963, pp. 317-336.

[5] R.D. DRIVER, Existence and continuous dependence of solutions of a neutral functional-differential equation, Arch. Rat. Mech. Anal. 19 (1965), 147-166.

[6] J.K. HALE and M.A. CRUZ, Existence, uniqueness and continuous dependence for hereditary systems, Annali di Matematica Pura ed Appl., 1970.

[7] J.K. HALE, A class of neutral equations with the fixed point property, Proc. Nat. Acad. Sci. 67, No. 1 (1970), 136-137.

[8] R.D. NUSSBAUM, The fixed point index for local condensing maps, Annali di Matematica Pura ed Appl., 1971.

University of Maryland, College Park, Maryland

KRONECKER INVARIANTS AND FEEDBACK*

R. E. Kalman

The purpose of this short note is to discuss the role of the so-called column indices of a singular pencil of matrices (introduced by Kronecker [8]) in the structure theory of linear constant dynamical systems, with special attention to the concept of "feedback". A complete exposition of the subject will appear in a future paper.

1. Definition of Invariants

We consider a finite-dimensional, linear, constant, discrete-time dynamical system Σ over an arbitrary field k,

$$x(t+1) = Fx(t) + Gu(t) ,$$

where $x(t) \in k^n = X$, $u(t) \in k^m = U$, and $t \in \mathbf{Z}$ = integers. (The output behavior of this system will not be of interest here.) We wish to associate with the pair of matrices (F, G) defining Σ certain invariants with respect to equivalence classes defined by the following collection of transformations:

(I) $(F, G) \mapsto (AFA^{-1}, AG)$, $\det A \neq 0$ (change of basis in X);
(II) $(F, G) \mapsto (F, GB)$, $\det B \neq 0$ (change of basis in U);
(III) $(F, G) \mapsto (F - GL, G)$, L = arbitrary $m \times n$ matrix (control law).

*Partially supported by the U. S. Air Force under Grant AFOSR 70-1898.

The family of all such transformations (applied in an arbitrary sequence) forms an abstract group **G**, which may be thought of as acting linearly on $k^{(m+n)n}$, as defined by the preceding formulas. The statement "(F, G) can be transformed to $(\hat{F}, \hat{G})$ via (I-III)" defines an equivalence relation $\sim$ on the set of (F, G)'s; the equivalence classes are usually called <u>orbits</u> of **G**. We want to find certain "entities" dependent on (F, G) which (i) are preserved under the action of the group **G** and (ii) allow us to decide whether or not (F, G) and $(\hat{F}, \hat{G})$ belong to the same orbit of **G**. Such "entities" constitute what is loosely called a "complete set of invariants" in the classical literature.

The construction of the invariants will require two steps.

<u>Step 1</u>. Consider the following blocks of column vectors:

$$(1) \qquad \begin{cases} g_1, \ldots, g_m & \text{(the columns of } G\text{);} \\ Fg_1, \ldots, Fg_m & \text{(the columns of } FG\text{);} \\ F^2g_1, \ldots, F^2g_m; & \\ \quad\ldots & \end{cases}$$

and so on. Let

$\ell_1 = m' = \text{rank } G \leqq m =$ the number of linearly independent vectors in block 1;

$\ell_2 =$ number of linear independent vectors in block 2, modulo (that is, independent also of) the vectors of block 1;

$\ell_3 =$ number of linearly independent vectors in block 3, modulo the vectors of the preceding blocks;

etc. There is an integer r such that <u>all</u> vectors of block $r+1$ are linearly dependent on the vectors in preceding blocks. (By the hypothesis of finite-dimensionality, r is finite; more precisely, the Hamilton-Cayley theorem shows that $r \leqq n$.) A direct computation shows that the numbers $\ell_1, \ldots, \ell_r$ <u>are invariant under</u> **G**. Note that $\ell_1 \geqq \cdots \geqq \ell_r$.

<u>Step 2</u>. Consider an $m \times r$ rectangular "crate" with mr compartments. Put ℓ_1 crosses $\times$ into ℓ_1 of the m compartments of the first row

of the crate (exactly one $\times$ per compartment but otherwise arbitrarily). Put ℓ_2 crosses in row 2 of the crate subject to the restriction that each cross in the second row lies below a cross of the first row. Continue. Add up the number of $\times$ in each column of the crate. This defines exactly m' positive integers $\kappa_1, \ldots, \kappa_{m'}$. Obviously,

$$(2) \qquad \begin{cases} \kappa_1 + \kappa_2 + \cdots + \kappa_{m'} \leqq n , \\ \kappa_i > 0, \quad i = 1, \ldots, m' . \end{cases}$$

The κ_i are invariant under **G**, because the ℓ_i are invariant. We shall call the κ_i the <u>control invariants</u> of $\Sigma = (F, G)$. (Later, we shall identify the κ_i (modulo ordering) with Kronecker's minimal column indices for the singular matrix pencil $[zI - F \;\; G]$. In the classical terminology, the κ_i are called <u>arithmetic invariants</u> of **G**.)

2. <u>Canonical Form</u>

To prove that the κ_i constitute a complete set of invariants for (F, G) under **G**, it is sufficient to find a special form for the matrices F and G having the properties that (i) every matrix pair (F, G) can be transformed into that form via (I-III), and (ii) the form is completely described by an arbitrary set of integers $\kappa_1, \ldots, \kappa_{m'}$ satisfying (2). When both requirements are satisfied, we will be justified in calling the special form <u>canonical</u>. (Compare Hodge and Pedoe [5, p. 328].)

To obtain such a canonical form, we begin by constructing a special basis in X, as follows. We list the vectors in (1) linearly by writing first block 1, then block 2, etc. Then we pick each vector which is not linearly dependent on the preceding ones, beginning with g_1. We put a $\times$ in the corresponding compartment of the crate (see step 2 above) whenever a vector is picked. If $F^i g_k$ is linearly dependent on the preceding ones (and is therefore not picked), then the same is true also for $F^{i+1} g_k$. Consequently the columns in the crate will contain no gaps in the sequence of $\times$'s, that is, satisfy the requirements stated in step 2.

We suppose further that (F, G) = completely reachable, which is equivalent to assuming that our list of vectors contains precisely n linearly independent ones. Then

$$(3) \qquad \sum_i \ell_i = \sum_i \kappa_i = n \, .$$

The vectors picked by the procedure just explained, namely

$$(4) \qquad \{g_j, \ldots, F^{\kappa_j - 1} g_j, \quad j \in M'\}$$

(where M' is some subset of $\{1, \ldots, m\}$ containing exactly m' elements), constitute a basis for X. By linear dependence, we get the relations

$$(5) \qquad F^{\kappa_i} g_i = - \sum_{j \in M'} \sum_{k=1}^{\kappa_j - 1} \alpha_{ijk} F^k g_j$$

for all $i \in M'$. We now construct a new basis defined by

$$(6) \qquad \begin{cases} e_{j1} = g_j \, , \\ e_{j2} = Fg_j + \alpha_{jj1} g_j \, , \\ \quad \cdot \\ \quad \cdot \\ \quad \cdot \\ e_{j\kappa_j} = F^{\kappa_j - 1} g_j + \alpha_{jj1} F^{\kappa_j - 2} g_j + \cdots + \alpha_{jj\kappa_j - 1} \, , \end{cases}$$

where $j \in M'$. This is a straightforward generalization of the procedure used in the construction of the so-called "control canonical form" in the case when $m = 1$. See Kalman, Falb, and Arbib [7, p. 55].

By direct computation it is easily verified that F, G have the following form with respect to the basis (6):

(7)

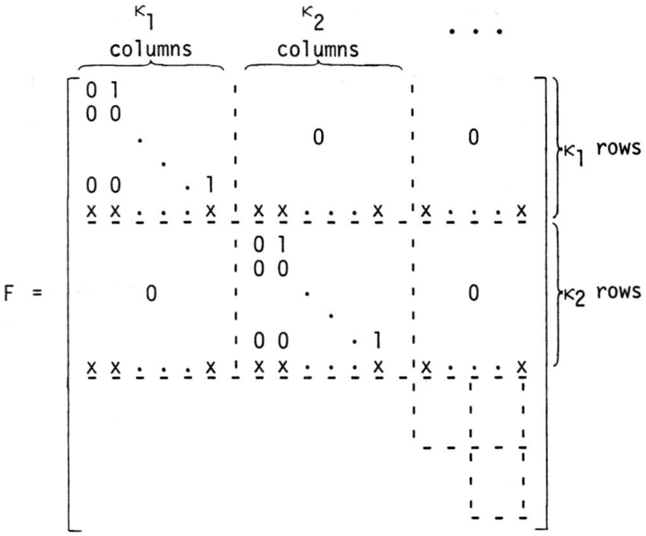

(8)

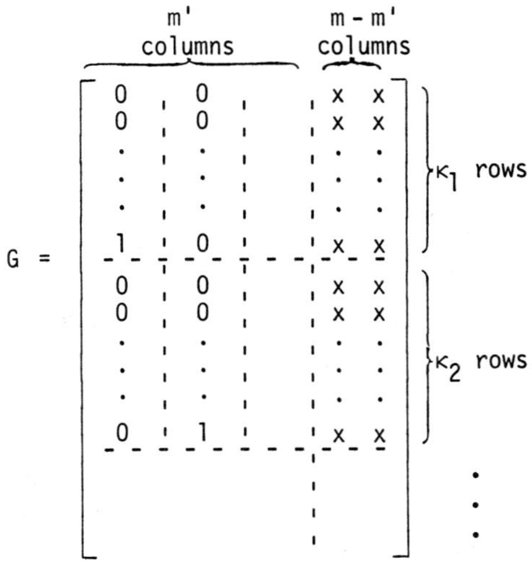

The elements marked "x" in F are given by the α_{ijk}. The m - m' columns of G marked "x" are linearly dependent on the preceding m' columns, by definition of the rank of G.

So far we have used only a transformation of type (I). Using a transformation of type (II) we can make the last $m - m'$ columns of G equal to 0. Finally, using a transformation of type (III), with the rows of L consisting of $-\alpha_{ijk}$, we can make 0 all the terms marked "x" in F.

We have now proved our

MAIN THEOREM. Let (F, G) be a completely reachable pair (over an arbitrary field k). Then there exist positive integers, $\kappa_1, \ldots, \kappa_{m'}$, uniquely determined by (F, G), such that (F, G) is equivalent via (I-III) to the matrices

$$F_c = \operatorname{diag}\,(E_{\kappa_1}, \ldots, E_{\kappa_{m'}}), \tag{9}$$

$$G_c = (\delta_{\kappa_1,1} \quad \delta_{\kappa_1+\kappa_2,2} \quad \cdots \quad \delta_{n,m'} \quad 0), \tag{10}$$

where

$$\sum_1^{m'} \kappa_i = n, \qquad m' = \operatorname{rank} G,$$

and

$$E_\kappa = \begin{pmatrix} 0 & 1 & 0 & \cdots & 0 \\ 0 & 0 & 1 & \cdots & 0 \\ \cdot & \cdot & \cdot & & \\ \cdot & \cdot & \cdot & & \\ \cdot & \cdot & \cdot & & \\ 0 & 0 & 0 & \cdots & 1 \\ 0 & 0 & 0 & \cdots & 0 \end{pmatrix} \quad (\kappa \times \kappa \text{ size}).$$

In other words, $\kappa_1, \ldots, \kappa_{m'}$ determine the orbit on which (F, G) lies; conversely, any two distinct sets of $\kappa_1, \ldots, \kappa_{m'}$ correspond to distinct orbits of $\mathbf{G}$.

We shall call (F_c, G_c) the generalized control canonical form of (F, G). It differs from the now classical scalar control canonical form in that all elements in the last row of E_κ are 0; this is the only way in which this canonical form can be made unique (modulo a permutation of the κ_i, of course). Notice that the basis (6) and in particular the α_{ijk} in (5) are not at all unique, but the nonuniqueness disappears after application of transformations of the type (II) and (III).

INTERPRETATION. The importance of the control invariants is that they give the smallest possible size of the cyclic blocks into which F can be decomposed via a transformation of type (III). As we shall see, this is the only "structural obstruction" to altering F by utilization of a control law L.

3. Minimal Column Indices

We can now easily demonstrate that the invariants of our Main Theorem are identical with those of Kronecker.

A pencil of matrices (archaic terminology) is simply a matrix-valued polynomial in z of degree 1 (= every component of the matrix is a polynomial of degree ≤ 1). A pencil is singular iff its determinant is the 0 polynomial. We shall examine the rectangular matrix $[zI - F \quad G]$ which is clearly a singular "pencil".

Kronecker's theorem (for which see Gantmakher [4, Chapter 12]) provides invariants for computing the equivalence of arbitrary pencils of matrices under the equivalence relation

$$(11) \qquad P(z) \sim Q(z) \quad \text{iff} \quad P(z) = AQ(z)C, \quad \det A \neq 0, \quad \det C \neq 0,$$

where A, C are constant matrices. The theory works over an arbitrary field k.

It is easy to see that

$$[zI - F \quad G] = A[zI - \hat{F} \quad \hat{G}]C$$

iff the matrix C has the special form

$$(12) \qquad C = \begin{bmatrix} A^{-1} & 0 \\ LA^{-1} & B \end{bmatrix}, \quad \det B \neq 0, \quad L = \text{arbitrary}.$$

Hence Kronecker's equivalence relation (11) applied to the singular pencil $[zI - F \quad G]$ is identical with the equivalence relation induced by the transformations (I-III) on pairs (F, G).

For the pencil $[zI - F \quad G]$ three of the four types of Kronecker invariants (invariant factors, infinite invariant factors, minimal row indices, minimal column indices) are trivial, and only the "minimal column indices" matter. Consequently these must be identical with the invariants constructed in Section 2. (This can be checked directly by comparing the canonical form (F_c, G_c) with Kronecker's canonical form for $[zI - F \quad G]$ given in Gantmakher [4, Chapter 12]. The minimal column indices are actually defined in such a way that $\kappa_1 \geqq \kappa_2 \geqq \cdots \geqq \kappa_{m'}$; this is irrelevant here because the ordering of the κ_i can always be achieved by a permutation of the columns of G, a transformation of type (II). The ordering is important, however, in computing the κ_i using Kronecker's definition.) The neat idea of using the pencil $[zI - F \quad G]$ in order to study the pair (F, G) is due to Rosenbrock [11, p. 95].

4. Applications

As an immediate consequence of the main theorem, we note the celebrated

GENERALIZED POLE SHIFTING THEOREM. Given any monic polynomial $\theta \in k[z]$ with $\deg \theta = n$, there exists a matrix L_θ such that $\det (zI - F + GL) = \theta(z)$ if and only if (F, G) is completely reachable.

Proof. The necessity is a well-known result in controllability theory, while the sufficiency follows immediately from the main theorem by applying the equally well-known scalar pole-shifting theorem. ¤

Remark. A quick proof of this result can be obtained even without the use of the control invariants; see Kalman [6, Chapter 6]. What is important here, however, is that the general case is really just a direct sum of the scalar (cyclic) case -- modulo the constraint represented on the size of the cyclic blocks by the values of the κ_i.

The main theorem also yields as an immediate corollary a result of Kalman (1966, unpublished), which reduces the general pole-shifting theorem to the scalar case through a different idea:

THEOREM. Let (F, G) be completely reachable and g_* any nonzero column of G. Then there is a matrix J such that $(F - GJ, g_*)$ is also completely reachable and moreover (i) $Jg_\star = 0$ and (ii) $\chi_F = \chi_{F-GJ}$. ($\chi_F(z) = \det(zI - F)$.)

Proof. Take (F, G) in canonical form. Modulo a transformation of type (II), $g_\star$ = last vector in the sequence $g_1, \ldots, g_m$ which is linearly independent of all the preceding vectors. Define J_c so that $F_c - G_cJ_c$ has all 1's in the diagonal above the main diagonal. (Draw a sketch to see that this is easily done.) Then $J_cg_\star = 0$, $F_c - G_cJ_c$ is a cyclic matrix with respect to the vector $g_\star$ and its characteristic polynomial is z^n, the same as that of F. ⌑

5. Rosenbrock's Theorem

The generalized pole shifting theorem gives only a partial picture as to what can be accomplished by a transformation of type (III) with regard to altering the algebraic structure of F. We owe to Rosenbrock [11, Chapter 5, Theorem 4.2] a remarkable result which settles the question, probably forever:

CONTROL STRUCTURE THEOREM. Let (F, G) be completely reachable, $m' = \text{rank } G$, and $\kappa_1 \geqq \cdots \geqq \kappa_{m'}$ the ordered control invariants of (F, G). Further, let $\psi_1, \ldots, \psi_q$ be arbitrary monic polynomials subject, however, to the following two conditions:

$$\text{(i)} \quad \psi_i | \psi_{i-1}, \quad i = 1, \ldots, q-1 \quad \text{and} \quad q \leqq m';$$

$$\text{(ii)} \quad \begin{cases} \deg \psi_1 \geqq \kappa_1, \\ \deg \psi_1 + \deg \psi_2 \geqq \kappa_1 + \kappa_2, \\ \qquad \text{etc.} \end{cases}$$

Then there is a matrix L such that $F - GL$ has the invariant factors $\psi_1, \ldots, \psi_q$; conversely, the invariant factors of $F - GL$ always satisfy (i)-(ii).

Since condition (i) is an intrinsic property always satisfied by the invariant factors, Rosenbrock's theorem says that $F - GL$ can have arbitrary structure except for a lower bound on the size of the cyclic components, as given via (ii) by the control invariants. (Here the ordering of the control invariants is essential, of course.)

Outline of Proof. Direct. Assume that (F, G) are given in control canonical form. For convenience of notation, consider the special case $\kappa_1 = 2$, $\kappa_2 = 2$ and $\deg \psi_1 = 3$, $\deg \psi_2 = 1$. The general case is similar. Examine the matrices

$$F = \left[\begin{array}{cc|cc} 0 & 1 & 0 & 0 \\ 0 & 0 & 0 & 0 \\ \hline 0 & 0 & 0 & 1 \\ 0 & 0 & 0 & 0 \end{array}\right], \qquad G = \left[\begin{array}{cc} 0 & 0 \\ 1 & 0 \\ \hline 0 & 0 \\ 0 & 1 \end{array}\right].$$

If

$$L = \begin{bmatrix} 0 & 0 & 0 & 0 \\ 0 & -1 & 0 & 0 \end{bmatrix}$$

then $(F - GL, \ G)$ is equivalent via (I-II) to

$$F = \left[\begin{array}{ccc|c} 0 & 1 & 0 & 0 \\ 0 & 0 & 1 & 0 \\ 0 & 0 & 0 & 0 \\ \hline 0 & 0 & 0 & 0 \end{array}\right], \qquad G = \left[\begin{array}{cc} 0 & 0 \\ 0 & 1 \\ 1 & 0 \\ \hline 0 & 1 \end{array}\right].$$

Now $\psi_1(z) = z^3$ and $\psi_2(z) = z$. The new pair (F, G) is not canonical, of course, because G is not canonical. However, the scalar pole-shifting trick still works because the first column of G is canonical and multiplying the second column by

$$L = \begin{bmatrix} 0 & 0 & 0 & 0 \\ 0 & 0 & 0 & \alpha \end{bmatrix}$$

puts the "wrong" term into the position f_{24} which has no effect on χ_F. Hence ψ_1 and ψ_2 can be arbitrarily chosen.

Converse. If we take (F, G) to be canonical, then (ii) is satisfied with the equal signs holding everywhere (obvious from (7)). Since every (F, G) can be obtained from the (unique) canonical pair (F_c, G_c) in its orbit by the construction outlined in the "direct" part of the proof, the inequalities (ii) must always hold. ¤

6. The Structure of L

It is not correct to call the transformation $F \mapsto F - GL$ "feedback" because the control law L may involve no "closed loops" at all -- this is the case, for example, for the J constructed in the proof of the second theorem in Section 4. This point has been generally overlooked in the literature. To remedy the situation, we propose the following new

DEFINITIONS. A control law J is called purely feedforward iff $\chi_F = \chi_{F-GJ}$; a control law K is called purely feedback iff $\deg \psi_i(F) = \deg \psi_i(F - GK)$ for all $i = 1, \ldots, q_F = q_{F-GK}$. A control law M is neutral iff it is both purely feedforward and purely feedback. (We shall employ the notations in the same sense below.)

In other words, a purely feedforward control law does not change the eigenvalues of F and a purely feedback law does not change the cyclic structure of F.

As a major new result we have the

THEOREM. Any feedback law L may be written $L = J + K$ where J, K are unique modulo M.

Outline of Proof. As in the proof of the control structure theorem, the prescribed $F - GL$ (within similarity) can be obtained by first setting up the required cyclic structure by means of a purely feedforward law J. Normalizing all eigenvalues to 0, it follows that J is unique modulo some M. The scalar pole-shifting trick applied to $F - GJ$ will produce the desired eigenvalues and (as the example used in the proof of the control structure theorem shows) this can always be done by a purely feedback type of control law, i.e. without interfering with the cyclic structure of $F - GJ$. ¤

Note that this theorem is a substantial generalization of our 1966 results quoted in Section 4.

7. Relation with Group Representations

The discussion of invariants via the "crate" of Section 1 is a device which has also been used in the theory of the representations of the symmetric group of m letters. This, in turn, is closely related to the theory of representations of the general linear group $GL(k^m)$. (In this literature, our "crate" is known as the "Young diagram". See Dieudonné [3].) In fact, each set of m positive integers κ_i satisfying

$$\kappa_1 + \cdots + \kappa_m = n, \qquad \kappa_1 \geqq \cdots \geqq \kappa_m \geqq 0 ,$$

that is, each possible generalized control canonical form, corresponds to a certain representation of $GL(k^m)$. It could be that this correspondence is more than a purely formal coincidence. In any case, this relationship between the Kronecker minimal indices and group representation theory (which was developed a few years later) does not seem to be worked out.

It is a well-known classical fact that the ring of algebraic invariants of F (polynomials in the elements of F invariant under (I)) has a basis consisting of the coefficients of the characteristic polynomial of F. These invariants do not form a complete system in the sense that their knowledge is not sufficient to determine the orbit of F under (I), that is, to classify the F's under similarity. (In fact, a complete system of invariants for F under similarity does not seem to be explicitly known at present.) The significance of the pole-shifting theorem is that it "destroys" all algebraic invariants and then the orbits can be completely classified by means of the Kronecker invariants; these can be explicitly computed via the procedure given in Section 2.

8. Historical Notes

There is much confusion concerning the results summarized here. Various authors have obtained parts of the overall picture but failed to notice further possibilities.

The canonical form (6) was (probably) first noted by Brunovský [1]. He was well aware of its invariant properties, unlike Luenberger [9]. (In fact, Luenberger's paper is an example of the confusion surrounding the word "canonical" in recent system-theoretic literature.) The Main Theorem itself was published recently by Brunovský [2, Theorems 1-2], but he did not know that his result is a special case of Kronecker's theorem. The Main Theorem also occurs in the work of Rosenbrock [11, p. 98]; he mentions Kronecker's invariants (but not Kronecker's canonical form) and his proof is circuitous, based on that of Luenberger [9]. Brunovský recognized at once the applicability of the Main Theorem to the pole shifting problem and also proved a result similar to the second theorem in Section 4 (see Brunovský [2, Corollary 2]), but he overlooked Rosenbrock's theorem -- surely the most striking application to date of the Main Theorem. Neither Brunovský nor Rosenbrock have looked at the question of "feedback" vs. "feedforward" analyzed in Section 6 here. The Main Theorem was also known to Popov (unpublished internal report, 1968), whose results will appear soon in a definitive form (Popov [10]).

REFERENCES

[1] P. BRUNOVSKÝ, O stabilizatsii lineinikh sistem pri opredelennom klasse postoyanno deistvuyushchikh vozmushchenii (in Russian), Differentsialnye upravneniya 2 (1966), 769-777.

[2] P. BRUNOVSKÝ, A classification of linear controllable systems, Kybernetika (Praha) 3 (1970), 173-187.

[3] J. DIEUDONNÉ, La théorie des invariants au XIXe siècle, Séminaire Bourbaki, exposé N° 395 (1971).

[4] F.R. GANTMAKHER, The Theory of Matrices and Its Applications. 2 volumes, Chelsea, 1959.

[5] W.V.D. HODGE and D. PEDOE, Methods of Algebraic Geometry. Vol. 1, Cambridge University Press, 1947.

[6] R.E. KALMAN, Lectures on Algebraic System Theory, Springer Lecture Notes in Mathematics, Vol. 2??, 1971.

[7] R.E. KALMAN, P.L. FALB and M.A. ARBIB, Topics in Mathematical System Theory. McGraw-Hill, 1969.

[8] L. KRONECKER, Algebraische Reduktion der Scharen quadratischen Formen, Sitzungsber. Berl. Akad. (1890), pp. 763-776, 1225-1375 (see collected works).

[9] D.L. LUENBERGER, Canonical form for linear multivariable systems, IEEE Trans. on Automatic Control, AC-12 (1967), 290-293.

[10] V.M. POPOV, Invariant description of linear, time-invariant controllable systems, to appear 1972.

[11] H.H. ROSENBROCK, Multivariable and State-Space Theory. John Wiley & Sons, 1970.

University of Florida, Gainesville, Florida, and
Stanford University, Stanford, California

LOWER BOUNDS AND UNIQUENESS FOR SOLUTIONS OF EVOLUTION INEQUALITIES IN A HILBERT SPACE

G. Ladas and V. Lakshmikantham

1. Introduction

Recently Ogawa [6] and Agmon and Nirenberg [1] obtained lower bounds and uniqueness theorems for the solutions of the differential inequalities

$$(1.1)\qquad \left\|\frac{du}{dt} - A(t)u(t)\right\|^2 \le \phi_1(t)\|u(t)\|^2 + \phi_2(t)a_t(u(t),u(t)), \quad t \in R$$

and

$$(1.2)\qquad \left\|\frac{du}{dt} - A(t)u(t)\right\|^2 \le \phi(t)[\|u(t)\|^2 + \int_t^T \omega(s)\|u(s)\|^2 ds]^{\frac{1}{2}}, \quad t \in [0,T)$$

where $A(t)$ is a linear operator with time-varying domain $D(A(t))$ in a Hilbert space H, a_t is a symmetric, positive semi-definite, bilinear functional defined on $D(A(t))$ and $\phi_1(t)$, $\phi_2(t)$, $\phi(t)$ and $\omega(t)$ are nonnegative scalar functions - under a variety of conditions. The operator $A(t)$ is supposed to admit a decomposition of the form

$$(1.3)\qquad A(t) = A_+(t) + \delta(A_-'(t) + A_-''(t))$$

where $A_+(t)$ is a linear symmetric operator over $D(A(t))$ while $A_-'(t)$ and $A_-''(t)$ are linear skew symmetric over $D(A(t))$. The constant δ of the decomposition is taken equal to zero in [6] and nonzero in [1].

The aim of this paper is to obtain similar results for the more general differential inequality

$$(1.4) \qquad \left\|\frac{du}{dt} - A(t)u(t)\right\| \leq \Phi(t, m(t), q(t)), \quad t \in J$$

where J is an interval on the real line R, $A(t)$ admits the decomposition (1.3), $m(t) = \|u(t)\|^2 + P(u(t))$ with $P(u(t))$ a nonnegative (nonlinear) functional defined on $D(A(t))$, $q(t) = (A_+(t)u(t), u(t))$ and Φ a function from $J \times R_+ \times R$ into the nonnegative real line R_+.

The form of the inequality (1.4) will offer a wider range of applicability of these results to partial differential equations but we shall be only concerned here with abstract differential inequalities.

In Section 2, we prove the main result of this paper, namely, Theorem 2.1, which gives lower bounds for the solutions of (1.4). The choice of hypotheses of this theorem was strongly suggested by the previous researches [1], [4], [5], [6] and [7]. The main ingredient in our proof is the theory of first order differential inequalities for which the reader is referred to [3, pp. 15-20]. As a consequence of Theorem 2.1, we also prove a uniqueness theorem analogous to Theorem 2(i) and (ii) in [1] and a "unique continuation at infinity" result analogous to that in [6].

In Section 3, we apply Theorem 2.1 to the differential inequality

$$(1.5) \quad \left\|\frac{du}{dt} - A(t)u(t)\right\| \leq [\phi_1(t)m(t) + \phi_2(t)a_t(u(t),u(t))]^{\frac{1}{2}}, \quad t \in R$$

where ϕ_1, ϕ_2 and a_t are as before, $m(t) = \|u(t)\|^2 + \int_t^T 2\omega(s)\|u(s)\|^2 ds$ and $A(t)$ admits the decomposition (1.3). Clearly (1.5) is more general than (1.1) or (1.2). Our hypotheses on (1.5) will be weaker than those imposed on (1.1) or (1.2) in the sense that when (1.5) reduces to (1.1) the hypotheses on (1.5) also reduce to the hypotheses on (1.1) imposed in [6]. The same is also true with respect to the hypotheses on (1.5) and those on (1.2) imposed in [1]. Our aim then will be to show that all the hypotheses of Theorem 2.1 are also satisfied for (1.5). Applying Theorem 2.1, we shall next compute explicit lower bounds for the solutions of (1.5). As particular cases of our estimates we shall obtain the Lemma in [6] and Theorem 1 in [1]. It should be noted that these results are the basic inequalities in [6] and [1] respectively from which all the other results follow.

2. Lower Bounds and Uniqueness

Let H be a Hilbert space with inner product $(,)$ and norm $\|\cdot\|$. Consider the evolution inequality (1.4) in H. By $C[J,H]$ we denote the space of continuous functions from the real interval J into the Hilbert space H. If $u \in C[J,H]$ then $\frac{du}{dt}$ denotes the derivative of u with respect to t in the strong topology of H.

Definition 2.1. A function $u \in C[J,H]$ is said to be a solution of inequality (1.4) if

(i) $u(t) \in D(A(t))$, $t \in J$, and $A(t)u(t)$ is continuous for $t \in J$;

(ii) $\frac{du(t)}{dt}$ exists and is continuous on J;

(iii) $u(t)$ satisfies the inequality (1.4) for all $t \in J$.

In this section we shall obtain general lower bounds and two uniqueness theorems for the solutions of (1.4) on an interval J of the real line R. We shall often use the inequality $m(t) \geq \|u(t)\|^2$ which is valid since $P(u(t)) \geq 0$. We also define $Q(t) = \frac{q(t)}{m(t)}$ as long as $m(t) \neq 0$. For easy references we state the following hypotheses:

(H_1) There exists a function $\phi \in C[J \times R, R_+]$ such that for all $t \in J$ with $m(t) \neq 0$

$$\Phi(t, m(t), q(t)) \leq m(t)^{\frac{1}{2}}\phi(t, Q(t)) .$$

(H_2) The functional $P(u(t))$ is differentiable with respect to t and for all $t \in J$

$$\left|\frac{d}{dt} P(u(t))\right| \leq 2\omega(t)m(t)$$

where $\omega \in C[J,R_+]$.

(H_3) There exist functions $\psi_i \in C[J \times R, R]$, $i = 1,2,3$ such that for any solution $u(t)$ of (1.4) with $m(t) \neq 0$ and for any number δ, $0 \leq \delta < \frac{2}{5}$, satisfying (1.3), the following estimates hold:

(i) $\delta \mathrm{Re}(A_+(t)u(t), A_-'(t)u(t)) \geq -\delta m(t)\psi_1(t,Q(t)) - \delta \|A_+(t)u(t) - Q(t)u(t)\|^2$;

(ii) $\delta \|A_-''(t)u(t)\|^2 \leq \delta m(t)\psi_2(t,Q(t)) + \delta \|A_+(t)u(t) - Q(t)u(t)\|^2$;

(iii) the function $(A_+(t)u(t), u(t))$ is differentiable on J and

$$\frac{d}{dt}(A_+(t)u(t),u(t)) - 2\mathrm{Re}(A(t)u(t),u'(t))$$
$$\geq -m(t)\psi_3(t,Q(t)) - \delta \|A_+(t)u(t) - Q(t)u(t)\|^2 .$$

It should be noted that when $A(t)$ is symmetric (as in [6]) then $\delta = 0$ and the hypotheses (i) and (ii) are automatically satisfied.

(H_4) There exists a function $\theta_1 \in C[J \times R, R]$ such that $\theta_1(t,y)$ is nondecreasing in y for each $t \in J$ and

$$2[y - \phi(t,y) - \omega(t)] \geq \theta_1(t,y), \quad t \geq t_0, \quad t \in J.$$

(H_5) There exists a function $\theta_2 \in C[J \times R, R]$ such that $\theta_2(t,y)$ is nondecreasing in y for each $t \in J$ and

$$2[y + \phi(t,y) + \omega(t)] \leq \theta_2(t,y), \quad t \leq t_0, \quad t \in J.$$

We define the function

$$w(t,y) \equiv \psi(t,y) + 2\omega(t)|y| + \frac{1}{2-5\delta}\phi^2(t,y) \tag{2.1}$$

where $\psi(t,y) \equiv 2\delta\psi_1(t,y) + \delta\psi_2(t,y) + \psi_3(t,y)$.

We also denote by $y_1(t)$ and $y_2(t)$ the right minimal and the left maximal solution, respectively, of the scalar initial value problem:

$$y' = -w(t,y) \ ; \quad y(t_0) = Q(t_0) . \tag{2.2}$$

We now pass to our main result of this section.

<u>THEOREM 2.1</u>. Let $u(t)$ be a solution of (1.4) and let t_0 be a point on J. Then,

(a) Under the hypotheses (H_1), (H_2), (H_3) and (H_4) the following lower bound is valid:

$$m(t) \geq m(t_0)\exp \int_{t_0}^{t} \theta_1(s, y_1(s))ds , \quad t \geq t_0, \quad t \in J ; \tag{2.3}$$

(b) Under the hypotheses (H_1), (H_2), (H_3) and (H_5) the following lower bound is valid:

$$(2.4)\qquad m(t) \geqslant m(t_0)\exp\int_t^{t_0} -\theta_2(s,y_2(s))ds, \quad t \leqslant t_0, \quad t \in J.$$

<u>Proof</u>. Suppose first that $m(t) > 0$ for all $t \in J$. Set

$$Lu(t) = u'(t) - A(t)u(t) \qquad (' = \frac{d}{dt})$$

Then, using the decomposition (1.3), the symmetry of A_+ and the skew symmetry of A'_- and A''_-, we obtain,

$$(2.5)\qquad \begin{aligned}\frac{d}{dt} m(t) &= 2\mathrm{Re}(u'(t),u(t)) + \frac{d}{dt} P(u(t) \\ &= 2Q(t)m(t) + 2\mathrm{Re}(Lu(t),u(t)) + \frac{d}{dt} P(u(t)).\end{aligned}$$

In view of (H_1) and (1.4) we have from (2.5)

$$(2.6)\qquad \left|\frac{d}{dt} m(t) - 2Q(t)m(t) - \frac{d}{dt} P(u(t))\right| \leqslant 2\Phi(t,m(t),q(t))\|u(t)\|$$

$$2\phi(t,Q(t)),m(t).$$

By (2.6) and (H_2) we are led to the inequalities:

$$(2.7)\qquad \frac{d}{dt} m(t) \geqslant 2[Q(t) - \phi(t,Q(t)) - \omega(t)]m(t), \quad t \in J$$

and

$$(2.8)\qquad \frac{d}{dt} m(t) \leqslant 2[Q(t) + \phi(t,Q(t)) + \omega(t)]m(t), \quad t \in J.$$

Since $A_+(t)u(t)$ is continuous in t and $A_+(t)$ is symmetric, it is easily seen by baking the diffeeence quotient and passing to the limit, that the function $(A_+(t)u(t),u(t))$ is differentiable and

$$(2.9)\qquad \frac{d}{dt}(A_+(t)u(t),u(t)) = (\dot{A}_+(t)u(t),u(t))+2\mathrm{Re}(A_+(t)u(t),u'(t))$$

where (as in [1])

$$\dot{A}_+(t)u(t),u(t)) \equiv \frac{d}{dt}(A_+(t)u(t),u(t))-2\mathrm{Re}(A_+(t)u(t),u'(t)).$$

It follows that (suppressing the variable t) and using (1.3)

$$(2.10)\qquad \frac{dQ}{dt} = \frac{(\dot{A}_+u,u)}{m} - \frac{Q}{m}\frac{d}{dt}P(u) + \frac{2}{m}[\|A_+u\|^2 - Q^2m + \mathrm{Re}(A_+u-Qu,Lu)]$$
$$+ \frac{2\delta}{m}\mathrm{Re}(A_+u,A'_-u) + \frac{2\delta}{m}\mathrm{Re}(A_+u,A''_-u)$$
$$\equiv I_1 + I_2 + I_3 + I_4 + I_5 .$$

We shall estimate the terms I_i, $i = 1,\ldots,5$. From (H_3)(iii) we get

$$(2.11)\qquad I_1 \geq -\psi_3(t,Q) - \frac{\delta}{m}\|A_+u-Qu\|^2 .$$

By (H_2) we see that

$$(2.12)\qquad I_2 \geq -2|Q|\omega(t) .$$

Notice that

$$(2.13)\qquad \begin{aligned}\|A_+u-Qu\|^2 &= \|A_+u\|^2 - 2Q(A_+u,u) + Q^2\|u\|^2\\ &\leq \|A_+u\|^2 - 2Q^2m + Q^2m\\ &= \|A_+u\|^2 - Q^2m .\end{aligned}$$

Next, using the arithmetic-geometric mean inequality on $2\mathrm{Re}(A_+u-Qu,Lu)$, we obtain, for any $a > 0$

$$(2.14)\qquad \begin{aligned}2\mathrm{Re}(A_+u-Qu,Lu) &\geq -2\|A_+u-Qu\|\,\|Lu\|\\ &\geq -a\|A_+u-Qu\|^2 - \frac{1}{a}\|Lu\|^2\\ &\geq -a\|A_+u-Qu\|^2 - \frac{m}{a}\phi^2(t,Q) .\end{aligned}$$

The estimates (2.13) and (2.14) yield

$$(2.15)\qquad I_3 \geq \frac{2-a}{m}\|A_+u-Qu\|^2 - \frac{1}{a}\phi^2(t,Q) .$$

In view of (H_3)(i), it follows that

$$(2.16)\qquad I_4 \geq -2\delta\psi_1(t,Q) - \frac{2\delta}{m}\|A_+u-Qu\|^2 .$$

By (H_3)(ii) and the skew symmetry of A''_- one gets

$$\begin{aligned}\delta|2\mathrm{Re}(A_+u,A''_-u)| &= \delta|2\mathrm{Re}(A_+u-Qu,A''_-u)|\\ &\leq \delta\|A_+u-Qu\|^2 + \delta\|A''_-u\|^2\\ &\leq 2\delta\|A_+u-Qu\|^2 + \delta m\psi_2(t,Q) ,\end{aligned}$$

which implies that

$$(2.17) \qquad I_5 \geq -\frac{2\delta}{m}\|A_+u-Qu\|^2 - \delta\psi_2(t,Q) .$$

Using (2.11), (2.12), (2.15), (2.16) and (2.17) in (2.10) and choosing $a = 2-5\delta$, we finally obtain

$$\frac{dQ}{dt} \geq -(2\delta\psi_1+\delta\psi_2+\psi_3) - 2|Q|\omega - \frac{1}{2-5\delta}\phi^2(t,Q)$$

i.e. (recalling (2.1))

$$(2.18) \qquad \frac{dQ}{dt} \geq -w(t,Q) .$$

The differential inequality (2.18) now yields ([3, pp. 15-20])

$$(2.19) \qquad Q(t) \geq y_1(t), \quad t \geq t_0, \quad t \in J$$

and

$$(2.20) \qquad Q(t) \leq y_2(t), \quad t \leq t_0, \quad t \in J .$$

Now we assume that (H_4) is satisfied. Then, using (2.7) and (2.19) we get

$$\frac{d}{dt}m(t) \geq \theta_1(t,y_1(t))m(t), \quad t \geq t_0, \quad t \in J$$

from which the estimate (2.3) follows and (a) is proved. Similarly, assuming that (H_5) is satisfied, we derive from (2.8) and (2.20) the inequality

$$\frac{d}{dt}m(t) \leq \theta_2(t,y_2(t))m(t), \quad t \leq t_0, \quad t \in J$$

from which the estimate (2.4) follows and (b) is proved.

We have established the desired lower bounds under the additional condition that $m(t) > 0$ for $t \in J$. We shall now remove this assumption. If $m(t_0) = 0$ the estimates (2.3) and (2.4) are clearly valid. Now assume that $m(t_0) > 0$. We shall prove that $m(t) > 0$ for all $t \in J$ and hence the previous arguments are valid. Otherwise there exists an interval with one end point t_0, say, $[t_0,t_1)$, such that $m(t) > 0$ on $[t_0,t_1)$ but $m(t_1) = 0$. Since (2.3) holds for all $t \in [t_0,t_1)$ it follows by continuity that the same bound holds also at $t = t_1$ contradicting the hypothesis $u(t_1) = 0$.

A similar argument, involving (2.4), is valid in case t_0 is a right-end point of the above interval. The proof is therefore complete.

An elementary, but very useful, application of Theorem 2.1 is the following:

<u>COROLLARY 2.1</u>. Let A be a linear symmetric operator in H with domain $D(A)$. Let $u(t)$ be a solution of the evolution inequality

$$\left\|\frac{du}{dt} - Au(t)\right\| \geq \phi(t)\|u(t)\|, \quad t \in J \tag{2.21}$$

where $\phi\colon J \to R_+$ is a given measurable and locally bounded function. Then, for $t, t_0 \in J$ the following lower bounds are valid:

$$\|u(t)\| \geq \|u(t_0)\| \exp\{Q(t_0)(t-t_0) - \int_{t_0}^{t} [\phi(s)+\tfrac{1}{2}(t-s)\phi^2(s)]ds\}, \quad t \geq t_0, \tag{2.22}$$

and

$$\|u(t)\| \geq \|u(t_0)\| \exp\{-Q(t_0)(t_0-t) - \int_{t}^{t_0} [\phi(s)+\tfrac{1}{2}(s-t)\phi^2(s)]ds\}, \quad t \leq t_0. \tag{2.23}$$

<u>Proof</u>. This is a very special case of Theorem 2.1. The hypotheses (H_1), (H_2), (H_3), (H_4) and (H_5) are satisfied with $\delta = 0$; $\Phi(t,m(t),q(t)) \equiv \phi(t)\|u(t)\|$; $P(u(t)) \equiv 0$; $\omega(t) \equiv 0$; $\psi_i \equiv 0$, $i = 1,2,3$; $\theta_1(t,y) = 2[y-\phi(t)]$; $\theta_2(t,y) = 2[y+\phi(t)]$; $w(t,y) \equiv \frac{1}{2}\phi^2(t)$;

$$y_1(t) \equiv Q(t_0) - \frac{1}{2}\int_{t_0}^{t} \phi^2(s)ds, \quad t \geq t_0$$

$$y_2(t) \equiv Q(t_0) - \frac{1}{2}\int_{t}^{t_0} \phi^2(s)ds, \quad t \leq t_0 .$$

By (2.3) it follows that

$$m(t) \geq m(t_0)\exp 2\{Q(t_0)(t-t_0) - \int_{t_0}^{t} [\phi(s)+\tfrac{1}{2}(t-s)\phi^2(s)]ds\}, \quad t \geq t_0$$

and since $m(t) = \|u(t)\|^2$ the estimate (2.22) results upon extracting square roots. Similarly (2.23) may be derived from (2.4). The proof is complete.

One can use the inequality (2.22) to establish the lower bounds in [2, p. 216].

A consequence of Theorem 2.1 is the following interesting uniqueness result, a special case of which is Theorem 2(i) and (ii) in [1].

THEOREM 2.2. Under the hypotheses (H_1), (H_2), (H_3), (H_4) and (H_5) for any solution $u(t)$ of (1.4) iehter $m(t) > 0$ for all $t \in J$ or $m(t) \equiv 0$ on J. In the special case when $P(u(t)) \equiv 0$, if $u(t_0) = 0$ for some $t_0 \in J$, then $u(t) \equiv 0$ on J.

Proof. It is clear from (2.3) and (2.4) that if $m(t_0) > 0$ then $m(t) > 0$ for all $t \in J$. Next, assume that $m(t_0) = 0$ for some $t_0 \in J$. We shall prove that $m(t) \equiv 0$ for all $t \in J$. If not, then $m(t)$ is not identically zero in an interval either to the left or to the right of t_0. Suppose that this happens to the left of t_0. Then, there must exist a subinterval $[t_1,t_2)$ of J such that $m(t) > 0$ for $t_1 \leq t < t_2$ and $m(t_2) = 0$. Applying the estimate (2.3) with $t = t_2$ and $t_0 = t_1$ we obtain a contradiction. Hence $m(t) \equiv 0$ to the left of t_0. Similarly, using the estimate (2.4) we obtain a contradiction unless $u(t) \equiv 0$ to the right of t_0. The proof is complete.

The following theorem shows that a solution of (1.4) (actually $m(t)$) cannot tend to zero too rapidly as $t \to \infty$ unless it is identically zero. This result is a "unique continuation at infinity" theorem analogous to Ogawa's theorem [6].

THEOREM 2.3. Let the hypotheses (H_1), (H_2), (H_3), (H_4) and (H_5) be satisfied on the whole real line R and $u(t)$ be a solution of (1.4). Assume that there exist constants k, ℓ, N such that $k \geq 0$, $\ell > 0$, and N depends on the solution satisfying the order relations:

$$m(t) = O(e^{-kt}) \quad \text{as} \quad t \to -\infty \tag{2.24}$$

$$m(t) = O(e^{-(k+\ell)t}) \quad \text{as} \quad t \to +\infty \tag{2.25}$$

$$\exp \int_{t_0}^{t} -\theta_1(s,y_1(s))ds = O(e^{-Nt}) \quad \text{as} \quad t \to +\infty \tag{2.26}$$

$$\exp \int_{t}^{t_0} \theta_2(s,y_2(s))ds = O(e^{-Nt}) \quad \text{as} \quad t \to -\infty . \tag{2.27}$$

Then, $m(t) \equiv 0$.

Proof. Let, for some t_0, $m(t_0) > 0$. Then the estimates (2.3) and (2.4) are valid and for convenience we write them in the form

$$m(t_0) \leq m(t)\exp \int_{t_0}^{t} -\theta_1(s,y_1(s))ds , \quad t \geq t_0 \tag{2.28}$$

and

$$m(t_0) \leq m(t)\exp \int_{t}^{t_0} \theta_2(s,y_2(s))ds , \quad t \leq t_0 . \tag{2.29}$$

From (2.28), (2.25) and (2.26) we obtain, with C standing for a generic constant, the inequality

$$m(t_0) \leq C \exp\{-(k+\ell+N)t\} , \quad \text{as} \quad t \to +\infty .$$

Since $m(t_0) > 0$, it is necessary that

$$k + \ell + N \leq 0 . \tag{2.30}$$

Similarly from (2.29), (2.24) and (2.27) we get

$$m(t_0) \leq C \exp\{-(k+N)t\} , \quad \text{as} \quad t \to -\infty ,$$

and therefore $k + N \geq 0$ which contradicts (2.30) and the proof is complete.

It is evident from the proof that there is an analogous theorem with the roles of $t = \infty$ and $t = -\infty$ interchanged. In the special case of equation (2.21) the hypotheses (2.26) and (2.27) are satisfied if we assume that $\phi \in L^2(R)$.

3. Here we shall apply Theorem 2.1 to study the inequality (1.5) which is still more general than (1.1) or (1.2). We shall impose the following hypotheses concerning (1.5).

There exist non-negative measurable functions α, γ_i, β_i, α_i, $i = 1,2,3$, bounded on every closed finite subinterval of R such that for any solution $u(t)$ of (1.5) with $m(t) \neq 0$ and for any number δ, $0 \leq \delta \leq \frac{1}{5}$, satisfying (1.3),

(I) $\delta \mathrm{Re}(A_+u,A_-'u) \geq -\delta\gamma_1\|A_+u\| \, \|u\| - \delta\beta_1\|u\|^2 - \delta\alpha_1 a_t(u,u)$;

(II) $\delta\|A_-''u\|^2 \leq \delta\gamma_2\|A_+u\|\ \|u\| + \delta\beta_2\|u\|^2 + \delta\alpha_2 a_t(u,u)$;

(III) The function $(A_+(t)u(t),u(t))$ is differentiable on J and

$$(\dot{A}_+u,u) \geq -\delta\gamma_3\|A_+u\|\ \|u\| - \beta_3\|u\|^2 - \alpha_3 a_t(u,u) .$$

If $\phi_2 + \alpha_1 + \alpha_2 + \alpha_3 \not\equiv 0$, we shall assume that

(IV) $-(A_+u,u) \geq a_t(u,u) - \alpha\|u\|^2$.

Notice that when $\delta = 0$ these hypotheses (I)-(IV) reduce to Ogawa's hypotheses imposed on (1.1) and when $\delta \neq 0$ and $\alpha_1, \alpha_2, \alpha_3, \alpha$ and ϕ_2 are identically zero, the same hypotheses reduce to Agmon-Nirenberg's hypotheses imposed on (1.2). In the last case (IV) is not assumed to hold because $\phi_2 + \alpha_1 + \alpha_2 + \alpha_3 \equiv 0$. First, our aim is to show that all the hypotheses (H_1)-(H_5) of Theorem 2.1 are also true for the equation (1.5). Then we shall obtain explicit lower bounds for the solutions of (1.5). As a byproduct of our estimates we shall obtain the inequalities of the lemma in [6] and Theorem 1 in [1] which are the key results there.

If $\phi_2(t) \not\equiv 0$, from the assumption (II) we get (suppressing t and using the same notation for q and Q as in Section 2)

$$\begin{aligned}[\phi_1 m + \phi_2 a_t(u,u)]^{\frac{1}{2}} &\leq [\phi_1 m + \phi_2 \alpha m - \phi_2 q]^{\frac{1}{2}} \\ &\leq m^{\frac{1}{2}}[\phi_1 + \phi_2(\alpha - Q)]^{\frac{1}{2}} \quad \text{when } m \neq 0.\end{aligned}$$

Hence (H_1) is satisfied with

$$(t,Q) = [\phi_1 + \phi_2(\alpha-Q)]^{\frac{1}{2}} \quad . \tag{3.1}$$

Notice also that because of (II), $\alpha - Q \geq 0$.

If $\phi_2(t) \equiv 0$ then $\phi(t,Q) = \sqrt{\phi_1}$. Hence in any case $\phi(t,Q)$ is given by (3.1).

Clearly, (H_2) is satisfied with

$$P(u(t)) = \int_t^T 2\omega(s)\|u(s)\|^2\, ds .$$

The proof that (H_3) is satisfied requires a trick used in [1]. Let θ denote the angle between the vectors A_+u and u in the Hilbert space H. Then

$$\|A_+u-Qu\|^2 \geq \|A_+u\|^2\sin^2\theta$$

and

$$|Q| = \frac{1}{m}|(A_+u,u)| = \frac{1}{m}\|A_+u\|\ \|u\|\ |\cos\theta| .$$

From these inequalities we obtain, using the δ of hypotheses (I)-(IV)

$$\begin{aligned}(3.2)\quad \delta\gamma_i\|A_+u\|\ \|u\| &= \delta\gamma_i\|A_+u\|\ \|u\|\ (\sin^2\theta + \cos^2\theta)\\ &\leq \delta\gamma_i\|A_+u\|\ \|u\|\ \sin^2\theta + \delta\gamma_i\|A_+u\|\ \|u\|\ |\cos\theta|\\ &\leq \delta\ [\|A_+u\|^2 + \gamma_i^2\ \|u\|^2]\sin^2\theta + \delta\gamma_i|Q|m\\ &\leq \delta\|A_+u-Qu\|^2 + \delta(\gamma_i^2 + \gamma_i|Q|)m , \quad i = 1,2,3 .\end{aligned}$$

In view of (I)-(IV) and (3.2) we get, as long as $m(t) \neq 0$,

(i) $\delta\mathrm{Re}(A_+u,A_-'u) \geq -\delta\|A_+u-Qu\|^2 - \delta m(\gamma_1^2 + \gamma_1|Q| + \beta_1 - \alpha_1 Q + \alpha\alpha_1)$;

(ii) $\delta\|A_-''u\|^2 \leq \delta\|A_+u-Qu\|^2 + \delta m(\gamma_2^2+\gamma_2|Q| + \beta_2 - \alpha_2 Q + \alpha\alpha_2)$;

(iii) $(\dot{A}_+u,u) \geq -\delta\|A_+u-Qu\|^2 - \delta m(\gamma_3^2 + \gamma_3|Q| + \beta_3 + \alpha_3 Q - \alpha\alpha_3)$.

Therefore (H_3) is established with

$$\psi_i(t,Q) = \gamma_i^2 + \gamma_i|Q| + \beta_i - \alpha_i Q + \alpha\alpha_i , \quad i = 1,2,3$$

and

$$\begin{aligned}(3.3)\quad \psi(t,Q) = &-(2\delta\alpha_1+\delta\alpha_2+\alpha_3)Q + (2\delta\gamma_1+\delta\gamma_2+\gamma_3)|Q| + (2\delta\gamma_1^2+\delta\gamma_2^2+\gamma_3^2)\\ &+ (2\delta\beta_1+\delta\beta_2+\beta_3) + \alpha(2\delta\alpha_1+\delta\alpha_2+\alpha_3) .\end{aligned}$$

Next we shall verify (H_4) and (H_5). Since the function $y-\phi(t,y) = y-[\phi_1+\phi_2(\alpha-y)]^{\frac{1}{2}}$, with $y \leq \alpha$ is increasing in y we could take for $\phi_1(t,y)$ the function

$$2[y - \phi(t,y) - \omega(t)] .$$

However a simpler function can be chosen as follows:

$$\begin{aligned}y - [\phi_1+\phi_2(\alpha-y)]^{\frac{1}{2}} &\geq y - \phi_1^{\frac{1}{2}} - [\phi_2(\alpha-y)]^{\frac{1}{2}}\\ &\geq y - \phi_1^{\frac{1}{2}} - \frac{\phi_2}{2\rho} - \rho(\alpha-y)\\ &= (1+\rho)y - \phi_1^{\frac{1}{2}} - \frac{\phi_2}{2\rho} - \alpha\rho .\end{aligned}$$

So for any $0 < \rho < 1$ we may take

$$\theta_2(t,y) \equiv 2[(1-\rho)y + \phi_1^{\frac{1}{2}} + \frac{\phi_2}{2\rho} + \alpha\rho + \omega]$$

and (H_5) is satisfied.

From (2.1), (3.3) and (3.1) we get

$$w(t,y) = -A(t)y + B(t)|y| + C(t) \tag{3.4}$$

where

$$A(t) = 2\delta\alpha_1 + \delta\alpha_2 + \alpha_3 + \frac{1}{2-5\delta}\phi_2$$

$$B(t) = 2\delta\gamma_1 + \delta\gamma_2 + \gamma_3 + 2\omega$$

$$C(t) = (2\delta\gamma_1^2 + \delta\gamma_2^2 + \gamma_3^2) + (2\delta\beta_1 + \delta\beta_2 + \beta_3) + \alpha(2\delta\alpha_1 + \delta\alpha_2 + \alpha_3) + \frac{1}{2-5\delta}(\phi_1 + \alpha\phi_2) .$$

In the special case of (1.1) studied by Ogawa [6], $\delta = 0$; $\omega(t) = 0$; $\gamma_i \equiv 0$, $i = 1,2,3$; $\beta_i = \alpha_i \equiv 0$, $i = 1,2$. In this situation

$$A(t) = \alpha_3 + \frac{1}{2}\phi_2; \quad B(t) \equiv 0; \quad C(t) = \beta_3 + \alpha\alpha_3 + \frac{1}{2}(\phi_1 + \alpha\phi_2).$$

Hence, from (3.4) and (2.18) we obtain

$$\frac{dQ}{dt} \geq (\alpha_3 + \frac{1}{2}\phi_2)Q - [\beta_3 + \alpha\alpha_3 + \frac{1}{2}(\phi_1 + \alpha\phi_2)] . \tag{3.5}$$

The inequality i3.5) is precisely the lemma in [6].

In the special case of (1.2) studied by Agmon and Nirenberg [1], $\phi_2 \equiv 0$, $\delta \neq 0$, e.g., $\delta = \frac{1}{5}$, $\alpha = \alpha_1 = \alpha_2 = \alpha_3 \equiv 0$, $\phi_1 = \phi^2$. Here

$$A(t) \equiv 0 ; \quad B(t) = \frac{2}{5}\gamma_1 + \frac{1}{5}\gamma_2 + \gamma_3 + 2\omega ;$$

$$C(t) = (\frac{2}{5}\gamma_1^2 + \frac{1}{5}\gamma_2^2 + \gamma_3) + (\frac{2}{5}\beta_1 + \frac{1}{5}\beta_2 + \beta_3) + \phi^2.$$

Hence,

$$\frac{dQ}{dt} \geq -B(t)|Q| - C(t) . \tag{3.6}$$

Notice that the function $\ell(t)$ in [1] satisfies the equality $\dot{\ell} = 2Q$ and therefore (3.6) implies

$$\ddot{\ell}(t) + B(t)|\ell(t)| + 2C(t) \geq 0$$

which is the equation (1.13) in [1] with coefficients having the same integral properties as the functions (1.13)' in [1]. Finally we must compute the solutions of (2.2). We first make the observation that the solution of the equation $y' = Ay - B|y| - C$ with $y(t_0) = y_0$ is given by the formula

$$y(t) = y_0 \exp \int_{t_0}^{t} (A \pm B)ds - \int_{t_0}^{t} \{C(s)\exp \int_{s}^{t} (A \pm B)d\xi\}ds$$

and therefore if $y_0 \leq 0$ then $y(t) \leq 0$ for $t \geq t_0$ and if $y_0 \geq 0$ then $y(t) \geq 0$ for $t \leq t_0$.

Set

$$\lambda_1 = \min(0, Q(t_0)) \quad \text{and} \quad \lambda_2 = \max(0, Q(t_0)).$$

By what we have above the solution $y_1^*(t)$ of the equation

$$y' = Ay - B|y| - C \ ; \qquad y(t_0) = \lambda_1, \qquad t \geq t_0$$

is negative for $t \geq t_0$ and therefore is given by

$$y_1^*(t) = \lambda_1 \exp \int_{t_0}^{t} (A+B)ds - \int_{t_0}^{t} \{C(s)\exp \int_{s}^{t} (A+B)d\xi\}ds, \quad t \geq t_0 .$$

From the theory of differential inequalities it follows that $y_1(t) \geq y_1^*(t)$ and the estimate (2.3) takes an explicit form with $y_1(t)$ replaced by $y_1^*(t)$.

Similarly, if $y_2^*(t)$ is the solution of the equation

$$y' = Ay - B|y| - C; \quad y(t_0) = \lambda_2, \qquad t \leq t_0$$

then $y_2^*(t)$ is positive for $t \leq t_0$ and is given by

$$y_2^*(t) = \lambda_2 \exp \int_{t_0}^{t} (A-B)ds - \int_{t_0}^{t} \{C(s)\exp \int_{s}^{t} (A-B)d\xi\}ds, \quad t \leq t_0 .$$

Also $y_2^*(t) \geq y_2(t)$ and the estimate (2.4) takes an explicit form with $y_2(t)$ replaced by $y_2^*(t)$.

It is clear from the foregoing explicit formulas that one could derive various lower bounds under appropriate conditions on the coefficients involved as it was done in [1] and [2].

REFERENCES

[1] S. AGMON and L. NIRENBERG, Lower bounds and uniqueness theorems for solutions of differential equations in a Hilbert space, Comm. Pure Appl. Math., 20 (1967), 207-229.

[2] A. FRIEDMAN, Partial Differential Equations. Holt, Rinehart and Winston, New York, 1969.

[3] V. LAKSHMIKANTHAM and S. LEELA, Differential and Integral Inequalities, Theory and Applications. Vol. I, Academic Press, New York, 1969.

[4] M. LEES, Asymptotic behavior of solutions of parabolic differential inequalities, Can. J. Math., 14 (1962), 626-631.

[5] H. OGAWA, Lower bounds for solutions of parabolic differential inequalities, Can. J. Math., 19 (1967), 667-672.

[6] H. OGAWA, On the maximum rate of decay of solutions of parabolic differential inequalities, Arch. Rational Mech. Anal., 38 (1970), 173-177.

[7] M.H. PROTTER, Properties of solutions of parabolic equations and inequalities, Can. J. Math., 13 (1961), 331-345.

University of Rhode Island, Kingston, Rhode Island

REMARKS ON LINEAR DIFFERENTIAL EQUATIONS WITH DISTRIBUTIONAL PERTURBATIONS

A. Lasota

1. Introduction

The purpose of this lecture is to present new theorems and open questions concerning the differential equation

$$x' = Ax + T \tag{1}$$

where A is a $d \times d$-matrix with C^∞ coefficients and T is a distribution. Section 2 contains basic notation and definitions. Section 3 is devoted to the study of boundary value problems and Sections 4 and 5 to the study of control problems.

2. Notation

For simplicity we restrict ourselves to the real case. As usual, by D we denote the space of all infinitely differentiable functions $\phi : R \to R$ with compact support and by D' the corresponding space of distributions. Both spaces are considered with the usual notation of the convergence. By $D_+(\Delta)$ we denote the subset of D which consists of all positive nontrivial functions with support in the interval Δ; that is

$$D_+(\Delta) = \{\phi : \phi(t) \geqslant 0, \ \phi(t) \not\equiv 0, \ \text{supp}\, \phi \subset \Delta\} .$$

The Cartesian products

$$\underbrace{D \times \cdots \times D}_{\text{d-times}}, \qquad \underbrace{D' \times \cdots \times D'}_{\text{d-times}}, \ldots$$

will be shortly denoted by $(D)^d$, $(D')^d$, ... , respectively. For $T \in (D')^d$ and $\phi \in (D)^d$, where $T = (T_1,\ldots,T_d)$ and $\phi = (\phi_1,\ldots,\phi_d)$, we write

$$\langle T,\phi \rangle = (T_1(\phi_1),\ldots,T_d(\phi_d)) .$$

The space of all $d \times d$-matrices with infinitely differentiable coefficients will be denoted $(C^\infty)^{d\times d}$. A sequence $\{A_n\} \subset (C^\infty)^{d\times d}$ will be called convergent to $A \in (C^\infty)^{d\times d}$ if for each integer $k \geqslant 0$

$$\lim_{n\to\infty} \frac{d^k A_n}{dt^k} = \frac{d^k A}{dt^k}$$

uniformly on compact sets. The same notion of convergence is introduced in the space $(C^\infty)^d$ of infinitely differentiable d-vector functions.

For any constant $d \times d$-matrix A, by $\|A\|$ we shall denote the norm of the linear operator $x \to Ax$ in the Euclidean space R^d; that is

$$\|A\| = \sup \{\|Ax\| : \|x\| = 1\} , \quad \|x\|^2 = \sum_{i=1}^{d} x_i^2 .$$

3. Boundary Value Problems

If T is a regular distribution (i.e. T corresponds to a locally integrable function), then all distributions satisfying equation (1) are regular and they correspond to the d-dimensional family of solutions in the Carathéodory sense. From this it follows easily the following

PROPOSITION 1. Suppose that for given $A \in (C^\infty)^{d\times d}$ and $\phi \in (D)^d$ the homogeneous problem

$$x' = Ax , \quad \langle x,\phi \rangle = 0 \tag{2}$$

has the unique $(C^\infty)^d$ solution $x \equiv 0$. Then for each $r \in R^d$ and

$T \in (D')^d$ there exists exactly one solution $x \in (D')^d$ of the boundary value problem

$$x' = Ax + T \;, \quad \langle x, \phi \rangle = r \;. \tag{3}$$

For the differential equation (1) the initial value (Cauchy) problem is meaningless. We can replace however the "sharp" Cauchy condition $x(0) = r$ by the "smooth" condition $\langle x, \phi \rangle = r$. The following theorem gives the exact estimation of the support of ϕ for which the boundary value problem (3) is correctly stated [4]:

<u>THEOREM 2</u>. Suppose that $A \in (C^\infty)^{d\times d}$, $T \in (D')^d$, $\phi \in (D_+(\Delta))^d$, $r \in R^d$ and suppose that

$$\int_\Delta A(t)\, dt < \frac{\pi}{2} \;. \tag{4}$$

Then there exists exactly one solution $x \in (D')^d$ of the boundary value problem (3).

<u>Proof</u>. By Proposition 1 it is sufficient to show that the homogeneous boundary value problem (2) admits only zero solution (in $(C^\infty)^d$). Suppose that x is such a solution of (2). Setting $x = (x_1,\ldots,x_d)$ and $\phi = (\phi_1,\ldots,\phi_d)$ we have

$$\int_{-\infty}^{+\infty} x_i(t)\phi_i(t)dt = 0 \;, \qquad i = 1,\ldots,d \;.$$

Since $\phi_i \in D_+(\Delta)$ there exist points $t_1,\ldots,t_d$ such that

$$x_i(t_i) = 0 \;, \qquad i = 1,\ldots,d \;. \tag{5}$$

Moreover, from the equation (1) it follows that

$$\|x'(t)\| \leqslant \|A(t)\| \, \|x(t)\| \;. \tag{6}$$

It is known [2] that (4), (5) and (6) imply $x \equiv 0$. This finishes the proof.

<u>THEOREM 3</u>. Suppose that

$$\lim_{n\to\infty} A_n = A \;, \quad \lim_{n\to\infty} \phi^n = \phi \;, \quad \lim_{n\to\infty} T_n = T \;, \quad \lim_{n\to\infty} r^n = r$$

where the sequences are convergent in the spaces $(C^\infty)^{d\times d}$, $(D)^d$, $(D')^d$ R^d respectively. Suppose, moreover, that

$$\phi \in (D_+(\Delta))^d \qquad \text{and} \qquad \int_\Delta \|A(t)\|dt < \frac{\pi}{2} .$$

Then for sufficiently large n the sequence $\{x^n\}$ of solutions of the boundary value problems

$$(x^n)' = A_n x^n + T_n , \qquad \langle x^n, \phi^n \rangle = r^n \tag{7}$$

is uniquely determined and converges (in $(D')^d$) to the solution x^o of the boundary value problem (3).

Proof. Denote by U and U_n the fundamental matrices of the equations $x' = Ax$ and $x' = A_n x$ respectively. We admit that $U(0) = U_n(0) = I$ (I = identity). By the usual continuous dependence argument we have

$$\lim_{n\to\infty} U_n = U$$

in the space $(C^\infty)^{d\times d}$. Since the homogeneous boundary value problem (2) has only the zero solution (according to Theorem 2), the matrix of the linear system

$$\langle Uc, \phi \rangle = 0$$

(with unknown c) is nonsingular. Consequently the matrices of the systems

$$\langle U_n c, \phi^n \rangle = 0$$

are nonsingular for sufficiently large n. This implies the existence and the uniqueness of the solutions x^n for sufficiently large n. Now, in order to complete the proof it is enough to write the explicit formulas for the solutions x^o and x^n respectively.

4. Control Theory (Controllability)

Consider now, the differential equation

$$x' = Ax + bu \tag{8}$$

with given $A \in (C^\infty)^{d\times d}$ and $b \in (C^\infty)^d$. Denote by D'_+ the set of all distributions with support in $[0,\infty)$. For any control $u \in D'_+$ there exists exactly one solution $x \in (D'_+)^d$ of (8). This solution will be called corresponding to u.

For $u = \delta^{(k)}$ (k^{th} derivative of δ - Dirac) the corresponding solution is given by the formula

$$(9) \qquad x = \sum_{i=1}^{k} (L^{k-i}b)\theta^{(i)} + U(L^k b)_0\theta$$

where U is the fundamental matrix of the homogeneous system $x' = Ax$ ($U(0) = I$) and L denotes the known in the optimal control theory operator

$$Lx = (A - \frac{d}{dt})x .$$

θ denotes the Heaviside function and the index zero in the second term denotes that the value is taken for $t = 0$.

The first term in (9) is a distribution with the support contained in the origin. The second is the infinitely differentiable curve starting from the point $(L^k b)_0$. Therefore any point

$$(10) \qquad x^k = \sum_{k=0}^{d-1} \lambda_k (L^k b)_0$$

of the subspace H of R^n spanned on the vectors

$$(b)_0 , (Lb)_0 , \ldots , (L^{d-1}b)_0$$

can be reached in the zero time by the control

$$(11) \qquad u^k = \sum_{k=0}^{d-1} \lambda_k \delta^{(k)} .$$

The classical result of the optimal control theory that any point of H is available can be proved for equation (8) in an extremely simple way. Namely for any such a point (10) we can show the required control (11). Moreover from this it follows that the available set for regular controls also contains H. In fact we can approximate (11) by regular distributions with support in any right-hand side neighbourhood of the origin. By continuous dependence argument, the corresponding solutions will approximate on $(0,\infty)$ (uniformly on compact sets) the regular part of (9). Therefore the available set for equation (8) with regular controls is dense in H. But this set is also a linear subspace of R^d and consequently it contains H.

5. Control Theory (Closedness of the Set of Solutions)

Let $F(t)$ be a set-valued mapping defined on the real line. We assume that for every $t \geqslant 0$ the set $F(t)$ is convex, closed and nonempty. For $t < 0$ we let $F(t) = \{0\}$. Consider the differential equation

$$x' = Ax + u \tag{12}$$

with given $A \in (C^\infty)^{d\times d}$. The control $u \in (D')^d$ will be called admissible if for each test function $\phi \in (D)^d$ there exists a locally integrable function $f : R \to R^d$ such that

$$\sum_{i=1}^{d} u_i(\phi_i) = \sum_{i=1}^{d} \int_{-\infty}^{+\infty} f_i(t)\phi_i(t)dt .$$

A regular control u is admissible if and only if $u(t) \in F(t)$ almost everywhere. A solution $x \in (D'_+)^d$ of (12) corresponding to an admissible (regular) control will be also called admissible (regular). Using the techniques developed in [3] we can prove the following

THEOREM 4. Suppose that F is upper semicontinuous in the sense of Cesari [1]; that is

$$F(t) = \bigcap_{\delta>0} \overline{\text{co}} \bigcup_{|s-t|<\delta} F(s) , \quad t \in R . \tag{13}$$

Then the set of all regular admissible solution of equation (12) is closed in $(D')^d$.

The theorem fails to be true if we omit assumption (13) but the following questions remain open:

i) Is the set of all admissible solutions closed for any $F(t)$ with convex closed values?

ii) Is the set of all admissible solutions closed for any F(t) satisfying the Cesari condition (13)?

Since any F satisfying (13) is convex closed valued the first question is more general.

REFERENCES

[1] L. CESARI, Existence theorems for optimal solutions in Pontryagin and Lagrange problems, J. SIAM Control, Ser. A, 3 (1966), 475-498.

[2] A. LASOTA and C. OLECH, An optimal solution of Nicoletti's boundary value problem, Ann. Polon. Math. 18 (1966), 131-139.

[3] A. LASOTA and C. OLECH, On the closedness of the set of trajectories of a control system, Bull. Acad. Polan. Sci., Ser. sci. math., astr. et phys. 14 (1966), 615-621.

[4] A. LASOTA and J. TRAPLE, Nicoletti boundary value problems for systems of linear differential equations with distributional perturbations, Zeszyty Naukowe U.J. (in press).

Jagellonian University, Krakow, Poland

COMPUTING BOUNDS FOR FOCAL POINTS AND FOR σ-POINTS FOR SECOND-ORDER LINEAR DIFFERENTIAL EQUATIONS*

Walter Leighton

1. Introduction

In this paper we shall be concerned with computing bounds for focal points and for σ-points for differential equations of the form

$$[r(x)y']' + p(x)y = 0 \ , \tag{1.1}$$

where $r(x) > 0$, $p(x) \geqslant 0$, and $r(x)$ and $p(x)$ are continuous, unless otherwise indicated, on a suitable interval I. The corresponding problem for eigenvalues and conjugate points was treated in [3].

If $y(x)$ is a nonnull solution of (1.1) that vanishes at a point $x = a \in I$, a point $x = \sigma \in I$ at which its derivative vanishes is called a σ-point of $x = a$. A point $x = c$ of I where $y(x)$ vanishes is, of course, a conjugate point of $x = a$. A focal point of a line $x = a$, $a \in I$, it will be recalled, is defined as follows. Let $y_1(x)$ be a solution of (1.1) such that $y_1'(a) = 0$, $y_1(a) \neq 0$. Any zero of $y_1(x)$ is called a focal point of the line $x = a$.

Because the methods to be described, when applied to the differential equation

$$y'' + p(x)y = 0 \ , \tag{1.2}$$

---

*This will acknowledge the partial support of the author by the U.S. Army Research Office (Durham) under the grant numbered DA-ARO-D-31-124-G1007.

are readily extended to the more general equation (1.1) we shall concern ourselves with differential equations of the form (1.2).

To develop the method, we consider an equation (1.2), where $p(x)$ is a step-function constant on subintervals. Specifically, let the interval $I:[a,b]$ be divided into n subintervals $I_1, I_2, \dots, I_n$, where $p(x) = c_i^2$ on I_i, $c_i \geqslant 0$. Assume first that all $c_i > 0$. The solution on the subinterval I_i $(i = 1, 2, \dots, n)$ and its derivative may be written

$$(1.3)\qquad \begin{aligned} y_i(x) &= \rho_i \sin(c_i x + \theta_i) , \\ y_i'(x) &= c_i[\rho_i \cos(c_i x + \theta_i)] , \end{aligned}$$

where

$$\rho_i = \sqrt{\alpha_i^2 + \beta_i^2} \neq 0, \quad \rho_i \sin\theta_i = \alpha_i, \quad \rho_i \cos\theta_i = \beta_i, \quad 0 \leqslant \theta_i \leqslant \pi ,$$

and α_i and β_i are constants $(i = 1, 2, \dots, n)$.

A solution of the equation (1.2) for such a function $p(x)$ must be of class C'. If I_i is the subinterval (a_{i-1}, a_i), this observation leads to the conditions

$$(1.4)\qquad \begin{aligned} \rho_{i+1}\sin(c_{i+1}a_i + \theta_{i+1}) &= \rho_i \sin(c_i a_i + \theta_i) , \\ c_{i+1}\rho_{i+1}\cos(c_{i+1}a_i + \theta_{i+1}) &= c_i \rho_i \cos(c_i a_i + \theta_i) , \end{aligned}$$

at $x = a_i$ $(i = 1, 2, \dots, n-1)$. Equations (1.4) lead to the equations

$$(1.5)\qquad \tan(c_{i+1}a_i + \theta_{i+1}) = \frac{c_{i+1}}{c_i}\tan(c_i a_i + \theta_i) \quad (i = 1, 2, \dots, n-1),$$

at least when both members of the equation are defined.

For convenience, we now assume that the subintervals are equal, that $a = 0$, and that $b/n = h$. Then, $a_i = ih$ $(i = 1, 2, \dots, n)$, and equation (1.5) becomes

$$(1.6)\qquad \tan(ic_{i+1}h + \theta_{i+1}) = \frac{c_{i+1}}{c_i}\tan(ic_i h + \theta_i) \quad (i = 1, 2, \dots, n-1) .$$

Finally, if we set

$$(1.7)\qquad \begin{aligned} z_1 &= c_1 h + \theta_1 \\ z_j &= (j-1)c_j h + \theta_j \quad (j = 2, 3, \dots, n) , \end{aligned}$$

equations (1.6) may be written in the form

$$\tan z_2 = \frac{c_2}{c_1} \tan z_1 \ ,$$

(1.8)

$$\tan z_k = \frac{c_k}{c_{k-1}} \tan(c_{k-1}h + z_{k-1}) \quad (k = 3, 4, \ldots, n) \ .$$

In the important case $n = 2$, the second line in (1.8) disappears, of course.

We have assumed that $p(x) \geqslant 0$ in (1.2). The form of solution of (1.2) on a subinterval where $c_i = 0$ is, of course, not trigonometric. The alteration in the method will be clear if we assume, for simplicity, that $c_1 = 0$ and $c_j > 0$ $(j = 2, 3, \ldots, n)$. In that case the first equation in (1.8) becomes simply

$$\tan z_2 = c_2 h + \beta_1 \ , \tag{1.9}$$

where β_1 is a suitably chosen constant.

2. σ-Points

Our present objective is the approximation of σ-points when $p(x)$ is continuous and nonnegative. Our method will involve approximating $p(x)$ by step-functions. For simplicity, we shall assume $p(x) > 0$. The last comment in Section 1 indicates the kind of alteration required when $p(x)$ is permitted to vanish - indeed, the kind of alteration, if $p(x)$ becomes actually negative.

Suppose that $p(x)$ is positive and continuous for $x \geqslant 0$ and let an interval containing $x = \sigma$ be divided into n equal parts (such an interval is, for example, $[0, \bar{b}]$, where $\bar{b}$ is any upper bound of σ [see 4]), $[a_0, a_1], [a_1, a_2], \ldots, [a_{n-1}, a_n]$, where $a_0 = 0$. Let p_i be the maximum of $p(x)$ on I_i: $[a_{i-1}, a_i]$, and consider the step-function defined by the p_i. It follows from an easy modification of an earlier result [2, Th. 1] that if σ exists, the differential equation (1.2) with $p(x)$ replaced by this step-function will have a σ-point $\sigma_0 \leqslant \sigma$, with strict inequality holding if $p(x) \not\equiv$ constant. In (1.7) we then take $\theta_1 = 0$, and there will then be a first index $i = m$ and a number $x = \sigma_0$ such that (see (1.3))

$$(2.1) \qquad c_m h + z_m = \pi/2 \ , \quad \sigma_0 = mh \ .$$

Equations (1.8) are the "connecting" conditions, it will be recalled, and we now take $n = m$. Condition (2.1) adjoined to (1.8) then determine σ_0, which is a lower bound for σ.

Similarly, if the p_i are taken to be the minimum values of $p(x)$ on the intervals I_i, conditions (1.8) with (2.1) adjoined will, in general,* determine an upper bound σ_1 for σ.

Example. Consider the differential equation

$$(2.2) \qquad y'' + (7 - x^2)y = 0 \ .$$

The σ-point of the origin is readily seen to be $\sigma = \sqrt{9 - \sqrt{57}}/2 = 0.6021$ (a solution of (2.2) is $(2x^3 - 3x)\exp(-x^2/2)$). If we take $m = 4$, our equations (1.8) and (2.1) become

$$(2.3) \qquad \begin{aligned} \tan z_2 &= \frac{c_2}{c_1} \tan c_1 h \ , \\ \tan z_3 &= \frac{c_3}{c_2} \tan (c_2 h + z_2) \ , \\ \tan z_4 &= \frac{c_4}{c_3} \tan (c_3 h + z_3) \ , \\ c_4 h + z_4 &= \pi/2 \ , \end{aligned}$$

where** $\sigma_j = 4h$ $(j = 0, 1)$. Equations (2.3) lend themselves readily to solution for h by successive approximations. We obtain $\sigma_0 = 0.60$, $\sigma_1 = 0.62$, to two decimal places.

---

* It is conceivable that $p(x)$, although possessing a σ-point, may be decreasing, for example, so rapidly that the step-function of minimum values of $p(x)$ may not possess a σ-point for any finite value of m. We exclude this case from consideration.

**Note that this condition, in effect, defines the interval we employ.

3. Convergence

The process that we propose involves subdividing an interval containing a σ-point into n equal subintervals and employing two step-functions -- one determined by the minimum values of $p(x)$ on the subintervals, and the other the maximum values. For definiteness, we shall assume that the nth step in the process involves subdividing the interval into n equal subintervals.

LEMMA 3.1. Let $p(x)$ be continuous for $x \geq a$, and let $y(x)$ be the solution of the system

$$(3.1)\qquad \begin{gathered} y'' + p(x)y = 0 \\ y(a) = \alpha\ , \quad y'(a) = \alpha' \end{gathered}$$

on $[a, g]$ $(g > a)$. If $q_n(x) \to p(x)$ uniformly on $[a, g]$, the solution $z_n(x)$ of the system

$$(3.2)\qquad \begin{gathered} z_n'' + q_n(x)z_n = 0\ , \\ z_n(a) = \alpha\ , \quad z_n'(a) = \alpha' \end{gathered}$$

approaches $y(x)$ uniformly on $[a, g]$.

This result is known (see for example, [1, p. 4]).

It follows from the lemma that $\lim \sigma_0 = \lim \sigma_1 = \sigma$, since the step-functions we employ clearly approximate the continuous function $p(x)$ uniformly. A similar analysis provides a proof of convergence for the approximation of conjugate points outlined in [3].

It should be pointed out that if the following subsequence of the process is employed, $\sigma_0 \to \sigma$ and $\sigma_1 \to \sigma$, monotonely. Let the first step be the subdivision of the original interval into m_1 equal subintervals and the second be the subdivision of each of these subintervals into m_2 equal parts, and so on. Then by Theorem 1 of [2] (which is readily shown to be valid for such step-functions), the lower bounds σ_0 never decrease, while the upper bounds never increase.

4. Focal Points

The method employed for determining bounds for the focal point f of the y-axis for the equation

$$y'' + p(x)y = 0 \,, \tag{4.1}$$

where $p(x)$ is positive and continuous for $x \geq 0$ will be seen to be general. In (1.7) we now take $\theta_1 = \pi/2$. Equations (1.8) are then unaltered, and equation (2.1) becomes

$$c_m h + z_m = \pi \,, \quad f_0 = mh \,, \tag{4.2}$$

where f_0 is the lower bound provided by the process for f. The process for obtaining an upper bound f_1 for f is strictly analogous.

We are here tacitly employing the following result that requires proof.

THEOREM 4.1. Suppose that $p(x)$ and $q(x)$ are piecewise continuous functions for $x \geq 0$ and that $q(x) \geq p(x)$. If the y-axis has a first (positive) focal point f with respect to equation (4.1), the first focal point f_1 of the y-axis exists with respect to

$$z'' + q(x)z = 0 \,, \tag{4.3}$$

and $f_1 \leq f$, with strict inequality, if $q(x) \not\equiv p(x)$.

To prove the theorem, let $y(x)$ and $z(x)$ be solutions of (4.1) and (4.3), respectively, such that

$$y(0) = z(0) = 1 \,,$$
$$y'(0) = z'(0) = 0 \,.$$

Then,

$$z(x)y'(x) - y(x)z'(x) \equiv \int_0^x [q(x) - p(x)]y(x)z(x)dx \quad (x \geq 0) \,.$$

Suppose that $z(x) > 0$ on $(0, f]$. Then,

$$z(f)y'(f) = \int_0^f [q(x) - p(x)]y(x)z(x)dx \,. \tag{4.4}$$

The right-hand member of (4.4) is nonnegative, and $y'(f) < 0$. It follows that $z(f) \leqslant 0$, and we have a contradiction. Further, if $z(f) = 0$, and if f is the first focal point of $z(x)$, the integral in (4.4) must be zero; that is, $q(x) \equiv p(x)$.

The proof of the theorem is complete, for if f is a zero of $z(x)$ other than the first, the result is obvious.

The method may also be employed for approximating the kth σ-point or the kth focal point by obvious modification [see [3]).

Example. Consider the differential equation

$$y'' + (7 - x^2)y = 0 . \tag{4.5}$$

If we take $m = 4$, our equations of approximation become

$$\tan z_2 = -\frac{c_2}{c_1}\cot(c_1 h) ,$$

$$\tan z_3 = \frac{c_3}{c_2}\tan(c_2 h + z_2) ,$$

$$\tan z_4 = \frac{c_4}{c_3}\tan(c_3 h + z_3) ,$$

$$c_4 h + z_4 = \pi ,$$

where $f_j = 4h$ $(j = 0, 1)$. A computation yields $f_1 = 0.60$, correct to two decimal places, while a lower bound $f_0 = 0.59$, correct to two decimal places. It is known [4] that $f < \sigma$, for (4.5).

REFERENCES

[1] PHILIP HARTMAN, Ordinary Differential Equations. Wiley, New York, 1964.

[2] WALTER LEIGHTON, Some elementary Sturm theory, J. Diff. Equations, 4 (1968), 187-193.

[3] __________, Upper and lower bounds for eigenvalues, J. Math. Anal. and Appl., to appear.

[4] __________, More elementary Sturm theory, to appear in the Journal of Applicable Analysis.

University of Missouri, Columbia, Missouri

GLOBAL CONTROLLABILITY OF NONLINEAR SYSTEMS

D. L. Lukes

This article deals with the controllability of nonlinear differential systems which arise when a linear system is perturbed. The sufficiency conditions presented insure that the nonlinear system will be (globally) controllable whenever its linear part is controllable. Moreover, the steering can be accomplished using continuous controls with arbitrarily prescribed initial and final values.

1. Introduction

The differential control system to be considered is of the type

$$\dot{x} = f(u,x,t) = Ax + Bu + h(u,x,t) \tag{1.1}$$

with the standing assumption that the perturbation $h(u,x,t)$ is continuous at all $(u,x,t) \in R^{r+n+1}$. The linear part is given in terms of the real matrices A, B of respective sizes $n \times n$, $n \times r$.

The aim of this article is to present conditions upon A, B and $h(u,x,t)$ which insure that for each $t_0, t_1 \in R^1$; $\alpha, \beta \in R^r$ and $x_0, x_1 \in R^n$ there exists a cortinuous control $u(\cdot)$: $[t_0, t_1] \to R^r$ for (1.1) with $u(t_0) = \alpha$, $u(t_1) = \beta$ which produces a response $x(t)$ satisfying the boundary conditions $x(t_0) = x_0$, $x(t_1) = x_1$.

The results generalize and strengthen the well known controllability theory of linear system (systems with $h \equiv 0$). By restricting our attention to systems (1.1) with a controllable linear part we are able to obtain global results for systems in which the state variable x and control variable u enter the system equation in a nonlinear fashion.

Complete proofs of all theorems and additional results may be found in [2].

To simplify the notation we make the preliminary translation of the origin $\tau = t - t_0$ which sends $t_0 \to 0$ and $t_1 \to T = t_1 - t_0$. This does not destroy the generality of the results.

2. Sufficient Conditions for Global Controllability

The next lemma will be vital to the perturbation analysis to follow.

LEMMA 2.1. Define $\Phi(\omega) = \int_0^\omega e^{A\sigma}B d\sigma$ and let $T > 0$. If rank $[B, AB, \ldots, A^{n-1}B] = n$ then the matrix

$$S_T = \int_0^T \Phi(\omega)\Phi^*(\omega)d\omega - \frac{1}{T}\left(\int_0^T \Phi(\omega)d\omega\right)\left(\int_0^T \Phi^*(\omega)d\omega\right) \tag{2.1}$$

is symmetric and positive definite.†

The following notations will prove convenient:

$$C_T(t) = \int_{T-t}^T \Phi^*(\omega)d\omega - \frac{t}{T}\int_0^T \Phi^*(\omega)d\omega \tag{2.2}$$

$$S_T(t) = \int_0^t e^{A(t-\tau)}BC_T(\tau)d\tau \quad . \tag{2.3}$$

(We will show $S_T(T) = S_T$.) Lemma 2.1 states a condition under which S_T is non-singular and hence we write down the following system of equations whose importance will become apparent in the next lemma.

---

†We denote the transpose of a matrix Φ by Φ^*.

$$(2.4) \qquad x(t) = e^{At}x_0 + \Phi(t)u_0 + \frac{1}{T}\int_0^t \Phi(\omega)d\omega(u_T - u_0) + S_T(t)y + \int_0^t e^{A(t-\omega)}h(u(\omega),\ x(\omega),\ \omega)d\omega$$

$$(2.5) \qquad u(t) = (1 - \frac{t}{T})u_0 + (\frac{t}{T})u_T + C_T(t)y$$

$$(2.6) \qquad y = S_T^{-1}[x_T - e^{AT}x_0 - \Phi(T)u_0 - \frac{1}{T}\int_0^T \Phi(\omega)d\omega(u_T - u_0)] - S_T^{-1}\int_0^T e^{A(T-\omega)}h(u(\omega),\ x(\omega),\ \omega)d\omega\ .$$

<u>LEMMA 2.2</u>. Any solution $x(t)$, $u(t)$ to the nonlinear functional equations (2.4)-(2.6) provides a solution to the boundary value problem

$$(2.7) \qquad \dot{x}(t) = Ax(t) + Bu(t) + h(u(t),\ x(t),\ t)$$

$$(2.8) \qquad u(0) = x_0\ , \qquad x(T) = x_T$$

$$(2.9) \qquad u(0) = u_0\ , \qquad u(T) = u_T\ .$$

<u>Remark 2.1</u>. Any solution $x(t)$, $u(t)$ to (2.4)-(2.6) will have $u(t)$ analytic in t and $x(t)$ continuous along with its first derivative. In general the differentiability will be limited only by the differentiability of h.

The following theorem presents our first sufficiency conditions which insure the global controllability of (1.1).

<u>THEOREM 2.1</u>. Consider the differential control system in R^n

$$(2.10) \qquad \dot{x} = f(u,x,t) = Ax + Bu + h(u,x,t)$$

with $h(u,x,t)$ continuous and $|h(u,x,t)|$ bounded on R^{r+n+1}. Further assume that $h(u,x,t)$ is periodic in t with period $T > 0$ for each fixed u,x. Finally assume that $\text{rank}\ [B,\ AB,\ \ldots,\ A^{n-1}B] = n$.

Then for each $\alpha,\beta \in R^r$; $x_0,x_1 \in R^n$ and $0 < t_1 < T$ there exists a control $u(\cdot) \in C^{\omega}([0,T],\ R^r)$ such that

(i) $u(0) = u(T) = \alpha$, $u(t_1) = \beta$ and

(ii) a corresponding solution of (2.10) for which $x(0) = x_0$ satisfies both $x(t_1) = x_1$, $x(T) = x_0$.

This control extends to a continuous periodic function on $(-\infty,\infty)$ with period T and its response agrees with the periodic extension of $x(t)$.

Remark 2.2. We do not require that the period T be a minimal period. Hence the theorem applies to the special case where $h = h(u,x)$ is independent of t (autonomous systems). In particular it applies to the case where $h \equiv 0$ (linear homogeneous systems). But even in the latter case it still says more than what is usually stated (see [1, p. 32]) - namely not only can the response be steered through any two states but moreover with any attainable prescribed velocities.

Remark 2.3. We may drop the periodicity requirement on $h(u,x,t)$ altogether. Then $T > 0$ is arbitrary. The conclusions of the theorem are all valid with the exception of the last sentence concerning the periodicity which is lost.

Remark 2.4. The class of systems covered by Theorem 2.1 is quite extensive. Any continuous $f(u,x,t)$ on a compact set in R^{r+n+1} will do as long as its extension to the complement of that set is of the type (2.10). Hence in particular it covers the situation where $h(u,x,t) = -(Ax + Bu)$ on a compact neighborhood of the origin, making $f(u,x,t) \equiv 0$ there, if outside some nonempty compact set $f(u,x,t)$ is of the form (2.10) with $h(u,x,t)$ continuous and $|h(u,x,t)|$ bounded on R^{r+n+1}. In such a case the control steering away from the origin would start out by "winding up" the value of $u(t)$ sufficiently large to get (u,x) out of the region where $f(u,x,t) \equiv 0$, etc.

Remark 2.5. When $f(u,x,t)$ has a representation of the type (2.10) described in Theorem 2.1 the linear parts can be computed by the formulas

$$Ax = \lim_{\varepsilon\to\infty} \frac{f(u,\varepsilon x,t)}{\varepsilon}$$

$$Bu = \lim_{\varepsilon\to\infty} \frac{f(\varepsilon u,x,t)}{\varepsilon} \quad .$$

Remark 2.6. We choose to avoid a formal definition of controllability. In the linear theory the usual definition as the capability to steer the response through any two states in finite time turns out to be equivalent to the condition rank $[B, AB, \ldots, A^{n-1}B] = n$. Hence we have shown that the controllability of a linear system is preserved when we add a perturbation $h(u,x,t)$ with $|h(u,x,t)|$ bounded.

Now we turn our attention to a different condition upon $h(u,x,t)$ which allows $|h(u,x,t)|$ to be unbounded.

THEOREM 2.2. Consider the one parameter family of differential control systems in R^n

$$\dot{x} = f_\varepsilon(u,x,t) = Ax + Bu + \varepsilon h(u,x,t) \tag{2.11}$$

with $h(u,x,t)$ continuous and

$$|h(u,x,t) - h(v,y,t)| \leqslant L[|u - v| + |x - y|] \tag{2.12}$$

on R^{r+n+1} for a constant L. Further assume that $h(u,x,t)$ is periodic in t with period $T > 0$ for each fixed u, x. Suppose that rank $[B, AB, \ldots, A^{n-1}B] = n$.

Then the conclusion in Theorem 2.1 and the remarks following it are valid for (2.11) for all ε in a neighborhood of $\varepsilon = 0$.

3. Comments About the Theorems and Their Proofs

Consider the two dimensional system

$$\dot{x}_1 = x_2 + \varepsilon(x_1^2 + x_2^2) \tag{3.1}$$

$$\dot{x}_2 = u \ . \tag{3.2}$$

By completing the square on the right hand side of (3.1) one can easily show this system is globally controllable only for $\varepsilon = 0$. Hence the requirement (2.12) in Theorem 2.2 is not superfluous.

The scalar system

$$\dot{x} = u + \varepsilon(u^2 + x^2)^{\frac{1}{2}} \tag{3.3}$$

is controllable for $\varepsilon = 0$ and satisfies an inequality of type (2.12).

Therefore Theorem 2.2 applies, giving global controllability for $|\varepsilon|$ small. However the requirement that $|\varepsilon|$ be small cannot be removed. This conclusion follows from the observation that $\varepsilon\dot{x}(t) \geq 0$ for all $t \in R^1$ if $|\varepsilon| \geq 1$. Estimates of the range of $|\varepsilon|$ over which the global controllability of (2.11) persists are given in [2].

The proofs of Theorems 2.1 and 2.2 are based upon establishing existence of solutions to (2.4)-(2.6) and application of Lemma 2.2. The boundedness of $h(u,x,t)$ allows the Arzela-Ascoli Theorem to operate in the proof of Theorem 2.1 and Theorem 2.2 is obtained by using (2.12) to develop a contractive mapping argument. In both cases the required controls are obtained as fixed points. Recall that the controllability of the linear part of (1.1) assured the nonsingularity of the matrix S_T in (2.4)-(2.6) and thereby entered the proof at an early stage.

4. An Example of a Globally Controllable Nonlinear System

This section demonstrates how Theorem 2.1 can be applied to detect controllability in certain nonlinear systems. The system is

$$\dot{x}_1 = D(x_2) \tag{4.1}$$

$$\dot{x}_2 = x_3 + u \tag{4.2}$$

$$\dot{x}_3 = -k_1x_1 + G(u - k_2x_2) + S(u) \tag{4.3}$$

where k_1, k_2 are parameters and D, G, S are the nonlinear functions

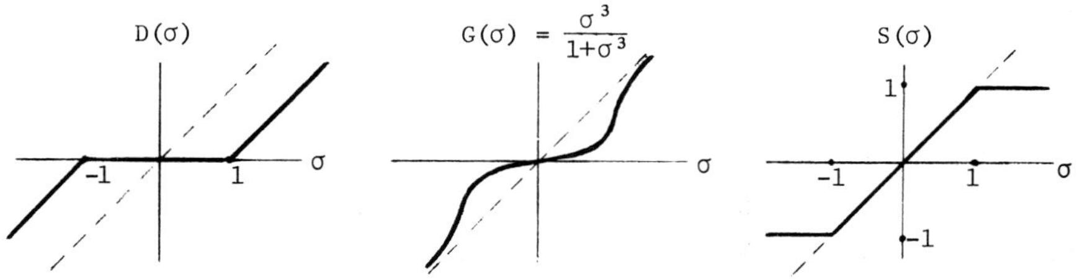

One can easily compute

$$\lim_{\varepsilon\to\infty} \frac{D(\varepsilon x_2)}{\varepsilon} = x_2 \tag{4.4}$$

$$\lim_{\varepsilon\to\infty} \frac{G(u - k_2\varepsilon x_2)}{\varepsilon} = - k_2x_2 \tag{4.5}$$

$$\lim_{\varepsilon\to\infty} \frac{G(\varepsilon u - k_2x_2)}{\varepsilon} = u \tag{4.6}$$

$$\lim_{\varepsilon\to\infty} \frac{S(\varepsilon u)}{\varepsilon} = 0 \ . \tag{4.7}$$

With $f(u,x,t)$ defined by (4.1)-(4.3) and the limits (4.4)-(4.7) we can apply Remark 2.5 to determine the linear part of f,

$$Ax = \lim_{\varepsilon\to\infty} \frac{f(u,\varepsilon x,t)}{\varepsilon} = \begin{pmatrix} x_2 \\ x_3 \\ -k_1x_1 - k_2x_2 \end{pmatrix} \tag{4.8}$$

$$Bu = \lim_{\varepsilon\to\infty} \frac{f(\varepsilon u,x,t)}{\varepsilon} = \begin{pmatrix} 0 \\ u \\ u \end{pmatrix} \ . \tag{4.9}$$

This shows that

$$A = \begin{pmatrix} 0 & 1 & 0 \\ 0 & 0 & 1 \\ -k_1 & -k_2 & 0 \end{pmatrix} \tag{4.10}$$

$$B = \begin{pmatrix} 0 \\ 1 \\ 1 \end{pmatrix} \ . \tag{4.11}$$

It follows that $\operatorname{rank} [B, AB, A^2B] = 3$ if and only if $k_1 \neq k_2 + 1$. Hence Theorem 2.1 applies and we conclude that the nonlinear system (4.1)-(4.3) is globally controllable in the sense described by the theorem.

REFERENCES

[1] E.B. LEE and L. MARKUS, Foundations of Optimal Control Theory. John Wiley, New York, 1967.
[2] D.L. LUKES, Global controllability of nonlinear systems, SIAM J. Control, to appear.

University of Virginia, Charlottesville, Virginia

THE PHRAGMÉN-LINDELÖF PRINCIPLE AND A CLASS OF FUNCTIONAL DIFFERENTIAL EQUATIONS

Grainger R. Morris
Alan Feldstein
Ernie W. Bowen

1. Introduction

This paper shows the relevance of applying complex-variable methods to the equation

$$\text{(A)} \qquad \dot{x}(t) = -x(\frac{t}{k}), \qquad k > 1$$

and to some of its generalizations. We shall show that (A) with $x(0) = 1$ has a unique solution, which is entire and, what is more interesting, show by use of the Phragmén-Lindelöf principle that this solution oscillates unboundedly as $t \to \infty$.

We present existence results as Theorems 1, 2, 3 and 4 for the equations

$$\text{(B)} \qquad \dot{x}(t) = \Sigma\, A_\nu(t)x(\frac{t}{k_\nu})$$

and

$$\text{(C)} \qquad \dot{x}(t) = \Sigma\, A_\nu(t)x(\frac{t-b_\nu}{k_\nu}) .$$

Theorems 5 and 6 give results on unboundedness and oscillation of solutions for (A), (B) and (C). As an example of the very different behavior that is possible for an advanced equation, we discuss in Section 7 the equation $\dot{x}(t) = x(t/k)$ when $k < 1$.

Although it is easy and natural to obtain some intermediate results for equations more general than (A), (B) or (C), we give, instead, examples in Section 8 to show that with more general equations one cannot always expect entire solutions and hence cannot always expect the interesting properties which such solutions would have. With the methods used here, we have to accept for (C) a restriction to real k_ν and b_ν, whereas for (B) the k_ν may be complex. This distinction arises because in discussing (B) only power series are used while in (C) we use results about equivalent real integral equations before we introduce power series.

In 1965 W. R. Utz [12] raised the question of determining conditions for the existence of real nontrivial functions $x(t)$ satisfying $\dot{x}(t) = Ax(\alpha(t))$ for a given real constant A and a given real function α. Utz commented about the corresponding equation for analytic functions. In 1967 Fred Gross [5] supplied some information about Utz' question by proving that nontrivial entire analytic solutions exist for complex constant A only if $\alpha(t)$ is a linear complex function; that is $\alpha(t) = \frac{(t-b)}{k}$. In [6], Gross and Yang give results on some interesting generalizations of this question.

Originally, the work in this paper was motivated by the results of systematic computer studies made of particular cases (reported in [2]). These experiments suggested the likelihood of unbounded oscillatory solutions to equation (A). That question is fully resolved affirmatively in the current paper. However, we still leave open the following related conjectures that were also suggested by those same numerical experiments (see [2] for a tabulation of the data):

Let the zeros of equation (A) with $x(0) = 1$ be denoted by $0 < z_1 < z_2 < \cdots$.

<u>CONJECTURE 1</u>. $\frac{z_{i+1}}{z_i}$ is monotonically decreasing.

<u>CONJECTURE 2</u>.

$$\lim_{i\to\infty} \frac{z_{i+1} - z_i}{z_i - z_{i-1}} = k .$$

CONJECTURE 3. Or more strongly,

$$\lim_{i\to\infty} \frac{z_{i+1}}{z_i} = k .$$

(In these conjectures, k is the same quantity that appears in the differential equation.)

2. Existence Theorems for (B)

The right hand side of equation (B) may have either a finite or an infinite series. We shall distinguish between these two cases by re-naming the equation as (1), or (3), respectively.

THEOREM 1. If for $\nu = 1, \ldots, n$, $A_\nu(t)$ are analytic in $|t| < R$ and $|k_\nu| > 1$, then

$$\text{(1)} \qquad \dot{x}(t) = \sum_{\nu=1}^{n} A_\nu(t)\, x\left(\frac{t}{k_\nu}\right)$$

has a unique solution with $x(0) = 1$, analytic in $|t| < R$. In particular, if $R = \infty$ the solution is entire, and, if further, $A_\nu(t)$ is of finite order ρ_ν and $\max \rho_\nu = \rho$, then the solution is also of finite order ρ at most.

If the k_ν are all real and I is a closed interval (symmetric about 0 if any k_ν is negative) for which

$$0 \in I \subset (-R,R) ,$$

then the analytic solution is the only solution which is continuous on I.

Proof. We work with dominant series. For any positive ε, each of the power series

$$A_\nu(t) = \sum_\mu a_{\nu\mu} t^\mu$$

is convergent when $t = R - \varepsilon$ and hence there is a $B_\nu = B_\nu(\varepsilon)$ such that

$$|a_{\nu\mu}| < (R - \varepsilon)^{-\mu} B_\nu \qquad \text{for } \nu = 1, \ldots, n .$$

A formal solution $\sum_\mu c_\mu t^\mu$ of equation (1) must satisfy

$$\sum_\mu (\mu+1) c_{\mu+1} t^\mu = \sum_{\nu=1}^{n} \left\{ \sum_\mu a_{\nu\mu} t^\mu \cdot \sum_\mu c_\mu \left(\frac{t}{k_\nu}\right)^\mu \right\} ,$$

in which it is evident that the successive coefficients can be obtained uniquely and that they satisfy

$$|c_\mu| \leqslant C_\mu ,$$

where $C_\mu = C_\mu(\varepsilon)$ are defined by

$$\sum_\mu (\mu+1)C_{\mu+1}t^\mu = \sum_\nu B_\nu \sum_\mu (R-\varepsilon)^{-\mu} t^\mu \sum_\mu C_\mu \left(\frac{t}{k}\right)^\mu$$

and $k = \min |k_\nu|$. The last equation is equivalent to

$$\left(1 - \frac{t}{R-\varepsilon}\right) \sum_\mu (\mu+1)C_{\mu+1}t^\mu = \sum_\nu B_\nu \sum_\mu C_\mu \left(\frac{t}{k}\right)^\mu$$

from which we can obtain the explicit result

$$\frac{C_{\mu+1}}{C_\mu} = \frac{\mu}{\mu+1} \cdot \frac{1}{R-\varepsilon} + \frac{\sum_\nu B_\nu}{(\mu+1)k^\mu} ,$$

which shows that $\Sigma\, C_\mu t^\mu$ has radius of convergence $R-\varepsilon$. Hence $\Sigma\, c_\mu t^\mu$ has radius of convergence at least $R-\varepsilon$ for each ε, which is to say it has radius of convergence R at least. If $R = \infty$, the above proof applies when $R-\varepsilon$ is replaced by any finite R_1, and we find that $\Sigma\, c_\mu t^\mu$ has radius of convergence ∞.

If the $A_\nu(t)$ are entire and of finite order ρ at most, then for any ε there is a K such that on $|t| = r$

$$|A_\nu(t)| \leqslant K \exp(r^{\rho+\varepsilon}) \qquad \text{for} \quad \nu = 1, \ldots, n .$$

Write $M(r) = \max_{|t|=r} |x(t)|$. If $M(r)$ is attained for $t = t^*$, then

$$M(r) = |x(t^*)| = \left|1 + \sum_{\nu=1}^{n} \int_0^{t^*} A_\nu(s)\, x\left(\frac{s}{k_\nu}\right) ds\right|$$

$$M(r) \leqslant 2nrK \cdot \exp(r^{\rho+\varepsilon}) M\left(\frac{r}{k}\right) . \tag{2}$$

Given $r > 1$, there is an N such that

$$\frac{r}{k^N} \leqslant 1 < \frac{r}{k^{N-1}}$$

and

$$N < 1 + \frac{\log r}{\log k} .$$

In equation (2) replace r successively by $k^{-1}r, k^{-2}r, \ldots, k^{-N+1}r$ and multiply the results. We obtain

$$M(r) = Q(r)M(1)\exp\left(\frac{k}{k-1} r^{\rho+\varepsilon}\right)$$

where $Q(r) = O\{(\log r)^2\}$. This implies that x is of finite order ρ at most.

Finally if the k_ν are real, suppose that $x(t)$ and $y(t)$ are two continuous solutions of (1) on I having $x(0) = y(0) = 1$. Write

$$L = \max_{\nu} \max_{t\in I} |A_\nu(t)| ,$$

$$I_r = I \cap [-r,r] ,$$

and

$$\delta = (2nL)^{-1} .$$

Then if $t \in I_\delta$

$$\begin{aligned} |x(t) - y(t)| &= \left|\Sigma \int_0^t A_\nu(s)\left[x\left(\frac{s}{k_\nu}\right) - y\left(\frac{s}{k_\nu}\right)\right]ds\right| \\ &\leqslant nL\delta \max_{u\in I_\delta} |x(u) - y(u)| \\ &= \frac{1}{2} \max |x(u) - y(u)| . \end{aligned}$$

This gives at once that $x(t) = y(t)$ on I_δ and we can in a finite number of steps extend this result to $I_{2\delta}, \ldots, I_{(m-1)\delta}, I_{m\delta} = I$. ∎

THEOREM 2. Let the following hold:

(i) $A_\nu(t)$ are analytic in $|t| < R$ for every ν.

(ii) $|k_\nu| \geqslant k > 1$.

(iii) For all $r < R$, $\sum_{\nu=1}^{\infty} \max_{|t|=r} |A_\nu(t)| < \infty$.

Then

$$\dot{x}(t) = \sum_{\nu=1}^{\infty} A_\nu(t)\, x\left(\frac{t}{k_\nu}\right) \tag{3}$$

has a unique solution with $x(0) = 1$, analytic in $|t| < R$. In particular, if $R = \infty$, the solution is entire, and, if further, for any $\varepsilon > 0$ there exists a K such that

$$(4) \qquad \sum_{\nu=1}^{\infty} |A_\nu(t)| \leqslant K \exp(r^{\rho+\varepsilon})$$

holds when $|t| = r$, then the solution is also of finite order ρ at most.

If the k_ν are all real and I is a closed interval (symmetric about 0 if any k_ν is negative) for which

$$0 \in I \subset (-R,R) ,$$

then the analytic solution is the only solution which is continuous on I.

Proof. We can use the proof of Theorem 1 with the following modifications.

First, we need some definiteness in the choice of B_ν since ΣB_ν is now an infinite series. Write

$$B_\nu(\varepsilon) = \max_{|t|=R-\varepsilon} |A_\nu(t)| ;$$

then we have $|a_{\nu\mu}| \leqslant (R-\varepsilon)^{-\mu} B_\nu$ as before, and (iii) guarantees that ΣB_ν is convergent.

Secondly, $|k_\nu|$ must be bounded uniformly away from 1, and this is guaranteed by (ii).

Thirdly, since $\Sigma |A_\nu(t)|$ is now an infinite series, we need the uniform growth condition in (4) in order to get the estimate in (2). ∎

3. Equivalent Integral Equations

When all the k_ν (and b_ν if they occur) are real our problems can be restated by using integral equations. Our arguments are straightforward, being concerned with the imposition of conditions and the introduction of notation which will allow us to quote standard work. We present these arguments rather discursively, and then state Lemma 1 to summarize what we have done. Although we are not striving for great generality, our notation will be more systematic if we work with

$$(5) \qquad \dot{x}(t) = \lambda \sum_{\nu=1}^{\infty} A_\nu(t) x(\alpha_\nu(t))$$

on a closed interval $I = [\gamma_1, \gamma_2]$ of the real axis.

We assume that each α_ν has a continuous non-vanishing derivative on I and that $\alpha_\nu(I) \subseteq I$; that is, that each α_ν is a diffeomorphism of I into itself. We write ψ_ν for the inverse of α_ν. We assume each $A_\nu(t)$ is a continuous function of $t \in I$; its values and the parameter λ may be complex.

The series on the right-hand side of (5) will be a finite sum if, for large ν, the A_ν vanish identically; if the series is genuinely infinite we write $C_\nu = \max_{t\in I} |A_\nu(t)|$ and require

$$\sum_{\nu=1}^{\infty} C_\nu < \infty$$

and

$$|\alpha_\nu'(t)| > \kappa^{-1} > 0 .$$

Suppose that x is a solution of (5) on I and that $t_0 \in I$. We must have

$$\text{(5a)} \qquad \begin{aligned} x(t) &= x(t_0) + \lambda \int_{t_0}^{t} \sum_{\nu=1}^{\infty} A_\nu(u)x(\alpha_\nu(u))du \\ &= x(t_0) + \lambda \sum_{\nu=1}^{\infty} \int_{t_0}^{t} A_\nu(u)x(\alpha_\nu(u))du . \end{aligned}$$

Since we shall need to say that our calculations are reversible, observe that the interchange above is valid if x is known to be bounded on I, even if it were not known to be continuous. If, working separately in each integral, we write $u = \psi_\nu(s)$, we get

$$\begin{aligned} x(t) &= x(t_0) + \lambda \sum_{\nu=1}^{\infty} \int_{\alpha_\nu(t_0)}^{\alpha_\nu(t)} A_\nu(\psi_\nu(s))\psi_\nu'(s)x(s)ds \\ &= x(t_0) + \lambda \sum_{\nu=1}^{\infty} \int_{\gamma_1}^{\gamma_2} K_\nu(t,s)x(s)ds , \end{aligned}$$

where

$$K_\nu(t,s) = \begin{cases} A_\nu(\psi_\nu(s))\psi_\nu'(s), & \text{if } \alpha_\nu(t_0) < s < \alpha_\nu(t) \\ -A_\nu(\psi_\nu(s))\psi_\nu'(s), & \text{if } \alpha_\nu(t) < s < \alpha_\nu(t_0) \\ 0, & \text{otherwise.} \end{cases}$$

(Observe that if $t > t_0$ then $K_\nu(t,s)$ has the sign of $A_\nu(\psi_\nu(s))$, whether α_ν is increasing or decreasing.) Now, since $|\psi_\nu'(s)|$ is uniformly bounded by κ, we have $\Sigma K_\nu(t,s)$ uniformly convergent and

$$x(t) = x(t_0) + \lambda \int_{\gamma_1}^{\gamma_2} \sum_{\nu=1}^{\infty} K_\nu(t,s)x(s)ds$$

which we shall write as

$$x(t) = x(t_0) + \lambda \int_{\gamma_1}^{\gamma_2} K(t,s)x(s)ds \ . \tag{6}$$

Thus we have shown that if $x(t)$ satisfies (5), it also satisfies (6). To look for solutions of (5), we first notice that K is bounded and a fortiori L^2 on $[\gamma_1,\gamma_2] \times [\gamma_1,\gamma_2]$. (We shall follow page 22 of Smithies [10] and avoid the word "eigenvalue" here. We use "characteristic" in his sense.) Except for the countable set of characteristic values of λ, equation (6) has a unique L^2 solution with $x(t_0) = 1$. If λ is a characteristic value, equation (6) with $x(t_0) = 0$ has nontrivial solutions and there is at least one such solution satisfying

$$\int_I |x(t)|dt = 1 \ .$$

In either case, (6) shows that $|x(t)|$ is bounded for $t \in I$ and we can reverse our previous steps and show that x satisfies (5a). Inserting x in the right-hand side of (5a) shows that $x(t)$ is continuous in t, and this gives that the integrand on the right-hand side of (5a) is continuous. Hence we can differentiate (5a) to show that x satisfies (5).

We can summarize this work as:

LEMMA 1. If, on the closed interval I, and for $\nu = 1, 2, \ldots$

(i) $A_\nu(t)$ is continuous,

(ii) α_ν is a diffeomorphism of I into itself,

(iii) $\Sigma \max |A_\nu(t)| < \infty$,

(iv) $|\alpha_\nu'(t)| > \kappa^{-1} > 0$,

then for all λ the problem

$$\dot{x}(t) = \lambda \Sigma A_\nu(t)x(\alpha_\nu(t))$$

with $\int_I |x(t)|dt = 1$ has a solution. Except for at most a countable set of λ, the solution is unique.

Further, if t_0 is given, then, except for at most a countable set of λ, there is a unique solution of the equation with $x(t_0) = 1$.

Remarks. (i) Since our K depended on t_0, a more complete notation would be K_{t_0}. Any other K_{t_1} will be simply related to it, but not the same and in general will have a different set of characteristic values. If λ_0 is a characteristic value of K_{t_0} but not of K_{t_1}, then the unique solution of (5) with $\lambda = \lambda_0$ determined by the condition $y(t_1) = 1$ must have $y(t_0) = 0$. If μ is a characteristic value of K_{t_0}, and if (5) with $\lambda = \mu$ has linearly independent solutions $u(t)$ and $v(t)$ with $u(t_0) = v(t_0) = 0$ (which requires that μ not be a simple characteristic value of K_{t_0}), then μ is a characteristic value of K_τ for every τ. Again if (5) with $\lambda = \mu$ has a nontrivial solution $u(t)$ with $u(t_0) = 0$ and a solution $w(t)$ with $w(t_0) = 1$ (which requires that the constant functions be orthogonal to every solution of the homogeneous adjoint integral equation), then μ is a characteristic value of K_τ for every τ.

(ii) The condition $|\alpha_\nu'(t)| > \kappa^{-1} > 0$ excludes some equations which are accessible to the methods of Section 2, and necessarily so, since we can construct examples where formal changes of variable will not lead to an equivalent integral equation. Consider, for instance, the equation

$$\dot{x}(t) = \Sigma \frac{1}{\nu^2} x\left(\frac{t}{\nu^5}\right) ,$$

and take $t_0 = \gamma_1 = 0$, $\gamma_2 = 1$. We find

$$K_\nu(t,s) = \nu^3 \quad \text{if} \quad 0 < s < t/\nu^5 .$$

and hence, in the triangle Δ_n: $t/(n+1)^5 < s < t/n^5$, we have

$$K(t,s) = \sum_{\nu=1}^{n} \nu^3 .$$

Clearly K is not bounded and since the contribution to $\iint K^2 dt\, ds$ from Δ_n is $5n^2/32 + O(n)$, K is not in L^2. ∎

4. Remarks on Volterra Kernels

In Section 5 we shall discuss (C), assuming analyticity of the functions A_ν and using complex-variable methods. Our results show, among other things, that the characteristic values of the corresponding K depend only on the values taken by the A_ν in a suitable interval I. Although it might seem that the analyticity is relevant to this property, it is a consequence of the properties of Volterra kernels. In this section, whose contents are not needed in establishing our later results, we shall indicate a line of argument leading to this assertion, but not give precise enunciations or proofs.

It is usual to say K is a Volterra kernel if it is defined on a square $I \times I$ and satisfies the condition

$$K(t,s) = 0 \quad \text{when} \quad s > t .$$

For our purposes it is desirable to work on both sides of the point t_0 and we shall say K is a Volterra kernel in the extended sense with respect to t_0 if

$$K(t,s) = 0 \quad \text{when} \quad |s-t_0| > |t-t_0| .$$

We shall denote this by

$$K \in VE(I \times I;\, t_0) .$$

If $K \in L^2(I \times I) \cap VE(I \times I;\, t_0)$ then, as in the more familiar case, K is free of characteristic values. This can be verified by examining the iterated kernels of K and showing that the Neumann series is always convergent. It can be shown more quickly by examining the modified Fredholm determinant

$$\delta(\lambda) = \sum_{n=0}^{\infty} \delta_n \lambda^n$$

of K. In this series, δ_n is given (see, for example, Smithies [10], page 99) by an n-fold integral of

$$\begin{vmatrix} 0 & K(u_1,u_2) & \cdots & K(u_1,u_n) \\ K(u_2,u_1) & 0 & \cdots & K(u_2,u_n) \\ \cdots & \cdots & \cdots & \cdots \\ K(u_n,u_1) & K(u_n,u_2) & \cdots & 0 \end{vmatrix} .$$

If $(u_1, u_2, \ldots, u_n)$ is a point in the region of integration and u_r is a coordinate for which $|u_i - t_0|$ is a minimum, then row r consists only of 0's, except when there is a $u_p \neq u_r$ such that

$$|u_p - t_0| = |u_r - t_0| .$$

The set of points with such a pair of coordinates is of n-dimensional measure 0 and hence $\delta_n = 0$ for all $n \geqslant 1$; that is, $\delta(\lambda) = 1$ for all λ.

Suppose now that I and J are two intervals of the real line, with $I \subset J$, and that K is defined on $J \times J$. Then it may well be that, outside $I \times I$, K satisfies the condition that $K(t,s) = 0$ when $|s - t_0| > |t - t_0|$. In this case we shall say that K is an <u>effectively Volterra</u> kernel outside $I \times I$. If K has this property and K restricted to $I \times I$ has Λ as its set of characteristic values, then Λ is exactly the set of characteristic values of K. This follows by showing that the formal work in the following can be justified.

Suppose $\lambda \notin \Lambda$ and that x_0, defined on I, is known to satisfy

$$x_0(t) = f(t) + \lambda \int_I K(t,s)x_0(s)ds .$$

Then any solution of

$$x(t) = f(t) + \lambda \int_J K(t,s)x(s)ds$$

must agree with x_0 on I and satisfy

$$x(t) = [f(t) + \lambda \int_I K(t,s)x_0(s)ds] + \lambda \int_{J \setminus I} K(t,s)x(s)ds .$$

In the square brackets is a known function, and the kernel of the integral over $J \setminus I$ is in $VE(J \times J; t_0)$. Hence the equation has a solution which does agree with x_0 on I.

Similarly, if $\lambda \in \Lambda$ and x_0 is a solution of the homogeneous integral equation on I, we can extend x_0 to a solution of the homogeneous equation on J.

5. Existence Theorems for (C)

The right hand side of equation (C) may have either a finite or an infinite series. We shall distinguish between these two cases by renaming the equation as (7) or (11), respectively. Since we appeal to the work of Section 3 we must restrict ourselves to the case when all k_ν and all b_ν are real. It will be convenient to insert a parameter λ, possibly complex, in the equation

$$\dot{x}(t) = \lambda \sum_{\nu=1}^{n} A_\nu(t) x\left(\frac{t-b_\nu}{k_\nu}\right) . \tag{7}$$

We note that $(t-b_\nu)/k_\nu$ is equal to t when

$$t = -\frac{b_\nu}{k_\nu - 1} = c_\nu \quad \text{say,}$$

and we shall often write $(t-b_\nu)/k_\nu$ as $c_\nu + (t-c_\nu)/k_\nu$. With this modification equation (7) will be denoted by (7').

If we write $\alpha_\nu(t) = c_\nu + (t-c_\nu)/k_\nu$, it is clear that, if $|k_\nu| > 1$, c_ν is an attractive fixed point of α_ν. Nevertheless, we must distinguish two cases. If $k_\nu > 1$, any interval containing c_ν is taken into itself by α_ν, but if $k_\nu < -1$, any interval $[\gamma_1,\gamma_2]$ containing c_ν is reversed in sense by α_ν, and will be taken into itself only if $c_\nu - \gamma_1 \leq |k_\nu|(\gamma_2 - c_\nu)$ and $\gamma_2 - c_\nu \leq |k_\nu|(c_\nu - \gamma_1)$. It is clear that any sufficiently large interval symmetrical about 0, namely any interval whose length is at least $2(|k_\nu|+1)|c_\nu|/(|k_\nu|-1)$, is taken into itself by α_ν.

If $k_\nu > 1$ for all ν, we shall work with the interval

$$I = [\min c_\nu, \max c_\nu] ,$$

which obviously satisfies the conditions of Lemma 1. If there are any k's less than -1 it will be convenient to work with intervals symmetrical about 0, although this does not give the best possible I. Write

$$d_\nu = \begin{cases} |c_\nu| & \text{if } k_\nu > 1 , \\ \dfrac{|k_\nu|+1}{|k_\nu|-1} \cdot |c_\nu| & \text{if } k_\nu < -1 . \end{cases}$$

Then the interval

$$I = [-\max d_\nu, \max d_\nu]$$

satisfies the conditions of Lemma 1.

THEOREM 3. Suppose that all k_ν in

$$\dot{x}(t) = \lambda \sum_{\nu=1}^{n} A_\nu(t)x(c_\nu + \frac{t-c_\nu}{k_\nu}) \tag{7'}$$

are real and satisfy $|k_\nu| > 1$ and that I is $[\min c_\nu, \max c_\nu]$ or $[-\max d_\nu, \max d_\nu]$ according as all k_ν are positive or some are negative. Then if

(i) $I \subset (-R,R)$, where R may be ∞,

(ii) $0 \in I$,

(iii) $A_\nu(t)$ are analytic in $|t| < R$,

there is, for every complex λ, at least one solution of (7') in I with $\int_I |x(t)|dt = 1$, and any solution of (7') in I can be extended to a function analytic in $|t| < R$. If $R = \infty$ and, further, $A_\nu(t)$ is of finite order ρ_ν and $\max \rho_\nu = \rho$, then the solution is of finite order ρ at most.

Proof. It is clear that the equation (7') and the interval I satisfy the conditions of Lemma 1, and hence there is, for every λ, at least one solution with the chosen normalization.

Since we may change t to $-t$ if necessary, there is no loss of generality if, in the case when all k_ν are positive, we assume $|\min c_\nu| \leqslant \max c_\nu$. After this, we write c^* for the right hand endpoint of I in either case; that is, we assume $|t| \leqslant c^*$ for all $t \in I$. Write $A_\nu(t) = \sum_{\mu=0}^{\infty} a_{\nu\mu}t^\mu$ and define $A_\nu{}^*(t) = \sum_{\mu=0}^{\infty} |a_{\nu\mu}|t^\mu$. Evidently $A_\nu{}^*(t)$ is analytic in $|t| < R$. By direct comparison of the expansions we see that for all $t \in I$

$$|A_\nu{}^{(s)}(t)| \leqslant A_\nu{}^{*(s)}(c^*)$$

for $\nu = 1, \ldots, n$ and $s = 0, 1, \ldots$.

Suppose now that $x(t)$ is any solution of (7') on I. By repeated differentiation of (7') we see that $x(t) \in C^\infty(I)$ and that

$$(8) \qquad x^{(s+1)}(t) = \lambda \sum_{\nu=1}^{n} \sum_{\mu=0}^{s} \binom{s}{\mu} A_\nu^{(s-\mu)}(t) k_\nu^{-\mu} x^{(\mu)}\left(c_\nu + \frac{t-c_\nu}{k_\nu}\right) .$$

There is a $w(t)$, analytic in $|t| < R$, for which

$$\dot{w}(t) = |\lambda| \sum_{\nu=1}^{n} A_\nu^*(t) w(t)$$

$$w(c^*) = \max_{t \in I} |x(t)| .$$

We shall use w and its derivatives to bound x and its derivatives. By repeated differentiation we have

$$w^{(s+1)}(t) = |\lambda| \sum_{\nu=1}^{n} \sum_{\mu=0}^{s} \binom{s}{\mu} A_\nu^{*(s-\mu)}(t) w^{(\mu)}(t) .$$

In particular

$$(9) \qquad w^{(s+1)}(c^*) = |\lambda| \sum_{\nu=1}^{n} \sum_{\mu=0}^{s} \binom{s}{\mu} A_\nu^{*(s-\mu)}(c^*) w^{(\mu)}(c^*) .$$

By the choice $w(c^*) = \max_{t \in I} |x(t)|$, we have for all $t \in I$

$$|x(t)| \leqslant w(c^*) .$$

Hence, if we put $s = 0$ in equations (8) and (9), a comparison of terms gives

$$|\dot{x}(t)| \leqslant \dot{w}(c^*) .$$

Induction yields

$$|x^{(s)}(t)| \leqslant w^{(s)}(c^*)$$

for all $t \in I$ and $s = 0, 1, \ldots$. Consider the power series expansion of $w(t)$ about c^*; namely,

$$\sum_{s=0}^{\infty} \frac{(t-c^*)^s}{s!} w^{(s)}(c^*) .$$

Consider, too, the formal power series expansion about some $\tau \in I$; namely,

$$\sum_{s=0}^{\infty} \frac{(t-\tau)^s}{s!} x^{(s)}(\tau) .$$

The first power series has positive radius of convergence at least equal to $R - c^*$ (which we interpret as ∞ if $R = \infty$) and its coefficients dominate those of the second series. Hence the second series has positive radius of convergence at least equal to $R - c^*$ and $x(t)$ can be extended to an analytic function in the region G, the union of discs of radius $R - c^*$ whose centers lie on I. If $R = \infty$, then G is the whole plane.

We now show for finite R that we can extend $x(t)$ to the whole disc $D(R)$ by using

$$(10) \qquad x(t) = x(0) + \lambda \int_0^t \sum_{\nu=1}^{n} A_\nu(u)\, x\left(c_\nu + \frac{u - c_\nu}{k_\nu}\right) du$$

a finite number of times. Write G_1 for the union of discs of radius $k(R - c^*)$ centered on I, where $k = \min |k_\nu|$. If $t \in G_1$, we assert that, for all ν, $\alpha_\nu(t) \in G$. This is immediate when $k_\nu > 1$ since c_ν is an attractive fixed point, α_ν takes any segment through c_ν into itself and G is obviously convex. When $k_\nu < -1$, we can write

$$t = u + v ,$$

where $u \in I$ and

(i) if $u \notin \partial I$, v is pure imaginary,

(ii) if u is the right-hand end-point of I, $|\arg v| \leq \frac{1}{2}\pi$,

(iii) if u is the left-hand end-point of I, $|\arg(-v)| \leq \frac{1}{2}\pi$.

We find

$$\alpha_\nu(t) = \alpha_\nu(u) + v/k_\nu \in G ,$$

and hence can use (10). Hence, for any $t \in G_1 \cap D(R)$, we may use (10) to define $x(t)$, provided that we choose a straight segment for the path of integration. For some finite m we have $k^m(R - c^*) \geq R$, and it suffices to have m applications of equation (10).

If now $R = \infty$ and the $A\ (t)$ are of finite order ρ at most, we can modify the relevant part of the proof of Theorem 1 so as to show that x is of finite order ρ at most. Instead of saying $|t/k_\nu| \leq |t|/k$ for all t, we now say there is an R_0 such that

$$|\alpha_\nu(t)| \leq \frac{|t|}{k'} \quad \text{for all } |t| > R_0 ,$$

where $k' = \frac{1}{2}(1+k)$. We end with

$$M(r) = Q(r)M(R_0)\exp\left(\frac{k'}{k'-1} r^{\rho+\varepsilon}\right),$$

where $\log Q(r) = O\{(\log r)^2\}$, which implies that x is of finite order ρ at most. ∎

THEOREM 4. If, in

$$\dot{x}(t) = \lambda \sum_{\nu=1}^{\infty} A_\nu(t)\, x\left(c_\nu + \frac{t-c_\nu}{k_\nu}\right), \tag{11}$$

the set $\{c_\nu\}$ is bounded above and below and all hypotheses of Theorem 3 hold (inf and sup replacing min and max in the definitions of I) and if, further,

$$1 < k \leqslant |k_\nu| < \kappa < \infty$$

and

$$\sum_{\nu=1}^{\infty} \max_{|t|=r} |A_\nu(t)| < \infty \quad \text{for all } r < R,$$

then the assertions of Theorem 3 hold, provided that in the case $R = \infty$ the hypothesis on finite order is strengthened to the uniform growth condition (4) of Theorem 2.

Proof. It is clear that Lemma 1 is applicable and that there is at least one solution with the chosen normalization.

If $x(t)$ is any solution on I we can follow the proof of Theorem 3 very closely, the only modifications being at the beginning and the end. If, as in the proof of Theorem 2, we write

$$B_\nu = \max_{|t|=R-\varepsilon} |A_\nu(t)|,$$

we deduce that $|a_{\nu\mu}| \leqslant (R-\varepsilon)^{-\mu} B_\nu$ and that $\sum_{\nu=1}^{\infty} A_\nu{}^*(t)$ is analytic in $|t| < R$. We can therefore define w as before. Since we cannot differentiate (11) arbitrarily often without justification, we must prove $x \in C^\infty(I)$ by induction using (8) and (9). The inductive proof that $|x^{(s)}(t)| \leqslant w^{(s)}(c^*)$ goes as before and the proof that $x(t)$ is analytic in $|t| < R$ needs no further adjustment.

Finally, if $R = \infty$ and the $A(t)$ are of finite order ρ at most, we can follow the discussion in Theorem 3 (which is a modification of that in Theorem 1) provided we introduce the modifications used in Theorem 2. When these changes are made we find that x is of finite order at most ρ. ∎

6. Unboundedness and Oscillation Theorems

In this section we shall be concerned with equations (B) and (C) when we know their solutions are of finite order ρ at most. We shall refer to those cases in which we know this occurs by saying that equation (B) or (C) satisfies condition E_ρ when the hypotheses of Theorem 1, 2, 3 or 4, as appropriate, are satisfied with $R = \infty$, including condition (4) on uniform growth (which for finite n is equivalent to the condition that each $A_\nu(t)$ should be of finite order ρ at most).

We shall describe (B) and (C) as real if all k_ν, b_ν (and hence c_ν) and λ are real and all $A_\nu(t)$ are real for real t.

THEOREM 5. Suppose (B) or (C) satisfies condition E_ρ with $\rho < \frac{1}{2}$. Then all solutions of the equation are unbounded on any ray.

If further,

(i) the equation is real,

(ii) $k_\nu > 1$ for all ν, and

(iii) there is an L such that $A_\nu(t) < 0$ for $t > L$ and for all ν,

then every solution oscillates unboundedly as $t \to +\infty$ through real values.

Proof. By applying the Phragmén-Lindelöf principle to a sector of opening 2π, we see (Titchmarsh [11], p. 253) that $x(t)$, or indeed any entire function of order less than $\frac{1}{2}$, is unbounded on any ray.

If the further conditions hold, suppose, if possible, that $x(t) > 0$ for $t > t_1 \geq L$. Then for $t > \max[k_\nu t_1 - (k_\nu - 1)c_\nu]$, it would follow that $\dot{x}(t) < 0$ and that $x(t)$ tended to a limit as $t \to \infty$. Since this is inconsistent with the unboundedness, $x(t)$ must become negative. By repeating the argument we see that $x(t)$ cannot remain negative, and so on. Thus $x(t)$ oscillates. ∎

Clearly Theorem 5 implies that any nontrivial solution to (A) oscillates unboundedly as $t \to \infty$. Although our main emphasis here has been on complex-variable methods, this seems a suitable place to give for equation (A) an independent and elementary real-variable proof of oscillation communicated to us at the Conference by Professor Wolfgang Hahn. It seems reasonable to expect that a proof of unboundedness by real-variable methods would require very detailed estimates.

THEOREM 6. [Hahn] If $k > 1$, every solution of

$$\dot{x}(t) = -x\left(\frac{t}{k}\right)$$

has an infinity of positive zeros.

Proof. Suppose, if possible, that $x(t) > 0$ for $t > t_0/k^2$. Then we have

$$x(t) > 0, \quad \dot{x}(t) < 0 \quad \text{and} \quad \ddot{x}(t) > 0 \quad \text{for} \quad t > t_0.$$

Consider the graph of the function in the interval $[t_0, kt_0]$.

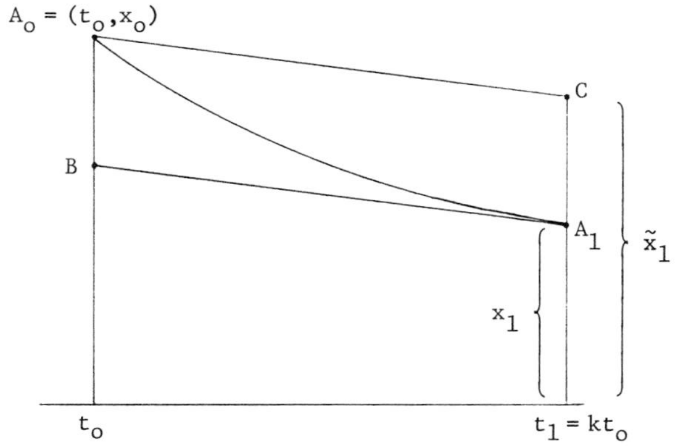

Graph of $x(t)$ in $[t_0, kt_0]$

Write $x_0 = x(t_0)$, $t_1 = kt_0$, $x_1 = x(t_1)$ and write A_0, A_1 for the points (t_0, x_0), (t_1, x_1). Then the arc A_0A_1 is convex and lies above the tangent at A_1 and below the line through A_0 parallel to this tangent. These parallel lines (which are evidently of slope $-x_0$) meet the ordinates $t = t_0$ and $t = t_1$ in points B and C respectively, and $\tilde{x}_1$, the ordinate of C, is given by

$$\tilde{x}_1 = x_0[1-(t_1-t_0)] .$$

Evidently $\tilde{x}_1 > x_1$.

We can repeat this in successive intervals $[t_1,kt_1], \ldots, [t_n,kt_n]$ where $t_n = kt_{n-1}$. We obtain

$$x_n < \tilde{x}_n = x_{n-1}[1-(t_n-t_{n-1})]$$

$$= x_{n-1}[1-k^{n-1}(t_1-t_0)] .$$

For sufficiently large n, the contents of the square brackets are negative and we cannot have both x_{n-1} and x_n positive. Hence the supposition that $x(t) > 0$ for sufficiently large t must be rejected, and similarly $x(t)$ cannot be negative for all large t. ∎

We have now shown by two methods that the solution of (A) with $x(0) = 1$; namely

$$\sum_{n=0}^{\infty} \frac{(-1)^n t^n}{n!k^{\frac{1}{2}n(n-1)}} ,$$

has an infinity of positive zeros. It is obvious that this function cannot have negative zeros. If we recall work of Laguerre, Pólya and Schur we can show that there are no zeros other than the positive ones.

Laguerre [7] considered infinite real sequences, and with

$$\alpha_0, \alpha_1, \ldots$$

he associated the operation of changing any polynomial

$$a_0 + a_1 t + \cdots + a_m t^m$$

into

$$\alpha_0 a_0 + \alpha_1 a_1 t + \cdots + \alpha_m a_m t^m .$$

He pointed out that there are sequences with the property that any polynomial, all of whose zeros are real, is taken into a polynomial, all of whose zeros are real; in particular the sequences

$$1, \frac{1}{\omega}, \frac{1}{\omega(\omega+1)}, \ldots \qquad \text{for} \quad \omega > 0$$

and

$$1, q, q^4, \ldots, q^{n^2} \qquad \text{for} \quad |q| \leqslant 1 .$$

This work was developed by Pólya and Schur [9] who called such sequences <u>factor sequences of the first kind</u>. They showed on page 104 that a necessary and sufficient condition for $\{\alpha_n\}$ to be a factor sequence of the first kind is that the series

$$\sum_{n=0}^{\infty} \frac{\alpha_n z^n}{n!}$$

should be everywhere convergent, in which case its sum would be expressible either as

$$\frac{\alpha_r}{r!} z^r e^{\gamma} \prod_{\nu=1}^{\infty} (1 + \gamma_\nu z)$$

or as

$$\frac{\alpha_r}{r!} (-z)^r e^{-\gamma z} \prod_{\nu=1}^{\infty} (1 - \gamma_\nu z)$$

with $\alpha_r \neq 0$, $\gamma \geqslant 0$ and $\gamma_\nu \geqslant 0$. To apply this result of Pólya and Schur to our $x(t)$, the solution of (A) with $x(0) = 1$, first notice that

$$x(t) = \sum_{n=0}^{\infty} \frac{(-k^{\frac{1}{2}}t)^n}{n!(k^{\frac{1}{2}})^{n^2}} .$$

This is precisely in the desired form where

$$z = -k^{\frac{1}{2}}t$$

$$\alpha_n = (k^{-\frac{1}{2}})^{n^2} = q^{n^2}, \text{ say.}$$

We see that $x(t)$ must be expressible in the form

$$e^{-\gamma k^{\frac{1}{2}} t} \prod_{\nu=1}^{\infty} (1 - \gamma_\nu k^{\frac{1}{2}} t)$$

with $\gamma \geqslant 0$ and $\gamma_\nu \geqslant 0$. We know that $\gamma = 0$, but what we desire is that $\gamma_\nu \geqslant 0$. This proves that all the zeros of $x(t)$ are positive.

7. An Advanced Equation

When the differential equation is advanced, the type of situation that may arise can be noted by a detailed study of

$$\dot{x}(t) = x\left(\frac{t}{k}\right) \quad \text{for} \quad k < 1 .$$

In this case there exist solutions C^{∞} on the real axis, analytic in the half planes $\mathrm{Re}(t) > 0$ and $\mathrm{Re}(t) < 0$. To see this one may proceed formally:

Let

$$x(t) = \sum_{\nu=-\infty}^{\infty} c_\nu e^{-\gamma_\nu t} .$$

For this to be a form solution we require

$$-\sum_{\nu=-\infty}^{\infty} c_\nu \gamma_\nu e^{-\gamma_\nu t} = \sum_{\nu=-\infty}^{\infty} c_\nu e^{-\gamma_\nu t/k}$$

$$= \sum_{\nu=-\infty}^{\infty} c_{\nu+1} e^{-\gamma_{\nu+1} t/k} .$$

This implies that

$$\gamma_{\nu+1} = k\gamma_\nu$$

$$c_{\nu+1} = -c_\nu \gamma_\nu .$$

The solution of these equations are

$$\gamma_\nu = \gamma_0 k^\nu$$

$$c_\nu = (-1)^\nu c_0 \gamma_0^{\ \nu} k^{\frac{1}{2}\nu(\nu-1)} .$$

Thus, a formal solution would be

$$x(t) = c_0 \sum_{\nu=-\infty}^{\infty} (-1)^\nu \gamma_0^{\ \nu} k^{\frac{1}{2}\nu(\nu-1)} e^{-\gamma_0 k^\nu t} .$$

Here γ_0 is a free parameter. To decide what values of γ_0 are appropriate we investigate the convergence of the above formal solution. Recall that $0 < k < 1$.

I. $\nu > 0$. Let

$$N_1 = 1 + \frac{2 \log (2|\gamma_0|)}{\log (1/k)} .$$

If $\nu \geqslant N_1$, then

$$|\gamma_0 k^{\frac{1}{2}(\nu-1)}| \leqslant \frac{1}{2}, \text{ and}$$

$$|\gamma_0^{\nu} k^{\frac{1}{2}\nu(\nu-1)}| \leqslant (\frac{1}{2})^{\nu} .$$

Since $0 < k < 1$, then for all γ_0 and t

$$e^{-\gamma_0 k^{\nu} t} \to 1 \quad \text{as} \quad \nu \to +\infty .$$

Therefore

$$\sum_{\nu=-N_1}^{\infty} (-1)^{\nu} \gamma_0^{\nu} k^{\frac{1}{2}\nu(\nu-1)} e^{-\gamma_0 k^{\nu} t}$$

is absolutely convergent for all real t and complex γ_0.

II. $\nu < 0$. Let

$$N_2 = -1 + \frac{2 \log (1/2|\gamma_0|)}{\log (1/k)} .$$

If $-\nu \geqslant N_2$, then

$$|\gamma_0 k^{\frac{1}{2}(\nu-1)}| \geqslant 2 , \text{ and}$$

$$|\gamma_0^{\nu} k^{\frac{1}{2}\nu(\nu-1)}| \leqslant (\frac{1}{2})^{-\nu} .$$

Since $0 < k < 1$, then $k^{\nu} \to \infty$ as $\nu \to -\infty$.
Therefore as $\nu \to -\infty$,

$$|e^{-\gamma_0 k^{\nu} t}| \to \begin{cases} \infty, & \text{Re } \gamma_0 t < 0 \\ 1, & \text{Re } \gamma_0 t = 0 \\ 0, & \text{Re } \gamma_0 t > 0 \end{cases} .$$

Furthermore, if $\text{Re } \gamma_0 t < 0$, then $\exp(-\gamma_0 k^{\nu} t)$ diverges to ∞ faster than $\gamma_0^{\nu} k^{\frac{1}{2}\nu(\nu-1)}$ converges to zero so that their product also diverges to ∞. This means that

$$\sum_{\nu=-\infty}^{-N_2} (-1)^{\nu} \gamma_0^{\nu} k^{\frac{1}{2}\nu(\nu-1)} e^{-\gamma_0 k^{\nu} t}$$

converges absolutely for $\operatorname{Re} \gamma_0 t \geqslant 0$ and diverges for $\operatorname{Re} \gamma_0 t < 0$. Hence the formal solution $x(t)$ is convergent if and only if $\operatorname{Re} \gamma_0 t \geqslant 0$. Since t is real and γ_0 is a free (complex) parameter, the formal solution converges if and only if

a. $t \geqslant 0$ when $\operatorname{Re} \gamma_0 > 0$

or b. $t \leqslant 0$ when $\operatorname{Re} \gamma_0 < 0$.

Hence all solutions in $t \geqslant 0$ are $C^\infty[0,\infty)$ and join on to the solutions in $t \leqslant 0$ which are $C^\infty(-\infty,0]$. However, there is no unique determination at the origin and we cannot uniquely continue any solution through the origin.

The question of analytic solutions to equations of the form

$$\dot{x}(t) = \sum_{\nu=1}^{n} A_\nu x\left(\frac{t}{k_\nu}\right) + bx(t) \tag{12}$$

for $k_\nu < 1$ and A_ν constant, has been studied by Frederickson [4]. He showed that the behavior exhibited above for $n = 1$ and $b = 0$ does hold in general, and he showed that under one set of conditions there are almost periodic solutions to (12) while under another set of conditions, there are totally monotone solutions.

In a paper on probability theory Ferguson [3] was led to an equation like (12); in particular, to

$$\dot{x}(t) = x\left(\frac{t}{k}\right) - x(t), \quad 0 < k < 1 .$$

His application had the boundary conditions

$$x(0) = 0 \quad \text{and} \quad \lim_{t\to\infty} x(t) = 1 .$$

Ferguson showed for $t \geqslant 0$ that this had a unique solution, which he gave as a power series.

Oberg [8] has made an interesting study of local existence and uniqueness of analytic solutions of

$$\dot{x}(t) = f(t,x(t), x(\alpha(t))) ,$$

with analytic f and α. His study concerns analyticity in neighborhoods of t_0, fixed points of $\alpha(t)$. Oberg showed that a unique local analytic solution always exists if $|\alpha'(t_0)| < 1$, "usually" exists if $|\alpha'(t_0)| = 1$, and "almost never" exists if $|\alpha'(t_0)| > 1$.

8. Examples

The question as to what might happen in Utz' [12] problem $\dot{x}(t) = Ax(\alpha(t))$, as mentioned in Section 1, when A is not a constant has been partly answered in this paper and is partly answered by

EXAMPLE 1.

$$x(0) = 1$$
$$\dot{x}(t) = A(t)x(te^{-t})$$

where $A(t) = a\exp\{at(1-e^{-t})\}$, a = complex constant. A solution is $x(t) = e^{at}$. In this case $A(t)$ is not constant and $\alpha(t) = te^{-t}$ is nonlinear, yet an entire solution exists. The next two examples show what can happen when $\alpha(t)$ is quadratic, though A is constant.

EXAMPLE 2.

$$\dot{x}(t) = \pm x(t^2), \qquad x(0) = 1 .$$

The solution is $x(t) = \sum_{\nu=0}^{\infty} (\pm 1)^{\nu} a_{\nu} t^{\lambda_{\nu}}$ where

$$\lambda_{\nu} = 2^{\nu} - 1$$
$$a_0 = 1$$
$$a_{\nu} = 1 / \prod_{i=1}^{\nu-1} \lambda_i \quad \text{for } \nu \geq 1 .$$

As noted in [1], the solution is a lacunary series. The unit circle is a natural boundary. No analytic continuation beyond $|t| = 1$ is possible.

EXAMPLE 3.

$$\dot{x}(t) = x(2t - t^2), \qquad x(0) = 1 .$$

Oberg [8] pointed out that this has a solution $x(t) = (1-t)^{-1}$ which is analytic in $|t| < 1$. However, for $0 \leq t < 1$ it has another solution, which is linearly independent from $(1-t)^{-1}$. Indeed, it is

$$x(t) = pg(1-t), \quad \text{with} \quad p^{-1} = g(1)$$

where $g(t)$ is the solution to

$$\dot{g}(t) = -g(t^2); \quad g(0) = 1 .$$

By Example 2 it is easy to see that $g(1) \neq 0$ (so p exists) and that $pg(1-t)$ is linearly independent from $(1-t)^{-1}$.

As has been mentioned in Section 1, experimental mathematics and the computer have played a role in motivating the work in this paper. One added fruitful result of using the computer is that many graphs were produced. Since graphs of solutions to retarded ordinary differential equations are rather rare, we include some of the more interesting oscillatory ones here. (See [1] and [2] for a discussion of the numerical analysis aspects.) Figures 1 through 5 graph the solutions to $y'(x) = -y(\alpha(x))$ with $y(x) = 1$ for $x \leqslant 0$ where $\alpha(x)$ is, respectively, as follows:

1. $\alpha(x) = x/k$ for $k = 3/2, 2, 5, \infty$
2. $\alpha(x) = (x \pm \sin x)/2$
3. $\alpha(x) = (x - b)/2$ for $b = 0, 1, 2$
4. $\alpha(x) = (3x + 2x \sin x)/5$
5. $\alpha(x) = (x + x \sin x)/2$.

Figure 6 graphs the functions $\alpha(x)$ used in Figures 4 and 5.

REFERENCES

[1] ALAN FELDSTEIN, Discretization methods for retarded ordinary differential equations, Ph.D. Dissertation, University of California, Los Angeles, 1964.

[2] ALAN FELDSTEIN and CAROLYN K. CRAFTON, Experimental mathematics: an application to retarded ordinary differential equations with infinite lag, Proc. 1968 ACM National Conf., Brandon Systems Press, 1968, pp. 67-71.

[3] THOMAS S. FERGUSON, Lose a dollar or double your fortune, Proc. 6th Berkeley Symposium on Mathematical Statistics and Probability, University of California Press, 1971.

[4] PAUL O. FREDERICKSON, Analytic Solutions for Certain Functional Differential Equations of Advanced Type, Mathematics Report #10, Department of Mathematics, Lakehead University, 1970.
(Also see his forthcoming paper: Dirichlet series solution for certain functional differential equations, to appear in Proc. of

the U.S.-Japan Seminar on Ordinary Differential and Functional Equations, Kyoto, 1971, Springer, 1972.)

[5] FRED GROSS, On a remark of Utz, Amer. Math. Monthly, Nov. 1967, pp. 1107-1108.

[6] FRED GROSS and CHUNG-CHUN YANG, On meromorphic solutions of a certain class of functional differential equations, submitted to Tohoku Math. Journal.

[7] EDMOND N. LAGUERRE, Oeuvres de Laguerre, Vol. 1. Gauthier-Villars, Paris, 1898, pp. 31-35 and 199-206.

[8] ROBERT J. OBERG, Local theory of complex functional differential equations, Trans. Amer. Math. Soc., Nov. 1971, pp. 302-327.

[9] GEORGE PÓLYA and I. SCHUR, Über zwei Arten von Faktorenfolgen in der Theorie der algebraischen Gleichungen, J. Reine Angew. Math. 144 (1914), pp. 89-113.

[10] F. SMITHIES, Integral Equations. Cambridge University Press, Cambridge, 1958.

[11] E.C. TITCHMARSH, Theory of Functions. 2nd edition, Oxford Press, London, 1958.

[12] W.R. UTZ, The equation $f'(x) = af(g(x))$, Bull. Amer. Math. Soc. 71, No. 694 (1965).

GRM: *University of New England, Australia, and Brown University, Providence, Rhode Island*

AF: *University of Arizona, Tempe, Arizona, and Naval Research Laboratory, Washington, D.C.*

EWB: *University of New England, Australia*

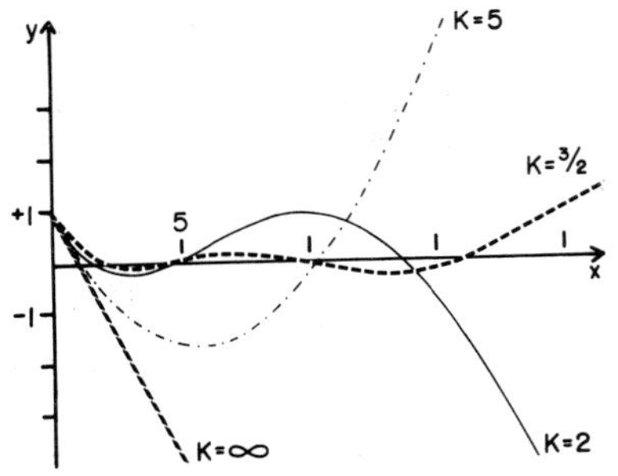

Fig. 1. Solution of $y'(x) = -y(x/k)$, $y(0) = 1$.

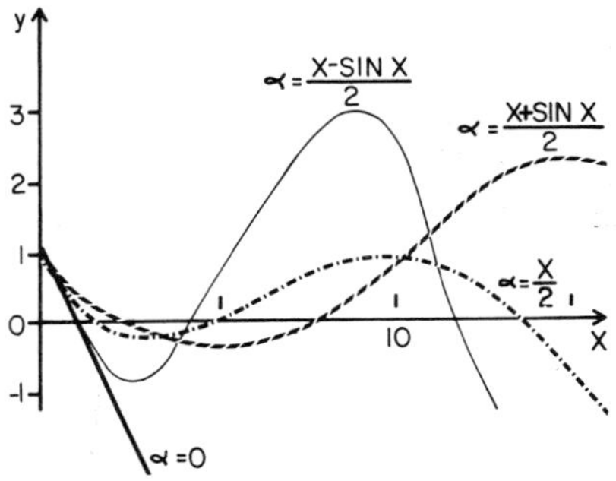

Fig. 2. Solution of $y'(x) = -y(\alpha(x)), y(x) = 1$ for $x \leqslant 0$.

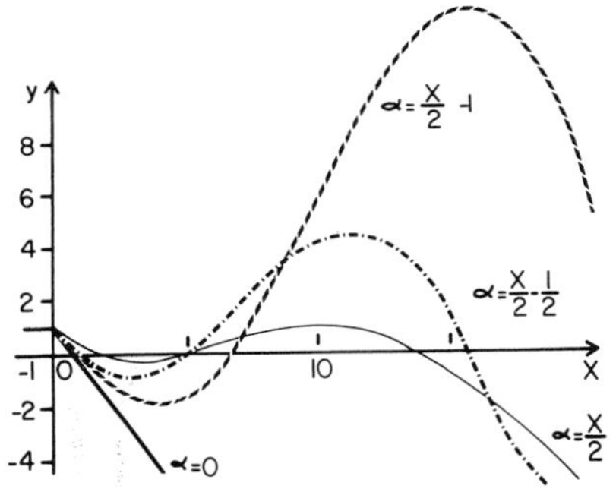

Fig. 3. Solution of $y'(x) = -y(\alpha(x)), y(x) = 1$ for $x \leqslant 0$.

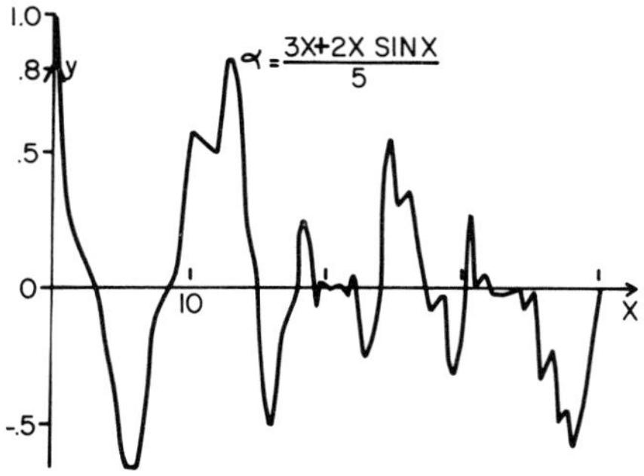

Fig. 4. Solution of $y(x) = -y(\alpha(x)), y(0) = 1$. See Fig. 6 for graph of α.

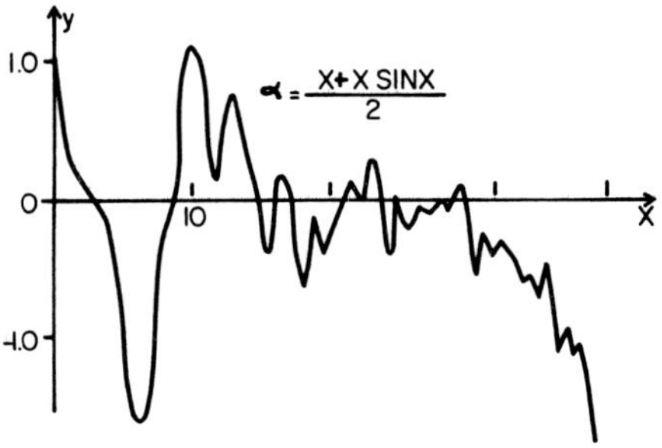

Fig. 5. Solution of $y(x) = -y(\alpha(x)), y(0) = 1$. See Fig. 6 for graph of α.

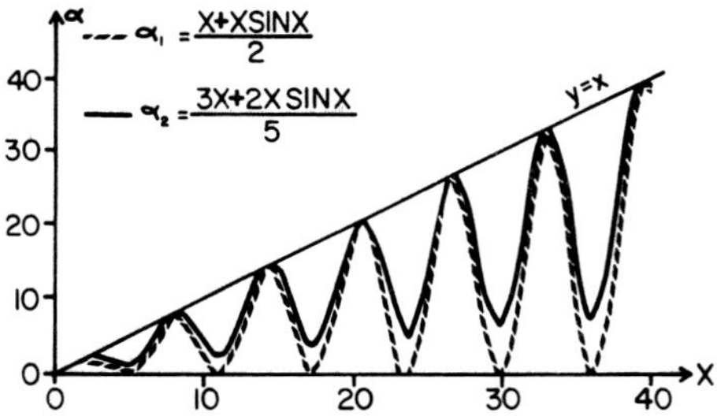

Fig. 6. Graph of α's. α_1 from Fig. 5, α_2 from Fig. 4.

SINGULAR PERTURBATIONS AND THE LINEAR STATE REGULATOR PROBLEM

R. E. O'Malley, Jr.

Consider the linear system

$$\begin{cases} \dfrac{dx}{dt} = a_1(\varepsilon)x + a_2(\varepsilon)z + b_1(\varepsilon)u \\[2ex] \varepsilon\,\dfrac{dz}{dt} = a_3(\varepsilon)x + a_4(\varepsilon)z + b_2(\varepsilon)u \end{cases}$$

of two scalar equations on the interval $t \geq 0$ with the initial states $x^0(\varepsilon)$ and $z^0(\varepsilon)$ prescribed where the control $u(t,\varepsilon)$ minimizes the quadratic cost

$$J(\varepsilon) = \int_0^\infty \left[\begin{pmatrix} x(t,\varepsilon) \\ z(t,\varepsilon) \end{pmatrix}' Q(\varepsilon) \begin{pmatrix} x(t,\varepsilon) \\ z(t,\varepsilon) \end{pmatrix} + r(\varepsilon)u^2(t,\varepsilon) \right] dt \, .$$

Here, the prime denotes transposition, ε is a small positive parameter, and the $a_i(\varepsilon)$'s, $b_i(\varepsilon)$, $x^0(\varepsilon)$, $z^0(\varepsilon)$, $r(\varepsilon)$ and $Q(\varepsilon)$ have asymptotic power series expansions as $\varepsilon \to 0$ such that $r(\varepsilon)$ is strictly positive and $Q(\varepsilon) = \begin{pmatrix} q_1(\varepsilon) & q_2(\varepsilon) \\ q_2(\varepsilon) & q_3(\varepsilon) \end{pmatrix}$ is positive semi-definite for ε sufficiently small.

---

*This research was supported in part by the National Science Foundation under grant number GP-27209.

Such problems are of considerable significance in practical situations where ε represents certain often-neglected "parasitic" parameters whose presence causes the order of the mathematical model to increase. P. V. Kokotović and his coworkers have recently studied these problems extensively (see Kokotović and Yackel (1970)).

By combining the Riccati matrix techniques of optimal control (cf. Kalman, Falb, and Arbib (1969)) with singular perturbation methods (cf. Hoppensteadt (1966) and O'Malley (1971)), we can obtain the asymptotic solution as $\varepsilon \to 0$. We shall assume that for ε sufficiently small

(i) $b_2(a_4b_1 - a_2b_2) \neq 0$

(ii) $a_4^2 + \frac{b_2^2}{r} q_3 \neq 0$, and

(iii) $(a_2a_3 - a_1a_4)^2 + \frac{1}{r}\begin{pmatrix} a_4b_1 - a_2b_2 \\ a_1b_2 - a_3b_1 \end{pmatrix}' Q(\varepsilon) \begin{pmatrix} a_4b_1 - a_2b_2 \\ a_1b_2 - a_3b_1 \end{pmatrix} \neq 0$.

Then, the optimal control, the corresponding trajectories $x(t,\varepsilon)$ and $z(t,\varepsilon)$, and the minimum cost $J^*(\varepsilon)$ will have unique asymptotic expansions of the form

$$u(t,\varepsilon) = \sum_{k=0}^{\infty} [u_k(t) + v_k(t/\varepsilon)]\varepsilon^k ,$$

$$x(t,\varepsilon) = X_0(t) + \sum_{k=1}^{\infty} [X_k(t) + m_{k-1}(t/\varepsilon)]\varepsilon^k ,$$

$$z(t,\varepsilon) = \sum_{k=0}^{\infty} [Z_k(t) + n_k(t/\varepsilon)]\varepsilon^k , \quad \text{and}$$

$$J^*(\varepsilon) = \sum_{k=0}^{\infty} J_k^* \varepsilon^k .$$

These expansions are uniformly valid for all $t \geq 0$. Further, the terms depending on t/ε tend to zero as t/ε tends to infinity. Thus, they are asymptotically negligible away from $t = 0$. Lastly, if $a_4(0) \neq 0$, the limiting solution $(U_0(t), X_0(t), Z_0(t), J_0^*)$ satisfies the reduced optimal control problem:

$$\begin{cases} \dfrac{dX_0}{dt} = a_1(0)X_0 + a_2(0)Z_0 + b_1(0)U_0 \\ 0 = a_3(0)X_0 + a_4(0)Z_0 + b_2(0)U_0 \end{cases}$$

for $t \geq 0$ with the initial value $x_0(0)$ prescribed where $U_0(t)$ minimizes the quadratic cost

$$J(0) = \int_0^\infty \left[\begin{pmatrix} X_0(t) \\ Z_0(t) \end{pmatrix}' Q(0) \begin{pmatrix} X_0(t) \\ Z_0(t) \end{pmatrix} + r(0)\, U_0^2(t) \right] dt .$$

To obtain the asymptotic solution, we first give the optimal control as a linear function of the states x and z through a matrix $K(\varepsilon)$. The asymptotic expansion for $K(\varepsilon)$ is obtained termwise under hypotheses (i)-(iii) as the symmetric, positive definite matrix solution of a quadratic, or Riccati, equation. Knowing K, the original system becomes a linear singular perturbation problem on the infinite interval $t \geq 0$.

Note:

(a) Hypothesis (i) implies that the original system is completely controllable for ε sufficiently small. This allows use of the Riccati matrix approach to the problem.

(b) Hypothesis (ii) implies that the boundary layer terms v_k, m_k, and n_k will decay exponentially as the stretched variable $t/\varepsilon \to \infty$.

(c) Hypothesis (iii) implies that the outer expansion terms U_k, X_k and Z_k will decay exponentially as $t \to \infty$. Such asymptotic stability requirements were shown to be necessary by Hoppensteadt [1].

(d) If the corresponding finite interval problem were considered, much different results and hypotheses would be appropriate since the calculus of variations would convert the original problem to a singularly perturbed two-point boundary value problem. Boundary layer behavior would then be generally required near both endpoints.

REFERENCES

[1] F.C. HOPPENSTEADT, Singular perturbations on the infinite interval, Trans. Amer. Math. Soc., 123 (1966), 521-535.

[2] R.E. KALMAN, P.L. FALB, and M.A. ARBIB, Topics in Mathematical System Theory. McGraw-Hill, New York, 1969.

[3] P.V.KOKOTOVIČ and R.A.YACKEL, Singular perturbation theory of linear state regulators, Proc. Eight Annual Allerton Conf. Circuit and System Theory, 1970, pp. 310-321.

[4] R.E. O'MALLEY, JR., Boundary layer methods for initial value Problems, SIAM Review, 13 (1971).

Courant Institute of Mathematical Sciences, New York, New York, and New York University, New York, New York

DELAY-FEEDBACK, TIME-OPTIMAL, LINEAR TIME-INVARIANT CONTROL SYSTEMS*

V. M. Popov

Summary

One considers the system

$$(1) \qquad dx(t)/dt = Ax(t) + Bx(t-h), \quad (t > 0, \quad h > 0)$$

where $x(t) \in R^n$ and A and B are real, $n \times n$, constant matrices. One defines the error $y(t) = q^T x(t)$, where $q \neq 0$, is a real, constant n-vector. One assumes that A, h and q are fixed, whereas B must be chosen so as to solve the following non-standard variant of the classical problem of time-optimal control: For a given B, denote by S_B the set of all continuous functions $x : [-h,\infty] \to R^n$ which satisfy (1) for $t > 0$. One must find B such that, for every solution of (1), ($x \in S_B$) the corresponding error vanishes in the minimal time, remaining equal to zero afterwards. (The minimal time must be finite and independent of $x \in S_B$.)

Using some recent results from the theory of delay-differential equations, one shows that the above problem has a simple solution.

* * *

*This research was supported in part by the Air Force Office of Scientific Research under Grant AF-AFOSR 69-1646.

In this note we study a non-standard variant of the problem of time-optimal control (see D. Bushaw [1] or L. S. Pontryagin, V. G. Boltyanskii, R. V. Gamkrelidze and E. F. Mishchenko [2] for the classical form of the problem). The new problem and its solution were inspired by some recent results in the study of delay-differential equations (see Remark 2 below).

Consider the delay differential equation

$$\frac{dx(t)}{dt} = Ax(t) + Bx(t-h) , \quad t > 0 , \tag{1}$$

where $x(t)$ is a real n-vector, A and B are $n \times n$ matrices with real, constant entries and h is a positive constant. One defines the "error"

$$y(t) = q^T x(t) , \quad q \neq 0 \tag{2}$$

where q is a real, constant n-vector. (The superscript T denotes transpose.)

Assume that A, h and q are given, whereas B can be freely chosen so as to ensure the following property: the error (2) vanishes in the minimal time, remaining identically zero afterwards.

For a given B, denote by S_B the set of all continuous functions x, defined for $t \geqq -h$, taking values $x(t)$ in R^n and satisfying Eq. (1) for $t > 0$. Our problem can be stated as follows:

<u>Delay-feedback, time-optimal problem</u>: To find $B : n \times n$, for which there exists a finite number $\tau > 0$ with the following properties:

(I) every solution $x \in S_B$ of Eq. (1) satisfies the condition $q^T x(t) = 0$ for every $t \geqq \tau$;

(II) τ is minimal; in other words, for every matrix $\tilde{B}$ and every number $\tilde{\tau}$ having the same property as in (I) (with S_B replaced by $S_{\tilde{B}}$), one has $\tilde{\tau} \geqq \tau$.

Before showing how this problem can be solved, let us make more precise the comments from the beginning of this note.

<u>Remark 1</u>. The above problem is closely related to a well known problem of optimal control for the equation

(3) $$\frac{dx}{dt} = Ax + u .$$

For a given initial condition, one must find u, in a given set of functions, such that the corresponding solution x of Eq. (3) reaches zero in the minimal time (see [1], [2]). After solving this problem, the time-optimal system is written as

(4) $$\frac{dx}{dt} = Ax + f(x)$$

and thus the "optimal control" u is secured by means of a "feedback function" f (having, in general, a complicated form).

In our case, the optimal system is required, from the beginning, to have a feedback of the delayed-type: $u(t) = Bx(t-h)$. Moreover, only the error $q^T x(t)$ is required to vanish in the interval $t \geq \tau$. These modifications of the classical problem of optimal control are, obviously, natural and meaningful.

Remark 2. It is easy to see that, for $h = 0$, there exists no number τ with property (I). As shown below, the existence of such a number for $h > 0$ can be proved in a few lines once one knows the form of the solution. However this form was not obvious from the beginning. The question is equivalent to a problem which was formulated by L. Weiss in [3]: Eq. (1) (with given A, B, and h) is called "pointwise complete" iff for every vector $q \neq 0$ and every $\tau > 0$ there exists a continuous function $x \in S_B$ - satisfying Eq. (1) for $t > 0$ - such that $q^T x(\tau) \neq 0$. Obviously, if Eq. (1) is pointwise complete property (I) cannot be satisfied. It was conjectured that all the systems of the form (1) were pointwise complete. Subsequent research, done by J. A. Yorke, J. Kato, E. B. Lee (in unpublished works) and R. M. Brooks and K. Schmitt [4], has partially confirmed the conjecture for some classes of equations of the form (1). However counterexamples to the conjecture were obtained recently by the author in [5] and by A. M. Zverkin [6].*

---

*At this moment, the only information we have about this work is a (rather detailed) summary, kindly communicated to the author by Professor A. Halanay.

A detailed study of the property of pointwise completeness for Eq. (1) was done in [7]. This study can now be used to obtain a solution of the problem from the present paper. However the proofs from [7] are more complicated than necessary for this purpose, since they also serve to obtain results which are not needed here. Therefore we prefer to give independent, shorter proofs, for the general result of this note, referring to [7] only for some more precise results, mentioned at the end of the paper.

Now we formulate the main result of the note:

<u>A solution of the delay-feedback, time-optimal problem</u>: Assume that $n \geqq 3$ and that the vectors q^T, q^Te^{Ah} and $q^Te^{Ah}A$ are linearly independent. Choose an n-vector r according to the equations:

$$(5) \qquad \begin{cases} q^Tr = 1 \\ q^Te^{Ah}r = 0 \\ q^Te^{Ah}Ar = 0 \; . \end{cases}$$

Then a solution of our problem is given by

$$(6) \qquad B = AZ - ZA$$

where

$$(7) \qquad Z = rq^Te^{Ah} \; .$$

Moreover the minimal time τ is equal to $2h$.

<u>Proof</u>. Take an arbitrary solution $x \in S_B$ of Eq. (1), where B is defined, as above, by Eq. (6). Consider the function v, defined for $t \geqq 0$ by

$$(8) \qquad v(t) = x(t) + Zx(t-h) \; , \quad t \geqq 0 \; .$$

Then, using Eq. (1), one obtains, for $t \geqq h$

$$(9) \qquad \frac{dv(t)}{dt} = Ax(t) + Bx(t-h) + Z(Ax(t-h) + Bx(t-2h)) \; .$$

The right hand member is equal to $Av(t)$, as one easily sees using

Eq. (8) and the relations $ZB = 0$ and $B + ZA = AZ$ (see Eqs. (5)-(7)). Hence (9) becomes $dv(t)/dt = Av(t)$ for $t \geqq h$ which implies $v(t) = e^{Ah}v(t-h)$ for $t \geqq 2h$. This further gives, using Eq. (8),

$$(10) \qquad x(t) + Zx(t-h) = e^{Ah}(x(t-h) + Zx(t-2h)), \quad \text{for} \quad t \geqq 2h .$$

Multiplying this equation, on the left, with q^T and using the relations $q^T Z = q^T e^{Ah}$ and $q^T e^{Ah} Z = 0$ (which follow from Eqs. (5) and (7)) one obtains $q^T x(t) = 0$ for $t \geqq 2h$. Therefore, for the number $\tau = 2h$, property (I) from the statement of our problem is satisfied.

It remains to show that τ is minimal (property II)), that is, that for $\tau < 2h$ property (I) cannot be satisfied. First one shows that for $0 < \tau \leqq h$ property (I) cannot be satisfied. Indeed, consider Eqs. (1), (2) - for an arbitrary $q \neq 0$ and an arbitrary B - and take τ arbitrarily in the interval $0 < \tau \leqq h$. Consider a continuous function x, satisfying Eq. (1) for $t > 0$ and having the following properties: 1° $x(0) = (q^T e^{A\tau})^T$, 2° $x(t) = 0$ in the interval $-h \leqq t \leqq \tau - h - \varepsilon$ (where ε is a small nonnegative number) and 3° x is linear in the interval $\tau - h - \varepsilon \leqq t \leqq 0$ (if $\tau < h$, ε may be taken equal to zero). Then from Eq. (1) one easily obtains

$$q^T x(\tau) = q^T e^{A\tau} x(0) + \mathcal{O}(\varepsilon) = \|q^T e^{A\tau}\|^2 + \mathcal{O}(\varepsilon)$$

where $\lim \mathcal{O}(\varepsilon) = 0$ as $\varepsilon \to 0$. Therefore $q^T x(\tau) \neq 0$ if ε is small enough. This shows that, if $0 < \tau \leqq h$, property (I) cannot be satisfied.

Finally consider $h < \tau < 2h$. Assume that there exists $q \neq 0$ and B such that property (I) is satisfied. We shall show that this assumption is contradictory. Given an arbitrary function $x \in S_B$, introduce the continuous function y, defined for $t \geqq 0$ as

$$(11) \qquad y(t) = \begin{pmatrix} x(t) \\ x(t-h) \end{pmatrix} .$$

If one introduces the 2n-vector $p^T = (q^T \quad 0)$ one has

$$(12) \qquad q^T x(t) = p^T y(t) ,$$

and our assumption that property (I) is true can be written as

$$p^T y(t) = 0 \quad \text{for} \quad t \geqq \tau \,. \tag{13}$$

From Eqs. (1) and (11) one obtains, for $t > h$

$$\frac{dy(t)}{dt} = F\,y(t) + G\,x(t-2h) \tag{14}$$

where

$$F = \begin{pmatrix} A & B \\ 0 & A \end{pmatrix}$$

$$G = \begin{pmatrix} 0 \\ B \end{pmatrix} .$$

Now one proves that, for every $x \in S_B$ the corresponding function y, given by (11) satisfies the equations

$$p^T \frac{d^k y(t)}{dt^k} = p^T F^k\, y(t)\,, \quad k = 0,1,\ldots\,, \quad t > h\,. \tag{15}$$

Indeed, the above equation is obviously true for $k = 0$. Moreover if (15) is true for $k = k_1$ (where k_1 is a nonnegative integer) then (15) is also true for $k = k_1+1$. Indeed, using (15) (for $k = k_1$) and (14), one obtains

$$p^T \frac{d^{k_1+1} y(t)}{dt^{k_1+1}} = p^T F^{k_1}(F\,y(t) + G\,x(t-2h))\,, \quad t > h\,. \tag{16}$$

From (13) it follows that the left hand member of (16) (for $t = \tau$) is equal to zero, which implies

$$0 = p^T F^{k_1+1} y(\tau) + p^T F^{k_1} G x(\tau-2h)\,. \tag{17}$$

In particular, (17) is true for the continuous function x which satisfies Eq. (1) for $t \geqq 0$ and has the following properties: 1° $x(\tau-2h) = (p^T F^{k_1} G)^T$, 2° $x(t) = 0$ for $-h \leqq t \leqq \tau-2h-\varepsilon$, 3° $x(t) = 0$ for $\tau-2h+\varepsilon \leqq t \leqq 0$ (where ε is a small positive number) and 4° x is linear in the intervals $\tau-2h-\varepsilon \leqq t \leqq \tau-2h$ and $\tau-2h \leqq t \leqq \tau-2h+\varepsilon$.

Since this function satisfies Eq. (1) for $t \geqq 0$ it is easy to see that $\|x(t)\|$ is of the order of ε for $t > 0$ and therefore the corresponding $\|y(t)\|$, given by (11), is also of the order of ε for $t > h$. Writing (17) for the just defined functions x and y one obtains

$$0 = p^T F^{k_1+1} y(\tau) + \|p^T F^{k_1} G\|^2$$

and since $y(\tau)$ is of the order of ε one obtains $p^T F^{k_1} G = 0$. Using this conclusion and Eq. (16) one obtains (15) for $k = k_1+1$. Thus Eq. (15) is proved for every integer k.

Using the theorem of Cayley-Hamilton, let us introduce the constants α_i, $i = 0,1,\ldots,2n$ ($\alpha_{2n} = 1$) such that $\alpha_0 F^0 + \alpha_1 F + \alpha_2 F^2 + \cdots + \alpha_{2n} F^{2n} = 0$. Then from (15) one obtains (since $\alpha_{2n} = 1$)

$$\frac{d^{2n}(p^T y(t))}{dt^{2n}} + \alpha_{2n-1} \frac{d^{2n-1}(p^T y(t))}{dt^{2n-1}} + \cdots + \alpha_0 p^T y(t) = 0$$

- an ordinary differential equation which is satisfied for $t > h$. From this equation and from (13) it follows that $p^T y(t) = 0$ for $t > h$. Using (12), one obtains (since x is continuous) $q^T x(h) = 0$, for every $x \in S_B$ - which means that, for $\tau = h$, property (I) is satisfied. This conclusion contradicts the previous section of the proof and shows that, for $h < \tau < 2h$ property (I) cannot be satisfied. Therefore, using the preceding sections of the proof, one sees that $\tau = 2h$ is the minimal time for which property (I) can be satisfied. This proves that Eqs. (5)-(7) give a solution of our problem.

Final Comments. Using [7] one can show that the problem considered in this paper cannot be solved for $n < 3$. Moreover, for $n = 3$ the problem can be solved if and only if the vectors q^T, $q^T e^{Ah}$ and $q^T e^{Ah} A$ are linearly independent. Then the solution of the problem is unique and is given by Eqs. (5)-(7).

If $n > 3$, it is clear that, in general, there exist several vectors r satisfying Eq. (5) and therefore the solution of the problem is not unique. The solution can be made unique if one introduces other conditions (for instance, one may require that several quantities of the form $q_i^T x(t)$ should vanish in the minimal time). It is clear that there exists a whole class of time-optimal problems of the type considered here. The discussion of these problems is beyond the scope of this note, whose only purpose was to signalize the existence of a class of time-optimal problems which have simple and general solutions and around which much interesting work can be developed.

REFERENCES

[1] D.W. BUSHAW, Differential equations with discontinuous forcing term, doctoral dissertation, Princeton University, 1952.

[2] L.S. PONTRYAGIN, V.G. BOLTYANSKII, R.V. GAMKRELIDZE and E.F. MISHCHENKO, The Mathematical Theory of Optimal Processes. Wiley (Interscience), New York, 1962.

[3] L. WEISS, On the controllability of delay-differential systems, SIAM J. Control 5, No. 4 (1967).

[4] R.M. BROOKS and K. SCHMITT, Pointwise completeness of differential difference equations (to be published).

[5] V.M. POPOV, Pointwise Complete and Pointwise Degenerate Linear, Time-Invariant, Delay-Differential Systems, Technical Report R-71-03, University of Maryland.

[6] A.M. ZVERKIN, Pointwise completeness of delay-differential equations, Conference of Lumumba-University, May 24-27, 1971, Moscow.

[7] V.M. POPOV, Pointwise degeneracy of linear, time-invariant, delay-differential equations (to be published).

University of Maryland, College Park, Maryland

THE FAMILY OF DIRECT PERIODIC ORBITS OF THE FIRST KIND IN THE RESTRICTED PROBLEM OF THREE BODIES

Dieter S. Schmidt

1. Description of Problem

The planar restricted problem of three bodies is defined in barycentric, synodic coordinates by the Hamiltonian function

$$h = \frac{1}{2}(p_1^2 + p_2^2) + q_2p_1 - q_1p_2 - (1-\mu)/\rho_1 - \mu/\rho_2 \tag{1}$$

where q_1, q_2 are position coordinates and p_1, p_2 the conjugate momenta. ρ_1 and ρ_2 are the distances of the third body of negligible mass to the primaries of mass $1-\mu$ and μ which are located at $(-\mu, 0)$ and $(1-\mu, 0)$ respectively.

For $\mu = 0$, the problem reduces to the central force problem in a rotating coordinate system. Its solutions are known [5]. For example, circular orbits of radius a are given by

$$q_1 = a \cos \omega t$$

$$q_2 = a \sin \omega t$$

with

$$\omega + 1 = \pm a^{-3/2} .$$

The upper (lower) sign corresponds to the so-called direct (retrograde) orbits, which have the same (opposite) orientation as the rotation of the synodic coordinate system around the origin of an inertial coordinate system.

The circular orbits can be arranged in three families which are parameterized by the energy

$$h = -1/(2a) \mp \sqrt{a}$$

1) direct circular orbits with $0 < a < 1$
2) direct circular orbits with $1 < a < \infty$
3) retrograde circular orbits $0 < a < \infty$.

A standard theorem asserts that if the nontrivial characteristic multipliers are not $+1$, then for small values of μ the circular orbit can be continued to an orbit of the restricted problem, which Poincaré calls of the first kind. The nontrivial characteristic multipliers of a circular orbit are $e^{\pm 2\pi i/\omega}$ and they are different from 1 if $\omega \neq 1/k$, k integer. This condition is usually given separately for the three families (i.e. see [5])

1) $a^{3/2} \neq k/(k+1)$ $k = 1, 2, \ldots$
2) $a^{3/2} \neq (k+1)/k$ $k = 1, 2, \ldots$
3) There is no restriction on retrograde circular orbits with the exception $a \neq 1$ due to a singularity in the perturbing function.

The purpose of this paper is to show what happens to direct circular orbits, which belong to these exceptional values of a.

To this end we have to mention a few more facts concerning the central force problem. If $a^{3/2} = k/\ell$ where k and ℓ are positive integers, relatively prime, then orbits which are elliptic in an inertial coordinate system will be periodic in the synodic system. In particular, for each such a there exist two natural families $E_a^{(i)}(h)$ $i = 1,2$ of periodic orbits which are symmetric to the q_1-axis and which can be parameterized by the energy h in $h_1 \leq h \leq h_2$, $h_1 = -1/(2a) - \sqrt{a}$, $h_2 = -1/(2a) + \sqrt{a}$. The orbits in these families have the following properties:

1) The period of each orbit is $2\pi k$.

2) For $h_1 < h < -1/(2a)$ the orbits are direct and have the winding number $\ell - k$ with respect to the origin of the synodic coordinate system.

3) As $h \to h_1+$ the orbits approach a direct circular orbit of radius a but with $|k - \ell|$ times its minimal period.

4) For $h = -1/(2a)$ the orbit consists of loops of collision orbits and with increasing h the orbits become retrograde and approach a retrograde circular orbit of radius a as $h \to h_2-$.

Poincaré calls an orbit of the second kind if it approaches an elliptic orbit as $\mu \to 0$. Their existence was first proven in [1]. The author showed in [4] that in case k and ℓ are not consecutive integers the two families $E_a^{(i)}(h)$ can be continued to two families $E_a^{(i)}(h,\mu)$ of periodic orbits of the second kind with equivalent properties.

In case $\ell = k \pm 1$ everything shown in [4] holds except near direct circular orbits. The difficulty arises because the orbits in $E_a^{(i)}(h)$ $i = 1,2$ have a winding number of $+1$ and their period is equal to the minimal period of the direct circular orbit of radius $a = (k/(k \pm 1))^{2/3}$ to which they tend as $h \to h_1+$.

2. Poincaré's Variables

We will use Poincaré's variables which are valid near direct orbits of the first kind. The Hamiltonian (1) becomes

$$h = \frac{1}{2} P_1^{-2} - P_1 + \frac{1}{2}(P_2^2 + Q_2^2) + \mu F(Q_1, Q_2, P_1, P_2) , \tag{2}$$

where μF describes the perturbation from the central force problem. Unfortunately the transforamtion from rectangular to Poincaré's coordinates cannot be given in closed form. Thus to describe them we can either give them in terms of the more well known variables of Delaunay ℓ, g, L and G

$$Q_1 = \ell + g \qquad Q_2 = -\sqrt{2(L-G)} \sin g$$

$$P_1 = G \qquad P_2 = \sqrt{2(L-G)} \cos g$$

or we can exhibit their meaning in the central force problem

$$\begin{aligned} Q_1 &= \omega t + \ell_0 + g_0 \\ P_1 &= \sqrt{a} \\ Q_2 &= \sqrt{2\sqrt{a}\,(1 - \sqrt{1-e^2})}\, \sin(t - g_0) \\ P_2 &= \sqrt{2\sqrt{a}\,(1 - \sqrt{1-e^2})}\, \cos(t - g_0) . \end{aligned} \tag{3}$$

The search for periodic orbits is simplified by the following

LEMMA (Birkhoff). An orbit satisfying

$$q_2(0) = q_2(T/2) = p_1(0) = p_1(T/2) = 0$$

is symmetric to the q_1-axis and is periodic with period $T > 0$.

In terms of Poincaré's variables the condition of the lemma takes on the following form, where we restrict ourselves to the case $a < 1$.

For direct orbits of the first kind

(4) $$\text{at } t = 0 \qquad Q_1 = Q_2 = 0$$
$$\text{at } t = T/2 \qquad Q_1 = \pi, \quad Q_2 = 0 .$$

For direct orbits of the second kind the condition depends on the values and parity of k and ℓ. For $a^{3/2} = k/(k+1)$ the condition is the same as given in (4) with the addition as $\mu \to 0$ $Q_2(t)$ will have opposite signs for orbits in the two distinct families $E_a^{(i)}$ $i = 1,2$.

The differential equations corresponding to (2) are

$$\frac{dQ_1}{dt} = P_1^{-3} - 1 + \mu \frac{\partial F}{\partial P_1} \qquad \frac{dP_1}{dt} = -\mu \frac{\partial F}{\partial Q_1}$$

$$\frac{dQ_2}{dt} = P_2 + \mu \frac{\partial F}{\partial P_2} \qquad \frac{dP_2}{dt} = -Q_2 - \mu \frac{\partial F}{\partial Q_2}$$

If the initial conditions at $t = 0$ are $Q_1 = Q_2 = 0$, $P_1 = P_{10}$, $P_2 = P_{20}$, then the solutions for Q_1 and Q_2 assume the form

$$Q_1 = (P_{10}^{-2} - 1)t + \mu Q_{11}(t, P_{10}, P_{20}, \mu)$$

$$Q_2 = P_{20} \sin t + \mu Q_{21}(t, P_{10}, P_{20}, \mu)$$

and the condition (4) for periodic orbits reads

(5a) $$\varphi = (P_{10}^{-3} - 1)T/2 + \mu Q_{11}(T/2, P_{10}, P_{20}, \mu) - \pi = 0$$

(5b) $$\psi = P_{20} \sin T/2 + \mu Q_{21}(T/2, P_{10}, P_{20}, \mu) = 0 .$$

For $\mu = 0$ the family of circular orbits is characterized by $P_{20} = 0$. The orbits have period $T = 2\pi/(P_{10}^{-3} - 1)$ where $P_{10} = \sqrt{a}$ is arbitrary but for the case under investigation near $(k/(k+1))^{1/3}$. On the other hand setting $P_{10} = (k/(k+1))^{1/3}$ and $T = 2\pi k$ satisfies both equations (5a) and (5b). The arbitrary parameter is now $P_{20} > 0$ or $P_{20} < 0$ which gives the two natural families of elliptic orbits $E_a^{(1)}$ and $E_a^{(2)}$ respectively, which bifurcate from the family of circular orbits when $P_{20} = 0$.

THEOREM 1. For a small $\mu > 0$ the family of direct orbits of the first kind breaks up near $P_{10} = (k/(k+1))^{1/3}$, $k = 1,2,\ldots$ and each part connects with one of the families of the second kind belonging to the above value of P_{10}.

Proof. It is beneficial to introduce a scale factor $\varepsilon > 0$ into equations (5a,b)

$$P_{10} \to \varepsilon \tilde{P}_{10} + (k/(k+1))^{1/3}$$
$$P_{20} \to \varepsilon \tilde{P}_{20}$$
$$\mu \to \varepsilon^2 \mu \ .$$

Equation (5a) can always be solved for $T = T(P_{10}, P_{20}, \mu)$ to give

$$T = 2\pi k + 2\varepsilon\alpha \tilde{P}_{10} + O(\varepsilon^2)$$

where

$$\alpha = 3\pi k^{2/3} (k+1)^{4/3} \ .$$

Using this value of T in (5b) we obtain

$$\psi = \varepsilon^2(\alpha \tilde{P}_{10} \tilde{P}_{20} + \mu\beta + O(\varepsilon)) \tag{6}$$

with

$$\beta = Q_{21}(\pi k, (k/(k+1))^{1/3}, 0, 0) \ .$$

Assuming $\beta \neq 0$ (say $\beta > 0$), which will be shown later, we introduce into equation (6) (see Fig. 1)

$$\tilde{P}_{10} = u - v$$
$$\tilde{P}_{20} = u + v$$

and are left with solving

$$\tilde{\psi} = \alpha(u^2 - v^2) + \mu\beta + O(\varepsilon) = 0 \ .$$

For $\varepsilon = 0$ this has two solutions

$$v = + \sqrt{u^2 + \mu\beta/\alpha}$$

or

$$v = - \sqrt{u^2 + \mu\beta/\alpha} \ .$$

For $\varepsilon \neq 0$ but sufficiently small we still can solve for v as a function of u since

$$\left.\frac{\partial\tilde{\psi}}{\partial v}\right|_{\varepsilon=0} = -2\alpha v = \mp 2\alpha\sqrt{u^2 + \mu\beta/\alpha} \neq 0 \ .$$

For P_{20} not near zero or P_{10} not near $(k/(k+1))^{1/3}$ the results of [4] apply, which complete the proof.

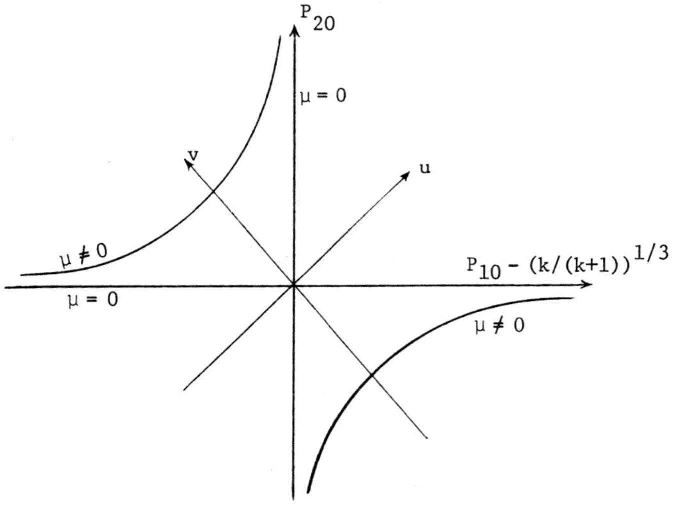

Fig. 1 Representation of families of periodic orbits for $\mu = 0$ and $\mu \neq 0$ $(\beta > 0)$.

3. Verification for the Restricted Problem

The verification of $\beta \neq 0$ requires the knowledge of the disturbing function F in (2). But since the eccentricity e of the orbits under consideration is small we can use an expansion of F into powers of Q_2 and P_2 (see eq. (3))

$$F = F_0 + F_1 Q_2 + F_2 P_2 + \cdots .$$

The functions F_i depend only on Q_1 and P_1. Following [2] one finds after some lengthy but elementary computations

$$F_0 = a \cos Q_1 + a^{-1} - \Delta$$

$$F_1 = \frac{1}{2} a^{3/4}(-\sin 2Q_1 - 2a^{-2} \sin Q_1 - \Delta^{-3}(2a \sin Q_1 + \sin 2Q_1))$$

$$F_2 = \frac{1}{2} a^{3/4}(-3 + \cos 2Q_1 + 2a^{-2} \cos Q_1 - \Delta^{-3}(3 - 2a \cos Q_1 - \cos 2Q_1))$$

with

$$\Delta^2 = a^2 - 2a \cos Q_1 + 1 \quad \text{and} \quad a = P_1^2 .$$

In order to find β we have to solve a system of the form

$$\dot{x} = f(x) + \mu g(x)$$

where

$$x = \begin{pmatrix} Q_1 \\ P_1 \\ Q_2 \\ P_2 \end{pmatrix} \qquad f = \begin{pmatrix} P_1^{-3} - 1 \\ 0 \\ P_2 \\ -Q_2 \end{pmatrix} \qquad \text{and} \quad g = \begin{pmatrix} F_{P_1} \\ -F_{Q_1} \\ F_{P_2} \\ -F_{Q_2} \end{pmatrix} .$$

Letting $x = x_0 + \mu x_1 + \cdots$ we know x_0 and obtain

$$x_1 = \phi(t) \int_0^t \phi^{-1}(s) g(x_0(s)) ds$$

where $\phi(t)$ is the fundamental matrix of $\dot{\phi} = f_x(x_0)\phi$. The third component of x_1 gives

$$Q_{21} = \int_0^t \{\cos(t-s)\, F_{P_2} - \sin(t-s)\, F_{Q_2}\} ds .$$

In particular

$$\beta = (-1)^k \int_0^{\pi k} (\cos s \;\; F_2 + \sin s \;\; F_1) ds$$

where the functions F_1 and F_2 are given above with setting $Q_1 = (P_{10}^{-3} - 1)s = s/k$ and $a = (k/(k+1))^{2/3}$. After more elementary computations

$$\beta = (-1)^k \frac{k a^{3/4}}{2} \int_0^{\pi} \frac{\cos(2+k)\sigma + 2a \cos(1+k)\sigma - 3 \cos k\sigma}{(a^2 - 2a \cos\sigma + 1)^{3/2}} d\sigma \; .$$

The three parts of the integral are also known as Laplace's coefficients, which are defined by

$$\frac{1}{2} B_n = \frac{1}{\pi} \int_0^{\pi} \frac{\cos n\sigma}{(a^2 - 2a \cos\sigma + 1)^{3/2}} d\sigma \; , \qquad n = 0,1,2,\ldots \; .$$

They can be expressed for $a < 1$ by means of the following infinite series ([2], Ch. 6.5)

$$\frac{1}{2} B_n = \frac{3\cdot5\cdots(2n+1)}{2\cdot4\cdots(2n)} a^n \left\{1 + \frac{3}{2}\,\frac{2n+3}{2n+2} a^2 + \frac{3\cdot5}{2\cdot4}\,\frac{(2n+3)(2n+5)}{(2n+2)(2n+4)} a^4 + \cdots\right\} .$$

Thus

$$\beta = (-1)^k \frac{k\; a^{3/4}\; \pi}{4} \{B_{k+2} + 2a\; B_{k+1} - 3\; B_k\}$$

and we can show that the expression in parentheses is always negative by using the infinite series and combining equal powers of a.

For $k = 1$ we find $\beta > 0$ and the situation is as assumed in the proof of Theorem 1. The family of periodic orbits of the first kind connects with the family $E_a^{(1)}(h,\mu)$, $a = (1/2)^{2/3}$ $(P_{20} > 0)$. As seen from (3), this requires that the initial conditions for the orbit are on the positive q_1-axis at a perigee $(g_0 = \ell_0 = 0)$.

Referring to [4] this entire family $E_a^{(1)}(h,\mu)$ exists with increasing h until its termination on the family of retrograde orbits of the first kind, since orbits in this particular family never come close to the singularity of the perturbing function F at $q_1 = 1 - \mu$ $q_2 = 0$. Thus we can state

THEOREM 2. For small mass ratios μ the family of direct periodic orbits of the first kind becomes with increasing h near $a = (1/2)^{2/3}$ a family of the second kind which terminates by triplication on an orbit in the family of retrograde periodic orbits of the first kind.

This is part of a conjecture of A. Deprit [3] which deserves recognition as being the starting point for these investigations.

REFERENCES

[1] R.F. ARENSTORF, Periodic solutions of the restricted three body problem representing analytic continuations of Keplerian elliptic motions, Amer. J. Math. 83 (1963), 27.

[2] C.L. CHARLIER, Die Mechanik des Himmels, Vol. 1. Walter de Gruyter, Berlin, 1927.

[3] A. DEPRIT and J. HENRARD, The Trojan Manifold - Survey and Conjectures, Mathematical Note No. 622, Boeing Scientific Research Laboratories, 1969.

[4] D.S. SCHMIDT, Families of periodic orbits in the restricted problem of three bodies connecting families of direct and retrograde orbits, to appear in SIAM J. Appl. Math. 21 (1971).

[5] V. SZEBEHELY, Theory of Orbits. Academic Press, New York, 1967.

University of Maryland, College Park, Maryland

A GENERAL THEORY OF LIAPUNOV STABILITY

P. Seibert

Introduction

The purpose of this paper is to sketch a theory of Liapunov stability based on the "second method", under minimal assumptions. The continuous flow is replaced by a preorder defined on an arbitrary set X. The metric or topological structure of the state space is replaced by two collections of subsets of X which play the role of "reference sets" for the definition of stability. A third structural element is introduced on X by means of a real-valued function which plays the role of a Liapunov function. Definitions and theorems are expressed in terms of relations between the four resulting collections of subsets of X. As immediate corollaries of the two principal theorems (2 and 3) one obtains criteria for uniform and non-uniform stability in the cases of autonomous and non-autonomous systems for eventual stability, and, using results of [7] and [8], for total stability.

The idea of Liapunov stability with respect to collections of sets (filters in that case) appears already in [1]. A similar theory, using uniformities, was presented by D. Bushaw in [3].

1. Stability of Succession Structures

Let X be an arbitrary set (the "state set" or "event set"), and denote by $\mathcal{X}$ the set of all nonempty subsets of X. We introduce a map

$$\sigma : X \to \mathcal{X}$$
$$x \to \sigma(x) =: x^\sigma \quad ,$$

which we call the "succession map". ($y \in x^\sigma$ means that the event y may succeed the event x.) If $A \subset X$, we write $A^\sigma := \cup \{x^\sigma \mid x \in A\}$. The pair (X,σ) we call a general succession structure.

Let $\mathcal{D}$ and $\mathcal{E}$ be nonempty subsets of $\mathcal{X}$. We say "$\mathcal{D}$ is finer than $\mathcal{E}$" iff every element of $\mathcal{E}$ contains an element of $\mathcal{D}$, notationally,

$$\mathcal{D} \rhd \mathcal{E}, \quad \text{or} \quad \mathcal{E} \lhd \mathcal{D} .$$

Moreover, we define

$$\mathcal{E}^\sigma := \{A^\sigma \mid A \in \mathcal{E}\} \quad .$$

Definition. The general succession structure (X,σ) is $\mathcal{E}$-stable iff $\mathcal{E} \lhd \mathcal{E}^\sigma$ ("$\mathcal{E}^\sigma$ is finer than $\mathcal{E}$"). In a more general sense, we say (X,σ) is $(\mathcal{D},\mathcal{E})$-stable iff $\mathcal{E} \lhd \mathcal{D}^\sigma$; in other words, given any set $A \in \mathcal{E}$, there exists a set $B \in \mathcal{D}$ such that $B^\sigma \subset A$.

THEOREM 1. If (X,σ) is $\mathcal{E}$-stable, then it is $(\mathcal{D},\mathcal{E})$-stable for any collection of sets $\mathcal{D}$ which is finer than $\mathcal{E}$.

The proof is left to the reader.

Examples. A) The classical case of a succession structure is a dynamical system (X,R,π). Here the succession relation is defined by means of the positive (or negative) semiorbits:

$$x^\sigma := \pi(x,R^\pm) \quad ,$$

where R^+ (R^-) denotes the nonnegative (nonpositive) reals. The generalization to dynamical polysystems (systems with multivalued phase map π) is obvious.

a) In Liapunov stability, the collections $\mathcal{D}$ and $\mathcal{E}$ are certain systems of neighborhoods of a given set M, or, if one wants to characterize boundedness properties, the complements of bounded sets. In the classical case, X is a metric space, $\mathcal{E}$ is the metric neighborhood filter, and $\mathcal{D}$ the complete neighborhood filter of M (which is assumed to be closed). Then $\mathcal{E}$-stability reduces to uniform stability (in the sense of Bhatia-Szegö [2], or "stability" in the sense of Zubov [11]), and $(\mathcal{D},\mathcal{E})$-stability becomes "stability" in the sense of [2].

b) Total stability or stability under persistent perturbations. Consider an autonomous differential equation which defines a dynamical system on X. In this case one can characterize total stability in terms of "contracting neighborhoods"* [7, 8]. A set A is called contracting under a given flow iff any point x in the closure of A has the property that xt, for any $t > 0$, is an interior point of A.

THEOREM [8]. A compact invariant set $M \subset X$ is totally stable iff it possesses a fundamental system of contracting neighborhoods.

This result enables us to extend the concept of total stability immediately to continuous flows, defining a set to be totally stable iff it possesses a fundamental system of contracting neighborhoods. Denote by $\mathcal{N}$ the neighborhood filter of M, and by $\mathcal{C}$ the system of contracting neighborhoods of M. Since the contracting property obviously implies positive invariance, it follows that $\mathcal{C}^{\sigma} = \mathcal{C}$, and consequently, $(\mathcal{C},\mathcal{N})$-stability is total stability. (Note that $\mathcal{N}$-stability (which is just Liapunov stability) does not imply $(\mathcal{C},\mathcal{N})$-stability. This, however, does not contradict Theorem 1, because $\mathcal{C}$ is in general coarser than $\mathcal{N}$.)

B) We now turn to space-time systems which are characterized by the two following properties: 1. The set X, which in this case we call "set of events", is a product set of the form

$$X = R \times Y$$
$$x = (t,y)$$

where R denotes the real line, and Y is an arbitrary set which we call the "set of states". 2. Given any point $x_0 = (t_0,y_0) \in X$, all t-sections of x_0, i.e.

*For a definition of total stability see, for instance, [4], § 28.

$$x_0 t = \pi(t_0, y_0, t) := \{(s,y) \in x_0^\sigma \mid s = t\} ,$$

with $t \geq t_0$, are non-empty.

All non-autonomous differential equations,

$$\dot{y} = f(t,y) ,$$

with global existence in the positive sense, define space-time systems. Stability of the origin in Y reduces to stability of the line $y = 0$ in the space time system. However, there is one concept of stability which has no equivalent in autonomous systems, namely <u>eventual stability</u> (introduced, independently, by LaSalle-Rath [5, 6] and by T. Yoshizawa [9, 10]). The origin of the state space Y is called eventually uniformly stable iff, given $\varepsilon > 0$, there exists a $\delta > 0$ and a $\tau \in R$ such that $\|y_0\| < \delta$, $\tau \leq t_0 \leq t$ together imply $\|y(t,y_0,t_0)\| < \varepsilon$; here $y(t,y_0,t_0)$ denotes the solution (assuming existence and uniqueness) satisfying the initial condition $y(t_0,y_0,t_0) = y_0$. This concept is easily seen to be equivalent to $(\mathcal{D},\mathcal{E})$-stability, with $\mathcal{D}$ and $\mathcal{E}$ defined as follows:

$$\mathcal{D} = \{S_Y(o,\varepsilon) \times [\tau,\infty) \mid \varepsilon > 0, \ \tau \in R\} ,$$

$$\mathcal{E} = \{S_Y(o,\varepsilon) \mid \varepsilon > 0\} ;$$

here $S_Y(o,\varepsilon)$ denotes the ε-neighborhood of the origin of the space Y.

2. <u>Liapunov Functions</u>

Consider functions

$$v: X \to R^+ ,$$

associate to each value $\lambda > 0$ the set

$$S_v^\lambda := \{x \in X \mid v(x) < \lambda\} ,$$

and denote the collection of all these sets by $\mathcal{S}_v$:

$$\mathcal{S}_v := \{S_v^\lambda \mid \lambda > 0\} .$$

We call v a <u>Liapunov function</u> for $(X,\sigma,\mathcal{D},\mathcal{E})$ iff it satisfies the following conditions:

(A) $$\mathcal{E} \lhd \mathcal{S}_v ,$$

(B) $$\mathcal{S}_v \lhd \mathcal{D} ,$$

(C) $$x \in S_v^\lambda \quad \text{implies} \quad x^\sigma \subset S_v^\lambda .$$

One verifies easily that in the case of a closed set M in a metric space, with $\mathcal{D}$ and $\mathcal{E}$ both equal to the metric neighborhood filter of M, the conditions (A), (B), and (C) are equivalent to Zubov's conditions for Liapunov stability [11]. Note also that in the case $\mathcal{D} = \mathcal{E}$, conditions (A) and (B) together state the equivalence of $\mathcal{E}$ and S_v (i.e., each element of one contains an element of the other).

<u>THEOREM 2</u>. If there exists a Liapunov function for $(X,\sigma,\mathcal{D},\mathcal{E})$, then the general succession structure (X,σ) is $(\mathcal{D},\mathcal{E})$-stable.

The proof is left to the reader.

2.1. In order to formulate an inverse theorem, certain conditions will be imposed on the succession relation σ and the collection $\mathcal{E}$.

<u>Definition 1</u>. The pair (X,σ) is a <u>succession structure</u> iff the relation between x and y defined by $y \in x^\sigma$ is reflexive and transitive.

<u>Definition 2</u>. The collection $\mathcal{E}$ is called <u>admissible</u> iff there exists a nested sequence of subsets of X, $\mathcal{B} = \{B_n \mid n = 1,2,\ldots\}$, such that $\cup B_n = X$, and $\mathcal{E}$ and $\mathcal{B}$ are equivalent, i.e., $\mathcal{E} \lhd \mathcal{B}$ and $\mathcal{B} \lhd \mathcal{E}$.

<u>THEOREM 3</u>. If (X,σ) is a $(\mathcal{D},\mathcal{E})$-stable succession structure, and $\mathcal{E}$ is admissible, then there exists a Liapunov function for $(X,\sigma,\mathcal{D},\mathcal{E})$.

The proof consists in verifying the conditions (A), (B), (C) for the function

$$v(x) = \inf \{\tfrac{1}{n} \mid x^\sigma \subset B_n\} .$$

REFERENCES

[1] J. AUSLANDER, On stability of closed sets in dynamical systems, Sem. Differential Equat. Dynam. Syst. II, Maryland 1969, Lecture Notes Math., 144 (1970), 1-4.
[2] N.P. BHATIA and G.P. SZEGÖ, Stability Theory of Dynamical Systems. Springer Verlag, 1970.
[3] D. BUSHAW, A stability criterion for general systems, Math. Systems Theory, 1 (1967).
[4] W. HAHN, Theory and Application of Liapunov's Second Method. Prentice-Hall, 1963.
[5] J.P. LASALLE, Recent advanced in Liapunov stability theory, SIAM Rev., 6 (1964), 1-11.
[6] J.P. LASALLE and R.J. RATH, Eventual stability, Proc. 2nd Congress IFAC, Part 2; Butterworths, London, 1964, 556-569.
[7] P. SEIBERT, A concept of stability in dynamical systems, Topological Dynamics, Internat. Sympos., Ft. Collins, 1967; Benjamin, New York and Amsterdam, 1968, 423-433.
[8] P. SEIBERT, Estabilidad bajo perturbaciones sostenidas y su generalización en flujos continuos, Acta Mexicana Ci. Tecn., 2 (1968), 154-165.
[9] T. YOSHIZAWA, Eventual properties and quasi-asymptotic stability of non-compact sets, Funkcial. Ekvac., 8 (1966), 79-90.
[10] T. YOSHIZAWA, Stability Theory by Liapunov's Second Method, Math. Soc. Japan, 1966.
[11] V.I. ZUBOV, The Methods of Liapunov and Their Application. Noordhoff, Groningen, 1964 [First publication: Leningrad, 1957].

Universidad Católica de Chile, Santiago, Chile

FINITE TIME STABILITY OF LINEAR DIFFERENTIAL EQUATIONS*

Leonard Weiss and Jong-Sen Lee

1. Introduction

"Finite time stability" differs from classical stability in that one is interested in the behavior of system trajectories which originate within an *a priori* fixed region in the state space over a given fixed time interval (which may be finite or infinite). In the past 25 years, exploration of various facets of this concept has been made by a number of workers (see [1] - [21]), many of them motivated by the fact that in a variety of practical situations, the "finite time" concept of stability is more pertinent than the classical concept. The theory, in its early stages, was mainly (though not entirely) concerned with linear differential equations ([1] - [8]), proceeded through some preliminary probing toward a qualitative nonlinear theory ([9] - [12]), and then underwent systematic development using a particular Lyapunov-like approach initiated in [13] - [15] and continued in [16] - [21].

In this paper, we bring the development full circle by presenting some new results in the theory of finite time stability for linear autonomous differential equations. Not only are these results computationally attractive, but the techniques used to obtain them allow

*The results in this paper were announced at the Fall SIAM Meeting, Boston, Oct. 1970. Dr. Weiss' research is supported, in part, by the Air Force Office of Scientific Research under Grant AFOSR 69-1646.

elementary and elegant proofs of known theorems to be given, and enable succinct characterizations of finite time stability to be made for certain classes of linear differential equations.

Extensions of the aforementioned results to the case of linear systems driven by "white noise" are made and also result in easily computable criteria for finite time stability.

2. Preliminaries

The symbol $\|\cdot\|$ denotes the euclidean norm on R^n. If A is an $n \times n$ real matrix, the set of eigenvalues of A is denoted by $\{\lambda(A)\}$. If the eigenvalues are real, $\hat{\lambda}(A) \triangleq \max \{\lambda(A)\}$. The spectral norm of A is denoted by $\|A\|^* \triangleq [\hat{\lambda}(A'A)]^{\frac{1}{2}}$ where A' is the transpose of A.

Consider the system of linear differential equations

$$\dot{x}(t) = Ax(t) \ , \quad t \geqslant 0 \tag{1}$$

where A is $n \times n$, real, and $x(0) \triangleq x_0$. The unique solution of (1) at time t is given by

$$x(t, x_0) \triangleq x(t) = e^{At} x_0 \ . \tag{2}$$

Our objective is to investigate (1) for the following type of behavior.

Definition 1. The system (1) is stable with respect to (α,β,T), $\alpha \leqslant \beta$, if $\|x_0\| < \alpha$ implies $\|x(t)\| < \beta$ for all $t \in [0,T)$.

3. A Fundamental Result

The theorem given below will be used in conjunction with the linear algebraic machinery developed in the next section to generate further stability results.

THEOREM 1. A necessary and sufficient condition for the system (1) to be stable (Def. 1) is

$$\|e^{At}\|^* \leqslant \frac{\beta}{\alpha} \ , \quad \forall t \in [0,T) \ . \tag{3}$$

Proof. For each fixed t,

$$\|x(t)\| = \|e^{At}x_0\| \leq \|e^{At}\|^* \|x_0\| \tag{4}$$

with equality achieved for some x_0 (independent of $\|x_0\|$). Hence, $\|x(t)\| < \beta$, $\forall t \in [0,T)$ if and only if $\|e^{At}\|^* \|x_0\| < \beta$, $\forall t \in [0,T)$ and the latter holds on $[0,T)$ with $\|x_0\| < \alpha$ if and only if (3) holds. Q.E.D.

Since our aim is to obtain results which are computationally tractable, our effort in the succeeding two sections is mainly devoted to translating (3) into computable conditions on the coefficient matrix A.

4. Some Useful Lemmas

The statements designated as Lemmas 1, 2, 4 below are well known. Lemma 3 is less well known (see Dahlquist [22]), and we therefore provide a proof for it.

LEMMA 1. If $f(A)$ is a well defined function of a matrix A, then $\lambda \in \{\lambda(A)\}$ implies $f(\lambda) \in \{\lambda(f(A))\}$.

Proof. See [23].

LEMMA 2. If A is a symmetric matrix, then $\|e^A\|^* = e^{\hat{\lambda}(A)}$.

Proof. Follows from Lemma 1.

LEMMA 3. For any $n \times n$ real matrix A,

$$\|e^A\|^* \leq e^{\hat{\lambda}(\frac{1}{2}(A' + A))} \tag{5}$$

with equality if and only if A is normal.

Proof. Any $n \times n$ matrix A has a unique decomposition $A = A_1 + A_2$ where $A_1 = \frac{1}{2}(A + A')$, $A_2 = \frac{1}{2}(A - A')$. Now, $e^A = e^{A_1 + A_2} = e^{A_1}e^{A_2}$ if and only if A_1 and A_2 commute (A_1 and A_2 commute if and only if A is normal). Hence, under such conditions $\|e^A\|^* = \|e^{A_1}e^{A_2}\|^*$ and since A_2 is skew-symmetric, $A_2' = -A_2$. Therefore, we have

$$(\|e^A\|^*)^2 = \hat{\lambda}(e^{-A_2}e^{A_1}e^{A_1}e^{A_2}) = \hat{\lambda}(e^{2A_1}) = e^{\hat{\lambda}(A'+A)} .$$

This proves the "equality" part. To prove the inequality, we start with the "Rayleigh quotient" [24] representation of $\|\cdot\|^*$, i.e.,

$$\begin{aligned}
\hat{\lambda}(A'+A) &= \max_{x\neq 0} \frac{x'(A'+A)x}{\|x\|^2} = \max_{x(\cdot)} \int_0^1 \frac{x'(t)(A'+A)x(t)}{\|x(t)\|^2}\, dt \\
&\geq \max_{\substack{x(\cdot)\\ \dot{x}=Ax}} \int_0^1 \frac{\dot{x}'(t)x(t) + x'(t)\dot{x}(t)}{\|x(t)\|^2}\, dt \\
&\geq \max_{x_0} \int_{x_0}^{x(1)} \frac{d(\|x(t)\|^2)}{\|x(t)\|^2} \geq \max_{x_0} \ln\left(\frac{\|x(1)\|^2}{\|x_0\|^2} \right) \\
&= \max_{x_0} \ln\left(\frac{\|e^A x_0\|^2}{\|x_0\|^2} \right) = \ln\left(\max_{x_0} \frac{\|e^A x_0\|^2}{\|x_0\|^2} \right) \\
&\geq \ln\, (\|e^A\|^*)^2 ,
\end{aligned}$$

which implies

$$\|e^A\|^* \leq e^{\hat{\lambda}(\frac{1}{2}(A'+A))} . \qquad \text{Q.E.D.}$$

Finally, we shall make use of the well known relationship (see [23]) between the spectral radius of an $n \times n$ matrix A and its spectral norm, namely

<u>LEMMA 4</u>. $\max\{|\lambda(A)|\} \leq \|A\|^*$.

5. <u>Applications of Theorem 1 and Lemmas 1-4</u>

The following result is new and characterizes finite time stability for a class of linear systems via an easily computable criterion.

<u>THEOREM 2</u>. Let A in (1) be normal. Then (1) is stable (Def. 1) if and only if

(6) $$\hat{\lambda}(\tfrac{1}{2}\,(A+A')) \leq \frac{1}{T} \ln \beta/\alpha .$$

<u>Proof</u>. By Theorem 1 and Lemmas 2, 3 we have

$$\text{Stability (Def. 1)} \iff \|e^{At}\|^* \leqslant \beta/\alpha \qquad \forall t \in [0,T)$$

$$\iff e^{\hat{\lambda}(\frac{1}{2}(A'+A)t)} \leqslant \beta/\alpha \quad \forall t \in [0,T)$$

From which the necessity and sufficiency of (6) follows.

This generalizes, via our simpler and more direct argument, an earlier result by Kaplan [21] for the case when A is symmetric. The machinery developed in Section 4 allows other known results in finite time stability to also be obtained directly. For example, as an immediate corollary of Lemma 3 and Theorem 1, we obtain

THEOREM 3 (Dorato [8]): A sufficient condition for (1) to be stable (Def. 1) is

$$\hat{\lambda}\left(\frac{1}{2}(A+A')\right) \leqslant \frac{1}{T} \ln \beta/\alpha \quad . \tag{7}$$

Even more striking is the simple and direct proof which can be given of the following necessary condition, originally established (see Kaplan [21] by means of a long induction argument.

THEOREM 4. The system (1) is stable (Def. 1) only if

$$\max \{\mathrm{Re}(\lambda(A))\} \leqslant \frac{1}{T} \ln \beta/\alpha \quad . \tag{8}$$

Proof. By Theorem 1 and Lemma 4 we obtain, as a necessary condition for stability (Def. 1),

$$\max \{|\lambda(e^{At})|\} \leqslant \beta/\alpha \qquad \text{for all} \quad t \in [0,T) \ . \tag{9}$$

By Lemma 1, this implies

$$e^{\max\{\mathrm{Re}(\lambda(A))\}t} \leqslant \beta/\alpha \qquad \text{for all} \quad t \in [0,T)$$

which implies, in turn, that

$$e^{\max\{\mathrm{Re}(\lambda(A))\}(T-\varepsilon)} \leqslant \beta/\alpha \quad \text{for all arbitrarily small} \quad \varepsilon > 0 \ . \tag{10}$$

Taking the logarithm and the limit as $\varepsilon \to 0$, yields (8).

6. Connections to Nonlinear Theory

Let $V: R^n \times [0,T) \to R$ be a continuous function with continuous first partial derivatives. Define

$$\dot{V}(x,t) = \text{grad } V \cdot \frac{dx}{dt} + \frac{\partial V}{dt}$$

$$V_m^a(t) \triangleq \min_{\|x\|=a} V(x,t)$$

$$V_M^a(t) \triangleq \max_{\|x\|=a} V(x,t) \ .$$

$B(\alpha) = \{x \in R^n : \|x\| < \alpha\}$, and $\overline{B}(\alpha)$ = closure of $B(\alpha)$.

Consider the following theorem*, which holds for certain nonlinear as well as linear systems.

THEOREM 5. The system (1) is (uniformly**) stable (Def. 1) if and only if there exists a real-valued function $V(x,t)$ as above and a real-valued integrable function $\phi(t)$ such that

(i) $\dot{V}(x,t) \leq \phi(t)$ for all $x \in (B(\beta) - \overline{B}(\alpha))$, all $t \in [0,T)$

(ii) $\int_{t_1}^{t_2} \phi(t)dt < V_m^{\beta}(t_2) - V_M^{\alpha-\varepsilon}(t_1)$, all $t_2 > t_1$, $t_1, t_2 \in [0,T)$.

To relate Theorem 5 to previous results, we note first that $V(x,t)$ can be chosen as $V(x)$ since (1) is autonomous. Consider

$$V(x) = \ln \|x\| \ . \tag{11}$$

Then

$$\dot{V}(x) = \frac{x'Ax}{\|x\|^2} \ .$$

---

* This is actually a trivial perturbation of Theorem 1 in [15], the difference being that $V_M^{\alpha-\varepsilon}$ appears in this case rather than V_M^{α} as in [15]. This occurs because the definition of finite time stability in [15] allowed an initial condition $\|x_o\| = \alpha$. The proof is exactly the same.

**Uniform finite time stability is defined in [15] and is equivalent to stability (Def. 1) for systems of the form (1).

Now, with A_1 and A_2 defined as in the proof of Lemma 3, we have

$$x'A_2x = 0 ,$$

and so

$$\dot{V}(x) = \frac{x'A_1x}{\|x\|^2} \leq \hat{\lambda}(A_1) .$$

Let $\phi(t) = \hat{\lambda}(A_1)$. Then Theorem 5 yields the result

$$\int_0^T \hat{\lambda}(A_1)dt < \ln \frac{\beta}{\alpha-\varepsilon}$$

or

$$\hat{\lambda}(A_1) < \frac{1}{T} \ln \frac{\beta}{\alpha-\varepsilon} \quad \text{for } \varepsilon > 0, \text{ arbitrarily small.}$$

Taking the limit as $\varepsilon \to 0$, we get

$$\hat{\lambda}(A_1) \leq \frac{1}{T} \ln \frac{\beta}{\alpha}$$

which is the relation given in Theorem 3.*

Remarks. 1. An extension of Theorem 5 to the case of nonuniform finite time stability has been made by various people working independently (c.f. (for sufficient conditions) [18], [19], and (for converse theorems), [20], [21]). The extension replaces hypothesis (ii) (assuming initial time 0) by

$$\int_0^T \phi(t)dt < V_m(T) - \max_{\|x\| \leq \alpha-\varepsilon} V(x,0) .$$

With this extension, the function $V(x) = \ln\|x\|$ yields a known sufficient condition (see [8]) for stability (Def. 1) of the linear time-varying system $\dot{x}(t) = A(t)x(t)$, $t \geq 0$, namely,

$$\int_0^T \hat{\lambda}\left(\frac{1}{2} (A'(s) + A(s)\right)ds \leq \ln \frac{\beta}{\alpha} .$$

*This connection has been noticed independently by A. Michel [27].

2. If A in (1) is normal, then the Lyapunov-like function (11) will <u>always</u> be conclusive in the test for stability (Def. 1).

3. It is interesting (though not surprising once one sees the pattern of results) that the natural Lyapunov-like function associated with finite time stability of linear systems is <u>not</u> a quadratic form, but the logarithm of the latter.

7. Linear Systems Driven by White Noise

In this section, we consider linear systems of the form

$$\dot{x}(t) = Ax(t) + Bu(t) , \quad t \geq 0 \tag{12}$$

where A is $n \times n$, B is $n \times m$, and $u(\cdot)$ is a vector white noise with zero mean and covariance matrix $Q(t)$ (i.e., $E\{u(t)\} = 0$ and $E\{u(t)u'(s)\} = Q(t)\delta(t-s)$, where E = Expectation, and δ = Dirac delta.)

For any $n \times n$ matrix G, let $tr(G)$ = trace of G. Then we have

<u>Definition 2</u>. The system (12) is mean square stable with respect to (α,β,γ,T), $\alpha \leq \beta$, if the conditions $E\{\|x(0)\|^2\} < \alpha^2$ and $tr(Q(t)) \leq \gamma^2$ for all $t \in [0,T]$ imply $E\{\|x(t)\|^2\} < \beta^2$ for all $t \in [0,T]$.

The main result in this section depends on the following Lemma.

<u>LEMMA 5</u>. Let F be an $n \times n$ symmetric matrix and let $\mathcal{P}$ denote the set of $n \times n$ nonnegative definite matrices. Then

$$\hat{\lambda}(F) = \max_{P \in \mathcal{P}} \frac{tr(PF)}{tr(P)} \quad . \tag{13}$$

<u>Proof</u>. Let S be an $n \times n$ orthogonal matrix such that $S'FS = \Lambda = \text{diag}(\lambda_i)$. Also let $S'PS = D$. Then

$$\begin{aligned} tr(PF) &= tr(PS\Lambda S') \\ &= tr(S'PS\Lambda) \\ &= tr(D\Lambda) \end{aligned}$$

$$= \lambda_1 d_{11} + \lambda_2 d_{22} + \cdots\cdots + \lambda_n d_{nn}$$

$$\leqslant \hat{\lambda}(F)(d_{11} + d_{22} + \cdots + d_{nn}), \quad d_{ii} \geqslant 0$$

$$\leqslant \hat{\lambda}(F)\,tr(D)\ .$$

Therefore it is possible to choose $\{d_{ii} | \ i = 1, .., n\}$ such that

$$\hat{\lambda}(F) = \max_{d_{ii}} \frac{tr(D\Lambda)}{tr(D)} = \max_{P \in \mathcal{P}} \frac{tr(PF)}{tr(P)}\ .$$

Q.E.D.

Now consider

THEOREM 6. The system (12) is mean square stable (Def. 2) if and only if

(14) $$\alpha^2(\|e^{At}\|^*)^2 + \gamma^2 \int_0^t (\|e^{A\xi}B\|^*)^2\,d\xi \leqslant \beta^2 \qquad t \in [0,T]\ .$$

Proof. Let $P(t) = E\{x(t)x'(t)\}$. Then $P(t) \in \mathcal{P}$ for each t, and differentiation yields

(15) $$\dot{P}(t) = AP(t) + P(t)A' + BQ(t)B'\ .$$

The solution to (15) is given by (see [26])

$$P(t) = e^{At}P(0)e^{A't} + \int_0^t e^{A(t-s)}BQ(s)B'e^{A'(t-s)}ds\ .$$

Then

$$tr(P(t)) = tr(P(0)e^{A't}e^{At}) + \int_0^t tr(Q(s)B'e^{A'(t-s)}e^{A(t-s)}B)ds$$

and, by Lemma 5,

(16) $$tr(P(t)) \leqslant (\|e^{At}\|^*)^2\,tr(P(0)) + \int_0^t (\|e^{A(t-s)}B\|^*)^2\,tr(Q(s))ds\ .$$

Sufficiency of (14) now follows by substituting, for any fixed $t \in [0,T]$, $tr(P(0)) < \alpha^2$, and $tr(Q(s)) \leqslant \gamma^2$ for all $s \in [0,t]$ into (16). To prove necessity, we first note that th re exists a $P(0)$ and a $Q(s)$ such that equality occurs in (16), and such that $tr(P(0)) = \alpha^2$ and $tr(Q(s)) = \gamma^2$ for all $s \in [0,t]$. Suppose (14) does not hold at some $t = t_1 \in [0,t]$. Then with $P(0)$ and $Q(s)$ chosen as indicated, (16) yields $tr(P(t_1)) > \beta^2$, in which case $tr(P(t_1)) \geqslant \beta^2$ as long as

$\mathrm{tr}(P(0)) < \alpha^2$ and $\mathrm{tr}(Q(s)) = \gamma^2$ for all $s \in [0,t_1]$. Hence the negation of (14) implies the negation of stability (Def. 3) and the theorem is proved.

We now develop results for the system (12) analogous to our earlier ones for (1). Let

$$\rho^2 = \frac{\gamma^2(\|B\|^*)^2}{\hat{\lambda}(A'+A)} . \tag{17}$$

COROLLARY 1. A sufficient condition for the system (12) to be mean square stable (Def. 2) is

$$\hat{\lambda}\left(\frac{1}{2}(A'+A)\right) \leq \frac{1}{2T} \ln \frac{\beta^2 + \rho^2}{\alpha^2 + \rho^2} . \tag{18}$$

Proof. Let

$$U(t) = (\|e^{At}\|^*)^2 \, \mathrm{tr}(P(0)) + \int_0^t (\|e^{A(t-s)}B\|^*)^2 \, \mathrm{tr}(Q(s))ds .$$

Then stability (Def. 2) occurs if $U(t) \leq \beta^2$, $\forall t \in [0,T]$, for $\mathrm{tr}(P(0)) < \alpha^2$ and $\mathrm{tr}(Q(s)) \leq \gamma^2$ for all $s \in [0,T]$. Now

$$U(t) < \alpha^2(\|e^{At}\|^*)^2 + \gamma^2 \int_0^t (\|e^{A\xi}B\|^*)^2 \, d\xi .$$

By Lemmas 2 and 3 and the fact that $\|GH\|^* \leq \|G\|^*\|H\|^*$,

$$U(t) < \alpha^2 e^{\hat{\lambda}(A'+A)t} + \gamma^2(\|B\|^*)^2 \int_0^t e^{\hat{\lambda}(A'+A)\xi} \, d\xi$$

or

$$U(t) < \alpha^2 e^{\hat{\lambda}(A'+A)t} + \frac{\gamma^2(\|B\|^*)^2}{\hat{\lambda}(A'+A)} [e^{\hat{\lambda}(A'+A)t} - 1] \triangleq V(t) .$$

Then stability (Def. 2) is implied by $V(T) \leq \beta^2$. Taking the log of both sides and using (17) yields (18). Q.E.D.

COROLLARY 2. Consider the system (12) and suppose $B = I$ and A is normal. Then (18) is a necessary and sufficient condition for (12) to be mean square stable (Def. 2).

Proof. From Theorem 6, we obtain the following necessary and sufficient condition.

$$\alpha^2(\|e^{At}\|^*)^2 + \gamma^2 \int_0^t (\|e^{A\xi}\|^*)^2 \, d\xi \leqslant \beta^2 \qquad \forall t \in [0,T] .$$

Applying Lemma 3 allows the integral to be evaluated, and taking the log of the resulting inequality with $t = T$, yields (18).

The close correspondence of these results with those for deterministic undriven systems is completed by giving the following necessary condition for stability (Def. 2). Let

$$\mathrm{Re}(\bar{\lambda}) = \max \{\mathrm{Re}(\lambda(A))\}$$

and let

$$\mu^2 = \frac{\gamma^2}{2(\mathrm{Re}(\bar{\lambda}))} . \tag{19}$$

Then we have

<u>THEOREM 7</u>. Let $B = I$ in (12). Then the system (12) is mean square stable (Def. 2) only if

$$\mathrm{Re}(\bar{\lambda}) \leqslant \frac{1}{2T} \ln \frac{\beta^2 + \mu^2}{\alpha^2 + \mu^2} . \tag{20}$$

<u>Proof</u>. Similar in structure to that of Theorem 4, and follows from applying Lemma 4 to the necessary condition (14).

REFERENCES

[1] B.V. BULGAKOV, On the accumulation of disturbances in linear oscillating systems with constant parameters (in Russian), Dokl. Akad. Nauk SSSR, Vol. 51, No. 5 (1946).

[2] G.V. KAMENKOV, On the stability of motion in a finite time interval, P.M.M. 17 (1953), 529-540.

[3] A.A. LEBEDEV, The problem of stability in a finite interval of time, P.M.M. 18 (1954), 75-94.

[4] A.A. LEBEDEV, On stability of motion during a given interval of time, P.M.M. 18 (1954), 129-148.

[5] SY-IN CHZHAN, Stability of motion during a finite time interval, P.M.M. 23 (1959), 333-343.

[6] SY-IN CHZHAN, On estimates of solutions of differential equations, accumulation of perturbation, and stability of motion during a finite time interval, P.M.M. 23 (1959), 640-649.

[7] L.S. GNOENSKII, On the accumulation of disturbances in linear systems, P.M.M. 25 (1961), 319-331.

[8] P. DORATO, Short time stability in linear time-varying systems, 1961 IRE Convention Record, Part 4, 83-87.

[9] J.P. LASALLE and S. LEFSCHETZ, Stability by Liapunov's Direct Method with Applications. Academic Press, New York, 1961.

[10] A.M. LETOV, Stability in Nonlinear Control Systems. Princeton University Press, Princeton, 1961.

[11] R. HAYART, Extension des théoremes de Liapounoff et de Chetayev relatifs à la stabilité, Analyse Mathématique, Compt. Rend. 259 (1964), 38-41.

[12] R. HAYART, Complements relatifs à la stabilité (α,β,τ), Equations Différentielles, Compt. Rend. 260 (1965), 1331-1333.

[13] L. WEISS and E.F. INFANTE, On the stability of systems defined over a finite time interval, Proc. Nat. Acad. Sci. 54 (1965), 44-48.

[14] L. WEISS and E.F. INFANTE, Finite time stability under perturbing forces and on product spaces, IEEE Trans. Auto. Cont. AC-12 (1967), 54-59.

[15] L. WEISS, Converse theorems for finite time stability, SIAM J. Appl. Math. 16 (1968), 1319-1324.

[16] R.W. GUNDERSON, On stability over a finite interval, IEEE Trans. Auto. Cont. AC-12 (1967), 634-635.

[17] A.A. KAYANDE and J.S.W. WONG, Finite time stability and comparison principles, Proc. Camb. Phil. Soc. 64 (1968), 749-756.

[18] J.A. HEINEN and S.H. WU, Further results concerning finite time stability, IEEE Trans. Auto. Cont. AC-14 (1969), 211-212.

[19] L. WEISS, An uniform and nonuniform finite time stability, IEEE Trans. Auto. Cont. AC-14 (1969), 313-314.

[20] A.A. KAYANDE, A theorem on contractive stability, 1970 (to appear).

[21] J. KAPLAN, Ph.D. dissertation, University of Maryland, 1970.

[22] G. DAHLQUIST, Stability and Error Bounds in the Numerical Integration of Ordinary Differential Equations. Almqvist & Wiksells Boktryckeri AB, Stockholm, 1958.

[23] F.R. GANTMAKHER, The Theory of Matrices, Vol. I. Chelsea, New York, 1959.

[24] R. BELLMAN, Introduction to Matrix Analysis. McGraw-Hill, New York, 1960.

[25] L. COLLATZ, Functional Analysis and Numerical Mathematics. Academic Press, New York, 1966.

[26] A.E. BRYSON and Y.C. HO, Applied Optimal Control. Blaisdell, New York, 1969.

[27] A. MICHEL, Quantitative analysis of system stability, boundedness, and trajectory behavior, Arch. Rat. Mech. Anal. 38, No. 2 (1970).

LW: University of Maryland, College Park, Maryland, and Naval Research Laboratory, Washington, D.C.

JSL: Naval Research Laboratory, Washington, D.C.

SECOND ORDER OSCILLATION WITH RETARDED ARGUMENTS

James S. W. Wong

1. Introduction

The study of oscillatory behavior of solutions of functional differential equations has received little attention as compared with the corresponding theory in the case of ordinary differential equations. There seem to be two reasons for this phenomenon. One is the fact that available techniques in the past to study oscillation require spontaneous descriptions of the solution and its derivative at the same instant t. This information, though embodied in the given equation, is much more difficult to extract in case of functional differential equations. Another obstacle in such a study is the lack of concrete examples on which one may formulate general results. It is well known that even in the simplest case of linear difference differential equations, the solutions are described in terms of roots of certain transcendental equations, solving which remains one of the most difficult mathematical problems, see e.g. Bellman and Cooke [2], Hale [8].

As early as 1921, Fite [6] has considered the problem of oscillation for solutions of certain functional differential equations. Since functional differential equations with retarded arguments find important applications in physical problems, most results on oscillation are obtained for this case. We refer to the monograph of Myshkis [16] and Norkin [18] for a discussion of oscillation results on first and second order delay differential equations. A systematic study on the oscillatory behavior of solutions of first order delay equations was initiated in a recent paper by Lillo [15].

We are here concerned with the study of oscillatory behavior of solutions of second order delay differential equations; in particular, our attention is focused on the following equation

$$(1.1) \qquad y''(t) + a(t)y^{\gamma}(\sigma(t)) = 0 , \qquad t \geq 0 ,$$

where $a(t)$, $\sigma(t)$ are real valued functions of t and $\gamma > 0$ is the quotient of two odd integers. We assume also that $a(t)$ and $\sigma(t)$ are piecewise continuous on $[0,\infty)$. To simplify our discussion, we consider only the case of oscillation at infinity, thereby dictating the assumption

$$(1.2) \qquad \lim_{t\to\infty} \sigma(t) = \infty .$$

For each $t_0 \geq 0$, denote by $E_{t_0} = \{t : \sigma(t_0) \leq t \leq t_0\}$. By a solution $y(t)$ of (1.1), we mean a real valued function satisfying (1.1) on $t \geq t_0$ and $y(t) = \phi(t)$ on E_{t_0} for some initial continuous function $\phi(t)$. A solution $y(t)$ of (1.1) is called oscillatory if $y(t)$ assumes both positive and negative values for arbitrarily large values of t. It is common to place a boundedness assumption on the delay function $\sigma(t)$, namely,

$$(1.3) \qquad 0 \leq t - \sigma(t) \leq M .$$

(The requirement that $\sigma(t) \leq t$ will be assumed throughout our discussion without further mention. However, we shall not assume in general that (1.3) holds.) Under the present hypothesis, it is well known that solution to the initial value problem for (1.1) with any continuous initial function $\phi(t)$ on E_{t_0} exists on $[t_0,\infty)$, (see El'sgol'ts [5], p. 16). Moreover, if $\gamma \geq 1$, then the solution is in fact unique.

Recent interest in the study of oscillation problems for equation (1.1) is evidenced by the works [4], [7], [13], [14], [15], [21], [22], [23], all of which are concerned with second order delay equations of a similar type. The purpose of this note is to show that a number of oscillation results in the ordinary differential equation case, i.e., $\sigma(t) = t$, can be carried over almost verbatim to this general situation. Following Myshkis [16] for the first order equation, we shall refer to equation (1.1) as the stable equation if $a(t) \geq 0$, the unstable equation if $a(t) \leq 0$, and the equation of mixed type if $a(t)$ assumes both positive and negative values for arbitrarily large values of t. Our results are mostly directed to the stable equation, but results in connection with the other two cases are also discussed.

In Section 2, we consider the question of the existence of nonoscillatory solutions to equation (1.1). Oscillation theorems for the stable equation are given in Sections 3 and 4, respectively, In Section 5, we discuss oscillatory behavior of solutions of the unstable equation.

2. Nonoscillation Theorems

We consider primarily the stable equation (1.1), namely, $a(t) \geq 0$, together with the assumption that $\sigma(t) \leq t$ and satisfies (1.2). The first result is an immediate extension of a necessary and sufficient condition for the existence of a bounded nonoscillatory solution in the ordinary case. (For $\sigma(t) \equiv t$, $\gamma > 1$, see Atkinson [1].)

THEOREM (2.1). Let $\gamma > 0$. Assume that $a(t) \geq 0$ and $\sigma(t)$ satisfies (1.2). Then equation (1.1) has a bounded nonoscillatory solution if and only if

$$\int^{\infty} ta(t)dt < \infty. \tag{2.1}$$

Proof. Let $y(t)$ be a bounded nonoscillatory solution. Since γ is the quotient of two odd integers, we may assume without loss of generality that $y(t) > 0$ for $t \geq T$. In this case, it is easy to see that $y'(t)$ must be positive and nonincreasing, so for $t \geq t_0 \geq T$, we have

$$y(t) - y(t_0) = \int_{t_0}^{t} y'(s)ds \geq y'(t)(t - t_0) . \tag{2.2}$$

Now consider the equivalent integral equation of (1.1) in the following form

$$y(t) = y(t_0) + y'(t)(t - t_0) + \int_{t_0}^{t} sa(s)y^{\gamma}(\sigma(s))ds . \tag{2.3}$$

Since $y'(t) > 0$, $\lim_{t\to\infty} y(t) = L$ exists. Using (2.2) and the boundedness of $y(t)$ in (2.3), we obtain the boundedness of

$$\int_{t_0}^{t} sa(s)y^{\gamma}(\sigma(s))ds < \infty. \tag{2.4}$$

Choose t_0 so large that $y(\sigma(t)) \geq \frac{L}{2}$ for $t \geq t_0$. It then follows from (2.4) that

$$\left(\frac{L}{2}\right)^{\gamma} \int_{t_0}^{\infty} sa(s)ds \leq \int_{t_0}^{\infty} sa(s)y^{\gamma}(\sigma(s))ds < \infty,$$

proving the necessity.

Conversely, suppose that (2.1) holds. One needs to establish the existence of a solution to the integral equation

$$y(t) = 1 - \int_{t}^{\infty}(s - t)a(s)y^{\gamma}(\sigma(s))ds . \tag{2.5}$$

We define the Caratheodory approximate solutions to (2.5) as follows:

$$\begin{cases} y_0(t) \equiv 1 , \\ y_n(t) = 1 - \int_{t}^{\infty}(s - t)a(s)y_{n-1}^{\gamma}(\sigma(s))ds . \end{cases} \tag{2.6}$$

Choose t_0 such that

$$\int_{t_0}^{\infty} sa(s)ds < \frac{1}{2} . \tag{2.7}$$

Using (2.7) in (2.6), we can show inductively that for all n, $t \geq t_0$,

$$\frac{1}{2} \leq y_n(t) \leq 1 . \tag{2.8}$$

From (2.6) we find

$$y_n'(t) = \int_{t}^{\infty} a(s)y_{n-1}^{\gamma}(\sigma(s))ds ,$$

hence

$$|y_n'(t)| \leq \int_{t}^{\infty} a(s)ds \leq \int_{t}^{\infty} sa(s)ds < \frac{1}{2} , \tag{2.9}$$

for $t \geq t_0$. It follows from (2.8) and (2.9) that the sequence $\{y_n(t)\}$ defines a uniformly bounded and equicontinuous family on $[t_0,\infty)$, hence it follows from the Arzela-Ascoli theorem that there exists a subsequence $\{y_{n_k}(t)\}$ uniformly convergent on every compact subinterval of $[t_0,\infty)$.

Now a standard argument, see for example [25], yields a function $y(t)$ which is a solution to (2.5).

The next result gives a necessary and sufficient condition for the existence of a nonoscillatory solution of a special kind to (1.1), namely, asymptotically linear ones. (For $\sigma(t) \equiv t$, $\gamma \geq 1$ see Nehari [17] and for $0 < \gamma < 1$, see Belohorec [3] and Heidel [10].)

Here by an asymptotically linear solution $y(t)$, we mean a solution for which there exists a constant $\alpha \neq 0$ such that

$$\lim_{t\to\infty} \frac{y(t)}{t} = \alpha . \tag{2.10}$$

THEOREM (2.2). Let $\gamma > 0$. Assume that $a(t) \geq 0$ and $\sigma(t)$ satisfies (1.2). Then a necessary and sufficient condition for equation (1.1) to possess an asymptotically linear solution is

$$\int^{\infty} a(t)\sigma^{\gamma}(s)ds < \infty . \tag{2.11}$$

Proof. Suppose that equation (1.1) has an asymptotically linear solution, then there exists $T \geq 0$ and a positive constant c such that for all $t \geq T$

$$0 < \frac{c}{2} \leq \frac{y(t)}{t} \leq c < \infty . \tag{2.12}$$

Since $a(t) \geq 0$, we have $y'(t) > 0$ for $t \geq t_0$. Thus, integrating (1.1) we find

$$y'(t) = y'(t_0) - \int_{t_0}^{t} a(s)y^{\gamma}(\sigma(s))ds > 0 . \tag{2.13}$$

Using (2.12) in (2.13), one obtains

$$(\frac{c}{2})^{\gamma} \int_{t_0}^{\infty} a(s)\sigma^{\gamma}(s)ds \leq \int_{t_0}^{\infty} a(s)y^{\gamma}(\sigma(s))ds < \infty ,$$

proving necessity.

Conversely, suppose that (2.11) holds. Consider the equivalent integral equation to (1.1) in the following form

$$y(t) = y(t_0) - y'(t_0)(t - t_0) - \int_{t_0}^{t} (s - t_0)a(s)y^{\gamma}(\sigma(s))ds . \tag{2.14}$$

We can estimate (2.14) as follows for $t \geq 1$,

$$(2.15) \qquad |y(t)| \leq t\left(C + \int_{t_0}^{t} a(s)|y(\sigma(s))|\,ds\right) = t\Phi(t) ,$$

where $C = |y(t_0)| + |y'(t_0)|$. Using the definition of $\Phi(t)$ defined in (2.15), we obtain

$$(2.16) \qquad \Phi'(t) = a(t)\,|y(\sigma(t))|^{\gamma} \leq a(t)\sigma^{\gamma}(t)\Phi^{\gamma}(\sigma(t)) .$$

Since $\Phi(t)$ is nondecreasing, (2.16) gives

$$(2.17) \qquad \Phi(t) \leq C + \int_{t_0}^{t} a(s)\sigma^{\gamma}(s)\Phi^{\gamma}(s)ds .$$

From the integral inequality (2.17), we can establish a bound for the function $\Phi(t)$ and hence a bound for the solution $y(t)$. Using the generalized Gronwall's inequality in [24], we obtain the following bounds for $\Phi(t)$. For $\gamma \neq 1$,

$$(2.18) \qquad \Phi(t) \leq \left\{C^{1-\gamma} + (1-\gamma)\int_{t_0}^{t} a(s)\sigma^{\gamma}(s)ds\right\}^{\frac{1}{1-\gamma}} ,$$

whenever the term inside the bracket above remains positive. For $\gamma = 1$, we apply simply the Gronwall's inequality and obtain

$$(2.19) \qquad \Phi(t) \leq C \exp\left(\int_{t_0}^{t} a(s)\sigma^{\gamma}(s)ds\right) = M_0 .$$

Thus, in view of (2.11), (2.18) and (2.19) show that for $0 < \gamma \leq 1$, $\Phi(t) \leq M_0$. For $\gamma > 1$, $\Phi(t)$ will be bounded, according to (2.18), provided that

$$(2.20) \qquad |y(t_0)| + |y'(t_0)| \leq \frac{1}{2}\left[(\gamma-1)\int_{0}^{\infty} a(s)\sigma^{\gamma}(s)ds\right]^{\frac{1}{1-\gamma}} = M_1 .$$

So in any case, there exist solutions of (1.1) satisfying

$$(2.21) \qquad |y(t)| \leq Mt , \qquad t \geq t_0 \geq 1 .$$

Using (2.21), we can estimate (2.14) from below, namely,

$$\frac{|y(t)|}{t} \geq |y'(t_0)| - \frac{1}{t}\int_{t_0}^{t} sa(s)y^{\gamma}(\sigma(s))ds + o(1) ,$$

or

(2.22) $$\frac{|y(t)|}{t} \geq |y'(t_0)| - M^\gamma \int_{t_0}^{\infty} a(s)\sigma^\gamma(s)ds \ .$$

Choose $y(t_0) = 0$ and $|y'(t_0)| \neq 0$ such that (2.20) is satisfied. Next, choose $t_0 \geq 1$ sufficiently large so that the right hand side of (2.22) is positive. Therefore, (2.22) shows that $y(t)$ is nonoscillatory, moreover it follows that $y(t)$ is asymptotically linear.

We remark that in the above two theorems the hypothesis that $a(t) \geq 0$ is used only in proving the necessity part. The sufficiency will hold provided that we replace $a(t)$ in conditions (2.1) and (2.11) by its absolute value $|a(t)|$. We state this explicitly in the following

THEOREM (2.3). Let $\gamma > 0$. Assume that $\sigma(t)$ satisfies (1.2). Then the conditions

$$\int^{\infty} t|a(t)|dt < \infty$$

and

$$\int^{\infty} |a(t)|\sigma^\gamma(t)dt < \infty$$

are sufficient for equation (1.1) to possess a bounded nonoscillatory and an asymptotically linear solution, respectively.

The above results can be extended to n^{th} order equations in much the same way, like the ordinary differential equation case. See e.g. Kartsatos [12], Onose [19] and Ladas [13].

3. Oscillation Theorems, $\gamma \neq 1$

In this section, we consider the counterparts to our results in the preceding section, namely, oscillation theorems for equation (1.1). As in the case without delay, $\sigma(t) \equiv t$, these results are only available for nonlinear equations, $\gamma \neq 1$, and also in the stable case, $a(t) \geq 0$.

THEOREM (3.1). Let $0 < \gamma < 1$. Assume that $a(t) \geq 0$ and $\sigma(t)$ satisfies (1.2). Then every solution of (1.1) is oscillatory if and only if

(3.1) $$\int^{\infty} a(t)\sigma^\gamma(t)dt = \infty \ .$$

Proof. The necessity part is contained in Theorem (2.2). To prove the sufficiency, we assume the existence of a nonoscillatory solution which can without loss of generality be assumed positive, say $y(t) > 0$ for $t \geq t_0$. Since $a(t) \geq 0$, equation (1.1) gives by a standard argument that $y'(t) > 0$ for $t \geq t_0$. It then follows from the fact that $y(t) > 0$, $y'(t) > 0$ and $y''(t) \leq 0$ that there exists a positive constant $K > 0$ such that

$$\frac{y(t)}{y'(t)} \geq Kt , \qquad t \geq t_0 , \tag{3.2}$$

(see Heidel [10]). Dividing (1.1) by $y^{-\gamma}(\sigma(t))$ and using (3.2), we find

$$y'^{-\gamma}(\sigma(t))y''(t) + a(t)(K\sigma(t))^{\gamma} \leq 0 . \tag{3.3}$$

Since $y'' \leq 0$ and $\sigma(t) \leq t$, so $y'(\sigma(t)) \geq y'(t)$. Thus, (3.3) yields

$$y'^{-\gamma}(t)y''(t) + K^{\gamma}a(t)\sigma^{\gamma}(t) \leq 0 .$$

Upon an integration of the above, we find

$$y'^{1-\gamma}(t) - y'^{1-\gamma}(t_0) + (1-\gamma)K^{\gamma} \int_{t_0}^{\infty} a(t)\sigma^{\gamma}(t)dt \leq 0 . \tag{3.4}$$

Since $\gamma < 1$, condition (3.1) together with (3.4) produce an immediate contradiction.

Remark (3.1). The above result was given in a weaker form by Gollwitzer [7] where it is assumed that $\sigma(t)$ satisfies the stronger assumption (1.3) instead of (1.2) and the condition (3.1) is replaced by the stronger assumption

$$\int^{\infty} t^{\gamma}a(t)dt = \infty .$$

However, our proof of the sufficiency patterns after that of Heidel [10] for the case without delay.

The analogous result in the case $\gamma > 1$ requires a stronger assumption on the delay function $\sigma(t)$. We propose the following condition that there exists a positive constant c such that for all large t

$$(3.5) \qquad ct \leq \sigma(t) \leq t\,, \qquad t \geq t_0\,.$$

It is clear that condition (3.5) is intermediate between (1.2) and (1.3). It turns out that this condition also plays an important role in our study of the linear equation in the next section.

THEOREM (3.2). Let $\gamma > 1$. Assume that $a(t) \geq 0$ and $\sigma(t)$ satisfies (3.5). Then every solution of (1.1) is oscillatory if and only if

$$(3.6) \qquad \int^{\infty} ta(t)dt = \infty\,.$$

Proof. The necessity part follows from Theorem (2.1). To prove the sufficiency, we assume that there exists a solution $y(t) > 0$ for $t \geq t_0$. Again, it follows from the fact that $a(t) \geq 0$ and $y(t) > 0$, $y''(t) \leq 0$ and $y'(t) > 0$ for all $t \geq t_0$. Dividing (1.1) through by $ty^{-\gamma}(\sigma(t))$ and integrating, we find

$$\int_{t_0}^{t} sy''(s)y^{-\gamma}(\sigma(s))ds + \int_{t_0}^{t} sa(s)ds \leq 0\,.$$

Since $y(t) \geq y(\sigma(t)) \geq y(ct)$ and $y'' \leq 0$, the above inequality reduces to

$$(3.7) \qquad \int_{t_0}^{t} sy''(s)y^{-\gamma}(cs)ds + \int_{t_0}^{t} sa(s)ds \leq 0\,.$$

Carrying out the integration in the first term of (3.7), we find

$$sy'(s)y^{-\gamma}(cs)\Big|_{t_0}^{t} - \int_{t_0}^{t} y'(s)\left\{y^{-\gamma}(cs) - \gamma csy'(cs)y^{-\gamma-1}(cs)\right\} ds\,.$$

The last term in the above integral is nonnegative. The first term can be integrated as follows

$$-\int_{t_0}^{t} y'(s)y^{-\gamma}(cs)ds \geq -\int_{t_0}^{t} y'(cs)y^{-\gamma}(cs)ds$$

$$= \frac{1}{c(\gamma-1)}\left\{y^{-\gamma+1}(ct) - y^{-\gamma+1}(ct_0)\right\}\,,$$

which is bounded below since $\gamma > 1$. These estimates together with (3.6) give a desired contradiction in (3.7).

Remark (3.2). Theorem (3.2) can be found in Gollwitzer [7] under the stronger assumption (1.3). The proof is based upon Heidel's proof of Atkinson's result.

In case $\gamma = 1$, Bradley [4] has shown that condition (3.6) implies all bounded solutions of (1.1) are oscillatory provided that $\sigma(t)$ satisfies (1.3). It is easy to see from the proof of Theorem (3.2) and that given by Bradley [4] that the same conclusion remains valid under the weaker assumption (3.5) on $\sigma(t)$. We omit the details.

4. Oscillation, $\gamma = 1$

In this section, we consider the linear equation

$$y''(t) + a(t)y(\sigma(t)) = 0 , \qquad t \geq 0 , \tag{4.1}$$

which is a special case of (1.1) when $\gamma = 1$. We shall assume also here that $a(t) \geq 0$ and that the delay $\sigma(t)$ satisfies (3.5). We are concerned here with showing how some of the oscillation results in the ordinary case can be carried over to this more general setup. Waltman [23] has shown that the classical Fite-Wintner oscillation criterion remains valid for (4.1), namely

$$\int^{\infty} a(t)dt = \infty . \tag{4.2}$$

In view of this, we confine ourselves to cases when the coefficient $a(t)$ does not satisfy (4.2), hence we may introduce the function

$$A(t) = \int_t^{\infty} a(s)ds . \tag{4.3}$$

Our first result is an extension of Fite-Wintner's oscillation criterion due to Hartman and Wintner in the ordinary case, namely,

THEOREM (4.1). Suppose that there exists a real number $\lambda < 1$, such that

$$\int^{\infty} t^{\lambda} a(t)dt = \infty , \tag{4.4}$$

then every solution of (4.1) is oscillatory.

Proof. Without loss of generality, we can assume $\lambda \geq 0$. Assume that $y(t) > 0$ on some half infinite interval $[t_0,\infty)$. Since $y''(t) \leq 0$, it follows that $y'(t) > 0$ on $[t_0,\infty)$. Condition (3.5) implies that $y(\sigma(t)) \geq y(ct)$, hence $y(t)$ satisfies the second order differential inequality

$$y''(t) + a(t)y(ct) \leq 0 . \tag{4.5}$$

Let $v(t) = \frac{y'(t)}{y(ct)}$ which is nonnegative on $[t_0,\infty)$. Using (4.5), it is easy to see that $v(t)$ satisfies the following first order Riccati inequality

$$v'(t) + cv^2(t) + a(t) \leq 0 . \tag{4.6}$$

Multiplying (4.6) through by t^λ and integrating from t_0 to t, we find

$$\int_{t_0}^{t} s^\lambda v'(s)ds + c\int_{t_0}^{t} s^\lambda v^2(s)ds + \int_{t_0}^{t} s^\lambda a(s)ds \leq 0 . \tag{4.7}$$

Integrating the first integral in (4.7), we obtain

$$t^\lambda v(t) - \lambda\int_{t_0}^{t} s^{\lambda-1}v(s)ds = k_0 + \int_{t_0}^{t} s^\lambda v'(s)ds , \tag{4.8}$$

where $k_0 = t_0^\lambda v(t_0)$. Applying Schwarz's inequality to the integral on the left of (4.8), we find

$$\left(\int_{t_0}^{t} s^{\lambda-1}v(s)ds\right)^2 \leq \left(\int_{t_0}^{t} s^{\lambda-2}ds\right)\left(\int_{t_0}^{t} s^\lambda v^2(s)ds\right) \tag{4.9}$$

$$\leq k_1\left(\int_{t_0}^{t} s^\lambda v^2(s)ds\right) ,$$

where $k_1 = \frac{1}{1-\lambda}t_0^{\lambda-1}$. Substituting (4.8) and (4.9) into (4.7), we obtain

$$-\lambda k_1^{\frac{1}{2}}\left(\int_{t_0}^{t} s^\lambda v^2(s)ds\right)^{\frac{1}{2}} + c\left(\int_{t_0}^{t} s^\lambda v^2(s)ds\right) \tag{4.10}$$

$$+ \int_{t_0}^{t} s^\lambda a(s)ds \leq k_0 .$$

Suppose that $\int^{\infty} s^{\lambda} v^2(s)ds$ is finite, then (4.4) gives an immediate contradiction in (4.10). Otherwise, when it is infinite, the first two terms in (4.10) become eventually nonnegative and we obtain a similar contradiction. This completes the proof.

The reduction of (4.1) in case of nonoscillation to the Riccati inequality (4.6) permits further refinements concerning oscillation criteria even when $a(t)$ satisfies (4.4) with $\lambda = 1$. These are known as pointwise conditions. The following result extends the well known oscillation criteria of Kneser [9], Hille [11] and Opial [20] in the ordinary differential equation case.

THEOREM (4.2). Suppose that for some large t_0, $a(t)$ satisfies for $t \geq t_0$ any one of the following three conditions

$$a(t) \geq \frac{1+\varepsilon}{4ct^2}, \qquad \varepsilon > 0, \tag{4.11}$$

$$A(t) \geq \frac{1+\varepsilon}{4ct}, \qquad \varepsilon > 0, \tag{4.12}$$

$$\int_t^{\infty} A^2(s)ds \geq \frac{1+\varepsilon}{4c} A(t) \neq 0, \qquad \varepsilon > 0, \tag{4.13}$$

where $A(t)$ is defined by (4.3) and ε is some positive number. Then every solution of (4.1) is oscillatory.

Proof. Proceed in the same manner as the proof of Theorem (4.1). Suppose that (4.11) holds. Introduce a new independent variable $\tau = ct$ and obtain from (4.6) and (4.11) the following Riccati inequality.

$$\frac{du}{d\tau}(\tau) + u^2(\tau) + \frac{1+\varepsilon}{4\tau^2} \leq 0, \tag{4.14}$$

from which it follows ([9], p. 362) that the second order equation

$$\frac{d^2z}{d\tau^2}(\tau) + \frac{1+\varepsilon}{4\tau^2} z(\tau) = 0$$

is nonoscillatory, which is a desired contradiction.

Next, suppose that (4.12) holds. From (4.6), we have

$$v'(t) + cv^2(t) \leq 0,$$

or

$$\frac{d}{dt}\left(-\frac{1}{v(t)} + ct\right) \leq 0 ,$$

which implies

$$(4.15) \qquad 0 < v(t) \leq \frac{1}{ct + d} ,$$

where $d = \frac{1}{v(t_0)} - ct_0$. From (4.15), it follows that $\lim_{t\to\infty} v(t) = 0$. So we can integrate (4.6) from t to infinity and obtain

$$(4.16) \qquad v(t) \geq c \int_t^\infty v^2(s)ds + A(t) .$$

The remainder of the proof starting from (4.6) is the same as in the ordinary case. We sketch the argument for the sake of completeness. Using (4.12) and (4.16), we can prove inductively

$$(4.17) \qquad v(t) \geq \alpha_n \frac{1 + \varepsilon}{4ct} ,$$

where $\alpha_0 = 1$ and $\alpha_n = \frac{1+\varepsilon}{4}\alpha_{n-1}^2 + 1$. Furthermore, $\alpha_n - \alpha_{n-1} = \frac{1+\varepsilon}{4}(\alpha_{n-1}^2 - \alpha_{n-2}^2)$ which shows that α_n is non-decreasing. Thus, $\lim_{n\to\infty} \alpha_n = L$ exists and satisfies $L = \frac{1+\varepsilon}{4} L^2 + 1$ which is impossible. Therefore $L = \infty$, but this produces a contradiction in (4.17).

Finally, suppose that (4.12) holds. A similar argument would lead to

$$(4.18) \qquad v(t) \geq \alpha_n A(t) ,$$

instead of (4.17) where α_n is the same as defined above. The contradiction is reached in exactly the same fashion as in the previous case.

Remark (4.1). We note that if $\sigma(t)$ satisfies the boundedness assumption (1.3) then conditions (4.11), (4.12), and (4.13) can be improved by replacing $c = 1$ in these inequalities. To see this, for a given $\varepsilon > 0$, we simply take $c = 1 - \frac{\varepsilon}{2}$, then $a(t) \geq \frac{1+\varepsilon}{4t^2}$ implies that $a(t) \geq \frac{1+\frac{\varepsilon}{4}}{4ct^2}$ for sufficiently small ε. On the other hand if $\sigma(t)$ satisfies (1.3) then we can choose t_0 sufficiently large so that $t - M \geq (1-\varepsilon)t$ for any $\varepsilon > 0$.

5. Unstable Case

In this section, we consider the oscillatory behavior of solutions of equation (1.1) when $a(t) \leq 0$. The main result here is an extension of a recent theorem of Ladas and Lakshikantham [14], where it was shown that if $a(t)$ is bounded away from zero by some negative constant depending on the delay then all bounded solutions of (1.1) must be oscillatory.

THEOREM (5.1). Let $0 < \gamma \leq 1$. Suppose that $a(t) \leq 0$, $\sigma(t)$ satisfies (1.2) and the following condition holds

$$\liminf_{t\to\infty} \int_{\sigma(t)}^{t} (t-\tau)a(\tau)d\tau \leq -1 . \tag{5.1}$$

Then all bounded solutions are oscillatory.

Proof. Let $y(t) > 0$, then $y''(t) \geq 0$. If $y(t)$ is to be bounded, then $y'(t) \leq 0$ and must also satisfy $\lim_{t\to\infty} y'(t) = 0$. We now claim that $\lim_{t\to\infty} y(t) = 0$. Suppose not, then there exists t_0 so that $y(t) \geq \alpha > 0$ for $t \geq t_0$. Integrating equation (1.1) from s to t, we have

$$y'(t) - y'(s) + \int_{s}^{t} a(\tau)y\,(\sigma(\tau))d\tau = 0 , \tag{5.2}$$

which implies

$$y'(s) \leq \alpha^{\gamma} \int_{s}^{t} a(\tau)d\tau . \tag{5.3}$$

Now integrate (5.3) from $\sigma(t)$ to t and obtain

$$y(t) - y(\sigma(t)) \leq \alpha^{\gamma} \int_{\sigma(t)}^{t} (t-\tau)a(\tau)d\tau . \tag{5.4}$$

By (5.1), there exists a sequence $\{t_n\}$ so that the right hand side of (5.4) tends to a number $\leq -\alpha^{\gamma}$. But the limit of the left hand side of (5.4) is zero as $t \to \infty$. This is a contradiction, so $y(t) \to 0$ as $t \to \infty$.

Choose t_0 so large that $y(\sigma(t)) \leq 1$ for $t \geq t_0$. Equation (5.2) now implies that

$$y'(s) \leq y(\sigma(t)) \int_s^t a(\tau)d\tau$$

$$\leq y(\sigma(t)) \int_s^t a(\tau)d\tau .$$

Integrating above again from $\sigma(t)$ to t, we obtain

$$-y(\sigma(t)) < y(t) - y(\sigma(t)) \leq y(\sigma(t)) \int_{\sigma(t)}^t (t-\tau)a(\tau)d\tau .$$

Applying (5.1) to the above estimate, we arrive at the desired contradiction.

Remark (5.1). Suppose that $a(t)$ is in addition nondecreasing, then condition (5.1) reduces to

$$(t-\sigma(t))^2 a(t) \leq -2 . \tag{5.5}$$

This is the result of Ladas and Lakshmikantham [14] proved under the additional assumption that $\sigma'(t) \geq 0$.

Remark (5.2). It seems desirable to leave condition (5.1) in its integral form for it has an advantage that it applies not only to those $a(t)$ which do not satisfy $a'(t) \geq 0$ but also to those $a(t)$ which are not bounded away from zero. For example, consider the simpler case when $\sigma(t) = t - M$ and the coefficient $a(t) = -K(1 + \sin \frac{2\pi t}{M})$ which has an infinity of zero. Condition (5.1) shows in this case that if $K \geq \frac{2}{M^2}(1 + \frac{1}{\pi})^{-1}$ then all bounded solutions of (1.1) are oscillatory.

REFERENCES

[1] F.V. ATKINSON, On second order non-linear oscillation, Pacific J. Math., 5 (1955), 643-647.

[2] R. BELLMAN and K.L. COOKE, Differential Difference Equations. Academic Press, New York, 1963.

[3] S. BELOHOREC, Oscilatoricke riesenia istej nelinearnej differencialnej rovnice druheho radu, Matematicky Casopis, 11 (1961), 250-255.

[4] J.S. BRADLEY, Oscillation theorems for a second order delay equation, J. Differential Equations, 8 (1970), 397-403.

[5] L.E. EL'SGOL'TS, Introduction to the Theory of Differential Equations with Deviating Arguments. Holden Day, San Francisco, 1966.

[6] W.B. FITE, Properties of the solutions of certain functional differential equations, Trans. Amer. Math. Soc., 22 (1921), 311-319.
[7] H.E. GOLLWITZER, On nonlinear oscillations for a second order delay equation, J. Math. Anal. Appl., 26 (1969), 385-389.
[8] J.K. HALE, Functional Differential Equations. Springer-Verlag, Berlin, 1971.
[9] P. HARTMAN, Ordinary Differential Equations. John Wiley, New York, 1964.
[10] J.W. HEIDEL, Rate of growth of nonoscillatory solutions for the differential equation $y'' + q(t)|y|^\gamma \operatorname{sgn} y = 0$, $0 < \gamma < 1$, Quart. Appl. Math., 28 (1971), 601-606.
[11] E. HILLE, Nonoscillation theorems, Trans. Amer. Math. Soc., 64 (1948), 234-252.
[12] A.G. KARTSATOS, On oscillation of solutions of even order nonlinear differential equations, J. Differential Equations, 6 (1969), 232-237.
[13] G. LADAS, Oscillation and asymptotic behavior of solutions of differential equations with retarded argument, (to appear).
[14] G. LADAS and V. LAKSHMIKANTHAM, Oscillation caused by retarded arguments, (to appear).
[15] J.C. LILLO, Oscillatory solutions of the equation $y'(x) = m(x)y(x - n(x))$, J. Differential Equations, 6 (1969), 1-35.
[16] A.D. MYSHKIS, Linear Differential Equations with Retarded Arguments, Deutscher Verlag der Wissenschaften, Berlin, 1955 (German).
[17] Z. NEHARI, On a class of nonlinear second order differential equations, Trans. Amer. Math. Soc., 95 (1960), 101-123.
[18] S.B. NORKIN, Differential Equations of Second Order with Deviating Arguments, Nauka, Moscow, 1965 (Russian).
[19] H. ONOSE, Oscillatory property of ordinary differential equations of arbitrary order, J. Differential Equations, 7 (1970), 454-458.
[20] Z. OPIAL, Sur les intégrales oscillantes de l'equation différentielle $u'' + f(t)u = 0$, Ann. Polon. Math., 4 (1958), 308-313.
[21] V.A. STAIKOS, Oscillatory property of certain delay differential equations, Bull. Soc. Math. Grece, 11 (1970), 1-5.
[22] V.A. STAIKOS and A.G. PETSOULAS, Some oscillation criteria for second order nonlinear delay-differential equations, J. Math. Anal. Appl., 30 (1970), 695-701.
[23] P. WALTMAN, A note on an oscillation criterion for an equation wtih a functional argument, Canad. Math. Bull., 11 (1968), 593-595.
[24] D. WILLETT and J.S.W. WONG, On the Discrete Analogues of Some Generalizations of Gronwall's Inequality, Monatshefte fur Math., 69 (1965), 362-367.
[25] J.S.W. WONG, On second order nonlinear oscillation, Funkcialaj Ekvacioj, 11 (1969), 207-234.

University of Iowa, Iowa City, Iowa

A NOTE ON MALMQUIST'S THEOREM ON FIRST-ORDER DIFFERENTIAL EQUATIONS

Chung-Chun Yang

Malmquist showed that if $R(z, y(z))$ is a rational function of z and $y(z)$ and if the equation (D): $y'(z) = R(z, y(z))$ has a solution $y(z)$ which is meromorphic in the whole plane, then either $y(z)$ is a rational function or $R(z, y(z))$ is a polynomial in $y(z)$ of degree ≤ 2.

From this and a growth argument it can be shown that if $y(z)$ is a transcendental meromorphic solution of (E) with finitely many poles, then $R(z, y(z))$ must be a linear function in $y(z)$.

In this paper, based on Nevanlinna's theory of meromorphic functions, it is shown that the above conclusion still holds when the coefficients in $R(z, y(z))$ are replaced by arbitrary meromorphic functions and the solutions $y(z)$ considered are meromorphic functions having relatively small number of poles, with growth rate much faster than all the coefficients.

1. Introduction

In 1913, J. Malmquist [3] proved the following important theorem:

THEOREM A. If $R(z, y(z))$ is a rational function of z and $y(z)$ and if the equation

$$y'(z) = R(z, y(z)) \tag{1.1}$$

has a solution $y(z)$ single-valued in its domain of existence, then either $y(z)$ is a rational function or (1.1) is a Riccati equation (equation (1.1) is said to be of Riccati's type if and only if $R(z, y(z))$ is a polynomial in $y(z)$ of degree ≤ 2).

Later a proof based on R. Nevanlinna's theory of meromorphic functions was given by K. Yosida [7]. In particular, Yosida's argument can give us the following result:

THEOREM B. Let $R(z, y(z))$ be a rational function of z and $y(z)$. Then if the following equation

$$y' = R(z, y(z)) \tag{1.2}$$

admits a transcendental meromorphic solution $y(z)$ which has only finitely many poles, then $R(z, y(z))$ must be a linear function in $y(z)$, i.e., $R(z, y(z)) = r_1(z) + r_2(z)y(z)$ (r_1, r_2 are rational functions).

Remark. Without the assumption that $y(z)$ has only a finite number of poles, the assertion is false. For example, $y = \tan z$ is a transcendental meromorphic solution of the equation $y' = 1 + y^2$.

The argument used in [7] relies heavily on the fact that all the coefficients in R9z, $y(z))$ are rational functions which enables one to use a result of Valiron [6]. Unfortunately, there is no corresponding result for a broader class of meromorphic functions. The main purpose of this note, among other things, it to employ techniques of Nevanlinna (see [2], Ch. 1, 2 and 3) to extend Theorem B (see Theorem 2 below) by allowing the coefficients in $R(z, y(z))$ to be arbitrary meromorphic functions. The methods developed are different from Yosida's. We shall focus our attention on the solutions which grow (in terms of Nevanlinna characteristic function) much faster than all the coefficients in the equation and the number of their poles are required to satisfy certain conditions.

2. Notation and Preliminary Lemmas

In the sequel, we shall employ the usual notation of Nevanlinna theory. $y(z)$ will always denote a function meromorphic in the whole complex plane. Following Nevanlinna, we define

$$(2.1) \qquad N(r,a) = N(r,a,y) = \int_0^r \frac{[n(t,a)-n(o,a)]}{t}\,dt + n(o,a)\log r ,$$

where $n(r,a) = n(r,a,y)$ denotes the number of roots of the equation $y(z) = a$ in $|z| \leqslant r$.

$$N(r,y) = N(r,\infty,y) .$$

We also define

$$(2.2) \qquad m(r,y) = m(r,\infty,y) = \frac{1}{2\pi}\int_0^{2\pi} \log^+ |y(re^{i\theta})|\,d\theta ,$$

where $\log^+ |x| = \max(\log|x|, 0)$,

$$m(r,a,f) = m(r,\infty,\frac{1}{y-a}), \qquad a \neq \infty$$

and

$$T(r,y) = m(r,y + N(r,y) .$$

$T(r,y)$ is called the characteristic function of y. By virtue of this we are able to give a measure of the growth rate of a meromorphic function.

We shall denote by $S(r,y)$ any function of r depending on $y(re^{i\theta})$ and satisfying

$$(2.3) \qquad S(r,y) = o\{T(r,y)\}$$

as $r \to \infty$, possibly outside a set of r of finite linear measure.

We denote by $P_n(y)$ a polynomial in y and its derivatives of degree at most n with coefficients $a(z)$ satisfying

$$T(r, a(z)) = S(r,y) ;$$

such polynomials will be called differential polynomials.

LEMMA 1 (Milloux [4]). Let ℓ be a positive integer and

$$(2.4) \qquad \Psi(z) = \sum_{i=o}^{\ell} a_i(z)y^{(i)}(z) \qquad (y^{(o)} \equiv y) ,$$

where $a_i(z)$ are functions meromorphic in the plane and satisfy

$$(2.5) \qquad T(r, a_i(z)) = S(r, y(z)) .$$

Then

$$(2.6) \qquad m(r, \frac{\Psi}{y}) = S(r,y) .$$

and

$$(2.7)\qquad T(r,\Psi) \leqslant (\ell + 1)\, T(r,y) + S(r,y)\ .$$

Remark. If the function y satisfies the condition:

$$(2.8)\qquad N(r,y) = S(r,y)\ ,$$

then (2.6) and (2.8) yield

$$(2.9)\qquad T(r, \frac{\Psi}{y}) = S(r,y)\ .$$

For our estimation, the following result will play a basic role.

LEMMA 2. Let $y(z)$ be a transcendental meromorphic function with $N(r,y) = S(r,y)$. Assume that $a_i(z)$ $(i = a, 1, 2, \ldots, \ell)$ be meromorphic function and satisfy conditions (2.5). Then

$$(2.10)\qquad T(r,y^{\ell} + a_1(z)\pi_{\ell-1}(y) + a_2(z)\pi_{\ell-2}(y) + \cdots + a_{\ell-1}(z)\pi_1(y) + a_{\ell}(z)$$

$$= \ell T(r,y) + S(r,y)\ ,$$

where $\pi_i(y)$ are homogeneous differential polynomial in y of degree i.

Proof. Set

$$(2.11)\qquad p_{\ell}(y) = y^{\ell} + a_1(z)\pi_{\ell-1}(y) + \cdots + a_{\ell-1}(z)\pi_1(y) + a_{\ell}(z)\ .$$

We rewrite (2.11) as

$$(2.12)\qquad p_{\ell}(y) = y^{\ell}(z)(1 + \frac{a_1(z)\pi_{\ell-1}(y)}{y^{\ell-1}}\frac{1}{y} + \cdots + \frac{a_{\ell-1}(z)\pi_1(y)}{y}\frac{1}{y^{\ell-1}}$$

$$+ a_{\ell}(z)\frac{1}{y^{\ell}})$$

$$= y^{\ell}(z)(1 + \frac{A_1(z)}{y} + \frac{A_2(z)}{y^2} + \cdots + \frac{A_{\ell}(z)}{y^{\ell}})\ ,$$

where $A_i(z) = a_i(z)\dfrac{\pi_{\ell-i}(y)}{y^{\ell-i}}$ $(i = 1, 2, \ldots, \ell)$.

Thus by (2.9) of Lemma 1 and by addition, we have

$$(2.13)\qquad T(r, A_i(z)) = S(r,y)\quad (i = 1, 2, \ldots, \ell)\ .$$

Now on the circle $|z| = r$, let

(2.14) $$A(re^{i\theta}) = \max_{1\leq i\leq \ell} |A_i(re^{i\theta})|^{\frac{1}{i}}, \quad i = 1, 2, \ldots, \ell.$$

Let E_1 by the set of θ in $o \leq \theta \leq 2$ for which $|y(re^{i\theta})| \geq 2A(re^{i\theta})$, and E_2 be the complementary set.

On E_1 we have

(2.15) $$\begin{aligned} |p_\ell(y)| &= |y^\ell(z)| \, |1 + \frac{A_1(z)}{y} + \frac{A_2(z)}{y^2} + \cdots + \frac{A_\ell(z)}{y^\ell}| \\ &\geq |y^\ell(z)| \, \{1 - |\frac{A_1}{y}| - |\frac{A_2}{y^2}| - \cdots - |\frac{A_\ell}{y^\ell}|\} \\ &\geq |y^\ell(z)| \, \{1 - \frac{1}{2} - \frac{1}{2^2} - \cdots - \frac{1}{2^\ell}\} \\ &= \frac{1}{2^\ell} |y^\ell(z)| . \end{aligned}$$

Hence, from this, (2.13) and (2.14), we have

(2.16) $$\begin{aligned} \ell\, m(r,y) &= m(r, y^\ell) \\ &= \frac{1}{2\pi}\int_{E_1} \log^+ |y^\ell(re^{i\theta})| d\theta + \frac{1}{2\pi}\int_{E_2} \log^+ |y^\ell(re^{i\theta})| d\theta \\ &\leq \frac{1}{2\pi}\int_0^{2\pi} \log^+ |2^\ell p_\ell(y)| d\theta + \frac{1}{2\pi}\int_{E_2} \log^+ |2A(re^{i\theta})|^\ell d\theta \\ &= m(r, p_\ell(y)) + \ell \log 2 + S(r,y) . \end{aligned}$$

Adding $\ell\, N(r,y)$ on both sides of the above inequality, we obtain

(2.17) $$\begin{aligned} \ell\, T(r,y) &\leq T(r, p_\ell(y)) + \ell\, N(r,y) + S(r,y) \\ &= T(r, p_\ell(y)) + S(r,y) . \end{aligned}$$

In the last step we made use of the hypothesis that $N(r,y) = S(r,y)$.

On the other hand, it is easily shown from the expression (2.12) and by induction on ℓ that

(2.18) $$T(r, p_\ell(y)) \leq \ell\, T(r,y) + S(r,y) .$$

Thus, combining (2.17) and (2.18), we obtain (2.10).

In a polynomial $p(z, y_0, y_1, \ldots, y_n)$, we shall denote $k_0 + k_1 + \cdots + k_n$ and $k_1 + 2k_2 + \cdots + nk_n$ as the dimension and weight of a term $a(z)y_0^{k_0} y_1^{k_1} y_2^{k_2} \cdots y_n^{k_n}$ $(a(z) \not\equiv 0)$ respectively. The degree of $p(y)$ is the maximal dimension among all its terms. We shall call $p(y)$ a polynomial of degree d as non-degenerate if

$$\Sigma a(z) \not\equiv 0 , \tag{2.19}$$

where $a(z)$ are coefficients in $p(y)$ and the summation is taken over all the terms with dimension d and the weight being maximal.

LEMMA 3. Let $y(z)$ be meromorphic transcendental with $N(r,y) = S(r,y)$. If $p(y)$ is a non-degenerate differential polynomial of degree d, then

$$T(r, p(y)) = d\, T(r,y) + S(r,y) . \tag{2.20}$$

Proof. By a result of Y. Tumura [5] one can express

$$y^{(n)} = y(F^n + p_{n-1}(F)) , \quad n = 1, 2, \ldots , \tag{2.21}$$

where $F = \frac{y'}{y}$. We note that $T(r,F) = N(r,F) + m(r,F) = S(r,y)$. Thus by substituting this into $p(y)$, $p(y)$ will have the form $A(z)y^d + p_{d-1}(y)$. The rest follows by Lemma 2.

LEMMA 4 (Clunie [1]). Suppose that $p(y)$ and $q_n(y)$ be two differential polynomials in y and that

$$y^n(z)\, p(y) = q_n(y) . \tag{2.22}$$

Then

$$m\{r, p(y)\} = S(r,y) \quad \text{as} \quad r \to \infty . \tag{2.23}$$

3. Main Results

We first prove a general result.

THEOREM 1. Let $R_i(z, y(z), y_1(z), \ldots, y_k(z))$, $i = 1, 2$, be given rational functions in $y_0(z), y_1(z), \ldots, y_k(z)$ $(y_i(z) \equiv y^{(i)}(z))$ with meromorphic functions as their coefficients. Assume that

$$(3.1)\qquad R_i(z, y_0(z), y_1(z), \ldots, y_k(z)) = \frac{p_{i1}(z,y_0,\ldots,y_k)}{q_{i1}(z,y_0,\ldots,y_k)} ,$$

where, $p_{i1}(z, y_0, y_1, \ldots, y_k)$, $q_{i1}(z, y_0, y_1, \ldots, y_k)$ are two relatively prime polynomials in $y_0, y_1, \ldots, y_k$ $(i = 1, 2)$ with degree n_i, m_i $(i = 1, 2)$ respectively.

Assume that

$$(3.2)\qquad p_{21}(z, y_0, \ldots, y_k) = y^{n_2}(z) + p_2(z, y_0, \ldots, y_k)$$

and

$$(3.3)\qquad q_{21}(z, y_0, \ldots, y_k) = y^{m_2}(z) + q_2(z, y_0, \ldots, y_k)$$

where $p_2(z, y_0, \ldots, y_k)$ and $q_2(z, y_0, \ldots, y_k)$ are two polynomials in $y_0, \ldots, y_k$ with degree less than n_2, m_2 respectively.

Further assume that p_{11} and q_{11} are non-degenerate. Then if the differential equation

$$(3.4)\qquad R_1(z, y_0, \ldots, y_k) = R_2(z, y_0, \ldots, y_k)$$

admits a transcendental meromorphic solution $y(z)$ with

$$(3.5)\qquad N(r,y) = S(r,y)$$

and such that

$$(3.6)\qquad T(r, a(z)) = S(r,y) \quad \text{as} \quad r \to \infty$$

hold for all the coefficients $a(z)$ in R_1 and R_2, we must have

$$n_1 + m_1 \geqq |n_2 - m_2| .$$

<u>Proof</u>. By assumption that p_{11} and q_{11} both are non-degenerate, and the property that $T(r, f_1f_2) \leqq T(r,f_1) + T(r,f_2)$ (see e.g. [2]) we have

$$T(r, R_1) \leqq T(r, p_{11}) + T(r, \frac{1}{q_{11}}) .$$

It follows by Lemma 1 and Nevanlinna's first fundamental theorem that

$$(3.7) \qquad T(r, R_1) \leqq n_1 T(r,y) + m_1 T(r,y) + S(r,y) = (n_1 + m_1) T(r,y) + S(r,y) ,$$

while

$$(3.8) \qquad T(r, R_2) \geqq T(r, p_{21}) - T(r, q_{21}) = n_2 T(r,y) - m_2 T(r,y) + S(r,y) = (n_2 - m_2) T(r,y) + S(r,y) .$$

Since p_{21} and q_{21} are interchangable, we thus get

$$T(r, R_2) \geqq |n_2 - m_2| T(r,y) + S(r,y) .$$

Our assertion follows from this, (3.7), and the fact that $T(r, R_1) = T(r, R_2)$.

Now let us consider the special case

$$R_1(z, y_o, y_i, \ldots, y_k) = y' \quad \text{and} \quad R_2(z, y, y_1, \ldots, y_k)$$

is a rational function of y only. Thus equation (3.4) assumes the form

$$(3.9) \qquad y' = \frac{a_1(z) + a_2(z)y + \cdots + a_{n_2}(z)y^{n_2}}{b_1(z) + b_2(z)y + \cdots + b_{m_2}(z)y^{m_2}} ,$$

$(a_{n_2} \neq 0, \ b_{m_2} \neq 0)$.

Now suppose that (3.9) has a transcendental meromorphic solution $y(z)$ satisfying

$$(3.10) \qquad N(r,y) = S(r,y) ,$$

$$(3.11) \qquad T(r, a_i(z)) = S(r,y) \quad (i = 1, 2, \ldots, n_2) ,$$

and

$$(3.12) \qquad T(r, b_j(z)) = S(r,y) \quad (j = 1, 2, \ldots, m_2) .$$

Then by (3.10) and (2.5) of Lemma 1 we have

$$(3.13)\qquad T(r, y') \leqq m(r, \frac{y'}{y}) + m(r, y) + N(r, y')$$
$$= S(r, y) + T(r, y) + S(r, y)$$
$$= T(r, y) + S(r, y)\ .$$

It follows from this and Theorem 1 that

$$(3.14)\qquad |m_2 - n_2| \leqq 1\ .$$

Now suppose that $m_2 \neq 0$. Then we have (i) $n_2 = m_2 - 1$, (ii) $n_2 = m_2$, (iii) $n_2 = m_2 + 1$.

In case (i) we have, according to (3.9), that

$$b_{m_2}(z)y^{m_2}y' + b_{m_2-1}(z)y^{m_2-1}y' + \cdots = a_{n_2}(z)y^{n_2} + \cdots\ .$$

Hence

$$(3.15)\qquad y^{m_2}(b_{m_2}y') + p_{m_2-1}(y) = q_{n_2}(y)$$

or

$$(3.16)\qquad y^{m_2-1}[b_{m_2}(z)yy' + b_{m_2-1}(z)y'] + p_{m_2-2}(y) = q_{n_2}(y)$$

where $q_{n_2}(y)$ is a differential polynomial in y of degree at most n_2.

Applying Lemma 4 to (3.15) and (3.16) we obtain

$$(3.17)\qquad m(r, b_{m_2}y') = S(r, y)$$

and

$$(3.18)\qquad m(r, b_{m_2}y'y + b_{m-1}(z)y') = S(r, y)\ .$$

By (3.10) we deduce

$$(3.19)\qquad T(r, b_{m_2}y') = S(r, y)$$

and

$$(3.20)\qquad T\{(r, y'(b_{m_2}y + b_{m_2-1})\} = S(r, y)\ .$$

Thus it follows from (3.19) and (3.20) that

$$(3.21)\qquad T(r,\ b_{m_2}y + b_{m_2-1}) \leqq T\{(r,\ y'(b_{m_2}y + b_{m_2-1})\} + T(r\ \frac{1}{b_{m_2}y'})$$
$$= S(r,\ y)\ .$$

This gives a contradiction, since

$$(3.22)\qquad T(r,\ b_{m_2}y + b_{m_2-1}) = T(r,\ y) + S(r,\ y)\ .$$

Case (ii) and case (iii) can be handled in a similar manner and will lead to the same contradiction. Thus we must have $m_2 = 0$.

Now according to (3.14) we conclude $n_2 = 1$. Thus the following theorem is proved.

THEOREM 2. If the differential equation

$$(3.23)\qquad y' = \frac{a_1(z) + a_2(z)y + \cdots + a_{n_2}(z)y^{n_2}}{b_1(z) + b_2(z)y + \cdots + b_{m_2}(z)y^{m_2}}$$

has a transcendental meromorphic solution $y(z)$ such that conditions (3.10), (3.11) and (3.12) are satisfied, then it is necessary that the right hand side of equation (3.23) is linear in y.

Remark. The above argument also shows that the same conclusion holds if one replaces y' by $y^{(n)}$ $(n \geqq 0)$ in the equation (3.23).

The following two results follow immediately from the proof of Theorem 2.

COROLLARY 1. Let $a_1(z)$ $(\not\equiv 0)$, $a_2(z)$ $(\not\equiv 0)$ and $a_3(z)$ be three meromorphic functions. Then the following differential equation

$$a_1(z)y(z)y'(z) + a_2(z)y'(z) + a_3(z) = 0$$

has no entire solution $y(z)$ which satisfies $T(r,\ a_i(z)) = S(r,\ y(z))$, $i = 1,\ 2,\ 3$.

COROLLARY 2. Let $a_1(z)$ $(\not\equiv 0)$, $a_2(z)$ $(\not\equiv 0)$ and $a_3(z)$ be three meromorphic functions. Then the following differential equation

$$a_1(z)y(z)y'(z) + a_2(z)y(z) + a_3(z) = 0$$

has no entire solution $y(z)$ which satisfies

$$T(r, a_i(z)) = S(r, y(z)), \quad i \equiv 1, 2, 3.$$

REFERENCES

[1] J. CLUNIE, On integral and meromorphic functions, Jour. Lond. Math. Soc., 37 (1962), p. 20.
[2] W.K. HAYMAN, Meromorphic Functions. Oxford, 1964, p. 5.
[3] J. MALMQUIST, Sur les fonctions a un rompre fini de branches definies par les equations différentielles dur premier order, Acta Math., 36 (1913), 297-343.
[4] H. MILLOUX, Les fonctions meromorphes et leurs derivées. Paris, 1940.
[5] Y. TUMURA, On the extensions of Borel's theorem and Saxer-Csillag's theorem, Proc. Phys-Math. Soc. Jap. III.s., 19 (1937), p. 32.
[6] G. VALIRON, Sur la derivée des fonctions algébroides, Bull. de la Soc. Math. de France (1931), p. 34.
[7] K. YOSIDA, A generalization of Malmquist's theorem, Jap. J. Math., 9 (1932), 253-256.

Mathematics Research Center, Naval Research Laboratory, Washington, D.C.